# NON-CLASSICAL LOGICS AND THEIR APPLICATIONS TO FUZZY SUBSETS

## A HANDBOOK OF THE MATHEMATICAL FOUNDATIONS
## OF FUZZY SET THEORY

# THEORY AND DECISION LIBRARY

*General Editors:* W. Leinfellner (*Vienna*) and G. Eberlein (*Munich*)

Series A: Philosophy and Methodology of the Social Sciences

Series B: Mathematical and Statistical Methods

Series C: Game Theory, Mathematical Programming and Operations Research

Series D: System Theory, Knowledge Engineering and Problem Solving

## SERIES B: MATHEMATICAL AND STATISTICAL METHODS

### VOLUME 32

*Scope:* The series focuses on the application of methods and ideas of logic, mathematics and statistics to the social sciences. In particular, formal treatment of social phenomena, the analysis of decision making, information theory and problems of inference will be central themes of this part of the library. Besides theoretical results, empirical investigations and the testing of theoretical models of real world problems will be subjects of interest. In addition to emphasizing interdisciplinary communication, the series will seek to support the rapid dissemination of recent results.

*The titles published in this series are listed at the end of this volume.*

# NON-CLASSICAL LOGICS
# AND THEIR APPLICATIONS
# TO FUZZY SUBSETS

*A Handbook of the
Mathematical Foundations
of Fuzzy Set Theory*

*edited by*

## ULRICH HÖHLE

*Fachbereich Mathematik,
Bergische Universität, Wuppertal, Germany*

and

## ERICH PETER KLEMENT

*Institut für Mathematik,
Johannes Kepler Universität, Linz, Austria*

## SPRINGER SCIENCE+BUSINESS MEDIA, B.V.

Library of Congress Cataloging-in-Publication Data

```
Non-classical logics and their applications to fuzzy subsets : a
  handbook of the mathematical foundations of fuzzy set theory /
  edited by Ulrich Höhle and Erich Peter Klement.
      p.   cm. -- (Theory and decision library. Series B,
  Mathematical and statistical methods ; v. 32)
    "Proceedings of the 14th Linz Seminar on Fuzzy Set Theory ... the
  second week of September 1992 at the Bildungszentrum St. Magdalena
  (Linz, Austria)"--P. 1.
    Includes bibliographical references and index.
    ISBN 978-0-7923-3194-0      ISBN 978-94-011-0215-5 (eBook)
    DOI 10.1007/978-94-011-0215-5
    1. Nonclassical mathematical logic--Congresses.  2. Fuzzy set
  theory--Congresses.   I. Höhle, Ulrich.  II. Klement, E. P. (Erich
  Peter)  III. Series.
  QA9.4.N64  1995
  511.3--dc20                                             94-35630
```

ISBN 978-0-7923-3194-0

*Printed on acid-free paper*

# CONTENTS

# Authors and Editors

**Lawrence Peter Belluce**
Department of Mathematics
University of British Columbia
121-1984 Mathematics Road
Vancouver, BC, Canada V6T 1Y4.

**Antonio Di Nola**
Istituto di Matematica
Facoltà di Architettura
Università di Napoli
Via Monteoliveto 3
I-80134 Napoli, Italy.

**Revaz Grigolia**  Institute of Cybernetics
Georgian Academy of Sciences
S. Euli str.,5
380086 Tbilisi, Republic of Georgia.

**Ulrich Höhle**
Fachbereich 7 Mathematik
Bergische Universiät Wuppertal
Gaußstraße 20
D-42097 Wuppertal, Germany.

**Frank Klawonn**  Institut für Betriebssysteme und Rechnerverband
Technische Universität Braunschweig
Bültenweg 74/75
D-38106 Braunschweig, Germany.

**Erich Peter Klement**  Institut für Mathematik
Johannes Kepler Unversität Linz
A-4040 Linz, Austria.

**Ladislav J. Kohout**
Department of Computer Science B–173
and The Institute for Cognitive Sciences
Florida State University
Tallahassee, Florida 32306-4019, USA.

**Christopher J. Mulvey**
Mathematics Division
School of Mathematical and Physical Sciences
University of Sussex, Falmer
Brighton, BN1 9QH, Great Britain.

**Daniele Mundici**
Dipartimento di Scienze della Informazione
Università di Milano
via Comelico, 39/41
I-20135 Milano, Italy.

**Mohammad Nawaz**
Bureau of Curriculum and Extension Centre
Quetta, Baluchistan, Pakistan.

**Vilém Novák**
Institute of Geonics
Academy of Sciences of Czech Republic
Studentská 1768
70800 Ostrava-Poruba, Czech Republic.

**Salvatore Sessa**
Istituto di Matematica
Facoltà di Architetture
Università di Napoli
Via Monteoliveto 3
I-80134 Napoli, Italy.

**Lawrence Neff Stout**
Department of Mathematics
Illinois Wesleyan University
Bloomington, IL 61702-2900, USA.

**Oswald Wyler**
Department of Mathematics
Carnegie Mellon University
Pittsburgh, PA 15213-3890, USA.

# Foreword

The present volume collects the proceedings of the 14th *Linz Seminar on Fuzzy Set Theory*, a research symposium dedicated exclusively to non-classical logics and their applications to fuzzy set theory. This conference took place in the second week of September 1992 at the Bildungszentrum St. Magdalena (Linz, Austria). It was sponsored by the Johannes Kepler Universität Linz and supported by the Austrian Bundesministerium für Wissenschaft und Forschung in Vienna and by the Linzer Hochschulfonds.

As in past years it was the aim of the organizers to bring together a small group of leading researchers from all over the world, for this time with special interests in the fields of intuitionistic logic, Lukasiewicz logic and fuzzy set theory. During this week 26 talks were presented, highlighted by three invited two hour–talks, covering topics reaching from monoidal, lattice–theoretical structures and categorical aspects of non–classical logics to epistomological problems of fuzzy set theory.

Due to the excellent conditions provided by the Bildungszentrum St. Magdalena there was much opportunity for informal and private discussions, used by most of the participants during the breaks or in the evenings. Judging from their statements after returning home, the participants regarded this Seminar as being most stimulating and successful.

An important result of these efforts is the present volume containing twelve selected papers. We express our deepest appreciation to all authors contributing to this book, who agreed to rewrite, polish, and in many cases to enlarge their original presentation in response to the discussions and peer review during the Seminar.

We are therefore convinced that this volume not only surveys the past, but also presents a comprehensive picture of the present work and indicates reasonable directions of future research work in the field. In this sense the present volume is not to be compared with usual conference proceedings — it can be viewed as a first handbook of the mathematical foundations of fuzzy set theory.

The financial support by the Austrian Bundesministerium für Wissenschaft und Forschung in Vienna and the Linzer Hochschulfonds is very gratefully acknowledged, as well as the practical support by the staff of the Bildungszentrum St. Magdalena. Finally, we acknowledge Kluwer Academic Publishers for their cooperation in bringing this project to completion, particularly to Ms. Marie M. Stratta for her capable assistance and patience. Our warmest thanks are due to H.J. Skala, who strongly supported this project from the very beginning.

July 1994                                                         The Editors

# Introduction

Non-classical logics emerged in the early twenties and thirties of this century and go back to the famous papers of J. Lukasiewicz [1920], E.L. Post [1921], A. Heyting [1930], G. Birkhoff and J.v. Neumann [1936]. From a syntactical point of view they are always fragments of classical logic. Disregarding for a moment quantum logic, non-classical logics can be characterized by the abandonment of the *law of the excluded middle* and the maintenance of the integrality, the exportation, importation and Duns Scotus law.

This abandonment can be expressed in various ways : Let $\otimes$ be the associative and commutative logical operation appearing in the exportation and importation law; then the requirement of the idempotency of $\otimes$ implies that the abandonment of the law of the excluded middle is equivalent to the abandonment of the law of *double negation*. Vice versa, if one insists in the law of double negation, then the abandonment of the law of the excluded middle forces the *non–idempotency* of $\otimes$. It is well known that the first case leads to *intuitionistic logic* [Heyting 1930], while the second case can be understood as a kind of integral, commutative *linear logic* (see also J.Y. Girard [1987]). If in the second case we add the axiom of *divisibility*

$$((\alpha \wedge \beta) \to (\alpha \otimes (\alpha \to \beta)))$$

then we obtain the so–called Wajsberg axioms and arrive at the infinite–valued *Lukasiewicz logic*. Each of these logics has its own significant meaning: Intuitionistic logic plays a fundamental role in the foundations of constructive mathematics, Lukasiewicz logic admits *antinomies* and throws a special light on the paradoxes of set theory.

Let us now turn to the semantical aspects of non–classical logics. The definitions of notions like *satisfiability* or *validity* require a *lattice-ordered monoid* $M$, the structure of which corresponds uniquely to the given logical axioms. In accordance with Tarski's matrix method (J. Lukasiewicz and A. Tarski [1930]), respectively the papers of A. Mostowski [1948] ,and H. Rasiowa and R. Sikorski (see also [1970]) we first interpret logical operations as algebraic operations on $M$ and subsequently $m$–ary predicate symbols as M–valued maps. As an immediate consequence of this approach we obtain the elements of *many–valued, non-classical model theory*.

Now we extend our considerations and cover also the theory of *fuzzy sets*, as it was initiated by L.A. Zadeh 1965. It is not difficult to see that fuzzy set theory is not a part of many–valued model theory, but fuzzy set theory is very close to it. Contrary to many–valued model theory, a very special lattice, namely the real unit interval [0,1], is taken as a basis; then once again unary predicate symbols are interpreted as lattice–valued (in this case as [0,1]–valued) maps, conjunction and disjunction are given by the lattice operations min and max, and finally the negation is interpreted by the involution $x \rightsquigarrow 1 - x$ on [0,1]. It is a characteristic of the situation in fuzzy set theory that an explicit use of

3

the implication seems to be abandoned. Therefore the logical background of fuzzy set theory is ambiguous – e.g. it is not clear whether the infinite–valued Lukasiewicz logic is meant or, after giving up the divisibility axiom, some kind of general, integral, commutative, linear logic is applied. The remarkable thing about fuzzy set theory is that contrary to the work of A. Mostowski [1948] lattice–valued maps are understood as *generalized characteristic functions* of a new kind of sets – so-called *fuzzy sets*. In the following years this idea turned out to be very attractive, especially in system theory, and entails a large variety of research papers.

Despite the differences explained above, many–valued model theory and fuzzy set theory have a fundamental, semantical problem in common, namely the interpretation of the underlying *lattice L* as set of *truth values*. Even though there is a general agreement to understand the universal bounds 1 and 0 of $L$ as TRUE and FALSE, the discussion of the semantical meaning of the remaining elements being strictly located between 0 and 1 is quite controversial ([Lukasiewicz 1930, Rosser and Turquette 1952, Gottwald 1989]). Without going into details we would like to take the opportuinity to point out that *local existence*, an idea arising from sheaf theory (see also Scott [1979]), may be the key for the resolution of this problem. In fact, if we admit local existence to our discourse, then each element $\lambda$ of the underlying lattice $L$ represents the *domain* of TRUE. In this context FALSE is TRUE with *empty* domain.

The papers collected in this volume represent fundamental contributions to the above topics. Part A deals with lattice–theoretical and monoidal foundations of non–classical logics. Important properties (e.g. semi-simplicity, free generation, etc.) of MV–algebras, Heyting algebras, integral, commutative Girard–monoids and their mutual relationships are investigated. Part B covers the categorical foundations of non–classical logics and their applications to fuzzy set theory. All papers in this part carry an overall topos-theoretical flavour and deal with the relations between lattice–valued maps viewed as generalized characteristic functions and subobjects in various categories. Part C presents general aspects of non–classical logics and their relevance for applications. This part comprises fundamental investigations in epistemological questions of many–valued logics, studies of interesting, model–theoretical properties of fuzzy logic (invented by J. Pavelka in the late seventies), and a development of PROLOG within the scope of many–valued logics.

The book closes with a bibliography and an index. Even though the bibliography is far from being complete, we are convinced that it covers all important references to the relations between non–classical logics and fuzzy set theory.

July 1994                                                    The Editors

# Part A

# Algebraic Foundations Of Non–Classical Logics

# I

# $\alpha$–Complete MV–Algebras

## L.P. Belluce

**MV**–algebras first appeared in 1958 and were introduced for the purpose of analysing algebraically the many–valued logics of Lukasiewicz [1, 2]. The theory of these algebras developed slowly at first but since the mid 1980's has undergone a rapid growth.

These algebras, we now know, can be studied from various viewpoints, for example as commutative BCK - algebras or as Wajsberg algebras or as certain intervals of lattice–ordered abelian groups with strong order unit ([5]). Of course they may be studied intrinsically, that is, from their defining axioms. This will be the point of view taken here.

Intrinsically we know that an **MV**–algebra becomes a Boolean algebra if we adjoin to the axioms the identity, $x \oplus x = x$. Thus, an **MV**–algebra is, in some intuitive sense, a "non–idempotent Boolean algebra". Hence the theory of Boolean algebras becomes one of the guides to the development of the theory of **MV**–algebras. It can show us what is possible on one hand and give us certain limits on the other hand. What fails in general in the Boolean case must fail in general in the **MV** case since each algebra of the former type is also an algebra of the latter type.

With the above in mind we shall consider herein some principal theorems in the theory $\alpha$ – complete Boolean algebras and their counterparts in the theory of **MV**–algebras.

**Preliminaries.** Recall that an **MV**–algebra is a system $< A, \oplus, \cdot, \bar{\ }, 0, 1 >$ where $< A, \oplus, 0 >$ is an abelian monoid and such that $x+1 = 1$, $\bar{\bar{x}} = x$, $\bar{0} = 1$, $x \cdot y = \overline{\bar{x} \oplus \bar{y}}$ (also written as xy); moreover we must have $x \oplus (\bar{x} \cdot y) = y \oplus (\bar{y} \cdot x)$ .Defining $x \vee y = x \oplus (\bar{x} \cdot y)$ and $x \wedge y = x \cdot (\bar{x} \oplus y)$ the induced system $< A, \vee, \wedge, 0, 1 >$ is a bounded distributive lattice with least element 0 and greatest element 1. The operations $\oplus, \wedge$ are related by $x \oplus (y \wedge z) = (x \oplus y) \wedge (x \oplus z)$.

Clearly there is a duality involving 0, 1, and $\oplus, \cdot$, and $\vee, \wedge$. For simplicity we shall refer to an **MV**–algebra $< A, \oplus, \cdot, -, 0, 1 >$ by its underlying set $A$.

An *ideal* $I$ of $A$ is a non–empty subset of $A$ closed under $\oplus$ and such that $x \le y, y \in I$ imply $x \in I$. Every ideal of $A$ is also an ideal of the lattice

7

$< A, \vee, \wedge, 0, 1 >$ but not conversely. An ideal $P \subseteq A$ is *prime* if $1 \notin P$ and $x \wedge y \in P$ implies $x \in P$ or $y \in P$; an ideal $M \subseteq P$ is *maximal* if $1 \notin M$ and $M$ is not contained in any other proper ideal. Each ideal $I$ gives rise to a quotient algebra $A/I$ via the congruence, $x \equiv y$ iff $d(x,y) = (x \cdot \bar{y}) \oplus (\bar{x} \cdot y) \in I$. If $P$ is prime then $A/P$ is linearly ordered and if $M$ is maximal then $A/M$ is essentially a subalgebra of the unit interval $[0,1]$ where the latter has the operations $x \oplus y = min(1, x+y)$, $\bar{x} = 1 - x$. In every **MV**–algebra $A$ we have ([2]), $\bigcap\{P \mid P$ is a prime ideal of $A\} = 0$. $A$ is called *semi–simple* if $\bigcap\{M \mid M$ is a maximal ideal of $A\} = 0$.

Moreover, each ideal $I$ is the intersection of the prime ideals that contain it. If $S \subseteq A$ then by $idS$ (or $idx$ if $S = \{x\}$) is meant the ideal of $A$ generated by $S$.

If $x \in A$ and for some (least) positive integer $n$ we have $nx = x \oplus x \oplus \cdots \oplus x = 1$ ($n$ summands) we say $x$ has *order* $n$ and write $ordx = n$. If no such $n$ exists we say $x$ has *infinite order* and write $ordx = \infty$. $A$ is said to be *locally finite* provided each $x \in A \setminus \{0\}$ has order $n$ for some $n < \infty$. Each locally finite $A$ is (isomorphic to) a sub–algebra of $[0,1]$. Also, if $M$ is a maximal ideal of an **MV**–algebra $A$ then the quotient algebra is locally finite.

An **MV**–algebra $A$ will be called $\alpha$–*complete*, where $\alpha$ is an infinite cardinal, provided the induced lattice $< A, \vee, \wedge, 0, 1 >$ is $\alpha$ - complete. We shall denote the least upper bound of a subset $S \subseteq A$ by $\sum S$ or by $\sum_i x_i$ if $S$ is indexed by $i$ and $x_i \in S$, and the greatest lower bound by $\prod S$ or $\prod_i x_i$.

Here we want to look at the **MV** analogue of what in Boolean theory is called a field of subsets of a set, or to be more precise, an $\alpha - field$ of subsets of a set, where $\alpha$ is an infinite cardinal.

Recall that every Boolean algebra is isomorphic to a field of subsets of a set and if $\alpha$ –complete is called an $\alpha - field$ of sets, provided the least upper bounds of at most $\alpha$ elements coincide with the set theoretical unions.

The analgoue for **MV**–algebras pursued here shall be called an $\alpha$–complete bold field of (fuzzy) sets, $(\alpha - bffs)$. Let $X$ be a non–empty set, let $A \subseteq [0,1]^X$ be an **MV**–subalgebra, $A$ $\alpha$ –complete. We say $A$ is an $\alpha$ –complete bold field of (fuzzy) subsets of $X$ provided, if $a_i \in A$, $i \leq \alpha$, and if $a = \sum_i a_i$ then for each $x \in X$, $a(x) = \sup_i a_i(x)$. That is, $\sup_i a_i = \sum_i a_i$.

Note that in the treatment of Boolean algebras, if one uses characteristic functions in place of sets, then the set theoretic union corresponds to the *sup*. For this reason, we shall write $\bigcup_i a_i$ in place of $\sup_i a_i$.

Clearly $[0,1]^X$ is an $\alpha - bffs$ for any cardinal $\alpha$ and any non–empty set $X$. Moreover, if $A$ is any $\alpha$ –complete **MV**–algebra, then $A$ is semi–simple, [1], hence $A \subseteq [0,1]^X$ as a subalgebra (up to isomorphism) for some set $X \neq \emptyset$. Of course $A$ need not be an $\alpha - bffs$. In other words not every $\alpha$ –complete bold algebra of (fuzzy) sets $(\alpha - bafs)$ is an $\alpha - bffs$.

We wish to determine in general when an $\alpha - bafs$ is an $\alpha - bffs$. For this purpose, let us call an ideal $I$ of an $\alpha$ –complete algebra $A$ $\quad \alpha - closed$ if whenever $a_i \in I, \quad i \leq \alpha$, then $\quad \sum_i a_i \in I$.

In the Boolean case a theorem of Sikorski [6 , 24.1] tells us: Let $B$ be an $\alpha$ –complete Boolean algebra. Then $B$ is an $\alpha$ –field of sets iff for each $b \in B, \quad b \neq 1 \quad$, there's an $\alpha$ –closed maximal ideal $M \subseteq B$ such that $b \in M$.

The straightforward analogue in the **MV** case would replace $"b \neq 1"$ by $"ord\, b = \infty"$. This, however, fails for let $A = [0,1]^{\mathbb{N}}, \quad \mathbb{N} = \{1, 2, \ldots\}$ and let $x \in A$ be such that $x_n = \frac{1}{n}$. Then $ord\, x = \infty$ but $x \notin I$ for any $\alpha$ - closed ideal $I$ since $\quad \sum_n nx = 1 \quad$. This example suggests an obvious remedy.

Call an element $x$ of an **MV**-algebra $A$ an element of *ultra-infinite order* if $\sum_n nx$ exists in $A$ and $\sum_n nx \quad < 1 \quad$. If we replace $"b \neq 1"$ by $"b$ has ultra–infinite order" then the above theorem on Boolean algebras extends to **MV**-algebras. Another possible analgue in the **MV**-case would be to replace $"b \neq 1"$ by $"b \neq 1, b^2 = b"$. We could also just ask that the Boolean subalgebra, $B(A)$, of idempotents of $A$ be an $\alpha$ –field of sets. All of these turn out to be equivalent; we shall show,

**Theorem I** *Let $A$ be an $\alpha$ –complete **MV**-algebra. Then following are equivalent:*

**i)** *$A$ is an $\alpha - bffs$;*

**ii)** *each $a \in A$, $\quad a$ with ultra-infinite order, belongs to some $\alpha$–closed maximal ideal;*

**iii)** *$Rad_\alpha A = \bigcap \{M \mid M$ an $\alpha$–closed maximal ideal of $\quad A\} = 0$;*

**iv)** *each idempotent $b \in B(A)$, $\quad b \neq 1$, belongs to some $\alpha$–closed maximal ideal of $A$;*

**v)** *$B(A)$ is an $\alpha$–field of sets.*

We shall begin our study with the linearly ordered case.
From Chang [1] we observe that if $A$ is locally finite and infinite then $A$ is atomless and densely ordered. From this it's straightforward to show $\sup\{ord\, x \quad | \quad x \in A, \; x \neq 0\} = \infty$ and, considered as a subalgebra of $[0,1]$, $A$ is a dense subset.

Next,

**Proposition 1** *Let $A$ be an infinite subalgebra of $[0,1]$. Suppose $A$ is $\aleph_0$–complete. Then $A = [0,1]$.*

**Proof.** Let $x \in [0,1] \quad, 0 < x < 1$. Let $L = \{y \in A \mid y \; \leq x\}$, and let $b = \bigcup L$. Then $b \leq x$. For $n = 1, 2, \ldots$, let $y_n \in L$ be such that $b - \frac{1}{n} < y_n$. Let $a = \sum y_n$;

then $a \in A$. Since each $y_n \le a$, we have $\bigcup_n y_n \le a$. But clearly $\bigcup_n y_n = b$, so $b \le a$. Assume $b < a$. Then $a\bar{b} \ne 0$. Since $A$ is infinite, the remarks above imply there is a $c \in A$ with $0 < c < a\bar{b}$. Hence $b + \bar{a} < \bar{c}$, and so $a \wedge b < a\bar{c}$. But $a \wedge b = b$ so $b < a\bar{c} < a$. Since each $y_n \le b$ we see that we must have $a = b$. Suppose $b < x$. Then $x\bar{b} \ne 0$ so for some $u \in A$ we have, $0 < u < x\bar{b}$. Then $b \oplus u < b \oplus x\bar{b} = b \vee x = x$. But $b \oplus u \in A$, so $b \oplus u \in L$. Then $b = b \oplus u$ so $u = 0$. Therefore $b = x$ and we see that $A = [0,1]$.
□

Implicit in the above is the conclusion that $A$ is complete and for any $X \subseteq A$, $\sum X = \bigcup X$.

**Corollary 1** *If $A \subseteq [0,1]$ is an $\alpha$ -complete* **MV** *- algebra, $\alpha \ge \aleph_0$, then $A = [0,1]$ or $A = \{0, \frac{1}{n}, \dots \frac{(n-1)}{n}, 1\}$ for some $n$.*

Now let $A$ be any $\alpha$ -complete **MV**–algebra, $\alpha \ge \aleph_0$ let $B(A)$ be the Boolean subalgebra of idempotents of $A$. Then it is easy to show,

**Proposition 2** $B(A)$ *is $\alpha$-complete.*

The following generalizes Theorem 5C of [3].

**Lemma 1** *Suppose $x$, $x_i$, $i \le \alpha$, are in $A$. Then $x \oplus \sum_i x_i = \sum_i (x \oplus x_i)$.*

**Proof.** For each $i$ we have $x \oplus x_i \le x \oplus \sum_i x_i$. Let $a \in A$ be such that $x \oplus x_i \le a$ for each $i$. Then $x \le a$. Also $\bar{x}(x \oplus x_i)) \le \bar{x}a$ , so $\bar{x} \wedge x_i \le \bar{x}a$ . Thus $\sum_i (\bar{x} \wedge x_i) \le \bar{x}a$ . By [3] we have $\bar{x} \wedge \sum_i x_i \le \bar{x}a$. Hence $x \oplus (\bar{x} \wedge \sum_i x_i) \le x \oplus \bar{x}a$ . Thus we have $(x \oplus \bar{x}) \wedge (x \oplus \sum_i x_i) \le x \vee a$ , that is, $x \oplus \sum_i x_i \le a$ . Hence the assertion follows.
□

**Proposition 3** *For each $x \in A$, $\sum_n nx \in B(A)$.*

**Proof.** Clearly $mx \oplus \sum_n nx = \sum_n nx$ . Thus

$$\sum_n nx \oplus \sum_n nx = \sum_m (mx \oplus \sum_n nx) = \sum_m (\sum_n nx) = \sum_n nx$$

. □

Proposition 3 will play a crucial role in our work. Recall that an **MV**–algebra $A$ is *stonian* if for each $x$, $x^\perp = id(e)$, $(id(S)$ is the ideal of $A$ generated by $S$), where $e \in B(A)$. An ideal $I$ of $A$ is a *stone* ideal if $I = id(I \cap B(A))$. We have;

**Proposition 4** *If $A$ is $\alpha$-complete then $A$ is stonian; if $I$ is an $\alpha$-closed ideal of $A$ then $I$ is a stone ideal.*

**Proof.** Let $x \in A$ and let $e = \sum_n nx$. Clearly $x^\perp = e^\perp$. Thus $\bar{e} \in x^\perp$. Moreover, if $x \wedge y = 0$ , then $e \wedge y = 0$ so $y \le \bar{e}$ . Hence $x^\perp = id(\bar{e})$. Now let $I$ be an $\alpha$-closed ideal and let $J = I \cap B(A)$. Clearly $id(J) \subseteq I$. Let $x \in I$; then $e = \sum_n nx \in I \cap B(A) = J$ . Thus, since $x \le e$, $x \in id(J)$ so $I = id(J)$.
$\square$

We shall show if $P$ is an $\alpha$-closed prime deal then $P$ is maximal. First note,

**Proposition 5** *Let $A$ be $\alpha$-complete, $I$ an $\alpha$-closed ideal. Then $A/I$ is $\alpha$-complete and if $a_i \in A$, $i \le \alpha$, $a = \sum_i a_i$ then $\sum_i(a_i/I) = a/I$.*

**Proposition 6** *Let $A$ be $\alpha$-complete, $P$ an $\alpha$-closed prime ideal. Then $P$ is maximal.*

**Proof.** By the above $A/P$ is $\alpha$-complete; it is also linearly ordered and semi-simple, hence is locally finite. Thus $P$ is a maximal ideal.
$\square$

If $W$ is any **MV**-algebra, let $Rad_\alpha(W)$ be the intersection of all $\alpha$-closed maximal ideals of $W$. We have,

**Proposition 7** *Let $A$ be $\alpha$-complete. Then $Rad_\alpha(A) = 0$ iff $Rad_\alpha(B(A)) = 0$.*

**Proof.** We know if $M$ is an $\alpha$-closed maximal ideal of $A$, then $M = id(M \cap B(A))$ where $M \cap B(A)$ is an $\alpha$-closed maximal ideal of $B(A)$. Moreover, if $N \subseteq B(A)$ is an $\alpha$-closed maximal ideal then $id(N)$ is a prime ideal of $A$ since $A$ is stonian. If $x_i \in id(N)$, $i \le \alpha$, then there are $e_i \in N$ with $x_i \le e_i$. Thus $\sum_i x_i \le \sum_i e_i \in N$. Hence $\sum_i x_i \in id(N)$ so $id(N)$ is $\alpha$-closed, hence maximal. This implies that $Rad_\alpha(B(A)) = (Rad_\alpha A) \cap B(A)$. So $Rad_\alpha(A) = 0$ implies that $Rad_\alpha(B(A)) = 0$. Suppose $Rad_\alpha B(A) = 0$. Let $x \in Rad_\alpha A$. Then $e = \sum_n nx \in Rad_\alpha A \cap B(A)$, so $e = 0$. Hence $x = 0$ and $Rad_\alpha A = 0$.
$\square$

We can now prove the theorem. So let $A$ be an $\alpha$-complete **MV**-algebra.
i) Suppose $A \subseteq [0,1]^X$ is an $\alpha - bffs$. Let $a \in A$ be such that $e = \sum_n na < 1$. By assumption, for each $x \in X$, $\bigcup_n na(x) = e(x)$. Thus for some $x_0 \in X$, $\bigcup_n na(x_0) \ne 1$. Since $a(x_0) \in [0,1]$ this can happen only if $a(x_0) = 0$. Let $M = \{c \in A \mid c(x_0) = 0\}$. Then $M$ is an $\alpha$-closed maximal ideal and

$a \in M$. Thus ii) holds.

Assume ii) holds. Let $a \in Rad_\alpha A$ and suppose $a \neq 0$. Then $e = \sum_n na \in Rad_\alpha A$ and $e \neq 0$. Thus $\bar{e} \neq 1$. Since $\bar{e} \in B(A)$ it has ultra–infinite order. Thus by ii) $\bar{e}$ belongs to some $\alpha$–closed maximal ideal which is impossible. Thus $Rad_\alpha A = 0$, and iii) holds.

Assume $Rad_\alpha A = 0$. By Proposition 7 we see that $Rad_\alpha B(A) = 0$. Let $b \in B(A)$, $b \neq 1$. If $b$ belongs to no $\alpha$–closed maximal ideal of $A$ then $\bar{b} \in Rad_\alpha A$ which implies $b = 1$. Thus iv) holds.

If iv) holds then by Sikorski's theorem v) holds.

If v) is true, then it's easy to see that $Rad_\alpha B(A) = 0$.

Thus by Proposition 7, we have $Rad_\alpha A = 0$. Let $X$ be the set of $\alpha$–closed maximal ideals of $A$. Since $\bigcap X = 0$ we have an embedding $A \longmapsto \prod_{M \in X} (A/M)$.

Each $A/M$ is $\alpha$–complete and a subalgebra of $[0,1]$. Thus we have $A \longmapsto [0,1]^X$ where if $M \in X$, $a(M) = a/M \in [0,1]$. Let $a_i \in A$, $i \leq \alpha$, and let $a = \sum_i a_i$. By Propositions 1, 5 we have $\sum_i a_i(M) = \bigcup_i a_i(M) = a(M)$ for $M \in X$. Thus $A$ is an $\alpha - bffs$.

In the above, we have freely identified $A/M$ with its image in $[0,1]$.

Another question of interest about $\alpha$–fields of sets is knowing which homomorphic images modulo an $\alpha$–closed ideal are again $\alpha$–fields of sets. We can raise the same question for $\alpha - bffs$.

For the Boolean case Sikorski [6] answers this as follows:

*Let $F$ be an $\alpha$–field of subsets of a set $X$, $I$ an $\alpha$–closed ideal of $F$. If there's an $e \in 2^X$ such that $I = e^\perp \cap F$ then $F/I$ is an $\alpha$–field of sets. Moreover, if each $\alpha$–closed maximal ideal $M \subseteq F$ has the form $M = \{a \in F \mid x_0 \notin a\}$ for some $x_0 \in X$ then the above condition is necessary .*

We shall show this theorem extends to the **MV** case.

Thus again, let $A$ be an $\alpha$–complete **MV**–algebra and $J$ an $\alpha$–closed ideal of $A$. Let $I = J \cap B(A)$. Then since $B(A)$ is $\alpha$–complete we see that $I$ is an $\alpha$–closed ideal of $B(A)$.

Let $\alpha - SpecW$ denote the set of $\alpha$–closed maximal ideals of $W$, $W$ an **MV**–algebra. Then,

**Proposition 8**   $J = \bigcap\{id(N) \mid N \in \alpha - SpecB(A), \quad I \subseteq N\}$
*iff*   $I = \bigcap\{N \mid N \in \alpha - SpecB(A), \quad I \subseteq N\}$.

**Proof.** Clearly, for $N \in \alpha - SpecB(A)$, $I \subseteq N$ iff $J = id(I) \subseteq id(N)$. Suppose then $J = \bigcap\{id(N) \mid I \subseteq N \in \alpha - SpecB(A)\}$. Let $x$ belong to $\bigcap\{N \mid I \subseteq N \in \alpha - SpecB(A)\}$. Then clearly $x \in J$ and so $x \in J \cap B(A) = I$. On the other hand, assume that $I = \bigcap\{N \mid I \subseteq N \in \alpha - SpecB(A)\}$. Let $x \in id(N)$, $I \subseteq N \in \alpha - SpecB(A)$. Then   $e = \sum_n nx \in id(N)$,

$I \subseteq N \in \alpha - SpecB(A)$. Then $e = \sum_{n} nx \in N$, $\quad I \subseteq N \in \alpha - SpecB(A)$ and so $e \in I$. Thus $x \in id(I) = J$.

□

Since every $\alpha$–closed maximal ideal $M \subseteq A$ that contains $J$ is of the form $id(N)$ for some $\alpha$ - closed maximal ideal $N \subseteq B(A)$, $\quad I \subseteq N$, we have,

**Proposition 9** $A/J$ *is an* $\alpha - bffs$ *iff* $B(A)/I$ *is an* $\alpha$-*field of sets* .

**Proof.** If $A/J$ is an $\alpha - bffs$ then $Rad_\alpha(A/J) = 0$. By the above proposition $Rad_\alpha(B(A)/I) = 0$ and so by Sikorski's theorem $B(A)/I$ is an $\alpha$–field of sets. The argument is reversible.

□

Suppose now $A \subseteq [0,1]^X$. If $u \in [0,1]^X$ is such that $J = u^\perp \cap A$ then let $e_0 = \sum_{n} nu$. Then $e_0^\perp = u^\perp$, $\quad e_0 \in B([0,1]^X)$. Thus $J = e_0^\perp \cap A$ and so $I = e_0^\perp \cap B(A)$. Thus by Sikorski [6], $B(A)/I$ is an $\alpha$–field of sets hence by Proposition 9, $A/J$ is an $\alpha - bffs$. Suppose now that if $M \in \alpha - SpecA$ then $M = \{a \in A \mid a(x_0) = 0\}$ for a fixed $x_0 \in X$. Let $N \in \alpha - SpecB(A)$. Then $N = M \cap B(A)$ for some $M \in \alpha - SpecA$. Thus $N = \{e \in B(A) \mid e(x_0) = 0\}$. If $A/J$ is an $\alpha - bffs$ then so is $B(A)/I$. Hence by [6] there is an $e_0 \in B([0,1]^X)$ such that $I = e_0^\perp \cap B(A)$. It follows that $J = e_0^\perp \cap A$. So we conclude

**Theorem II** *Let* $A$ *be an* $\alpha$ - *complete subalgebra of* $[0,1]^X$ *and let* $J$ *be an* $\alpha$–*closed ideal. If for some* $x_0 \in [0,1]^X$*, we have* $J = x_0^\perp \cap A$ *then* $A/J$ *is an* $\alpha - bffs$*. Moreover if each* $M \in \alpha - SpecA$ *is determined (in the manner above) by some* $x_0 \in X$ *then the condition is necessary.*

The general question as to which $\alpha$–complete **MV**–algebras are representable in the form $F/J$, $\quad F$ an $\alpha - bffs$, $\quad J$ an $\alpha$–closed ideal now arises. In the Boolean case, this is answered by the Loomis–Sikorski theorem:
*If* $B$ *is an* $\aleph_0$–*complete Boolean algebra then there exists an* $\aleph_0$–*complete field of sets* $\Re$ *and an* $\aleph_0$–*ideal* $\wp$ *such that* $B \cong \Re/\wp$*. If* $\alpha > \aleph_0$ *then this statement is false.*

The corresponding question for **MV**–algebras is open.

Another problems of interest along these lines is to determine when an algebra is the "set of all subsets" of a given set. In the Boolean case a necessary and sufficient condition for this to happen is that the algebra be complete and atomic.

By contrast, the algebra $[0,1]^X$ of all (fuzzy) subsets of a non–empty set $X$ is atomless. It is clear that $B([0,1]^X)$ is complete and atomic. Moreover, for each atom $e \in B([0,1]^X)$ there are $a \in [0,1]^X$ such that $0 < a < e$. These conditions, as we shall show, are sufficient though we shall phrase them somewhat differently.

Let $A$ be an **MV**algebra, $a \in A$,  $a \neq 0$. We shall say $a$ is a *sub–boolean atom* (*subb - atom*) if $a < e$ for some atom $e \in B(A)$. We shall call $A$ *subb–atomic* if for each $x \in A$, $x \neq 0$, there is a subb–atom $a < x$, $a \in A$ (note the strict inequality).

We shall prove

**Theorem III** *A is isomorphic to the set of all (fuzzy) subsets of some non–empty set $X$ iff $A$ is complete and subb–atomic.*

Clearly $[0,1]^X$ is complete and subb–atomic for any non–empty set $X$. Now let $A$ be a complete and subb–atomic **MV**–algebra. Then $B(A)$ is a complete sub–algebra of $A$.

**Proposition 10** $B(A)$ *is atomic.*

**Proof** Let $b \in B(A)$,  $b \neq 0$. There's subb–atom $a \in A$,  $a < b$. By definition, $a < e$ for some atom $e \in B(A)$. Hence $a \leq b \wedge e = be$. Thus $be \neq 0$ and since $e$ is an atom of $B(A)$ we must have $e \leq b$.
□

Let $X$ be the set of atoms of $B(A)$. If $e \in X$ then from [3] we know that the ideal of $A$ generated by $e$, $id(e)$, can be made into an **MV**–algebra $A_e = < id(e), \oplus, \cdot, \sim, 0, e >$ where $\tilde{x} = \bar{x}e$. Moreover, if $M_e = e^\perp$ we have an isomorphism of $A/M_e$ onto $A_e$.

**Lemma 2** $M_e$ *is a closed maximal ideal of $A$.*

**Proof.** $M_e$ is clearly a closed ideal. Let $N = M_e \cap B(A)$, that is, $N = e^\perp \cap B(A)$. If $b \in B(A)$,  $b \notin N$, then $be \neq 0$. Hence, since $e \in X$,  $be = e$. Thus $\bar{b}e = 0$ and so $\bar{b} \in N$. So $N$ is a closed maximal ideal of $B(A)$. $A$ is complete, thus stonian, so $id(N)$ is a prime [4] and closed ideal of $A$, hence is maximal. But $id(N) \subseteq M_e$ and so $M_e$ is a closed maximal ideal.
□

Since $A$ is atomless it is clear that $A_e$ is infinite thus by Proposition 1, $A_e \cong [0,1]$. So we can identify $\prod_e A_e$ with $[0,1]^X$. Moreover, $\bigcap_e M_e = 0$. To see this suppose $x \in A$ is such that $xe = 0$ for all $e \in X$. Then $e \leq \bar{x}$ and so, $\sum X \leq \bar{x}$. But $\sum X = 1$ since $B(A)$ is atomic and so we way infer $x = 0$. From this we have that $A$ is a subdirect product of the $A_e$. We claim,

**Proposition 11** $A \cong \prod_e A_e$.

**Proof.** Let $< a_e > \in \prod_e A_e$, $a_e \in A_e$. Then $\{a_e \mid e \in X\} \subseteq A$. Let $a = \sum \{a_e \mid e \in X\}$. Thus, $a \in A$. The subdirect embedding $A \longmapsto \prod_e A_e$ is given by $u \longmapsto < ue >$. Hence $a \longmapsto < ae >$. Let $e_0 \in X$. Then $e_0 a = \sum_{e \in X} a_e e_0$. For $a_e \in A_e$ we have $a_e \leq e$, hence $a_e e_0 = 0$ for $e \neq e_0$ and $a_e e_0 = a_e$ for $e = e_0$. It follows that $ae = a_e$ for each $e \in X$, so $a \longmapsto < a_e >$ and the proposition follows.
□

From the remarks preceeding the proposition and the proposition itself we may infer that $A \cong [0, 1]^X$, thus the proof of Theorem III is complete.

By definition a subb–atomic algebra is atomless. The converse, however, is false. For let $B$ be a complete and atomless Boolean algebra and let $A = B \times [0, 1]$. Then $A$ is complete and atomless. The only atom of $B(A)$ is $(0, 1)$ and so the only subb–atoms of $A$ are the elements of the form $(0, x)$, $x \neq 1$. It follows that $A$ is not subb–atomic.

Another example of interest along these lines is the algebra $A = \mathbb{Q}^X$ where $X \neq \emptyset$ and $\mathbb{Q}$ is the **MV**–algebra of rational numbers in $[0, 1]$. Here $A$ is subb-atomic, $B(A)$ is complete and atomic, but $A$ is not complete.

The above results shed some light on the general representation theorem of complete $MV$-algebras. From [3] we know that if $A$ is complete then $A \cong A_1 \times A_2$ where $A_1$ is complete and atomic and $A_2$ is complete and atomless. The structure of $A_1$ is known; $A_1$ is a direct product of a family of finite locally finite **MV**–algebras. The structure of $A_2$ is more troublesome.

We have,

**Proposition 12** *If $A$ is complete and atomless and $B(A)$ is atomic then $A$ is subb-atomic.*

    **Proof.** Let $x \in A$, $x \neq 0$. If $x \wedge e = 0$ for all atoms $e$ of $B(A)$ then $x = 0$ since the supremum of all the atoms of $B(A)$ is 1. Thus for some atom $e \in B(A)$ we have $x \wedge e \neq 0$. Since $A$ is at0mless there exists an $a \in A$, $0 < a < x \wedge e$. Hence $a$ is a subb–atom and $a < x$.
□

From this we obtain

**Proposition 13** *If $A$ is a complete $MV-$algebra and $B(A)$ is atomic then $A \cong A_1 \times A_2$ where $A_1$ is complete and atomic and $A_2$ is complete and subb-atomic.*

**Proof.** It suffices to show that $B(A_2)$ is atomic. To this end let $e \in B(A_2)$, $e \neq 0$. Then $(0, e) \in B(A)$. Let $(a, b)$ be an atom of $B(A)$, $(a, b) \leq (0, e)$. Then $a = 0$ and $b \leq e$. Clearly $b$ is an atom of $B(A_2)$ hence $B(A_2)$ is atomic.
□

The converse of the above is also true. That is, if $A \cong A_1 \times A_2$ where $A_1$ is complete and atomic and $A_2$ is complete and subb-atomic then $A$ is complete and $B(A)$ is atomic.

To see this we observe that the direct product of two complete algebras is complete so we need only show that $B(A)$ is atomic. By Proposition 10 we have that $B(A_2)$ is atomic. Since the product of two atomic Boolean algebras is atomic and since $B(A) \cong B(A_1) \times B(A_2)$ it will suffice to show $B(A_1)$ atomic. Let $e \in B(A_1)$, $e \neq 0$. Since $A_1$ is atomic there's an atom $a \in A_1$, $a \leq e$. By Proposition 3, $e_0 = \sum_n na \in B(A_1)$. Since $na \leq e$ for all $n$, $e_0 \leq e$. Now let $f \in B(A_1)$; then $f \wedge e_0 = \sum_n f \wedge na = \sum_n n(f \wedge a)$ . (see [3]; we are also using the fact that idempotents distribute over $\oplus$). If $e_0 \not\leq f$ then $a \not\leq f$ so $f \wedge a = 0$, thus $f \wedge e_0 = 0$. If $e_0 \leq f$ then $a \leq f$ and so $f \wedge e_0 = e_0$. Thus $e_0$ is an atom of $B(A_1)$, hence $B(A_1)$ is atomic.

We would like to add that complete and atomless $MV$−algebras are part of the way to being injective. Complete and atomless does not imply injectivity, since there exist complete and atomless Boolean algebras and no Boolean algebra is injective in the category of $MV$−algebras. For a complete $MV$−algebras to be injective it must satisfy a divisibility condition; these divisibility conditions force the algebra to be atomless in a strong sense. We shall be precise.

Call an $MV$−algebra $A$ *strongly-atomless* (or simply, *st-atomless*) if for each $a \in A$, $a \neq 0$, there exists a $P \in SpecA$, $a \notin P$, and $a/P$ <u>not</u> an atom of $A/P$.

First we observe,

**Proposition 14** *If $A$ is st-atomless then $A$ is a atomless.*

**Proof.** For suppose $A$ has an atom $a$. Let $P \in SpecA$, $a \notin P$. Suppose for some $y \in A$ we have $0 < y/P \leq a/P$. Then $\bar{a}y \in P$.    $a$ is an atom, so $a\bar{y} = 0$ or $a\bar{y} = a$. If $a\bar{y} = a$ ,then by 1.15 of [1], $a \oplus \bar{y} = \bar{y}$. Hence $a \wedge y = y(a \oplus \bar{y}) = 0$.   $a \notin P$, thus $y \in P$ so $0 = y/P$. Therefore we have $a\bar{y} = 0$. Thus $d(a, y) \in P$ so $y/P = a/P$ and we see that $a/P$ is an atom. Since $A$ is st-atomless this is absurd. Hence $A$ is an atomless.
□

If $A$ is linearly ordered and atomless then it is trivially st-atomless.

We have the characterization.

**Proposition 15** *A is st-atomless iff for each $a \in A$, $a \neq 0$, there is a $z \in A$, $0 < z < a$, with $z \wedge a\bar{z} \neq 0$.*

**Proof.** Suppose $A$ is st-atomless but for some $a \in A$, $a \neq 0$, we have $z \wedge a\bar{z} = 0$ for all $z \leq a$. Let $y \in A$. Then for $z = a \wedge y$ we have $z \wedge a\bar{z} = 0$. Now $a\bar{z} = a(\bar{a} \vee \bar{y}) = a\bar{y}$ so $z \wedge a\bar{z} = y \wedge a\bar{y} = 0$. Let $P \in SpecA$, $a \notin P$. Suppose for some $y \in A$ that $0 < y/P < a/P$. Since $y \wedge a\bar{y} = 0$, $y \wedge a\bar{y} \in P$. Thus $y \in P$ or $a\bar{y} \in P$. $y \in P$ implies $y/P = 0$ and $a\bar{y} \in P$ implies $a/P \leq y/P$. Thus $a/P$ is an atom and this is impossible. Conversely, suppose the condition holds. Let $a \in A$, $a \neq 0$. Then there is a $z \in A$, $0 < z < a$, with $z \wedge a\bar{z} \neq 0$. Let $P \in SpecA$ be such that $z \notin P$. Then $a \notin P$ and $0 < z/P \leq a/P$. If $z/P = a/P$ then $a\bar{z} \in P$, thus $z \wedge a\bar{z} \in P$. So $z/P < a/P$ and $a/P$ is not an atom. Ergo $A$ is st-atomless.
□

Let us say that an *MV*-algebra $A$ is *sub-divisible* if for each $a \in A$, $a \neq 0$ and each integer $n > 0$, there exists an $x \in A$, $x \neq 0$, with $nx \leq a$.

The following extends 3.20 of [1].

**Proposition 16** *If A is st-atomless then A is sub-divisible.*

**Proof.** We induct on $n$, the statement being obvious for $n = 1$. Assume true for $n$. Let $a \in A$, $a \neq 0$. There exists an $x \in A$, $x \neq 0$, such that $nx \leq a$. Since $A$ is st-atomless there is a $z$, $0 < z < x$ with $z \wedge x\bar{z} \neq 0$. Let $w = z \wedge x\bar{z}$. Let $P \in SpecA$. Then $0 \leq w/P \leq z/P \leq x/P$. By [1, pg. 480] either $(z \oplus z)/P \leq x/P$ or $(x\bar{z} \oplus x\bar{z})/P \leq x/P$. In either case $(w \oplus w)/P \leq x/P$. Since $P$ is arbitrary we may infer $w \oplus w \leq x$. Now we have, $(n+1)w \leq 2nw = n(w \oplus w) \leq nx \leq a$. Since $w \neq 0$ the statement follows for $n + 1$.
□

Define $A$ to be *weakly-divisible* if $a \in A$, $a \neq 0$, and integer $n > 0$ there exists a unique, least element $b \in A$ with $nb = a$. Call $A$ *divisible* if in addition $b$ also satisfies $a\bar{b}^{n-1} = b$.

It is known that $A$ is injective iff $A$ is complete and divisible ([7]). Also, it is clear that if $A$ is weakly divisible then $A$ need not be atomless, as every Boolean algebra is weakly divisible. But, by contrast,

**Proposition 17** *If A is divisible then A is st-atomless.*

**Proof.** Let $a \in A$, $a \neq 0$. Let $b \in A$ be such that $2b = a$ and $a\bar{b} = b$. Then $0 < b < a$. Let $P \in SpecA$, $b \notin P$. Then $a \notin P$ and $0 < b/P \leq a/P$. If $b/P = a/P$ then $a\bar{b} \in P$. But $a\bar{b} = b$. Hence $0 < b/P < a/P$ and $a/P$ is not an atom.
□

**Corollary 2** *If $A$ is injective then $A$ is complete and st-atomless.*

This raises the question of the converse of the above to which we have no answer. In this regard we do have,

**Theorem IV** *If $A$ is complete and st-atomless then $A$ is weakly divisible.*

**Proof.** Let $a \in A$, $a \neq 0$ and let $n$ be a positive integer. Let $\Gamma = \{y \mid ny \leq a\}$. By Proposition 16 $\Gamma \neq \emptyset$ and $\Gamma \neq 0$. Index $\Gamma$ so that $\Gamma = \{x_\gamma \mid \gamma \in \Gamma\}$. Let $u = \sum \Gamma = \sum_\gamma x_\gamma$. Clearly $\sum_\gamma nx_\gamma \leq a$. We claim $nu = \sum_\gamma nx_\gamma$. It is clear that $nx_\gamma \leq nu$ for each $\gamma \in \Gamma$. Suppose $nx_\gamma \leq z$ for each $\gamma \in \Gamma$. Let $P \in SpecA$, $\gamma_1, \gamma_2, ..., \gamma_n \in \Gamma$. Then for $1 \leq i \leq n$, $x_{\gamma_i}/P \leq z/P$. Let $t = (x_{\gamma_1} \vee ... \vee x_{\gamma_n})/P$; as $A/P$ is linearly ordered, $x_{\gamma_1}/P \oplus ... \oplus x_{\gamma_n}/P \leq nt \leq z/P$.

Since $P$ is arbitrary we infer $x_{\gamma_1} \oplus ... \oplus x_{\gamma_n} \leq z$ and so $\sum_{\gamma_i \in \Gamma} (x_{\gamma_1} \oplus ... \oplus x_{\gamma_n}) \leq z$. By an obvious extension of Lemma 1, we obtain $u \oplus u = \sum_\gamma (u \oplus x_\gamma) = \sum_{\gamma, \gamma'} (x_\gamma \oplus x_{\gamma'})$ or in general, $nu = \sum_{\gamma_i} (x_{\gamma_1} \oplus ... \oplus x_{\gamma_n})$. Thus $nu \leq z$. Now we have $nu = \sum_\gamma nx_\gamma \leq a$. Suppose then that $nu < a$. Let $s = a(\overline{nu})$. So $s \neq 0$. For some $w \neq 0$ we have $nw \leq s$. Hence, $n(u \oplus w) = nu \oplus nw \leq nu \oplus s = nu \oplus a(\overline{nu}) = nu \vee a = a$. Therefore $u \oplus w \in \Gamma$ so $u \oplus w = u$. Hence $0 = \bar{u}u = \bar{u}(u \oplus w) = \bar{u} \wedge w$. Now $w \leq nw \leq s \leq \overline{nu}$, so $w \leq \bar{u}^n \leq \bar{u}$. Thus $\bar{u} \wedge w = w$ so $w = 0$. This shows that $s = 0$ and so $nu = a$. Consider now the set $\Delta = \{y \mid ny = a\}$. Let $b = \prod \Delta$. If $ny = a$ then $b \leq y$; so $nb \leq a$. By an extension of Theorem 5c [3] we may obtain $nb = \prod \{y_1 \oplus ... \oplus y_n \mid y_c \in \Delta\}$. Choose any $y_1, ..., y_n \in \Delta$ and any $P \in SpecA$. We can assume $y_1/P \leq y_i/P$, $1 \leq i \leq n$. Then $a/P = ny_1/P \leq y_1/P \oplus ... \oplus y_n/P$. Since $P$ is arbitrary, $a \leq y_1 \oplus ... \oplus y_n$. Thus $a \leq nb$ and we have $nb = a$. Clearly $b$ is the least such and must be unique.
□

We point out that complete, atomless and weakly divisible does not imply injectivity since the complete atomless Boolean algebras are weakly divisible. Observe, however, that no Boolean algebra is st-atomless since the quotient of a Boolean algebra by a prime ideal is always a 2 element algebra. With this in mind we have,

**Conjeture** If $A$ is complete and st-atomless then $A$ is injective.

In the final part of this paper we want to present some results concerning $SpecA$ when $A$ is complete, $SpecA = \{P \mid P$ a prime ideal of $A\}$.

Recall that a topological space is *extremally disconnected* iff the closure of each open set is open. For Boolean spaces it is well known that such a space is extremally disconnected iff the corresponding Boolean algebra is complete.

For **MV**−algebras, the analogue of the above mentioned result is much stronger insofar as completeness of the entire algebra is not necessary.

An **MV**−algebra $A$ called stonian provided for each $x \in A$, $x^{\perp} = id(e)$, $e \in B(A)$. Extending this notion, let us call an **MV**−algebra A *strongly stonian* (or just *sstonian*) iff for each ideal $I \subseteq A$, $I^{\perp} = id(e)$ for some $e \in B(A)$. (Clearly sstonian implies stonian since $x^{\perp} = id(x)^{\perp}$.).

With this concept the situation for **MV**−algebras with regard to extremally disconnected spectrum is,

**Theorem V** *Let $A$ be an* **MV**−*algrbra. Then, $SpecA$ is extremally disconnected iff $A$ is sstonian.*

The relation to complete **MV**-algebras is expressed in the following proposition

**Proposition 18** *If $A$ is a complete MV-algebra then $A$ is sstonian.*

**Proof.** Let $I$ be an ideal of $A$. Let $e = \sum I^{\perp}$. If $x \in I$ then $x \wedge e = \sum\{x \wedge y \mid y \in I^{\perp}\} = 0$. Thus $e \in I^{\perp}$. Hence $e \oplus e \in I^{\perp}$ so $e = e \oplus e$ and we may conclude that $e \in B(A)$ and $I^{\perp} = id(e)$.
□

In order to prove Theorem V we need several preliminary results. To this end let $A$ be a given **MV**−algebra and let $X = SpecA$. If $S \subseteq X$ let $\overline{S}$ denote the closure of $S$. We know that if $U \subseteq X$ is open then for some ideal $I \subseteq A$, $U = U(I) = \{P \in X \mid I \not\subseteq P\}$. Similarly if $V \subseteq X$ is closed then for some ideal $J \subseteq A$, $V = V(J) = \{P \in X \mid J \subseteq P\}$.

**Lemma 3** *Let $I \subseteq A$ be an ideal. Then $\overline{U(I)} = V(I^{\perp})$.*

**Proof.** We know $\overline{U(I)} = V(J)$ for some ideal $J \subseteq A$. Thus , for $P \in X$, $I \not\subseteq P$ implies $J \subseteq P$. Assume that $I \cap J \neq 0$. Then for some $P \in X$, $I \cap J \not\subseteq P$. Hence $I \not\subseteq P$, but then we have $J \subseteq P$ and so $I \cap J \subseteq P$. Therefore we have $I \cap J = 0$. Let $I \wedge J = id\{x \wedge y \mid x \in I, y \in J\}$. Since $I \wedge J \subseteq I \cap J$ we see that $J \subseteq I^{\perp}$. Hence $V(I^{\perp}) \subseteq V(J)$. Now $I \cap I^{\perp} = 0$ so $I \not\subseteq P$ implies $I^{\perp} \subseteq P$. Thus $U(I) \subseteq V(I^{\perp})$. It follow that $\overline{U(I)} \subseteq V(I^{\perp})$, so $V(J) \subseteq V(I^{\perp})$. Thus, $\overline{U(I)} = V(I^{\perp})$.
$\square$

**Proposition 19** *If $X$ is extremally disconnected then $A$ is sstonian.*

**Proof.** Let $I \subseteq A$ be an ideal. From Lemma 3 we know $\overline{U(I)} = V(I^{\perp})$ . Since $X$ is extremally disconnected we have $\overline{U(I)}$ is open. Thus for some ideal $J \subseteq A$ we obtain $V(I^{\perp}) = U(J)$. Hence for $P \in X$, $I^{\perp} \subseteq P$ iff $J \not\subseteq P$. $I^{\perp} \cap J \neq 0$ implies $I^{\perp} \cap J \not\subseteq P$ for some $P \in X$; but $J \not\subseteq P$ implies $I^{\perp} \subseteq P$ and so $I^{\perp} \cap J \neq 0$ is not possible. Thus, $I^{\perp} \wedge J = 0$ , and so $I^{\perp} \subseteq J^{\perp}$. If $I \not\subseteq P$, $P \in X$, then $I^{\perp} \subseteq P$ so $J \not\subseteq P$. Hence $J \subseteq P$ implies $I \subseteq P$. That is, $\{P \in X \mid J \subseteq P\} \subseteq \{P \in X \mid I \subseteq P\}$. Thus, $\bigcap\{P \in X \mid I \subseteq P\} \subseteq \bigcap\{P \in X \mid J \subseteq P\}$. But $I = \bigcap\{P \in X \mid I \subseteq P\}$ and $J = \bigcap\{P \in X \mid J \subseteq P\}$, so $I \subseteq J$ and we infer $J^{\perp} \subseteq I^{\perp}$. Now we have $I^{\perp} = J^{\perp}$. Continuing, we also have that $X = U(I^{\perp}) \cup V(I^{\perp})$ clearly so we also have $X = U(J^{\perp}) \cup U(J) = U(J \oplus J^{\perp})$ where $J \oplus J^{\perp}$ is $id\{x \oplus y \mid x \in J, \ y \in J^{\perp}\}$. Therefore we must have $J \oplus J^{\perp} = A$. So for some $x \in J$, $y \in J^{\perp}$ we have $x \oplus y = 1$, $x \wedge y = 0$. From this we obtain an $e \in B(A)$ such that $J = id(\bar{e})$ and $J^{\perp} = id(e)$. That is, $I^{\perp} = id(e)$ and we conclude that $A$ is sstonian.
$\square$

To complete the proof of Theorem V let us suppose that $A$ is sstonian. Let $U = U(I)$ be open in $X$, $I$ an ideal of $A$. We know $\overline{U(I)} = V(I^{\perp})$ .Since $A$ is sstonian ,there exists an $e \in B(A)$ with $I^{\perp} = id(e)$. So $\overline{U(I)} = V(e)$ . But $V(e) = U(\bar{e})$, thus $\overline{U(I)} = U(\bar{e})$ . Hence $\overline{U(I)}$ is open and $X$ is extremally disconnected.

We saw above that $A$ complete implies $A$ sstonian. Thus,

**Proposition 20** *If $A$ is a complete $MV$–algebra then $SpecA$ is extremally disconnected.*

Note that the converse of the above is false. For consider the algebra $C$ in [1]. Then $C$ is sstonian since $I^{\perp} = 0$ for all ideals $I \subseteq C$ but $C$ is not complete. The same is true for any cartesian power of $C$.

There is, however, a partial completeness imposed on $A$ when $SpecA$ is extremally disconnected namely,

**Proposition 21** *If $A$ is sstonian then $B(A)$ is complete.*

**Proof.** Let $\alpha$ be a cardinal number and assume that $e_i \in B(A)$, $i \leq \alpha$. Let $I = id\{e_i \mid i \leq \alpha\}$ be the ideal in $A$ generated by the $e_i$, $i \leq \alpha$. For some $e \in B(A)$, $I^{\perp} = id(\bar{e})$. Thus $e_i \wedge \bar{e} = 0$ for each $i \leq \alpha$ so $e_i \leq e$, $i \leq \alpha$. Suppose $f \in B(A)$ and $e_i \leq f$ for each $i \leq \alpha$. Then $e_i \wedge \bar{f} = 0$, $i \leq \alpha$, so $\bar{f} \in I^{\perp}$, so $\bar{f} \leq \bar{e}$, hence $e \leq f$. It follows that $B(A)$ is complete.
□

We note that the converse of the above is not true. Just take $A$ to be the subalgebra of $C \times C$ consisting of those pairs of the form $(nc, mc)$ or $(1 - nc, 1 - mc)$ where $n, m$ are non–negative integers and $c$ is the atom of $C$. Then $B(A)$ is finite hence complete but $A$ is not stonian (therefore not sstonian) since $(c, 0)^{\perp} = \{(0, nc) \mid n = 0, 1, 2, ...\}$ is not principal.

# References

1 C.C. Chang, *Algebraic- analysis of many valued logics*, Trans. Amer. Math. Soc. 88(1958) 467 - 490.

2 _____ ,*A new proof of the completeness of the Lukasiewicz axioms*, Trans. Amer. Math. Soc. 93 (1959) 74-80.

3 L.P. Belluce,*Semi-simple and complete* **MV**–*algebras*, Algebra Universalis, 29(1992), 1 - 9.

4 _____ , A. Di Nola, S. Sessa, *The prime ideal space of an* **MV**–*algebra*, Math. Logic Quartely **40** (1994).

5 D. Mundici,*Interpretation of AF $C^{*}$–algebras in Lukasiewicz sentential logic*, J. Functional Anal. 65(1986) 15 - 63.

6 R. Sikorski,*Boolean Algebras*, Springer - Verlag, (1960).

7 F. Lacava, *Sulle L-algebre iniettive*, Bolletino UMI (7) 3-A (1989), 319 - 324.

# II

# On MV–algebras of continuous functions

## A. Di Nola and S. Sessa

Abelian lattice ordered groups with strong unit and MV-algebras are categorically equivalent. The restriction of this equivalence to archimedean lattice ordered groups gives exactly the semisimple MV-algebras. Consequently, an MV-algebra is semisimple iff it is (up to isomorphisms) subalgebra of the MV-algebra of all $[0,1]$-valued continuous functions defined on a compact Hausdorff topological space. $\sigma$-complete, in general hypernormal MV-algebras of continuous functions, defined on suitable spaces, are also investigated.

## 1 Introduction

An MV-algebra is a system $A = (A, \oplus, \odot, *, 0, 1)$ such that $(A, \oplus, 0)$ is an Abelian monoid, $x \oplus 1 = 1$, $x^{**} = x$, $0^* = 1$, $x \odot y = (x^* \oplus y^*)^*$, $(x^* \oplus y)^* \oplus y = (y^* \oplus x)^* \oplus x$ for all $x, y \in A$. In accordance to [8, 1.11], by setting $x \vee y = (x \odot y^*) \oplus y$, $x \wedge y = (x \oplus y^*) \odot y$ and $x \leq y$ iff $x \wedge y = x$ for all $x, y \in A$, the system $L(A) = (A, \vee, \wedge, \leq, *, 0, 1)$ is a Kleene algebra, i.e., a bounded distributive lattice such that $(x \vee y)^* = x^* \wedge y^*$, $(x \wedge y)^* = x^* \vee y^*$, $(x \wedge x^*) \leq (y \vee y^*)$ for all $x, y \in A$. Thus A is said complete, $\sigma$-complete, Stonian, etc. iff its underlying lattice $L(A)$ has the respective same property. For all the unexplained notions on MV-algebras, we refer to [8] and [23].

The unit real interval $[0, 1]$ becomes an MV-algebra [8, p.473] by setting $x \oplus y = \min\{1, x + y\}$, $x^* = 1 - x$, and $x \odot y = \max\{0, x + y - 1\}$ for all $x, y \in [0, 1]$ and the natural order coincides with the lattice order of this MV-algebra.

Considering any topological space (shortly in the sequel, space) and $[0, 1]$ endowed with the natural topology, the family $C(X)$ of all $[0, 1]$-valued continuous functions defined on $X$ has a structure of MV-algebra, induced pointwise by the MV-operations on $[0, 1]$. The same operations induce on $[0, 1]^X$, if $X$ is a nonempty set, the MV-algebra of all the fuzzy sets of $X$, called usually Bold algebra of fuzzy sets of $X$ [2, p. 1358].

An MV-algebra $A$ is locally finite [8, 3.10] iff each $x \in A \setminus \{0\}$ has finite order $n \in \mathbb{N}$ (the positive integers), i.e.

$$\oplus nx = \underbrace{x \oplus \ldots \oplus x}_{n-\text{times}} = 1$$

$A$ is locally finite iff $A$ is subalgebra of $[0, 1]$ [9, Lemma 6]. We shall denote with $\text{Min}A$, $\text{Max}A$ and $\text{Spec}A$ the spaces (endowed with the Zariski topology) of the minimal, maximal and prime ideals of $A$, respectively. The radical of $A$, $\text{Rad}A$, is the intersection of all the maximal ideals of $A$ and $A$ is semisimple iff $\text{Rad}A = \{0\}$ [8, Theorem 4.9].

Well known concepts and theorems from the theory of rings of continuous functions shall be adapted in Section 4 for MV-algebras of $C(X)$ type without particular restrictions. However, we point out the following facts: let $X$ be a space and, as stressed in [12], it can be shown that there exists a Tychonoff (i.e., Hausdorff and completely regular) space $Y$ such that $C(Y)$ is isomorphic to $C(X)$ [14, 3.9]. Without loss of generality, we shall assume $X$ to be a Tychonoff space. Furthermore, we have that (the proof is similar to that of [14, 4.9]):

**Theorem 1.1** *Two compact spaces $X$ and $Y$ are homeomorphic iff $C(X)$ and $C(Y)$ are isomorphic MV-algebras.*

For all the notions on lattice ordered groups here recalled, we refer to [1] and [6]. Let $G$ be a lattice ordered abelian group (shortly in the sequel $\ell$-group) and let $u \in G$, $u > 0$. The unit interval $[0, u] = \{a \in G : 0 \leq a \leq u\}$ of $G$ is equipped with a structure of MV-algebra by means of the operations $a^* = u - a$, $a \oplus b = (a + b) \wedge u$, $a \odot b = 0 \vee (a + b - u)$ for all $a, b \in [0, u]$ (here, $\vee, \wedge, +$ are operations in $G$). In [20, Proposition 6], the author showed (the details appear in [21, Proposition 1.1]) that any MV-algebra $A$ can be considered like a positive segment, equipped with the above operations, of a suitable $\ell$-group $G$. An exact functorial representation of this result is established in [23, Theorem 3.9]. Strictly speaking, a categorical equivalence $\Gamma$ is defined between the category of MV-algebras (whose morphisms are the MV-homomorphisms [8, p.48]) and the category of $\ell$-groups with strong unit (whose morphisms are the unital $\ell$-homomorphisms [23, 1.4]). Thus, by [23, 3.10], for any MV-algebra $A$, there is exactly, (up to isomorphisms), an $\ell$-group $G$ with strong unit $u$ such that $A = \Gamma(G, u) = [0, u]$. By [23, 2.7(ii)], the lattice operations on $G$ agree with the lattice operations on $A$. Further, by [20, Corollary 1], the $\ell$-ideals of $G$ are in one-one correspondence with the ideals of $A$ and it is easily seen that this correspondence is an homeomorphism between $\text{Spec}G$ (the spectral space of prime $\ell$-ideals of $G$ with the Zariski topology) and $\text{Spec}A$. Consequently, $\text{Max}G$ and $\text{Min}G$ are homeomorphic to $\text{Max}A$ and $\text{Min}A$, respectively (here $\text{Max}G$ and $\text{Min}G$ have the usual meaning). For sake of completeness, we recall that [6, 2.2.12] $u > 0$ is a strong unit of $G$ iff for each $a \in G$, $nu \geq a$ for some $n \in \mathbb{N}$ and that $u > 0$ is a weak unit of $G$ if $u \wedge |x| = 0$ implies $x = 0$. In the

$\ell$-group of the real-valued bounded continuous functions on a topological space, the strong units are the functions less than a positive constant [6, p.41]. This $\ell$-group is Archimedean, i.e. for all $a, b \in G_+$ ($G_+$ is the positive cone of $G$), $na \leq b$ for all $n \in \mathbb{N}$ implies $a = 0$ [6, 11.1.1]. It is seen that the following result holds:

**Theorem 1.2** *Let $G$ be an $\ell$-group with strong unit $u$. Then $A = \Gamma(G, u)$ is semisimple iff $G$ is Archimedean.*

We supply a short proof of Theorem 1.2, then we deduce that

**Theorem 1.3** *An MV-algebra $A$ is semisimple iff $A$ is (up to isomorphisms) subalgebra of $C(X)$ for some compact Hausdorff space $X$.*

As in [12], we also characterize some classes of topological F-spaces (in particular, basically and extremally disconnected spaces) using MV-algebras of continuous functions and their associated $\ell$-groups in the functor $\Gamma$.

## 2   Semisimple MV-Algebras

Using [6, 2.6.3] we deduce the following:

**Lemma 2.1** *Let $G$ be a totally ordered $\ell$-group with strong unit $u$. Then $G$ is Archimedean iff $A = \Gamma(G, u)$ is locally finite.*

**Proof of Theorem 1.2** Let $G$ be Archimedean. Then, by [6, 4.3.9 (iii)], the intersection of all the maximal $\ell$-ideals of $G$ is $\{0\}$. Since Max$A$ and Max$G$ are in one-one correspondence, we deduce that Rad$A = \{0\}$ and $A$ is semisimple. Vice versa, let $A$ be semisimple. Then, by [8, Theorem 4.9], $A$ is representable as subalgebra of a direct product of locally finite MV-algebras $A_i, i \in I$. By [9, Lemma 6], there exists a totally ordered $\ell$-group $G_i$ with strong unit $u_i$ such that $A_i = \Gamma(G_i, u_i)$ and hence

$$A \subseteq \Pi_{i \in I} A_i \subseteq \Pi_{i \in I} G_i = G'. \tag{1}$$

Now, recalling [23, Theorem 3.8] or [20], $G$ coincides with the $\ell$-subgroup generated by $A$ in $G'$ and $u = \{u_i\}_{i \in I}$. By Lemma 2.1, each $G_i$ is Archimedean and hence $G'$ is also Archimedean. Then we deduce the thesis since $G$ is $\ell$-subgroup of $G'$.
$\square$

**Proof of Theorem 1.3** We only prove the nontrivial implication. Let $A$ be semisimple and $G$ be an $\ell$-group with strong unit $u$ such that $A = \Gamma(G, u)$. By Theorem 1.2, $G$ is Archimedean and by [6, 13.2.6], there is an $\ell$-isomorphism $\psi$ of $G$ onto an $\ell$-subgroup $H$, containing the constant function 1, of the $\ell$-group $G'$ of all the real–valued continuous functions defined on Max$G$ (these

functions are necessarily bounded since Max$G$ is a compact Hausdorff space
by [6, 10.2.2]), moreover $\psi(u) = 1$. Then $A = \Gamma(G, u)$ is isomorphic to the
subalgebra $\Gamma(\psi)(A) = \Gamma(H, 1)$ of the MV-algebra $\Gamma(G', 1) = C(\mathrm{Max}G)$. Since
Max$A$ and Max$G$ are homeomorphic spaces (cfr.Sec.1), $C(\mathrm{Max}G)$ and $C(\mathrm{Max}A)$
are isomorphic MV-algebras by Theorem 1.1 and Max$A$ meets the requirements
of the thesis.
□

**Remark 2.2** *In literature, other representation theorems for semisimple MV-
algebras are established. In [2, Theorem 4], it is proved, expliciting Theorem
4.9 of [8], that an MV-algebra $A$ is semisimple iff $A$ is subalgebra of some Bold
algebra of fuzzy sets, strictly speaking $[0, 1]^{\mathrm{Max}A}$. In [16, Corollary 3.7], the
author shows that $A$ is semisimple iff $A$ is (up to isomorphisms) subalgebra of
$C(X)$ for some Boolean space $X$.*

Now we recall some definitions. An MV-algebra $A$ is divisible [21, Definition
1.1] if for each $a \in A$ and for each $n \in \mathbb{N}$, there exists $x \in A$ such that $\oplus nx = a$
and $a^* \oplus \underbrace{(x \oplus \ldots \oplus x)}_{(n-1)-\text{times}} = x^*$.

The following is already known in literature, e.g. [21, Proposition 1.2] or
[16, Proposition 2.13].

**Lemma 2.3** *Let $G$ be an $\ell$-group with strong unit $u$. Then $A = \Gamma(G, u)$ is
divisible iff $G$ is divisible.*

Let $A$ be a subalgebra of an MV-algebra $B$. Following [22, Lemma 4], we
say that $A$ is large in $B$ iff for any ideal $I$ of $B$, $I \neq \{0\}$ implies $I \cap A \neq \{0\}$.
In other words, $A$ is large in $B$ iff for each $0 \neq b \in B$, there exists $0 \neq a \in A$
such that $a \leq \oplus nb$ for some $n \in \mathbb{N}$. Furthermore, we say that $A$ is dense in $B$
iff for each $0 \neq x \in B$, there exists $0 \neq a \in A$ such that $a \leq b$.

**Lemma 2.4** *Let $G$ be an $\ell$-subgroup of an $\ell$-group $G'$ with strong unit $u$ and
$u \in G$. Then $A = \Gamma(G, u)$ is large (resp. dense) in $B = \Gamma(G', u)$ iff $G$ is a large
(resp. dense) $\ell$-subgroup of $G'$.*

**Proof.** Since the $\ell$-ideals of $G'$ are in one-one correspondence with the ideals
of $B = \Gamma(G', u)$ (cfr. Sec.1), the proof is immediate bearing in mind Definition
11.1.14 (resp. 7.6.5) of [6].
□

**Theorem 2.5** *Let $A$ be a divisible and semisimple MV-algebra. Then $A$ is
(up to isomorphisms) a dense subalgebra of $C(\mathrm{Max}A)$. Moreover, any element
$f \in C(\mathrm{Max}A)$ is supremum of all the elements of $A$ majorized by $f$.*

**Proof.** Let $H \subseteq G'$ be the $\ell$-groups of the proof of Theorem 1.3, then $A = \Gamma(H, 1) \subseteq \Gamma(G', 1) = C(\text{Max}A)$. $H$ is a large $\ell$-subgroup of $G'$ by [6, 13.2.6], thus $A$ is large in $C(\text{Max}A)$ by Lemma 2.4. Since $A$ is divisible, then $A$ is dense in $C(\text{Max}A)$ by [22, Lemma 4].

In order to prove the second part of the thesis, we observe that $H$ is dense in $G'$ by Lemma 2.4. Now $G'$ is Archimedean and by [6, 11.3.6], any $f \in C(\text{Max}A) \subseteq G'$ is supremum of elements of $H$ majorized by $f$, i.e., $f = \sup H_f$ where $H_f = \{g \in H : g \leq f\}$. Now let $A_f = \{a \in A : a \leq f\}$, clearly $A_f$ is nonempty since $0 \in A_f$ and it is easily seen that $A_f = H_f \cap G'_+$. A trivial computation shows that $f = \sup A_f$, i.e., the thesis.
□

In the previous theorem, the density of $A$ is consequence of its divisibility. However, using Theorem 1.2 and [6, 11.3.6], it is seen, with similar proof to the Theorem 2.5, that the following more general result holds:

**Theorem 2.6** *Let $A$ be a dense subalgebra of a semisimple MV-algebra $B$. Then any element $b \in B$ is supremum of all the elements of $A$ majorized by $b$.*

**Remark 2.7** *This theorem parallels its analog one for Boolean algebras (e.g., cfr.[24, 23.1]).*

# 3  $\sigma$-Complete MV-algebras

In this Section we establish some results on $\sigma$-complete MV-algebras, starting with the following

**Theorem 3.1** *Let $A$ be an MV-algebra and $G$ be an $\ell$-group with strong unit $u$ such that $A = \Gamma(G, u)$. Then $G$ is a $\sigma$-complete (resp. complete) $\ell$-group iff $A$ is $\sigma$-complete (resp. complete).*

**Proof.** Let $A$ be $\sigma$-complete and $S$ be a bounded countable subset of $G$. Without loss of generality, assume $x \geq 0$ for each $x \in S$. By [23, Proposition 3.1(i)], we have $x = x_1 + x_2 + \ldots + x_n$ for some $x_1, x_2, \ldots, x_n \in A$, where $x_1 = x \wedge u$, $x_2 = (x - x_1) \wedge u, \ldots, x_n = (x - x_1 - x_2 - \ldots - x_{n-1}) \wedge u$. Since $A$ is $\sigma$-complete, $s_k = \sup\{x_k : x \in S\}$ exists in $A$ for each $k = 1, 2, \ldots, n$. Since the lattice operations of $A$ agree with those of $G$, $s_k$ is the supremum in $G$ as well. Let $s = s_1 + s_2 + \ldots + s_n$, hence $s \geq x$ for all $x \in S$ since $s_k \geq x_k$ for each $k = 1, 2, \ldots, n$. We claim $s = \sup S$ and indeed, let $t \geq 0$ be an upper bound for $S$, i.e., $t \geq x$ for all $x \in S$. Applying again [23, Proposition 3.1(i)], we have that $t = t_1 + t_2 + \ldots + t_n$ for some $t_1, t_2, \ldots, t_n \in A$, where $t_1 = t \wedge u, \ldots, t_n = (t - t_1 - t_2 - \ldots - t_{n-1}) \wedge u$. Then, by [23, Proposition 3.1(ii)], $x_k \leq t_k$ and hence $s_k \leq t_k$ for each $k = 1, 2, \ldots, n$ and therefore $t \geq s$. Since the converse implication is trivial, the theorem is proved.
□

**Theorem 3.2** *Let $A$ be a $\sigma$-complete (resp. complete) MV-algebra. Then $A$ is (up to isomorphisms) subalgebra of the $\sigma$-complete (resp. complete) MV-algebra $C(MaxA)$. In addition, if $A$ is divisible, then $A$ is dense in $C(MaxA)$ (resp. $A = C(MaxA)$).*

**Proof.** By [3, Corollary 1] or [10, Lemma 2.1], $A$ is semisimple and thus the thesis follows from Theorems 1.2 and 2.5. Later we shall show that $C(MaxA)$ is $\sigma$-complete (resp. complete) (cfr. Remark 4.8).
□

**Remark 3.3** *Divisible and complete MV-algebras coincide with the injective MV-algebras as proved in [22, Theorem 1] and [16, Theorem 2.14]. Of course, an $\ell$-group $G$ with strong unit $u$ is injective iff the MV-algebra $\Gamma(G, u)$ is injective as also claimed in [19, Theorem 4.2]. The injectivity of the $\ell$-groups with strong unit is also characterized in [15].*

# 4   Hypernormal MV-algebras

In this Section we always assume $X$ to be a Tychonoff space. As in [12], we borrow some well known definitions from rings of continuous functions [14]. $X$ is an $F$-space iff any two disjoint cozero-sets are completely separated, i.e., contained in disjoint zero-sets [14, 14N]. $X$ is basically (resp. extremally) disconnected iff the closure of any cozero-set (resp. open) is open [14, 1H]. Clearly any extremally disconnected space is basically disconnected, every basically disconnected space is an $F$-space, the converse implications fail [14, 4N and 14N].

Let $\beta X$ be the Stone-Cech compactification of a space $X$ and it is seen, as in [14, 6.6], that $C(\beta X) = \mu(C(X))$ for some MV-isomorphism $\mu$ mapping the constant function $1_X$ , defined on $X$, in the constant function $1_{\beta X}$ defined on $\beta X$. Let $G(X)$ (resp. $G(\beta X)$) be the $\ell$-group of all the real-valued continuous bounded functions defined on $X$ (resp. $\beta X$). If $1_X$ and $1_{\beta X}$ are assumed as strong units, we obviously have that $C(X) = \Gamma(G(X), 1_X)$ and $C(\beta X) = \Gamma(G(\beta X), 1_{\beta X})$. Since the functor $\Gamma$ preserves unital isomorphisms by [23, Proposition 3.5], we deduce that

**Theorem 4.1** $\lambda = \Gamma(\mu)$ *for some unital $\ell$-isomorphism $\lambda$, i.e., such that* $\lambda(G(X)) = G(\beta X)$ *and* $\lambda(1_X) = 1_{\beta X}$.

Following [12] or [5], we recall that an MV-algebra $A$ is hypernormal iff every prime ideal contains an unique minimal prime ideal. The same property characterizes semiprojectable $\ell$-groups [6, 7.5.1] and, since the functor $\Gamma$ maps one-one $SpecA$ onto $SpecG$ (cfr. Section 1), the following theorem is immediate:

**Theorem 4.2** *Let $G$ be an $\ell$-group with strong unit $u$. Then $A = \Gamma(G, u)$ is hypernormal iff $G$ is semiprojectable.*

**Corollary 4.3** *For any space $X$, the following are equivalent:*

*(i) $X$ is an F-space,*

*(ii) $C(X)$ is a hypernormal MV-algebra,*

*(iii) $G(X)$, with strong unit $1_X$, is a semiprojectable $\ell$-group.*

**Proof**. (i) and (ii) are equivalent by [12, Theorem 5.3]. (ii) and (iii) are equivalent by Theorem 4.2.
□

Following [5, Theorem 33], an MV-algebra $A$ is Stonian iff $L(A)$ is a Stone lattice. Furthermore, we recall that [1, p.118] an $\ell$-group (not necessarily abelian) is projectable (resp. strongly projectable) iff every principal polar (resp. polar) is a cardinal summand. It is well known that a strongly projectable $\ell$-group is projectable and a projectable $\ell$-group is semiprojectable, the converse implications fail [1, p.118]. In accordance to [13, Theorem 2.2], we say that an $\ell$-group $G$ (not necessarily abelian) is complemented iff Min$G$ is compact.

**Theorem 4.4** *Let $G$ be an $\ell$-group with strong unit $u$ and $A = \Gamma(G, u)$. Then the following are equivalent:*

*(i) $A$ is a Stonian MV-algebra,*

*(ii) $G$ is complemented and semiprojectable,*

*(iii) $G$ is projectable.*

**Proof**. Since Min$A$ = Min$L(A)$ by [11, Lemma 1.2 (ii)], i.e. Min$A$ can be seen like the minimal prime spectrum of $L(A)$, Min$A$ is a spectral space iff Min$A$ is compact by [25, Lemma 3.1]. Thus $A$ is Stonian iff $A$ is hypernormal and Min$A$ is compact by [5, Theorem 35]. Since Min$A$ is homeomorphic to Min$G$ (cfr. Section 1), then (i) and (ii) are equivalent by Theorem 4.2. Any strong unit is a weak unit [7, p.308] and a projectable $\ell$-group with weak unit is complemented [13, p.295]. Thus (ii) and (iii) are equivalent by Propositions 3.6 and 4.4 of [13].
□

**Theorem 4.5** *For any space $X$, the following are equivalent:*

*(i) $X$ is basically (resp. extremally) disconnected,*

*(ii) $C(X)$ is a $\sigma$-complete (resp. complete) MV-algebra,*

*(iii) $G(X)$, with strong unit $1_X$, is a $\sigma$-complete (resp. complete) $\ell$-group,*

*(iv) $G(X)$, with strong unit $1_X$, is a projectable (resp. strongly projectable) $\ell$-group.*

**Proof.** (ii) and (iii) are equivalent by Theorem 3.1. Each condition (i), (iii), (iv) is equivalent to the following ones: (j) $\beta X$ is basically (resp. extremally) disconnected by [14, 6M]. (jjj) $G(\beta X)$, with strong unit $1_{\beta X}$, is a $\sigma$-complete (resp. complete) $\ell$-group by Theorem 4.1. (jv) $G(\beta X)$, with strong unit $1_{\beta X}$, is a projectable (resp. strongly projectable) $\ell$-group by Theorem 4.1. (j), (jjj) and (jv) are well known to be equivalent (e.g., cfr.[1, p.127]).
□

In [5, Theorem 34], it is proved that a complete MV-algebra is Stonian. Now we generalize this result proving the following:

**Theorem 4.6** *A $\sigma$-complete MV-algebra A is Stonian.*

**Proof.** Let $G$ be an $\ell$-group with strong unit $u$ such that $A = \Gamma(G, u)$. Then $G$ is $\sigma$-complete by Theorem 3.1 and hence projectable by [6, 11.2.3]. Thus $A$ is Stonian by Theorem 4.4.
□

Let $B(A)$ be the Boolean subalgebra of an MV-algebra $A$. If $A$ is complete, then $B(A)$ is complete [10, Corollary 1.4]. However, by using the method of proof in [10, Corollary 1.4], we can prove that

**Theorem 4.7** *Let A be a $\sigma$-complete MV-algebra. Then $B(A)$ is a $\sigma$-complete Boolean algebra and MaxA, MinA are basically disconnected spaces.*

**Proof.** Let $\{a_n : n \in \mathbb{N}\}$ be a family of elements of $B(A)$ and let $a = \wedge_{n \in \mathbb{N}} a_n$ in $A$. Then $a^* = \vee_{n \in \mathbb{N}} a_n^*$ and $a \wedge a^* = \vee_{n \in \mathbb{N}} (a \wedge a_n^*) = 0$, i.e., $a \in B(A)$. Then $B(A)$ is a $\sigma$-complete Boolean algebra and it is well known that (e.g., [18, p. 99]) $\text{Spec} B(A)$ is a basically disconnected space. By Proposition 26 (resp. Proposition 35) of [5], MaxA (resp. MinA) and $\text{Spec} B(A)$ are homeomorphic spaces and hence the thesis.
□

**Remark 4.8** *In Theorem 4.7, $C(MaxA)$ is a $\sigma$-complete MV-algebra by Theorem 4.5. If A is a complete MV-algebra, then, proceeding as in the proof of Theorem 4.7, it is easily seen that MaxA and MinA are extremally disconnected spaces. Thus $C(MaxA)$ is a complete MV-algebra by Theorem 4.5 (this result is also obtained in [16]).*

**Remark 4.9** *The Stonian MV-algebras are called projectable in [17], where another categorical equivalence between these algebras and certain cyclic ordered groups is established. Furthermore, Theorem 4.6 and the fist part of Theorem 4.7 can be also seen as Corollaries and Propositions 4 and 2 of [4], respectively.*

**Acknowledgment** Thanks are due to Prof. D. Mundici for some useful suggestions received during the preparation of this paper.

# References

[1] M. Anderson and T. Feil, *Lattice Ordered Groups: an Introduction*, D. Reidel Publishing Company, Dordrecht, Holland, 1988.

[2] L. Belluce, *Semisimple algebras of infinite-valued logic and Bold fuzzy set theory*, Canad. J. Math. **38** (1986), 1356–1379.

[3] _____, *Semisimple and complete MV-algebras*, Algebra Universalis **29** (1992), 1–9.

[4] _____, *α-complete MV-algebras*, Tech. report, 1994, to appear.

[5] L. Belluce, A. Di Nola, and S. Sessa, *The prime spectrum of an MV-algebra*, Math. Log. Quart., to appear.

[6] A. Bigard, K. Keimel, and S. Wolfenstein, *Groupes et Anneaux Reticules*, Lecture Notes in Mathematics, no. 608, Springer-Verlag, Berlin, 1977.

[7] G. Birkhoff, *Lattice Theory*, 3 ed., Coll. Publ., no. 25, Amer. Math. Soc., New York, 1973.

[8] C. Chang, *Algebraic analysis of many-valued logics*, Trans. Amer. Math. Soc. **88** (1958), 467–490.

[9] _____, *A new proof of the completeness of the Lukasiewicz axioms*, Trans. Amer. Math. Soc. **93** (1959), 74–80.

[10] R. Cignoli, *Complete and atomic algebras of the infinite-valued Lukasiewicz logic*, Studia Logica **50** (1991), 375–384.

[11] R. Cignoli, A. Di Nola, and A. Lettieri, *Priestly duality and quotient lattices of many-valued algebras*, Rend. Circ. Mat. Palermo (2) **40** (1991), 371–384.

[12] R. Cignoli and A. Torrens, *Ordered spectra and Boolean product of MV-algebras*, submitted.

[13] P. Conrad and J. Martinez, *Complemented lattice-ordered groups*, Indag. Math. **3** (1990), 281–298.

[14] L. Gillman and M. Jerison, *Rings of Continuous Functions*, Graduate Texts in Mathematics, no. 43, Springer-Verlag, New York, 1976.

[15] D. Gluschankof, *Prime ideals and Sikorski extension theorem for some ℓ-groups*, Ordered Algebraic Structures (J. Martinez, ed.), Kluwer Acad. Publ., 1989, pp. 113–122.

[16] _____, *Prime deductive systems and injective objects in the algebras of Lukasiewicz infinite valued calculi*, Algebra Universalis **29** (1992), 354–377.

[17] _____, *Cyclic ordered groups and MV-algebras*, Czechoslovak Math. J. **43** (1993), 249–263.

[18] P. Halmos, *Lectures on Boolean Algebras*, D. Van Nostrand Co., Princeton, 1963.

[19] C.S. Hoo, *Unitary extensions of MV and BCK-algebras*, Math. Japon. **37** (1992), 585–590.

[20] F. Lacava, *Alcune proprietà delle L-algebre esistenzialmente chiuse*, Boll. Un. Mat. Ital. A (5) **16** (1979), 360–366.

[21] _____, *Sulle classi delle L-algebre e degli ℓ-gruppi abeliani algebricamente chiusi*, Boll. Un. Mat. Ital. B (7) **1** (1987), 703–712.

[22] _____, *Sulle L-algebre iniettive*, Boll. Un. Mat. Ital. A (7) **3** (1989), 319–324.

[23] D. Mundici, *Interpretation of AF C\*-algebras in Łukasiewicz sentential calculus*, J. Funct. Anal. **65** (1986), 15–63.

[24] R. Sikorski, *Boolean Algebras*, Ergebnisse der Math. und ihrer Grenzgeb., no. 25, Springer-Verlag, Berlin, 1960.

[25] H. Simmons, *Reticulated rings*, J. Algebra **66** (1980), 169–192.

# III

# Free and projective Heyting and monadic Heyting algebras

## R. Grigolia

Finitely generated Heyting algebras and monadic Heyting algebras are described in terms of perfect Kripke models using colouring technique. Defining projective algebras as retract of free algebras, the characteristic of finitely generated projective algebras is given in varieties of Heyting algebras and monadic Heyting algebras. By means of projective algebras, using an algebraic proof, the conditions of Friedman's conjecture (Problem 41 [8]) are confirmed for well-known Medvedev's logic.

## 1 Introduction

It is well known that Heyting algebras are algebraic models of intuitionistic propositional logic IL, which is properly contained in the classical propositional logic KL. In between IL and KL lie the intermediate propositional logics, such as Medvedev's logic ML [13]. The variety (equational class) of all Heyting algebras we shall denote by **HA**. **HA** may understand as a category where the objects of **HA** are Heyting algebras and morphisms are Heyting homomorphisms.

The category **K** of perfect Kripke models, which is dually equivalent to the category of **HA**, was constructed by L.Esakia [4]. The objects of **K** are partially ordered sets $(X; R)$ where $X$ is a Stone space, for any $x \in X$ $R(x)$ and $R^{-1}(x)$ are closed sets, the least closed sets which contains cone is a cone itself (subset $Y \subseteq X$ is a *cone* if for any $x \in Y$ and $xRy$ implies $y \in Y$). The morphisms of **K** are strongly isotone continuous mappings :

$$f : X \to X' \quad \text{is strongly isotone if} \quad f(y)Rx \Leftrightarrow (\exists y')(yRy' \& f(y') = x)$$

for any $y \in X$, $x \in X'$.

Quotient algebras correspond to closed cone and Heyting subalgebras correspond to correct partitions [4]. A partition $E$ of a space $X$ is called by *correct* if corresponding to it equivalence relation (which we shall denote by the same

33

symbol) is closed, $E$-saturation of any cone is a cone and there is perfect Kripke model $(X'; R')$ and strongly isotone mapping $f : X \to X'$ such that $E = \text{Ker } f$.

A *Heyting algebra* (pseudo-Boolean algebra, relatively pseudo-complemented distributive lattice with 0) is an algebra $(A; \vee, \wedge, \to, 0, 1)$ subject to certain polynomial identities [20]. Although $\neg$ has not been included here as a basic operation, we can regard $\neg p$ as an abbreviation for $p \to 0$, where $p$ is any Heyting-algebra polynomial.

Let $(X; R)$ be partially ordered set. Say that $Y \subseteq X$ *totally* covers the element $x \in X$, in notation $x \prec Y$ or $Y \succ x$, if $Y$ coincides with the set of all elements which cover $x$. If $Y$ is a singleton then we say that the element totally covers $x \in X$. Let $X$ be a topological space. A set $A \subseteq X$ is called **dense** in $X$ if $clA = X$, where $cl$ is closure operator of the space $X$.

**Theorem 1.1** *[3]. If a continious mapping $f$ of a dense subset $A$ of a topological space $X$ to a Hausdorff space $Y$ is continuously extendable over $X$, then the extension is uniquely determined by $f$.*

**Theorem 1.2** *[3]. Let $A$ be dense subspace of a topological space $X$ and $f$ is a continuous mapping of $A$ to a compact space $Y$. The mapping $f$ has a continuous extension over $X$ if and only if for every pair $B_1$, $B_2$ of disjoint closed subsets of $Y$ the invers images $f^{-1}(B_1)$ and $f^{-1}(B_2)$ have disjoint closures in the space $X$.*

The concept of monadic algebras was introduced by P.Halmos [10] for modal logic S5 and algebraic semantics for intuitionistic modal logic by R.Bull [12] and H.Ono [15] who named those algebras as bi–topological Boolean algebras (see also N.Y.Suzuki [22])

Note that intuitionistic modal logic IML is modal counterparts of intermediate predicate logic IPL. More precisely : we fix a first-order language $\mathcal{L}$ and modal propositional language $\mathcal{L}^\mathcal{M}$. Note that we use both $\Box$ (necessity) and $\Diamond$ (possibility) as primitive symbols in $\mathcal{L}^\mathcal{M}$. We fix one-to-one correspondence between propositional variables of $\mathcal{L}^\mathcal{M}$ and monadic predicate variables of $\mathcal{L}$. For each propositional variable $p$, $p^*$ denotes the monadic predicate variable corresponding to $p$. Let $x$ be a fixed individual variable. We define a translation $\psi$ of modal formulas into first-order formulas inductivaly as follows :

(1) $\psi(p) = p^*$ if $p$ is a propositional variable,
(2) $\psi(\alpha \wedge \beta) = \psi(\alpha) \wedge \psi(\beta)$
(3) $\psi(\alpha \vee \beta) = \psi(\alpha) \vee \psi(\beta)$
(4) $\psi(\alpha \to \beta) = \psi(\alpha) \to \psi(\beta)$
(5) $\psi(\neg\alpha) = \neg\psi(\alpha)$
(6) $\psi(\Box\alpha) = \forall x \psi(\alpha)$
(7) $\psi(\Diamond\alpha) = \exists x \psi(\alpha)$

Intuitionistic modal logic IML is obtained from the intuitionistic propositional logic IL by adding several rules [2] or axioms [15]. In between IML and S5 lie the intermediate intuitionistic modal logics L: IML$\subseteq$L$\subseteq$S5.

A *monadic Heyting algebra* (bi-topological pseudo-Boolean algebra) is an algebra $(A; \vee, \wedge, \rightarrow, I, C, 0, 1)$ satisfying the following conditions :

(1) $(A; \vee, \wedge, \rightarrow, 0, 1)$ is a Heyting algebra,

(2) for each $x, y \in A$

(i) $I(x \wedge y) = I(x) \wedge I(y)$ (i') $C(x \vee y) = Cx \vee Cy$

(ii) $Ix \leq x$ (ii') $x \leq Cx$

(iii) $IIx = Ix$ (iii') $Cx = CCx$

(iv) $I1 = 1$ (iv') $C0 = 0$

(v) $Ix \rightarrow Iy \leq I(Ix \rightarrow Iy)$ (v') $I(x \rightarrow y) \leq Cx \rightarrow Cy$

(vi) $CIx = Ix$

(vii) $ICx = Cx$

As we see from the axioms, the elements of the kind $Ia$, $Ca$ $(a \in A)$ form subalgebra of Heyting algebra $(A; \vee, \wedge, \rightarrow, 0, 1)$. So monadic Heyting algebra can represent as pair of Heyting algebras $(A; A_0)$ [5], where $A_0$ is relativaly complete subalgebra, i.e. there exist

$$\bigcap_{\substack{a \leq b \\ b \in A_0}} \quad \text{and} \quad \bigcup_{\substack{a \geq b \\ b \in A_0}} \quad \text{for any} \quad a \in A.$$

The dual category for Heyting algebras was constructed by L.Esakia [5,6]. In dual category we have $(X; R, E)$, where $(X; R)$ is perfect Kripke model, $E$ is equivalence relation on $X$ such that $E$−saturation of every clopen is clopen. Here $CY = E(Y)$, $IY = -R^{-1}(E(-Y))$ for $Y \subseteq X$.

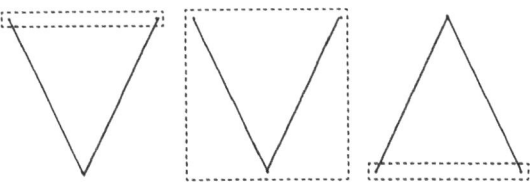

Fig. 1

On Fig.1 the equivalence classes is outlined by dotted line.
Homomorphisms corresponds to $E$−saturated closed cones.
Note that the descriptions of finitely generated Heyting algebras and projective Heyting algebras is given in [9] in Russian. Here we reproduce this description in English.

## 2    Finitely generated free Heyting algebras

For description of construction of free algebras the important role plays the criterion of Heyting algebras to be finitely generated. For this aim we introduce the notion of colouring.

Let $(X; R)$ be perfect Kripke model and $J$ be a non empty finite set. The partition $\{K_p : p \in J\}$ of a set $X$, every element of which is clopen, is said to be *colouring*.

**Theorem 2.1** *[7]. Heyting algebra $A$ is finitely generated if and only if corresponding to it perfect Kripke model $(X; R)(= A^*)$ admits colouring $\{K_p : p \in J\}$ such that any non trivial correct partition $E$ contains many-coloured class, i.e. there exests $q \in E$ such that $x, y \in q \Rightarrow x \in K_{p_1} \& y \in K_{p_2}(p_1 \neq p_2)$.*

Finitely generated Heyting algebras was described by A.Urquhart [22]. But this description is not effective. We use constructively given Kripke models for description of finitely generated Heyting algebras using colouring technique.

For colouring of $X_{\mathbf{HA}}(n)$, corresponding to $n$-generated free Heyting algebra, we shall use "colours" which is denoted by subsets of set $\{1, \ldots, n\}$ as follows : $P_0 = \emptyset$, $P_1 = \{1\}, \ldots, P_n = \{n\}, \ldots, P_{2^n - 1} = \{1, \ldots, n\}$.

For convinience sake suppose $n = 2$. In this case we have four colours : $\emptyset(= P_0)$, $\{1\}(= P_1)$, $\{2\}(= P_2)$, $\{1, 2\}(= P_3)$. Now we determine model $(X_{\mathbf{HA}}(2); R)$ by levels. $(X_{\mathbf{HA}}(2); R)$ has four maximal elements (that is the elements of the first level):$m_0, m_1, m_2, m_3$. . Let us colour these four elements of the first level in $4(= 2^2)$ distinct colours. Then for every set of incomparable elements $c \subseteq \{m_0, m_1, m_2, m_3\}(= \theta_1)$ there are $2^{|\cap P_{i_j}|}$ elements on the second level $(\theta_2)$, every of which is covered only by the elements of $c$ (that is $c$ totally covers $2^{|\cap P_{i_j}|}$ elements of the second level), coloured in $P_s \subseteq \cap_{j=1}^{k} P_{i_j}$ colours, where $k = |c|$, $P_{i_j}$ are colours of elements from $c$. Every element $m_i(i = 0, 1, 2, 3)$ of the first level, which has colour $P_i$, covers $2^{|P_i|} - 1$ elements of the second level, coloured in $P_j \not\subseteq P_i$ colours, and what's more any of these elements of the second level is covered by the only element $m_i$. The second level are exhausted by these elements. For description of the following levels we act analogically for incomparable elements for preceding levels, containing elements of the immediately preceding level, and for every element of the immediately preceding level. Let us denote the set of all elements of $n-$th level by $\theta_n$. We shall denote by $X_{\mathbf{HA}}(2)$ the set $\cup_{i=1}^{\infty} \theta_i$. The relation $R$ on $X_{\mathbf{HA}}(2)$ is determined by the construction.

Note that if $n = 1$ then $X_{\mathbf{HA}}(1)$ is the model described by L.Rieger [21]. Corresponding to it lattice is known by name Rieger-Nishimura ladder. This lattice (ladder) was constructed by I.Nishimura in [14]. Let $G_i(i = 1, 2)$ be the set of all elements of $X_{\mathbf{HA}}(2)$ having $P_j$ colour, such that $i \in P_j$. Let $(F_{\mathbf{HA}}(2); \cup, \cap, \rightarrow, X_{\mathbf{HA}}(2), \emptyset)$ be Heyting algebra generated by elements $G_1$ and

$G_2$, where $\cup, \cap$ are set-theoretical operation of union and intersection and $x \rightharpoonup y = -R^{-1}(x \cap -y)$.

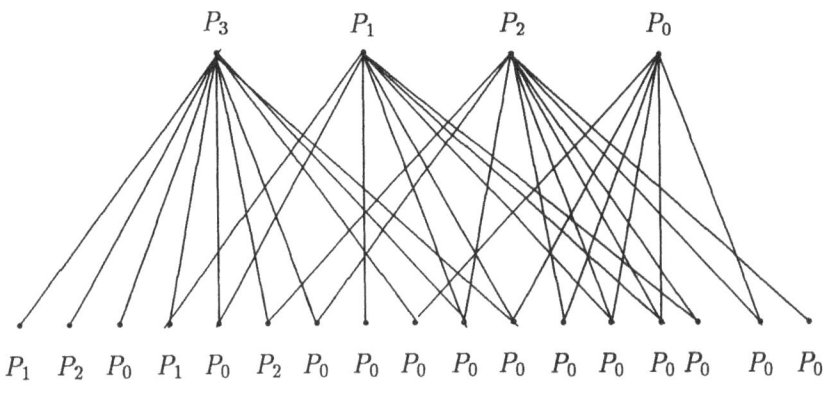

Fig. 2

We have such colouring of $X_{\mathbf{HA}}(2)$ that for any finite cone $Y$ of $X_{\mathbf{HA}}(2)$ and any correct partition $E$ on $Y$ $E$ contains many-coloured class. Note that a finite Kripke model, corresponding to any 2-generated Heyting algebra $A$ with generators $g_1$ and $g_2$, is a cone of $X_{\mathbf{HA}}(2)$ such that $G_i \cap Y = g_i^*$, where $g_i^*$ is a cone corresponding to $g_i (i = 1, 2)$. Now we can replace 2-generated Heyting algebra with $n-$generated Heyting algebra $(n > 2)$. Hence we have

**Lemma 2.2** *Any $n$-generated Heyting algebra is homomorphic image of algebra $F_{\mathbf{HA}}(n)$.*

**Theorem 2.3** *Algebra $F_{\mathbf{HA}}(n)$ is free $n$-generated Heyting algebra with free generators $G_1, \ldots, G_n (n \in \omega)$.*

**Proof.** It is suffucent to show that if polynomial identity $p = q$ on $n$ variables does not hold in variety **HA** then $p = q$ does not hold in $F_{\mathbf{HA}}(n)$. Let us suppose that $p = q$ does not hold in **HA**. Then $p = q$ does not hold in some finite Heyting algebra $A$ on some elements $a_1, \ldots, a_n$, where $a_1, \ldots, a_n$ generate $A$. Then, according Lemma 2.2, $A$ is a homomorphic image of $F_{\mathbf{HA}}(n)$. Denote this homomorphism by $\phi$, and besides $\phi(G_i) = a_i (i = 1, \ldots, n)$. Hence $p = q$ does not hold in $F_{\mathbf{HA}}(n)$.

Let $(\kappa X_{\mathbf{HA}}(n); \kappa R)(= \quad F_{\mathbf{HA}}^*(n) \quad )$ be the perfect Kripke model corresponding to the algebra $F_{\mathbf{HA}}(n)$, i.e. $\kappa X_{\mathbf{HA}}(n)$ is the set of all prime filters of algebra $F_{\mathbf{HA}}(n)$, and $\kappa R$ is the inclusion relation between filters. Let $U$ be the set of all principal prime filters of algebra $F_{\mathbf{HA}}(n)$. Then $U \subseteq \kappa X_{\mathbf{HA}}(n)$ and $(U; \kappa R)$ is isomorphic to $(X_{\mathbf{HA}}(n); R)$. Principal prime filters are generated by elements of kind $R(x)$, $x \in X_{\mathbf{HA}}(n)$. Identify $U$ with $X_{\mathbf{HA}}(n)$. Let us note that $R(x)$, where $x \in X_{\mathbf{HA}}(n)$, is an element of $F_{\mathbf{HA}}(n)$. In fact, $R(x)$ is a cone of $X_{\mathbf{HA}}(n)$. So $R(x)$ is an element of Heyting algebra $A_{\theta_m}(= R(\theta_m)^*)$, corresponding to Kripke model $R(\theta_m)$, for some $m \in \omega$. Hence there exists polynomial $k(x_1, \ldots, x_n)$ such that $R(x) = k(G_1^{m+1}, \ldots, G_n^{m+1}) \in A_{\theta_{m+1}} (\cong R(\theta)^*)$ (since $R(x) \in A_{\theta_{m+1}}$ and $G_1^{m+1}, \ldots, G_n^{m+1}$ are generators of $A_{\theta_{m+1}}$). Then $h(k(G_1, \ldots, G_n)) = k(h(G_1), \ldots, h(G_n)) = k(G_1^{m+1}, \ldots, G_n^{m+1}) = R(x)$, where $h$ is homomorphism from $F_{\mathbf{HA}}(n)$ onto $A_{\theta_{m+1}}$. But since $R(x) \cap \theta_{m+1} = \emptyset$ and $R(x)$ is a cone then $k(G_1, \ldots, G_n) = R(x)$. Hence we conclude that all finite cones of $X_{\mathbf{HA}}(n)$ are elements of $F_{\mathbf{HA}}(n)$. Since for every $x \in X_{\mathbf{HA}}(n)$ $R(x) \in F_{\mathbf{HA}}(n)$ then $R(x)$ is clopen, i.e. $x$ is isolated point of space $\kappa X_{\mathbf{HA}}(n)$. Note that non principal prime filters are "below" of elements from $X_{\mathbf{HA}}(n)$ with respect to relation $\kappa R$. $X_{\mathbf{HA}}(n)$ is a dense subset of space $\kappa X_{\mathbf{HA}}(n)$.

# 3   Finitely generated projective Heyting algebras

Projective Heyting algebra are defined as retracts of free algebras. R.Balbes and A.Horn [1] are described projective Heyting algebras, which are finite or chain. We characterize finitely generated projective Heyting algebras. Say that element $a \in A$ has *height* $n \in \omega$ if the greatest chain subalgebra of algebra $(a]$ contains n+1 elements.

**Theorem 3.1** *A n-generated Heyting algebra $A(= [g_1, \ldots, g_n]$ where $g_1, \ldots, g_n$ are generators of A) is projective if and only if A is finitely approximated and for every incomparable elements $z_1, \ldots, z_k$ of finitely height there exists join indecomposable element $z \in A$ such that $z$ covers $z_1 \vee \ldots \vee z_k$. and beginning from some $t \in \omega$ for any incomparable elements $u_1, \ldots, u_s$ or $s = 1$, the heights of which more or equal to $t$, if $u_1 \vee \ldots \vee u_s \leq g_{i_1} \wedge \ldots \ldots g_{i_r}$, and for any $j \in 1, \ldots, n - i_1, \ldots, i_r$ $u_1 \wedge \ldots \wedge u_s \nleq g_j$ then for any $a \subseteq i_1, \ldots, i_r$ there exists join indecomposable element $u_a$ such that $u_a$ covers $u_1 \vee \ldots \vee u_s$ and $u_a \leq \wedge_{i \in a} g_i$.*

**Proof.** Let $A$ be finitely generated and finitely approximated Heyting algebra which obeys the following conditions : for any incomparable elements $z_1, \ldots, z_k$ of finitely height there exists join indecomposable elment $z \in A$ such that $z$ covers $z_1 \vee \ldots \vee z_k$. Let $G'_1, \ldots, G'_n$ be generators of $A$. Then $(X_A; R_A)(= A^*)$ is isomorphic to a closed cone of $(\kappa X_{\mathbf{HA}}(n)(n); \kappa R)$. Identify $X_A$ with corresponding set of elements of $\kappa X_{\mathbf{HA}}(n)$, and let us consider dense part $X_{\mathbf{HA}}(n)$ of the space $\kappa X_{\mathbf{HA}}(n)$. Let $M_A = \{m_1, \ldots, m_k\}$ be the set of all maximal elements of $X_A$, and let $M$ be the set of all maximal elements of $X_{\mathbf{HA}}(n)$. Let $u_1^j, \ldots, u_{i_j}^j \in \theta_2 \cap X_A, j = 1, \ldots, k$, such that $u_t^j \prec \{m_j\}$ (i.e. $\{m_j\}$ totally covers $u_t^j$), $t = 1, \ldots, i_j$ (if they exist). Then let $-R^{-1}((M - \{m_j\}) \cup \{u_1^j, \ldots, u_{i_j}^j\})(j = 1, \ldots, k - 1)$ be the equivalence class of $m_j$ (which is the class of all elements which is equivalent to $m_j$). Let

$-R^{-1}(\{m_1, \ldots, m_{k-1}, u_1^k, \ldots, u_i^k\})$ be the equivalence class of $m_k$. Corresponding partition (and equivalence relation as well) we shall denote by $E_1$. Note that $q \cap X_A$, where $q \in E_1$, , is one element set, namely $q \cap X_A = \{m_j\}$ for $j \in \{1, \ldots, k\}$. In addition to every $q \in E_1$ is a cone. If there exists $a \in X_{HA}(n) - X_A$ such that $a \prec \{m_j\}, j < k$, then $a \in E_1(m_j)$; if there does not exist $a \in X_{HA}(n) - X_A$ such that $a \prec \{m_j\}, j < k$, then $E_1(m_j) = \{m_j\}$. Let $X_A \cap \theta_2 = \{n_1, \ldots, n_p\}$ and $v_1^s, \ldots, v_{i_s}^s \in X_A \cap \theta_3 (s = 1, \ldots, p)$ such that $v_t^s \prec n_s, t = 1, \ldots, i_s$. Let $\{l_1, \ldots, l_r\} \subseteq \theta_2 - X_A$ such that $l_i \prec B_i$, where $B_i \subseteq \theta_1, |B_i| > 1, i = 1, \ldots, r$, and $E_1(B_i) = E_1(B_j) = \bigcup_{m_i \in E_1(B_j)} E(m_i), i, j \in \{1, \ldots, r\}$. Since $A$ satisfies the condition of the theorem there is $n_s \in X_A \cap \theta_2$, where $n_s \prec \{m_i : m_i \in E_1(B_s)\} (= B), E_1(B) = E_1(B_s)$. Then

$$(-R^{-1}(M_{n_s} \cup \{v_1^s, \ldots, v_{i_s}^s\}) \cup \{l_1, \ldots, l_r\}) - E_1(M_A),$$

where $M_{n_s}$ is the set of all elements of $\theta_1 \cup \theta_2$ which are incomparable with $n_s$, is the equivalence class of $n_s (s = 1, \ldots, p)$. The partition (and the corresponding equivalence relation), the representatives of classes of which are $n_s$, we shall denote by $E_2$. Accordingly every element of $\theta_1 \cup \theta_2$ is contained in equivalence class of $E_1 \cup E_2$, the representative of which is element of $(\theta_1 \cup \theta_2) \cap X_A$. Every equivalence class of $E_2$ contains the only element from $\theta_2 \cap X_A$.

Let us assume that $E_1, \ldots, E_k$ are determined. Let us determine $E_{k+1}$. Let $X_A \cap \theta_{k+1} = \{r_1, \ldots, r_s\}$ and $w_1^j, \ldots, w_{i_j}^j \in \theta_{k+2} \cap X_A, j = 1, \ldots, s$, such $w_t^j \prec \{r_j\}, t = 1, \ldots, i_j$. Let $\{e_1, \ldots, e_t\} \subseteq \theta_{k+1} - X_A$ such that $e_1 \prec D_j, D_i \subseteq \bigcup_{i=1}^k \theta_i, |D_i| > 1, i = 1, \ldots, t$;

$$(E_1 \cup \ldots \cup E_k)(D_i) = (E_1 \cup \ldots \cup E_k)(D_j) = (E_1 \cup \ldots \cup E_k)(D),$$

$D = X_A \cap (\bigcup_{i=1}^k E)(D_j)$. Since $A$ satisfies the condition of the theorem, then there is $r_j$ such that $r_j \prec D$. Let $(-R^{-1}(M_{r_j} \cup \{w_1^j, \ldots, w_{i_j}^j\} \cup \{e_1, \ldots, e_t\}) - (\bigcup_{i=1}^k E_i)(\bigcup_{i=1}^k \theta_i)$ be the equivalence class of $E_{k+1}$, the representative of which is $r_j$, where $M_{r_j}$ is the set of all elements of $\bigcup_{i=1}^k \theta_i$ which are incomparable with $r_j$. Let $E = \bigcup_{i=1}^\infty E_i$. The equivalence relation $E$ has the following properties : $(\forall x \in X_{HA}(n))(\exists y \in X_A)(x \in E(y)), (\forall x, y)(x, y \in X_A^0 \& x \neq y) \Rightarrow E(x) \neq E(y))$, where $X_A^0 = X_A \cap X_{HA}(n)$, the saturation of any cone of $X_{HA}(n)$ with respect to $E$ is a cone.

The equivalence relation $E$ induces the mapping $f_0 : X_{HA}(n) \rightarrow X_A$, where $f_0(x) = y \in X_A, E(x, y)$. Let $A' = \cap A_i$ where $A_i$ is an algebra corresponding to $X_{HA}(n)/E_i$. It is obvious that $A_{i+1}$ is a subalgebra of $A_i$. Since $A_i$ is a subalgebra of $F_{HA}(n)$, then A' is also a subalgebra of $F_{HA}(n)$. Hence there exists a strongly isotone, continuous mapping $f : \kappa X_{HA}(n) \longmapsto X_A$ extending the continuous map $f_0$. In so far as $f(\kappa X_{HA}(n))$ is closed in $X_A$ and $X_A^0 (= X_A \cap X_{HA}(n)) \subseteq f(\kappa X_{HA}(n))$ is dense subset of $X_A$ (since $A$ is finitely approximated), the mapping $f$ is onto.

Identify the elements of $A$ with corresponding clopen cones of $X_A$. The Heyting algebra $A_0$ of cones $Y_0 = Y \cap X_A^0$ of the set $X_A$, where $Y \in A$, is isomorphic to $A$ (since $X_A^0 (= X_A \cap X_{HA}(n))$ is a dense subset of the space $X_A$). Let us consider sets $cl_{\kappa X_{HA}(n)} f_0^{-1}(Y), cl_{\kappa X_{HA}(n)} f_0^{-1}(X_A - Y)$. Since $f$ is continuous extention of $f_0$, then according to Theorem 1.3

$$cl_{\kappa X_{HA}(n)} f_0^{-1}(Y) \cap cl_{\kappa X_{HA}(n)} f_0^{-1}(X_A - Y) = \emptyset.$$

On the other hand

$$cl_{\kappa X_{HA}(n)} f_0^{-1}(Y) \cup cl_{\kappa X_{HA}(n)} f_0^{-1}(X_A - Y) =$$
$$= cl_{\kappa X_{HA}(n)}(f_0^{-1}(Y) \cup f_0^{-1}(X_A - Y)) = cl_{\kappa X_{HA}(n)} X_{HA}(n) = \kappa X_{HA}(n).$$

Hence $cl_{\kappa X_{HA}(n)} f_0^{-1}(Y) = -cl_{\kappa X_{HA}(n)} f_0^{-1}(X_A - Y)$. Therefore $cl_{\kappa X_{HA}(n)} f_0^{-1}(Y)$ is clopen of the space $\kappa X_{HA}(n)$. If $Y$ is a cone then $f_0^{-1}(Y)(= E(Y_0))$ is a cone and $f_0^{-1}(Y) \in F_{HA}(n)$ (since $cl_{\kappa X_{HA}(n)} f_0^{-1}(Y) \cap X_{HA}(n) = f_0^{-1}(Y))$.
The set of all saturated cones of $\kappa X_{HA}(n)$ form subalgebra of algebra $F_{HA}(n)$, which we shall denote by $A'$. Between $A$ and $A'$ there is isomorphism $\phi : \phi(Y) = Y \cap X_A^0, Y \in A'$.

Let us determine the embeding $\epsilon : A \to F_{HA}(n)$ as follows: $\epsilon(Y) = E(Y)$. The mapping $h : F_{HA}(n) \to A$, where $h(Y) = Y \cap X_A^0$, is homomorphism, and $h(E(Y)) = E(Y) \cap X_A^0 = Y$, where $E(Y) \in F_{HA}(n)$. Consequently $A_0$ is a retract of algebra $F_{HA}(n)$. Hence we conclude that algebra $A$ is projective. Note that if $A$ is projective then $A$ is a subalgebra of $F_{HA}(n)$ and, consequently, is finitely approximate.
Now let us suppose that for some incomparable elements $a_1, \ldots, a_k \in X_A \cap \bigcup_{i=1}^k \theta_i (\subseteq X_{HA}(n))$ there is not exist element $z \in \theta_{m+1} \cap X_A$ such that $z \prec \{a_1, \ldots, a_k\}$. But there is an element $z_0 \in \theta_{m+1}$ such that $z_0 \prec \{a_1, \ldots, a_k\}$. There does not exist a correct partition $E$ such that $x E z_0$ for some $x \in X$. In fact $z_0$ may not "paste together" with any element from $X_A \cap \theta_{m+1}$, since in this case the partition $E$ is not correct, namely : the saturation of the cone $R(z_0)$ is not a cone. By the same reason it is impossible to "paste together" with element $a_i$ for some $i \leq k$. Hence there does not exist a correct partition of $\kappa X_{HA}(n)$ such that for every equivalent class $q$ there is the only element $a \in X_A$ such that $a \in q$. If the last condition of the theorem does not hold, then $A'$ is not isomorphic to $A$, moreover $A'$ is not finitely generated. Therefore $A$ is not a retract of $F_{HA}(n)$. $\square$

A finite projective Heyting algebras was described as star sums of two- and four element Boolean algebras in [1]. The same algebras are the finite Heyting algebras satisfying the condition : for any incomparable elements $a_1, \ldots, a_n$ there exists join indecomposable element $z$ such that $z$ covers $a_1 \vee \ldots \vee a_n$. More precisely it holdes the following

**Corollary 3.2** *Finite Heyting algebra $A$ is projective if and only if $A$ is isomorphic to star sum of Boolean algebras $B_i$, $i = 1, \ldots, m$, where $B_i (i < m)$ is*

isomorphic to either two- or four element Boolean algebra, and $B_m$ is isomorphic to two element Boolean algebra.

**Proof.** If $A$ is isomorphic to star sum $B_1 \oplus \ldots \oplus B_m$, where $B_m \cong \mathbf{2}$, for $1 \leq i < m$ either $B_i \cong \mathbf{2}$ or $B_i \cong \mathbf{2}^2$, then for any incomparable elements $a_1, \ldots, a_n \in A$ there exists join indecomposable element $z \in A$ such that $z$ covers $a_1 \vee \ldots \vee a_n$ (note that $n \leq 2$). Hence $A$ projective.

Now let us suppose that $A$ is finite projective Heyting algebra. Therefore for any incomparable elements $a_1, \ldots, a_n \in A$ there exists join indecomposable element $z \in A$ such that $z$ covers $a_1 \vee \ldots \vee a_n$. Then $A$ is subdirect irreducible. In fact, if we suppose opposite then for some elements $b_1, \ldots, b_k \in A$, which distinct from 1, we have $1 = b_1 \vee \ldots \vee b_k$. Then there is join indecomposable element $z \in A$ such that $z$ covers $b_1 \vee \ldots \vee b_k (= 1)$, that is we get contradiction. Consequently $1 \in A$ is join indecomposable, that is $A$ is subdirect irreducible.

Let $(X; R) = A^*$. Since $A$ is subdirect irreducible then $(X; R)$ has the smallest element, that is the last level of $(X; R)$ contains only one element. Every level of $(X; R)$ contains not more than two elements. In fact, suppose that some level contains more than two elements, say $s(\geq 3)$. Then the following level must contain not less than $2^s - (s + 1)$ elements. That is the number of elements of the following levels must increase that contradicts the finiteness of $A$. Hence the number of incomparable elements (the cardinality of antichain) not more than two. From here we have that algebra $A$ has not more than two atoms, that is equivalent to that $(X; R)$ has not more than two maximal elements. Let $\{m_1, m_2\}$ be the set of maximal elements of $(X; R)$. Let us suppose that $m_1 \neq m_2$ and $\{m_1, m_2\}$ totally covers one element, say $a \in X$, and $\{m_1\}$ totally covers one element, say $b \in X$. Then $\{b, m_2\}$ and $\{a, b\}$ totally cover $b'$ and $a'$ respectively. Analogically, $\{a, b'\}$ and $\{a', b'\}$ must totally cover the elements of the following level and so on. Then we get the model $(X_{\mathbf{HA}}(1); R)$, that corresponds to free cyclic Heyting algebra. But this contradicts the finiteness of algebra $A$. Therefore the second level contains either two, and these two elements are totally covered by $\{m_1, m_2\}$, or one element, and this element is totally covered by $\{m_1, m_2\}$. If $m_1 = m_2$ then $\{m_1\}$ totally covers either one or two elements. In common case if $\{a, b\}$ is the set of elements of some level and $a \neq b$, then the following level contains either two elements, and these elements are totally covered by $\{a, b\}$, or one element, and this element is totally covered by $\{a, b\}$. If $a = b$ then $a$ totally covers either one or two elements. In the long run the last level contains one element.

It follows as mentioned above that $A \cong B_1 \oplus \ldots \oplus B_k$, where $\forall i < k$ either $B_i \cong \mathbf{2}$ or $B_i \cong \mathbf{2}^2$, and $B_k \cong \mathbf{2}$.

From Theorm 3.1 we obtain directly

**Corollary 3.3** *Let $D \subseteq Max\kappa X_{\mathbf{HA}}(n)$ and $A_D (= (-R^{-1}(D))^*)$ be the Heyting algebra corresponding to the perfect Kripke model $\kappa X_{\mathbf{HA}}(n) - R^{-1}(D)$. Then $A_D$ is projective Heyting algebra.*

Let **V** be subvariety of **HA**. Then a homomorphism onto $\alpha : A \to B$ is **V**−universal for $A \in$ **HA** if and only if $B \in$ **V** and for every $B' \in$ **V** and every homomorphism $\beta : A \to B'$ there exists a unique homomorphism $\gamma : B \to B'$ in **V** such that $\gamma\alpha = \beta$.

**Theorem 3.4** *Let* **V₂** *is a subvariety of* **V₁**. *Then* **V₂**−*universal homomorphic image of projective algebra $A$ in* **V₁** *is projective algebra in* **V₂**.

**Proof.** Let $\alpha : A \to A'$ be **V₂**−universal homomorphism, $h : B \to B'$ be homomorphism onto $(B, B' \in$ **V₂**$)$, and $\beta : A' \to B$ be any homomorphism. Since $A$ is projective then for $\beta\alpha$ and $h$ there exists $\gamma : A \to B$ such that $h\gamma = \beta\alpha$. Then there exists a unique homomorphism $\tau : A' \to B$ such that $\gamma = \tau\alpha$. From here we have $h\tau\alpha = \beta\alpha$. Since $\alpha$ is homomorphism onto, we have $h\tau = \beta$.

**Corollary 3.5** *Let* **V** *be any subvariety of* **HA** *and* $D \subsetneq MaxX_{\mathbf{V}}(n)$, *and* $A_D (= (-R^{-1}(D))^*)$ *be the Heyting algebra corresponding to* $\kappa X_{\mathbf{V}}(n) - R^{-1}(D)$. *Then* $A_D$ *is projective algebra in* **V**.

# 4   On the structural completeness of superintuitionistic logic (Algebraic approach)

H.Friedman in his work [8] formulated among others the following hypothesis :
There exist sets $T \subseteq F$ which satisfy the following conditions :
(i) $0 \notin T$, (ii) $\alpha \& \beta \in T \Leftrightarrow \alpha \in T$ and $\beta \in T$,
(iii) $\alpha \lor \beta \in T \Leftrightarrow \alpha \in T$ or $\beta \in T$,
(iv) $\alpha \to \beta \in T \Leftrightarrow (\forall e : V \to F)[h^e(\alpha) \in T \Rightarrow h^e(\beta) \in T]$,
where $h^e$ is the extension of the function $e : V \to T$ to an endomorphism of the algebra $(F; \&, \lor, \to, 0, 1)$ which is a free algebra in the class of algebras of the type (2,2,2,0,0). Futhemore this $T$ is not unique.

T.Prucnal in [17] has shown that intermediate Medvedev's logic ML satisfies these four conditions. It was carried out a syntactic proof. In the present paper we suggest an algebraic proof. Note that intuitionistic logic Int has disjunction property. Hence Int satisfies the clauses (i) - (iii). However this calculus does not obey the condition (iv) [11]. If $T$ satisfies (i) - (iv), then $T$ is intermediate logic [18]. Let us note that the condition (iv) is equivalent to the notion of structural completeness [16] in the sense of Pogorzelski, i.e. any structural permissible rule of a logic is derivable.

Any intermediate logic, which satisfies the condition (iv), contains logic KP [18], where KP is an intermediate logic obtained by adding to Int the axiom $(\neg p \to q \lor r) \to (\neg p \to q) \lor (\neg p \to r)$ (Kreisel-Putnam [12]).

It is well known that Medvedev's logic of finite problems [13] ML is approximated by finite Kripke models $M_n = (\mathcal{P}'(X); \leq)$, where $X(= \{1, \ldots, n\})$ is non empty finite set, $\mathcal{P}'(X)$ is the set of all non empty subsets of $X$, for any $Y_1, Y_2 \in \mathcal{P}'(X)$ $Y_1 \le Y_2 \Leftrightarrow Y_2 \subseteq Y_1$. In other words ML is approximated by partially ordered sets which are isomorphic to finite Boolean algebras without the greatest element. We shall denote by $A_n (= M_n^*)$ the Heyting algebra of cones of model $M_n$. The variety $\mathbf{M}$, generated by family $\{A_n\}_{n \in \omega}$, corresponds to Medvedev's logic ML.

**Theorem 4.1** *Let $\mathbf{V}$ be subvariety of $\mathbf{HA}$, such that $X_{\mathbf{V}}(n)$, $n \in \omega$, has the smallest element. Then, if identity $(\neg p \to (q \vee r)) = 1$ holds in $\mathbf{V}$, then $(\neg p \to q) \vee (\neg p \to r) = 1$ holds in $\mathbf{V}$, where $p, q, r$ are any Heyting polynomials.*

PROOF. Let us suppose that $p, q$ and $r$ have $n$ variables. Then $p(G_1, \ldots, G_n)$, $q(G_1, \ldots, G_n)$, $r(G_1, \ldots, G_n) \in F_{\mathbf{V}}(n)$, i.e. $p(G_1, \ldots, G_n)$, $q(G_1, \ldots, G_n)$ and $r(G_1, \ldots, G_n)$ are clopen cones of $X_{\mathbf{V}}(n)$ $(G_1, \ldots, G_n$ are free generators of $F_{\mathbf{V}}(n))$. If $p(G_1, \ldots, G_n) = 1$ then the theorem is trivial. Let us suppose that $p(G_1, \ldots, G_n) \ne 1$. Then $\neg p(G_1, \ldots, G_n)$ $(= -R^{-1}(p(G_1, \ldots, G_n)))$ has the smallest element, because of projectivity of $(-R^{-1}(p(G_1, \ldots, G_n)))^*$ (Corollary 3.5). Therefore $q(G_1, \ldots, G_n) \subseteq \neg p(G_1, \ldots, G_n)$ or $r(G_1, \ldots, G_n) \subseteq \neg p(G_1, \ldots, G_n)$. Hence $(\neg p \to q) \vee (\neg p \to r) = 1$ holds in $\mathbf{V}$.

Let L be any intermediate logic satisfying the conditions (i) - (iv). We shall denote by $\mathbf{L}$ the variety corresponding to logic L.

From THEOREM 4.1 we obtain directly the following

**Corollary 4.2** *Intermediate logic L contains a formula $((\neg p \to (q \vee r)) \to ((\neg p \to q) \vee (\neg p \to r))$, i.e. $KP \subseteq L$.*

Let us note that if we have a Heyting algebra $A$ and $X_A$ is Kripke model corresponding to $A$, then the identity $((\neg p \to (q \vee r)) \to ((\neg p \to q) \vee (\neg p \to r)) = 1$ holds in $A$ if for every clopen cone $Y \subseteq X_A$ $X_A - R^{-1}(Y)$ is either empty or has the smallest element. Therefore we have the following

**Lemma 4.3** *Intermediate logic KP is contained in ML.*

**Theorem 4.4** *Any finite subdirect irreducible algebra $A \in \mathbf{L}$ is a subalgebra of $F_{\mathbf{L}}(m)$ for some $m \in \omega$.*

**Proof.** Let $X_A$ be a Kripke model corresponding to $A$. Since $A$ is subdirect irreducible then $X_A$ has the smallest element, say $x_0 \in X_A$. $X_A$ is isomorphic to a cone of $X_{\mathbf{L}}(n)(\cong F_{\mathbf{L}}^*(n))$ for some $n \in \omega$. We identify the elements $X_A$ with the corresponding elements of $X_{\mathbf{L}}(n)$. Since $X_A$ is a finite cone of $X_{\mathbf{L}}(n)$, then $x_0$ is isolated point of the space $X_{\mathbf{L}}(n)$. Therefore $-R^{-1}(x_0) \cup \{x_0\}$ and $-R^{-1}(x_0)$ are clopen cones of $X_{\mathbf{L}}(n)$ and, hence, the elements of $F_{\mathbf{L}}(n)$. From here we obtain that there exist polynomials $p_1$ and $p_2$ on $n$ variables such that $p_1(g_1, \ldots, g_n) = -R^{-1}(x_0) \cup \{x_0\}$ and $p_2(g_1, \ldots, g_2) = -R^{-1}(x_0)$, where $g_1, \ldots, g_n$ are free generators of $F_{\mathbf{L}}(n)$. It is obvious that $p_1(g_1, \ldots, g_n) \to p_2(g_1, \ldots, g_n) \ne 1$ in $F_{\mathbf{L}}(n)$, i.e. $p_1 \to p_2 \notin L$. Then there exists endomorphism $s$ of algebra $F$ such that

$p_1(s(g_1), \ldots, s(g_n)) = 1$ and $p_2(s(g_1), \ldots, s(g_n)) \neq 1$ in $F_{\mathbf{L}}(m)$ on generators $g'_1, \ldots, g'_m \in F_{\mathbf{L}}(m)$, where $m$ is the number of variables in $s(p_1 \to p_2)$. The algebra $A' = [s(g_1), \ldots, s(g_n)]_{F_{\mathbf{L}}(m)}$ is $n$-generated (by $s(g_1), \ldots, s(g_n)$) subalgebra of $F_{\mathbf{L}}(m)$. Therefore $X_{A'}(= A'^*)$ is a cone (up to isomorphism), with the smallest element, of $X_{\mathbf{L}}(n)$. Let us note that if $Y_1, Y_2$ are distinct cones of $X_{\mathbf{L}}(n)$ and $Y_1 \cong Y_2$, then for some $i \leq n$ $g_i \cap Y_1$ is not isomorphic to $g_i \cap Y_2$. Therefore the only $n$-generated Heyting algebra, in which $p_1 = 1$ and $p_2 \neq 1$, is the algebra which is isomorphic to $A(= X_A^*)$.

**Lemma 4.5** *A variety* **L** *contains the algebra* $A_2$.

**Proof.** Since **L** satisfies condition (iii),then for every $n$ elements $x_1, \ldots, x_n \in \kappa X_{\mathbf{L}}(n)$ there exists an element $y \in X_{\mathbf{L}}(n)$ such that $y \leq x_i$ for any $i \leq n$. Let us construct correct partition $E$ on $X_{\mathbf{L}}(n) : -R^{-1}(A), -R^{-1}(B), R^{-1}(A) \cap R^{-1}(B)$, where $A, B$ are non empty subsets of $\mathrm{Max}(\kappa X_{\mathbf{L}}(n))$, such that $A \cup B = \mathrm{Max}(\kappa X_{\mathbf{L}}(n))$. and $A \cap B = \emptyset$. Then we have $\kappa X_{\mathbf{L}}(n)/E \cong X_{A_2}$.

**Lemma 4.6** *1-generated Heyting algebra* $Z_7$, *which contains 7 elements, is not contained in* **L**.

**Proof.** Let us suppose that $Z_7 \in \mathbf{L}$. According to Theorem 4.4 $Z_7$ is a subalgebra of $F_{\mathbf{L}}(m)$ for some $m \in \omega$. Therefore there exists correct partition $E$ on $\kappa X_{\mathbf{L}}(m)$ such that $\kappa X_{\mathbf{L}}(m)/E \cong X_{Z_7}$. Let $f$ be strongly isotone mapping from $X_{\mathbf{L}}(m)$ onto $X_{Z_7}$, such that $\mathrm{Ker} f = E$. Then we have four classes of partition $E : f^{-1}(i), i = 1, 2, 3, 4$. Let $m_1 \in \mathrm{Max} f^{-1}(1)$ and $m_2 \in \mathrm{Max} f^{-1}(2)$. Then there exists $m_{12} \in X_{\mathbf{L}}(m)$ such that $\{m_1, m_2\}$ totally covers $m_{12}$ (Lemma 4.5). Hence $E(m_{12})$ coincides with one of the classes of partition $E$. But, in this case, correctness of the partition of $E$ is breaked.

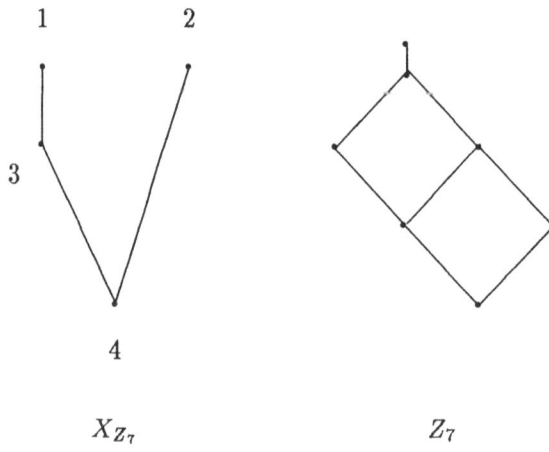

$$X_{Z_7} \qquad\qquad\qquad\qquad Z_7$$

Fig. 3

**Theorem 4.7** *The variety* **L** *contains algebras* $A_n (n \in \omega)$.

PROOF. Since **L** satisfies disjunction property then there exists subdirect irreducible algebra $A \in$ **L** with $k$ atoms and height $t \geq k$. Hence, according to Theorem 4.4 $A$ is subalgebra of $F_{\mathbf{L}}(m)$ for some $m \in \omega$. Therefore there exists strongly isotone mapping $f$ from $\kappa X_{\mathbf{L}}(m)$ onto $A$, that is there exists correct partition $E$ on $\kappa X_{\mathbf{L}}(m)$ such that $\kappa X_{\mathbf{L}}(m)/E \cong X_A$. Since $A$ has $k$ atoms then $X_A$ has $k$ maximal elements, say $a_1, ..., a_k$. Then from every $f^{-1}(a_i), i = 1, ..., k$, , we choose one element which is maximal element of $\kappa X_{\mathbf{L}}(m)$. Let $i \in f^{-1}(a_i)$ be such element from $\mathrm{Max}(\kappa X_{\mathbf{L}}(m))$. For $i \in f^{-1}(a_i)$ and $j \in f^{-1}(a_j)$ there exists $(i,j) \in q_{ij}(\in E)$ such that $i, j$ totally covers $(i,j)$ (Lemma 4.5). There does not exist an element $x$ from the second level of $\kappa X_{\mathbf{L}}(m)$ such that $x$ is totally covered by more than 2 maximal elements of $\kappa X_{\mathbf{L}}(m)$, because otherwise does not hold the following identity

$$(*) \qquad ((\neg p \rightarrow (q \vee r)) \rightarrow ((\neg p \rightarrow q) \vee (\neg p \rightarrow r)) = 1,$$

corresponding to Kreisel-Putnam formula (Corollary 4.2). Further, for every $i_j \in f^{-1}(a_{i_j}), j = 1, 2, 3$, there exists $(i_1, i_2, i_3) \in q_{i_1 i_2 i_3}(\in E)$ such that $(i_1, i_2, i_3)$ is totally covered by $(i_1, i_2), (i_1, i_3), (i_2, i_3)$. In general, for every $i_j \in f^{-1}(a_{i_j}), j = 1, ..., t(\leq k)$, there exists $(i_1, ..., i_t) \in q_{i_1 ... i_t}(\in E)$ such that $(i_1, ..., i_t)$ is totally covered by $\{(i_{j_1}, ..., i_{j_{t-1}}) : \{i_{j_1}, ..., i_{j_{t-1}}\} \subseteq \{i_1, ..., i_t\}\}$. Otherwise does not hold the identity $(*)$. In fact, let us note that $X_{\mathbf{L}}(m)$ contains a cone wich is isomorphic to $M_k$. Let $\{b_1, ..., b_n\}$ be the set of incomparable elements of $M_k$ such that there is not exist an element $x \in M_k$ which totally covered by $b_1, ..., b_n$. We add an element $a_0$ to the cone $R(\{b_1, ..., b_n\})(R =\leq)$ such way that $a_0 \leq x$ for every $x \in R(\{b_1, ..., b_n\})$. Thus we obtain a model $R(a_0)$ with the smallest element $a_0$. Let $A_{a_0}$ be the algebra corresponding to model $R(a_0)$. Algebra $A_{a_0}$ does not belong to variety **L**. In fact, if $\bigcap\limits_{i=1}^{n} b_i = b \neq \emptyset$, then $-R^{-1}(\{i_0\})(i_0 \in b)$ contains more than two minimal elements: $\{b_1 - \{i_0\}\}, ..., \{b_n - \{i_0\}\}$. If $\bigcap\limits_{i=1}^{n} b_i = \emptyset$, then for $j_0 \in b_i(i \leq n) -R^{-1}(\{j_0\})$ contains at least two minimums. Therefore in $A_{a_0}$ does not hold the identity $(*)$. Hence $A_{a_0} \notin$ **L**. Let $b$ be an element of $M_k$ which is totally covered by $\{b_1, ..., b_n\} \subseteq M_k$. Now let us suppose there exists an element $a$ such that $a$ is covered by $b_i(i = 1, ..., n)$ and $a_1, ..., a_t \notin M_k$, and $a \prec \{b_1, ..., b_n, a_1, ..., a_t\}$, $R(\{a_1, ..., a_t\}) - \mathrm{Max} M_k \neq \emptyset$. Then in algebra $A_a$, corresponding to the model $R(a)$, does not hold identity $(*)$. In fact, $R(a) - R^{-1}(x)$ contains at least two minimums, where $x \in R(\{a_1, ..., a_t\}) - \mathrm{Max} M_k$. If $R(\{a_1, ..., a_t\}) - \mathrm{Max} M_k = \emptyset$, then there exists correct partition $E$ on $R(a)$ such that $R(a)/E \cong R(b) \cong M_{k'}$ for some $k' \leq k$. Therefore $R(b)$ is a cone of $X_{\mathbf{L}}(m)$ (up to isomorphism). Let us consider else one case. Let $b$ be an element of $M_k$ which is totally covered by $\{b_1, ..., b_n\} \subseteq M_k$. Let us suppose that there exist elements $a_1, ..., a_t, a$, such that $a_i$ totally covered by $\{b_i\}(i = 1, ..., t)$, and $a$ is totallly covered by $\{a_1, ..., a_t, b_{t+1}, ..., b_n\}$. Then there exists correct partition $E$ $(a_i E b_i, i = 1, ..., t)$ such that $R(a)/E \cong R(b)(\cong M_{k'}, k' \leq k)$.

Hence we conclude that there exists a correct partititon $E$ on $\kappa X_{\mathbf{L}}$ such that $\kappa X_{\mathbf{L}}/E \cong M_k$. Since $k \in \omega$, we complete the proof. $\square$

**Theorem 4.8** *For every* $n \in \omega$ *the algebra* $A_n$ *is projective in variety* **L**.

**Proof.** Since an algebra $A_n$ belong to variety **L** , then $A_n$ is a subalgebra of $F_{\mathbf{L}}(m)$ for some $m \in \omega$ (Theorem 4.7), that is there exists correct partition $E$ on $X_{\mathbf{L}}(m)$ such that $X_{\mathbf{L}}(m)/E \cong M_n$. Let $f : X_{\mathbf{L}}(m) \rightarrow M_n$ be strongly isotone mapping such that $\mathrm{Ker} f = E$. Let us denote by the same symbols $\{i_1, ..., i_k\}(\subseteq \{1, ..., n\})$ the elements from $X_{\mathbf{L}}(m)$ the image $f(\{i_1, ..., i_k\})$ of which coincides with $\{i_1, ..., i_k\} \in M_n$ provided for every $k \leq n$ $\{k\} \in \mathrm{Max} X_{\mathbf{L}}(m)$. Then, as we see, there exists embeding $\epsilon : M_n \rightarrow \kappa X_{\mathbf{L}}(m)$, such that $f\epsilon = Id_{M_n}$. Hence $M_n$ is a retract of $\kappa X_{\mathbf{L}}(m)$. Therefore $A_n$ is a retract of $F_{\mathbf{L}}(m)$, that is $A_n$ is projective in **L**.

**Theorem 4.9** *Medvedev's logic ML satifies the conditions (i) - (iv).*

**Proof.** It is well known that ML satisfies conditions (i) - (iii). It is sufficient to show that if $\alpha \rightarrow \beta \notin$ ML then there is endomorphism $s$ of algebra $F$ such that $s(\alpha) \in$ ML and $s(\beta) \notin$ ML. Let us suppose that $\alpha \rightarrow \beta \notin$ ML. Then understanding $\alpha \rightarrow \beta$ as a Heyting polynomial, the identity $\alpha \rightarrow \beta = 1$ does not hold in $A_n$ for some $n \in \omega$ on some elements $a_1, ..., a_k \in A_n$, where $k$ is the number of variables of $\alpha \rightarrow \beta$. Then there exists join indecomposable element $a \in A_n$ such that $\alpha = 1$ and $\beta \neq 1$ in $(a](\cong A_{n_1}, n_1 \leq n)$ on the elements $f(a_1), ..., f(a_k) \in (a]$, where $f$ is homomorphism from $A_n$ onto $(a](\cong A_{n_1})$. But the algebra $(a]$ is projective in variety **M**, which corresponds to logic ML and, consequently, is a retract of $m-$generated free algebra $F_{\mathbf{M}}(m)$ for some $m \in \omega$. That is there exist homomorphisms $h : F_{\mathbf{M}}(m) \rightarrow (a]$ and $\delta : (a] \rightarrow F_{\mathbf{M}}(m)$ such that $h\delta = Id$. Since $\delta$ is embeding, then $\delta(\alpha(f(a_1), ..., f(a_k))) = 1$ and $\delta(\beta(f(a_1), ..., f(a_k))) \neq 1$. The element $\delta(f(a_i)) \in F_{\mathbf{M}}(m)$ can obtain by some polynomial $q_{a_i}$ on the free generators $g_1, ..., g_m \in F_{\mathbf{M}}(m)$, that is $q_{a_i}(g_1, ..., g_m) = \delta(f(a_1)), i = 1, ..., k$. Then the mapping $e : x_i \rightarrow q_{a_i}(x_1, ..., x_m)$ generates endomorphism $s$ of algebra $F$ such that $s(\alpha) \in ML$ and $s(\beta) \notin ML$. $\square$

From theorems 4.7 and 4.9 we obtain directly the following

**Corollary 4.10** *Medvedev's logic ML is the greatest intermediate logic among of all logics satisfying conditions (i) - (iv).*

**Theorem 4.11** *1-generated free algebra* $F_{\mathbf{L}}(1)$ *of the variety* **L** *is isomorphic to 9 element Heyting algebra* $Z_9$.

PROOF. It is clear that $F_\mathbf{L}(1)$ is a homomorphic image of 1-generated free Heyting algebra $F_\mathbf{HA}$. Because of the condition (iii) $F_\mathbf{L}(1)$ must be subdirectly irreducible. $Z_9$ belongs to variety $\mathbf{L}$. Actually, $Z_9$ is a subalgebra of the algebra $A_3$. In fact, the partition $E$ on $M_3$, the only non trivial classes of which are $\{\{1\},\{2\}\}$ and $\{\{1,3\},\{2,3\}\}$, is correct. Hence $Z_9$ is a homomorphic image of $F_\mathbf{L}(1)$. According to Lemma 4.6 1-generated Heyting algebra $Z_7$ does not belong to a variety $\mathbf{L}$. Then the only 1-generated Heyting algebra, homomorphic image of which is $Z_9$ and is not $Z_7$, is the algebra $Z_9$. Hence $F_\mathbf{L}(1) \cong Z_9$.

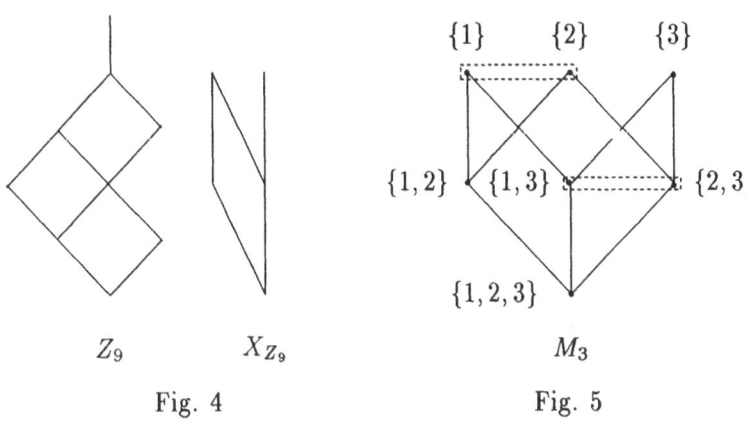

$Z_9 \qquad\qquad X_{Z_9} \qquad\qquad\qquad M_3$

Fig. 4 $\qquad\qquad\qquad\qquad$ Fig. 5

T.Prucnal in [18] formulated the hypotheses:

Let $D = \{(\alpha \rightarrow \beta) \rightarrow \beta : \alpha \in ML \& \beta \in B\}$, where $B$ is the smallest set containing $\{\neg p : p \in At\}$ and closed with respect to $\rightarrow, \vee, \wedge, \neg$ and let $H_0$ be the least intermediate logic containing $KP \cup D$. Then $H_0$ sastisfies conditions (i) - (iv).

**Assertion 4.12** $H_0$ *does not satisfy the conditions (i) - (iv).*

**Proof.** Let us denote by $\mathbf{H_0}$ the variety corresponding to $H_0$. Let us consider the algebra $A_0$, the Kripke model $X_{A_0}$ of which is represented on Fig.6. We shall show that $A_0 \in \mathbf{H_0}$. The intuitionistic formulas we shall understand as Heyting polynomials. Then the elements of $A_0$ obtained by polynomials from $B$ are $\{1,3\}$, $\{2,5\}$, $\{1,2,3,5\}$, $\emptyset$, $1_{A_0}$. If $\alpha \in ML$, then the elements obtained by polynomials is either $\{1,2,3,4,5\}$ or $1_{A_0}$. Then it is easy to check, that (*) and $\alpha = 1$ hold in $A_0$, where $\alpha \in H_0$. Hence $A_0 \in \mathbf{H_0}$.

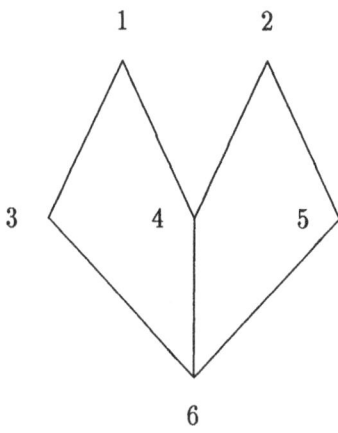

Fig. 6

Let us suppose that $H_0$ satisfies conditions (i) -(iv). Then $A_0$ is a subalgebra of $F_{H_0}(m)$ for some $m \in \omega$, (Theorem 4.4). Hence there exists strongly isotone mapping $f$ from $X_{H_0}(m)(\cong F^*_{H_0}(m))$ onto $X_{A_0}$. Let us suppose that $b \in f^{-1}(1) \cap \text{Max} X_{H_0}(m)$ such that $b$ has $\emptyset$ colour. Then

$$Y(= X_{H_0} - R^{-1}(f^{-1}(1) - \{b\}))$$

is a Kripke model corresponding to an algebra, say $A'$, which is projective in $H_0$ (Corollary 3.5). Hence $Y$ has the smallest element. Let us note that for $b$ and $b'$ from $f^{-1}(2) \cap \text{Max} X_{H_0}(m)$ there exists an element $c \in f^{1-}(4)$ such that $c$ is totally covered by $b, b'$. Hence $f^{-1}(4) - R^{-1}(f^{-1}(1) - \{b\}) \neq \emptyset$. But

$$f^{-1}(3) - R^{-1}(f^{-1}(1) - \{b\}) = \emptyset, \quad f^{-1}(6) - R^{-1}(f^{-1}(1) - \{b\}) = \emptyset.$$

Consequently $Y$ has at least two minimums. And we obtained a contradiction. Hence $H_0$ does not satisfy conditions (i) -(iv).

PROBLEM. If itermediate logic L satisfies the conditions (i) - (iv) and L finitely approximated, then L coincides with Medvedev's logic ML.

## 5    Free and projective monadic Heyting algebras

Let $(X; R, E)$ be the perfect Kripke model corresponding to monadic Heyting algebra $A$.

The partition $E_0$ of $X$ is said to be *correct*, if (1) $E_0$ is closed equivalence relation, (2) The $E_0$-saturation of a cone is cone, (3) $I(E_0(E(R(x)))) = E_0(E(R(x)))$ for every $x \in X$.
It is easy to proof the following.

**Theorem 5.1** *There is one-to-one correspondence between correct partitions of a model $(X; R, E)$ and subalgebras of algebra $A(= (X; R, E)^*$.*

Let us denote by **MHA** the variety of monadic Heyting algebras.

What kind of Kripke models we have for $n$-generated free monadic Heyting algebras?

Among the cones of $X_{\mathbf{MHA}}(1)$ we have finite cones $Z$ of model $X_{\mathbf{HA}}(1)$ whith the relation $E : xEy$ for every $x, y \in Z$. Hence $X_{\mathbf{MHA}}(1)$ has infinite maximum. Further, by levels (as in the case of Heyting algebras) we can construct $X_{\mathbf{MHA}}(1)$ and determine equivalence relation $E$ such way, that to keep correctness of corresponding partition.

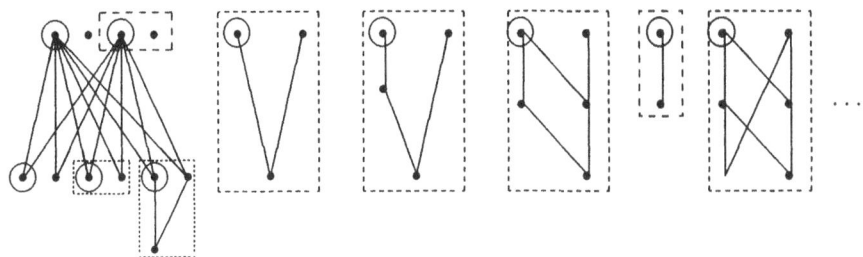

Fig. 7

In the Fig.7 is given the part of $X_{\mathbf{MHA}}(1)$. The generator is noted by circle. We see that we have complex picture.

**Theorem 5.2** $F_{\mathbf{MHA}}(1)$ *has infinitely many atoms.*

**Proof.** Let $g$ be free generator of $F_{\mathbf{MHA}}(1)$. Then $I(g)$ is atom. The element $I(\neg g)(= I(-R^{-1}(g)))$ is also atom. There are Heyting polynomials $p_n(x)$ such that $p_n(g)$ is an element of $F_{\mathbf{HA}}(1)$ corresponding to some cone of $X_{\mathbf{HA}}(1)$. Hence for some $n \in \omega$ we have $p_n(g) = \mathrm{Max}X_{\mathbf{HA}}(1)$. Then $(I(p_n(g)) \rightharpoonup (I(g) \vee I(\neg g))) \wedge I(p_n)$ is a cone cosisting of a cone as given in Fig. 8.

Fig. 8

In general case we can obtain any cone $Z \subsetneq X_{\mathbf{HA}}(1)$ with $E : xEy$ for every $x, y \in Z$. From here we conclude that $F_{\mathbf{MHA}}(1)$ has infinitely many atoms.

**Theorem 5.3** *There exists an element* $a \in F_{\mathbf{MHA}}(m)$ *such that the ideal* $(a]$ *is a homomorphic image of* $F_{\mathbf{MHA}}(m)$ *and* $(a]$ *is isomorphic to* $F_{\mathbf{HA}}(m)$.

**Proof.** Let $G_1, ..., G_m$ be free generators of $F_{MHA}(m)$, which are clopen cones of $\kappa X_{MHA}(m)$. Let us note, besides that $\kappa X_{MHA}(m)$ contains $\kappa X_{HA}(m)$ as a closed cone, and $\kappa X_{MHA}(m)$ containes finite cone $Z \subseteq X_{HA}(m)$ with equivalence relation $E : xEy$ for every $x, y \in Z$. An element $d(= I(G_1 \vee ... \vee G_m) \vee I(\neg(G_1 \vee ... \vee G_m)))$ coincides with the set all maximal elements $x$ of $X_{MHA}(m)$ such that $E(x) = x$. Then $\neg\neg d(= -R^{-1}(-R^{-1}(d)))$ is isomorphic to $\kappa X_{HA}(m)$, i.e. $\kappa X_{HA}(m)$ is a clopen cone of $X_{MHA}(m)$.

**Theorem 5.4** *Monadic Heyting algebra* $(A; \vee, \wedge, \rightarrow, I, C, 0, 1)$ *is projective in* **MHA** *if* $(A; \vee, \wedge, \rightarrow, 0, 1)$ *is isomorphic to* $F_{HA}(m)$ *and* $I(x) = C(x) = x$ *for every* $x \in A$.

**Proof.** The theorem has the same proof as Theorem 3.1.

**Corollary 5.5** *A finitely generated monadic Heyting algebra* $(A; \vee, \wedge, \rightarrow, I, C, 0, 1)$ *is projective in* **MHA** *if* $(A; \vee, \wedge, \rightarrow, 0, 1)$ *is projective Heyting algebra and* $I(x) = C(x) = x$ *for every* $x \in A$.

Friedman's problem 41 can reformulate for propositional calculus based on

$$\vee, \&, \rightarrow, \exists, \forall, 0$$

as follows:

There is a set of formulas $T$ obeying (i) $0 \notin T$, (ii)$\alpha \& \beta \in T \Leftrightarrow \alpha, \beta \in T$, (iii)$\alpha \vee \beta \in T \Leftrightarrow \alpha \in T$ or $\beta \in T$, (iv)$\alpha \vee \beta \in T \Leftrightarrow$ for any substitution $s$ $s(\alpha) \in T \Rightarrow s(\beta) \in T$, (v)$\alpha \in T \Rightarrow \exists \alpha \in T \Leftrightarrow \exists \exists \alpha \in T$, (vi)$\forall \alpha \in T \Leftrightarrow$ for any substitution $s$ $s(\alpha) \in T$. Furthermore this $T$ is not unique.

**Theorem 5.6** *Among intuitionistic modal logic there is not exist a logic* $T$ *satifying conditions (i) - (vi).*

**Lemma 5.7** *If intuitionistic modal logic* $L$ *satisfies conditions (iii) and (iv), then every finite subdirect irreducible algebra* $A \in L$ *is a subalgebra of* $F_L(m)$ *for some* $m \in \omega$, *where* **L** *is variety corresponding to logic* $L$.

**Proof.** The Lemma has the same proof as Theorem 4.4.

**Proof of Theorem 5.6.** Let $A$ be monadic Heyting algebra, the Kripke model of which is given on Fig.9.

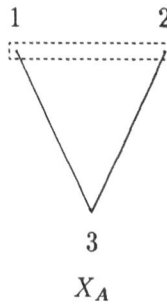

$$X_A$$

Fig. 9

It is clear that $A$ is a homomorphic image of $F_L(m)$ for some $m \in \omega$ where $L$ is any subvariety of the variety **MHA**. Let us suppose that $A$ is a subalgebra of $F_L(m)$ for some $m \in \omega$. Then there exists strongly isotone mapping $f$ from $X_L(m)$ onto $X_A$. Hence there is a correct partition $E_0$ of $X_L(m)$ such that $\operatorname{Ker} f = E_0$. $E_0$ has 3 classes. Then $f^{-1}(1)$ and $f^{-1}(2)$ contain all maximal elements of $X_L(m)$. One of them, say $f^{-1}(1)$, contains maximal element $x \in X_L(m)$ such that $E(x) = x$. Then $E_0(E(x)) = f^{-1}(1)$. But $I(f^{-1}(1)) = \emptyset \neq f^{-1}(1)$, and we have contradiction. From here we conclude that $A$ is not subalgebra $F_L(m)$.

Let us denote by $\mathbf{M}^+$ a subvariety of variety **MHA** such that $A \in \mathbf{M}^+$ if $(A; \vee, \wedge, \rightarrow, 1, 0) \in \mathbf{M}$ and for every $x \in A$, $I(x) = C(x) = x$, where **M** is a variety corresponding to Medvedev's logic ML. Let $\mathrm{ML}^+$ be the logic corresponding to variety $\mathbf{M}^+$. Then we have

**Theorem 5.8** *The logic $ML^+$ obeys conditions (i) - (vi).*

Let us note that $\mathrm{ML}^+$ is normal extension of intiuitionistic modal logic IML, but it is not intuitionistic modal logic, since $\mathrm{ML}^+ \not\subset$ S5.

# References

[1 ] R.Balbs and A.Horn, Injective and projective Heyting algebra, *Trans. Amer. Soc.* **148** (1970), 549-559

[2 ] R.Bull, MPS as the formalization of an intuitionistic concept of modality, *Journal of Symbolic Logic* **31** (1966), 609-616

[3 ] R.Engelking, General topology, Polish Scientific Publishers, Warszawa, 1977

[4 ] L.Esakia, Topological Kripke model, *Dokl. Sov. Akad.* **214**, 2, 298-301

[5 ] L.Esakia, The provability status of intuitionistic logic with the maximality principle, Preprint, 1989.

[6 ] L.Esakia, The provability logic with quantifier modalities, *IY Soviet-Finnish Symposium on Intensional Logic and Logical Structure of Theory* "Metsniereba", Tbilisi, (1988), 4-10

[7 ] L.Esakia, R.Grigolia, The criterion of Brouwerian and closure algebras to be finiteliy generated, *Bull. Sect. Logic* **6**, 2, (1973), 46-52

[8 ] H.Friedman, One hundred and two problems in mathematical logic, *Journal of Symbolic Logic* **40**(1975), 113-129

[9 ] R.Grigolia, Free algebras of non classical logics, Monograph, "Metsniereba", Tbilisi, 1987

[10 ] P.Halmos, Algebraic logic, Chesea Publ. Comp., N. Y., 1962

11 ] R.Harrop, Concerning formulas of types $A \rightarrow B \vee C, A \rightarrow (\exists x)B(x)$ in intuitionistic formal system, *Journal of Symbolic Logic* **25**(1960), 27-32

[12 ] G.Kreisel and H.Putnam, Eine Unableitbarkeitsbeweismethode für den ituitionistischen Aussagenkalkül, *Arch. Math. Logic* (Archiv für Math. Logik und Grundlagenforsch.) **3** (1957), 74-78.

[13 ] JV.Medvedev, Finite problems, *Dokl. Sov. Acad.* **142**, 5 (1962), 1015-1018

[14 ] I.Nishimura, On formulas on one variable in intuitionistic propositional calculus, *Journal of Symbolic Logic* **25**(1960), 327-331

[15 ] H.Ono, On some intuitionistic modal logic, *Publication of the Research Institute for mathematical sciences*, Kyoto university, **13**(1977), 687-722

[16 ] W.A.Pogorzelski, Structural completeness of the propositional calculus, *Bull. Acad. Polon. Sci. Ser. Math. Astr. Phys.* **19**(1971), 349-351

[17 ] T.Prucnal, Structural completeness of Medvedev's propositional calculus, *Reports on Math. Log.* **6** (1976)

[18 ] T.Prucnal, On two problems of Harvey Friedman, *SL* **38**,3 (1979), 247-262

[19 ] T.Prucnal, On the structural completeness of some pure implicational propositional calculus, *SL* **30**(1972), 45-52

[20 H.Rasiowa and R.Sikorski, The mathematics of metamathematics , (3rd ed.) Monografie mathematyczne, Warszawa, 1970

[21 ] L.Rieger, Zametki o tak nazyvaemykh algebrakh s zamykaniami, *Chekh. Math. Journ.* **7**, 16 (1957)

[22 ] N.I.Suzuki, An algrbraic approach to intuitionistic modal logic in connections with Intermediate predicate logic, *SL* **48**(1989), 141-155

[23 ] A.Urquhart, Free Heyting algebras, *Alg. Univ.* **3**,1, (1973), 94-97

# IV

# Commutative, residuated l–monoids

## U. Höhle

## 1  Introduction

The purpose of this paper is to outline a common framework for a diversity of
monoidal structures which constitute the basis of various papers in fuzzy set
theory. The most frequent structures we encounter in the literature are given
by Heyting algebras, MV–algebras and semigroup structures on the real unit
interval (so–called t–norms ([27]). Heyting algebras appear in papers looking
from an intuitionistic point of view at fuzzy set theory, MV–algebras form the
base for a positivistic approach to fuzzy set theory (cf. Poincaré's paradox and
related topics in [15]), and finally t–norms are prefered by statisticians working with a probabilistic understanding of fuzzy set theory. All these monoidal
structures have in common the following basic properties : Integrality, commutativity of the semigroup operation ∗ and the existence of a binary operation →
which is adjoint to the given operation ∗. Therefore we claim that the structure
of integral, commutative, residuated l–monoids forms the appropriate level of
generality for our intension.

Due to the limited space being available we do not attempt to develop the whole
theory of integral, commutative, residuated l–monoids; rather we concentrate
on those fundamental properties being important for the mathematical foundations of fuzzy set theory – e.g. algebraic strong de Morgan law, divisibility,
law of double negation, the existence of square roots etc. . We do not only
underline the importance of these axioms, but we also try to follow their consequences in various directions. For instance, if we compare from the point
of view of *semi–simplicity* Heyting algebras with MV–algebras, then it might
be interesting to note that divisibility and semi–simplicity imply already the
MV–algebra structure of the given monoid. On the other hand, if we study the
*MacNeille completion* of integral, commutative, residuated l–mnonoids, then we
see that Heyting algebras and (commutative) Girard–monoids behave well (i.e.
their structure is preserved under the MacNeille completion), while divisibility is an axiom which seems to be incompatible with this (minimal) extension
construction – e.g. the MacNeille completion of an MV–algebra M is again an

MV–algebra iff M is semi–simple. Finally in the case of left–continuous t–norms divisibility means continuity; this situation provokes of course the question : To which extent does fuzzy set theory require continuity ?

Further in the case of non–idempotent semigroup operations *square roots* can carry an important part of information. In this paper we make the first attempt to explore the significance of this concept in the scope of integral, commutative, residuated l-monoids. In particular we can easily build from square roots and the given semigroup operation further binary (bisymmetric) operations –e.g. if the underlying lattice is given by the real unit interval,then the *arithmetic*, respectively *geometric mean* arises in this way. Moreover we prove among others that every complete MV–algebra with square roots is either a complete Boolean algebra or a *Boolean valued model* of the real unit interval viewed as an MV–algebra or a product of both (i.e. of a complete Boolean algebra and some Boolean valued model of the real unit interval provided with Lukasiewicz's arithmetic conjunction).

The paper is organized as follows : Starting from general properties of integral, commutative, residuated l-monoids we give some mainstream examples including *Lindenbaum algebras* as well as t–norms. Subsequently, we develop a filter theory, an important prerequisite for the study of semi–simplicity. In section 5 we recall the MacNeille completion and its applications to integral, commutative, residuated l-monoids. Finally we present in section 6 various results on $\sigma$–complete (resp. complete) MV–algebras including a version of *Tarski's Lemma*.

Even though the paper is in large parts tutorial, we do hope that the expert in the field finds the orgaization stimulating and encouters at one or another place of this paper some additional information totally new also to him. In particular we apologize that we do not give credit at all those places where it might seem to be necessary; in this case the reader is referred to the papers quoted in the list of references.

# 2    Integral, residuated, commutative l-monoids

Let $(L, \leq)$ be a lattice – i.e. $(L, \leq)$ is a partially ordered set such that joins and meets of *finite* subsets of L exists. . In paricular $\bigwedge \emptyset$    (resp. $\bigvee \emptyset$ ) is the universal upper (resp. lower) bound 1 (resp. 0) in L .Throughout this paper we assume that L contains at least two elements – i.e. $0 \neq 1$ . Further let $(L, *)$ be a commutative monoid. Then the triple $(L, \leq, *)$ is called a residuated, commutative, l-monoid (cf. [4]) if and only if there exists a further binary operation $\rightarrow$ on L such that the condition

(AD)        $\alpha * \beta \; \leq \; \gamma \quad \Longleftrightarrow \quad \alpha \; \leq \; \beta \rightarrow \gamma$

holds for all $\alpha, \beta, \gamma \in L$ . It is easy to see that $\rightarrow$ is uniquely determined by (AD) .

A *homomorphism* between residuated, commutative l-monoids is a structure preserving map –i.e. $h : (L_1, \leq_1, *_1) \longmapsto (L_2, \leq_2, *_2)$ is a homomorphism if and only if h is a lattice-homomorphisms and a monoid-homomorphism satisfying the following additional condition

$$h(\alpha \rightarrow_1 \beta) \quad = \quad h(\alpha) \rightarrow_2 h(\beta)$$

Residuated commutative l-monoids and homomorphisms in the preceding sense form in an obvious way a category.

**Proposition 2.1 (General properties)** *Let* $(L, \leq, *)$ *be a residuated, commutative l-monoid. Then the following assertions are valid*

(i) $\quad \alpha * (\alpha \rightarrow \beta) \quad \leq \quad \beta \quad , \quad \quad \beta \quad \leq \quad \alpha \rightarrow (\alpha * \beta) \quad .$

(ii) $\quad (L, \leq, *)$ is a partially ordered monoid.

(iii) $\quad \alpha \rightarrow (\beta \rightarrow \gamma) \quad = \quad (\alpha * \beta) \rightarrow \gamma \ .$

(iv) $\quad \alpha * (\beta \vee \gamma) \quad = \quad (\alpha * \beta) \vee (\alpha * \gamma) \ .$

(v) $\quad \alpha \rightarrow (\beta \wedge \gamma) \quad = \quad (\alpha \rightarrow \beta) \wedge (\alpha \rightarrow \gamma) \ .$

$\quad \quad (\alpha \vee \beta) \rightarrow \gamma \quad = \quad (\alpha \rightarrow \gamma) \wedge (\beta \rightarrow \gamma) \ .$

(vi) $\quad \alpha * (\alpha \rightarrow \beta) \quad = \quad \beta \quad \Longleftrightarrow \quad \exists \gamma \in L \ \text{s.t.} \ \alpha * \gamma = \beta.$

(vii) $\quad \alpha \rightarrow (\alpha * \beta) \quad = \quad \beta \quad \Longleftrightarrow \quad \exists \gamma \in L \ \text{s.t.} \ \alpha \rightarrow \gamma = \beta \ .$

**Proof.** The assertion (i) follows from Axiom (AD) and the reflexivity of $\leq$ . In the case of $\alpha \leq \beta$ we derive from Assertion (i) and the transitivity of $\leq$ the following inequality

$$\alpha \quad \leq \quad \gamma \rightarrow (\beta * \gamma) \quad .$$

Applying (AD) we obtain $\alpha * \gamma \leq \beta * \gamma$ ; hence $(L, \leq, *)$ is a partially ordered monoid ([4]) . Because of (i) and (ii) the inequalities

$$\alpha * \beta * (\alpha \rightarrow (\beta \rightarrow \gamma) \quad \leq \quad \gamma \quad , \quad \quad \alpha * \beta * ((\alpha * \beta) \rightarrow \gamma) \quad \leq \quad \gamma$$

hold for all $\alpha, \beta, \gamma \in L$ . Now we invoke (AD) and obtain

$$\alpha \rightarrow (\beta \rightarrow \gamma) \quad \leq \quad (\alpha * \beta) \rightarrow \gamma \quad \leq \quad \alpha \rightarrow (\beta \rightarrow \gamma) \quad ;$$

hence (iii) is verified. If we put $\delta = (\alpha * \beta) \vee (\alpha * \gamma)$ , then we infer from (AD) $(\beta \vee \gamma) \leq \alpha \rightarrow \delta$ ; therefore the inequality $\alpha * (\beta \vee \gamma) \leq \delta$ follows again from (AD) . Because of (ii) the converse inequality $\delta \leq \alpha * (\beta \vee \gamma)$ also holds ; therewith the assertion (iv) is established. Finally it is not difficult to derive from (i) – (iv) the remaing relations (v) – (vii) . □

A residuated,commutative l-monoid $(L, \leq, *)$ is called *integral* if and only if the universal upper bound 1 acts as unit element w.r.t. $*$ . Integrality of residuated l-monoids monoids can be characterized as follows

**Lemma 2.2** *Let $M = (L, \leq *)$ be a commutative, residuated l-monoid, and $\top$ be the unit element of M. Further let 1 be the universal upper bound in $(L, \leq)$ and $\iota$ be an arbitrary element of L.*
*(a) The following assertions are equivalent*

(a1)     $\iota = \top$ .

(a2)     $\iota \leq \alpha \to \beta \quad \Longleftrightarrow \quad \alpha \leq \beta$ .

(a3)     $\alpha = \iota \to \alpha \quad$ for all $\alpha \in L$ .

*(b) The following assertions are equivalent*

(b1)     $(L, \leq, *)$   is integral .

(b2)     $1 = \alpha \to \beta \quad \Longleftrightarrow \quad \alpha \leq \beta$ .

(b3)     $\alpha = 1 \to \alpha \quad$ for all $\alpha \in L$ .

*(c) If there exists $\gamma \in L$ with $1 * \gamma = \top$ , then $1 = \top$ .*

**Proof.** The assertion (b) is an immediate consequence of the assertion (a) . In order to verify (a) we proceed as follows: Obviously (a1) implies (a2) . By virtue of Proposition 2.1 (iii) we derive from (a2) the following relation

$$\alpha \leq \iota \to \beta \quad \Longleftrightarrow \quad \iota \leq \alpha \to (\iota \to \beta) = \iota \to (\alpha \to \beta) \quad \Longleftrightarrow \quad \alpha \leq \beta \ ;$$

hence (a2) implies (a3) . If the assertion (a3) holds , then we infer from Proposition 2.1 (vii)

$$\iota * \beta \quad = \quad \iota \to (\iota * \beta) \quad = \quad \beta \quad ;$$

hence $\iota$ is the unit element of M.
Because of Proposition 2.1(ii) the following relation holds

$$1 = 1 * \top \leq 1 * 1 \leq 1 \quad .$$

On the other hand we infer from the hypothesis of Assertion (c) :

$$1 * (1 \to \top) = \top \quad .$$

Multiplying both sides by 1 the relation

$$1 = \top * 1 = 1 * 1 * (1 \to \top) = 1 * (1 \to \top) = \top$$

follows.
□

In any integral, residuated, commutative l-monoid the following relation holds

(viii)     $\alpha * \beta \quad \leq \quad \alpha \wedge \beta$ .

**Proposition 2.3** *In any integral, residuated, commutative l-monoid the following assertions are equivalent*

(ix)     $(\alpha \rightarrow \beta) \vee (\beta \rightarrow \alpha) \quad = \quad 1$     (Algebraic Strong De Morgan Law)

(x)     $\alpha \rightarrow (\beta \vee \gamma) \quad = \quad (\alpha \rightarrow \beta) \vee (\alpha \rightarrow \gamma)$ .

(xi)     $(\alpha \wedge \beta) \rightarrow \gamma \quad = \quad (\alpha \rightarrow \gamma) \vee (\beta \rightarrow \gamma)$ .

**Proof.** Referring to (AD), (i), (ii), (iii) and (viii) we obtain

$$\alpha * (\beta \rightarrow \gamma) \quad \leq \quad \beta \rightarrow (\alpha * \gamma) \quad ,$$

$$\begin{aligned}((\alpha \wedge \beta) \rightarrow \gamma) * (\alpha \rightarrow \beta) \quad &\leq \quad [(\alpha * (\alpha \rightarrow \beta)) \rightarrow \gamma] * (\alpha \rightarrow \beta)\\ &\leq \quad \alpha \rightarrow \gamma \quad .\end{aligned}$$

If (ix) holds , then we derive from (iv) and the preceding relations the following estimations

$\alpha \rightarrow (\beta \vee \gamma) \quad \leq$

$(\alpha \rightarrow [(\beta \vee \gamma) * (\beta \rightarrow \gamma)]) \bigvee (\alpha \rightarrow [(\beta \vee \gamma) * (\gamma \rightarrow \beta)]) \quad \leq$

$(\alpha \rightarrow \gamma) \vee (\alpha \rightarrow \beta) \quad ,$

$(\alpha \wedge \beta) \rightarrow \gamma \quad =$

$[((\alpha \wedge \beta) \rightarrow \gamma) * (\alpha \rightarrow \beta)] \bigvee [((\alpha \wedge \beta) \rightarrow \gamma) * (\beta \rightarrow \alpha)] \quad \leq$

$(\alpha \rightarrow \gamma) \vee (\beta \rightarrow \gamma) \quad ;$

hence (ix) implies (x) as well as (xi) . On the other hand, we deduce from (x) (resp. (xi)) and Lemma 2.2(b2) the following relations

$$\begin{aligned}(\alpha \rightarrow \beta) \vee (\beta \rightarrow \alpha) \quad &\geq \quad ((\alpha \vee \beta) \rightarrow \alpha) \quad \vee \quad ((\alpha \vee \beta) \rightarrow \beta) \quad = \quad 1\\ (\alpha \rightarrow \beta) \vee (\beta \rightarrow \alpha) \quad &\geq \quad (\alpha \rightarrow (\alpha \wedge \beta)) \quad \vee \quad (\beta \rightarrow (\alpha \wedge \beta)) \quad = \quad 1 \quad .\end{aligned}$$

Hence the equivalence between (ix), (x) and (xi) is established.
□

**Lemma 2.4** *Let $(L, \leq, *)$ be an integral, residuated, commutative l−monoid satisfying the algebraic strong de Morgan law (cf. (ix)) . Then the following assertions are valid*

(1)     $\alpha * \beta \quad \leq \quad (\alpha * \alpha) \vee (\beta * \beta)$ ,     $(\alpha * \alpha) \wedge (\beta * \beta) \quad \leq \quad \alpha * \beta$ .

(2)     $\alpha * (\beta \wedge \gamma) \quad = \quad (\alpha * \beta) \wedge (\alpha * \gamma)$ .

(3)        $\alpha \wedge (\beta \vee \gamma) \;=\; (\alpha \wedge \beta) \vee (\alpha \wedge \gamma)$        .

**Proof.** Combining the algebraic strong de Morgan law with the integrality of $(L, \leq, *)$ we obtain

$$\alpha * \beta \;\leq\; (\alpha * \beta * (\beta \to \alpha)) \vee (\alpha * \beta * (\alpha \to \beta)) \;\leq\; (\alpha * \alpha) \vee (\beta * \beta) \quad,$$

$$(\alpha * \alpha) \wedge (\beta * \beta) \;\leq\; (\alpha * \alpha * (\alpha \to \beta)) \vee (\beta * \beta * (\beta \to \alpha)) \;\leq\; \alpha * \beta \quad,$$

$$(\alpha * \beta) \wedge (\alpha * \gamma) \leq\; (\alpha * \beta) * (\beta \to \gamma)) \vee (\alpha * \gamma * (\gamma \to \beta)) \;\leq\; \alpha * (\beta \wedge \gamma) \quad,$$

$$(\alpha \wedge (\beta \vee \gamma)) \to ((\alpha \wedge \beta) \vee (\alpha \wedge \gamma)) \quad=$$

$$[(\alpha \wedge (\beta \vee \gamma)) \to (\alpha \wedge \beta)] \bigvee [(\alpha \wedge (\beta \vee \gamma)) \to (\alpha \wedge \gamma)] \quad\geq$$

$$((\beta \vee \gamma) \to \beta) \vee ((\beta \vee \gamma) \to \gamma) \quad=\quad (\gamma \to \beta) \vee (\beta \to \gamma) \;=\; 1 \quad;$$

hence the relations (1) – (3) are established.
□

An integral, residuated, commutative l-monoid $(L, \leq, *)$ is called *divisible* if and only if for every pair $(\alpha, \beta) \in L \times L$ with $\beta \leq \alpha$ there exists $\gamma \in L$ s.t. $\beta = \alpha * \gamma$. With regard to Lemma 2.2(c) we note that divisibility makes only sense in the scope of *integral* l-monoids.
A characterization of divisibility is given in the following

**Lemma 2.5** *Let* $(L, \leq *)$ *be an integral, residuated, commutative l-monoid. Then the following assertions are equivalent*

(1)        $(L, \leq, *)$   *is divisible.*

(2)        $\alpha \wedge \beta \;=\; \alpha * (\alpha \to \beta)$   .

(3)        $\alpha \to (\beta \wedge \gamma) \;=\; (\alpha \to \beta) * ((\alpha \wedge \beta) \to \gamma)$   .

**Proof.** If $(L, \leq, *)$ is divisible, then there exists $\gamma \in L$ s.t. $\alpha \wedge \beta = \gamma * \alpha$ . Because of $\gamma \leq (\alpha \to \beta)$ we obtain

$$\alpha \wedge \beta \;=\; \gamma * \alpha \;\leq\; \alpha * (\alpha \to \beta) \;\leq\; \alpha \wedge \beta \;;$$

i.e. (1) implies (2) . If (2) holds, then we infer from (iii) and (v)

$$(\alpha \to \beta) * ((\alpha \wedge \beta) \to \gamma) \quad=\quad (\alpha \to \beta) * ((\alpha \to \beta) \to (\alpha \to \gamma))$$
$$=\quad (\alpha \to \beta) \wedge (\alpha \to \gamma) \;=\; \alpha \to (\beta \wedge \gamma) \;;$$

hence (2) implies (3) . In the case of $\alpha = 1$ and $\gamma \leq \beta$ we apply Lemma 2.2(b3) and relation (3)

$$\beta * (\beta \to \gamma) \;=\; (1 \to \beta) * (\beta \to \gamma) \;=\; 1 \to \gamma \;=\; \gamma \;;$$

hence (3) implies (1) .
□

**Proposition 2.6** *Let* $(L, \leq, *)$ *be an integral, divisible, residuated, commutative l-monoid. Then the following relations are valid*

(xii)     If $\alpha$ is idempotent w.r.t. $*$ , then $\alpha \wedge \beta \;=\; \alpha * \beta$ for all $\beta \in L$.

(xiii)    $\alpha * (\beta \wedge \gamma) \;=\; (\alpha * \beta) \wedge (\alpha * \gamma)$ .

(xiv)    $\alpha * \beta \;\leq\; (\alpha * \alpha) \vee (\beta * \beta)$.

(xv)    $\alpha \wedge (\beta \vee \gamma) \;=\; (\alpha \wedge \beta) \vee (\alpha \wedge \gamma)$ .

**Proof.** A verification of (xii) and (xiii) is given in the proof of Lemma 2.2 in [17] . In order to prove (xiv) we first consider an arbitrary pair $(\alpha, \beta) \in L \times L$ . Since $(L, \leq, *)$ is divisible , there exist elements $\gamma_1, \gamma_2 \in L$ such that $\alpha \;=\; (\alpha \vee \beta) * \gamma_1$ and $\beta \;=\; (\alpha \vee \beta) * \gamma_2$ . Because of (iv) the equation $\alpha \vee \beta \;=\; (\alpha \vee \beta) * (\gamma_1 \vee \gamma_2)$ holds. Hence we obtain

$$
\begin{aligned}
\alpha * \beta \;&=\; \gamma_1 * \gamma_2 * (\alpha \vee \beta) * (\alpha \vee \beta) \\
&=\; \gamma_1 * \gamma_2 * (\gamma_1 \vee \gamma_2) * (\alpha \vee \beta) * (\alpha \vee \beta) \\
&\leq\; ((\gamma_1 * \gamma_1) \vee (\gamma_2 * \gamma_2)) * (\alpha \vee \beta) * (\alpha \vee \beta) \\
&=\; (\alpha * \alpha) \vee (\beta * \beta) \qquad ;
\end{aligned}
$$

hence the relation (xiv) is established. Further we deduce from Lemma 2.5(2)

$$
\begin{aligned}
\alpha \wedge (\beta \vee \gamma) \;&=\; (\beta \vee \gamma) * ((\beta \vee \gamma) \to \alpha) \\
&\leq\; (\beta * (\beta \to \alpha)) \vee (\gamma * (\gamma \to \alpha)) \;=\; (\alpha \wedge \beta) \vee (\alpha \wedge \gamma) \quad ;
\end{aligned}
$$

hence (xv) is verified .
□

**Corollary 2.7** *Let* $M = (L, \leq, *)$ *be an integral, residuated, commutative l-monoid. If $M$ is divisible and satisfies the algebraic, strong de Morgan law, then the subset $H_M$ of all idempotent elements w.r.t. $*$ forms a Heyting algebra, and the implication in $H_M$ coincides with the implication based on $*$ .*

**Proof.** Let us recall that $\to$ denotes the implication based on $*$ . Further let $e_1$ and $e_2$ , respectively, be an idempotent element. The inequality $e_1 \wedge (e_1 \to e_2) \;\leq\; e_2$ follows from (i) and (xii) . Now we invoke (xi) and obtain

$$ 1 \;\leq\; (e_1 \to e_2) \vee ((e_1 \to e_2) \to e_2) \qquad ; $$

hence the inequality

$$ e_1 \to e_2 \;\leq\; ((e_1 \to e_2) * (e_1 \to e_2)) \vee ((e_1 \to e_2) * (e_1 \to e_2) \to e_2)) $$

holds. Since $e_2$ is idempotent, we infer from Lemma 2.2(b3) and Lemma 2.5(2)

$$
\begin{aligned}
(e_1 \to e_2) * ((e_1 \to e_2) \to e_2) \;&=\; e_2 \wedge (e_1 \to e_2) \;=\; e_2 * (e_1 \to e_2) \\
&\leq\; (e_1 \to e_2) * (e_1 \to e_2)
\end{aligned}
$$

hence we obtain    $(e_1 \to e_2)$  $\leq$   $(e_1 \to e_2) * (e_1 \to e_2)$    – i.e.
$e_1 \to e_2$   is idempotent.
□

An integral, residuated, commutative l-monoid $(L, \leq, *)$ is called an integral, commutative *Girard*-monoid if and only if the "negation" is an involution – i.e.

(xvi)      $(\alpha \to 0) \to 0$   =   $\alpha$

**Proposition 2.8** *Let $(L, \leq, *)$ be an integral, commutative, Girard-monoid. Then the following relations are valid*

(xvii)      $\alpha \to \beta$   =   $(\alpha * (\beta \to 0)) \to 0$

(xviii)      $(\alpha \wedge \beta) \to 0$   =   $(\alpha \to 0) \vee (\beta \to 0)$    .

**Proof.** The relation (xvii) follows immediately from (iii) and (xvi) . Now we invoke (v) and (xvi) and obtain

$$((\alpha \to 0) \vee (\beta \to 0)) \to 0  =  ((\alpha \to 0) \to 0) \wedge ((\beta \to 0) \to 0)  =  \alpha \wedge \beta  ;$$

hence (xviii) follows again from (xvi) .
□

**Lemma 2.9** *Let $(L, \leq *)$ be an integral, commutative Girard-monoid. Then the following assertions are equivalent*

(1)      $(L, \leq, *)$ satisfies the algebraic strong de Morgan law.

(2)      $\alpha * (\beta \wedge \gamma)$   =   $(\alpha * \beta) \wedge (\alpha * \gamma)$    .

**Proof.** The implication (1) $\Longrightarrow$ (2) follows from Lemma 2.4 . On the other hand, if we use (2), (xvii) and (xviii) , then we obtain

$$\begin{aligned}
\alpha \to (\beta \vee \gamma)  &=  (\alpha * [(\beta \to 0) \wedge (\gamma \to 0)] \to 0) \\
&=  [(\alpha * (\beta \to 0)) \wedge (\alpha * (\gamma \to 0))] \to 0 \\
&=  [(\alpha * (\beta \to 0)) \to 0] \vee [(\alpha * (\gamma \to 0)) \to 0] \\
&=  (\alpha \to \beta) \vee (\alpha \to \gamma)  ;
\end{aligned}$$

hence we infer from Proposition 2.3 that the algebraic strong de Morgan law holds.
□

**Lemma 2.10** *Let $(L, \leq, *)$ be an integral, commutative Girard-monoid satisfying the algebraic strong de Morgan law. Then the "negation" has at most a fixpoint – i.e.*

$\alpha$   =   $\alpha \to 0$   *and*   $\beta$   =   $\beta \to 0$   *implies*   $\alpha = \beta$    .

**Proof.** If $\alpha$ and $\beta$ are fixpoints of the "negation" , then the algebraic strong de Morgan law implies

$$
\begin{aligned}
\alpha \quad &= \quad (\alpha * (\alpha \to \beta)) \vee (\alpha * (\beta \to \alpha)) \\
&= \quad (\alpha * (\alpha \to \beta)) \vee (\alpha * ((\beta \to 0) \to (\alpha \to 0))) \\
&= \quad \alpha * (\alpha \to \beta) \quad \le \quad \beta \quad ;
\end{aligned}
$$

hence the assertion follows.
□

An integral, residuated, commutative l–monoid $(L, \le *)$ has *square roots* if and only if there exists a unary operation $S : \quad L \longmapsto L \quad$ provided with the following properties

(S1) $\qquad S(\alpha) * S(\alpha) \quad = \quad \alpha$

(S2) $\qquad \beta * \beta \quad \le \alpha \quad \Longrightarrow \quad \beta \le S(\alpha)$ .

Obviously S is uniquely determined by (S1) and (S2) . Therefore we write instead of $S(\alpha)$ also $\alpha^{1/2}$ .

**Proposition 2.11 (General properties of square roots)** *Let* $(L, \le, *)$ *be an integral, residuated, commutative l-monoid with square roots. Then the following relations hold*

(xix) $\qquad \alpha \quad \le \quad \alpha^{1/2} \quad , \quad \alpha \le \beta \quad \Longrightarrow \quad \alpha^{1/2} \le \beta^{1/2}$

(xx) $\qquad \alpha^{1/2} * \beta^{1/2} \quad \le \quad (\alpha * \beta)^{1/2}$

(xxi) $\qquad (\alpha \to \beta)^{1/2} \quad = \quad \alpha^{1/2} \to \beta^{1/2}$

(xxii) $\qquad (\alpha \wedge \beta)^{1/2} \quad = \quad \alpha^{1/2} \wedge \beta^{1/2}$

(xxiii) $\qquad \alpha * \beta \quad \le \quad (\alpha * \alpha) \vee (\beta * \beta)$

(xxiv) $\qquad \alpha \wedge \beta \quad \le \quad \alpha^{1/2} * \beta^{1/2} \quad \le \quad \alpha \vee \beta$ .

**Proof.** (xix) and (xx) follow immediately from (S1) and (S2) . Now we invoke again (S2) and derive from

$$
(\alpha^{1/2} \to \beta^{1/2}) * (\alpha^{1/2} \to \beta^{1/2}) \quad \le \quad \alpha \to \beta
$$

the inequality $\quad \alpha^{1/2} \to \beta^{1/2} \quad \le \quad (\alpha \to \beta)^{1/2}$ . On the other hand, we infer from (i), (xix) and (xx) $\quad \alpha^{1/2} * (\alpha \to \beta)^{1/2} \quad \le \quad \beta^{1/2} \quad$ ; hence the converse inequality $\quad (\alpha \to \beta)^{1/2} \quad \le \quad \alpha^{1/2} \to \beta^{1/2} \quad$ follows. Therewith the relation (xxi) is established. Because of (xix) and $(\alpha^{1/2} \wedge \beta^{1/2}) * (\alpha^{1/2} \wedge \beta^{1/2}) \quad \le \quad \alpha \wedge \beta \quad$ the relation (xxii) holds. Further the inequality $\alpha \vee \beta \leq ((\alpha * \alpha) \vee (\beta * \beta))^{1/2}$ follows from (S2) . Taking on both sides the square we can apply (iv) and obtain

$$(\alpha * \alpha) \vee (\alpha * \beta) \vee (\beta * \beta) \quad \leq \quad (\alpha * \alpha) \vee (\beta * \beta) \quad ;$$

hence relation (xxiii) is valid. Finally we use (S1), (xix) and (xxiii)

$$\alpha \wedge \beta \quad = \quad (\alpha \wedge \beta)^{1/2} * (\alpha \wedge \beta)^{1/2} \leq \alpha^{1/2} * \beta^{1/2} \quad ;$$

thus the relation (xxiv) is established.
□

**Corollary 2.12** *Let $(L, \leq, *)$ be an integral, divisible, residuated, commutative l-monoid with square roots. Then the implication*
$\beta \leq \alpha^{1/2} * \beta^{1/2} \implies \beta \leq \alpha$ *holds.*

**Proof.** If $\beta \leq \alpha^{1/2} * \beta^{1/2}$ , then we derive from Lemma 2.5(2) and (xiii) the following estimation

$$
\begin{aligned}
\beta &= (\alpha^{1/2} \wedge \beta^{1/2}) * \beta^{1/2} = \beta^{1/2} * \beta^{1/2} * (\beta^{1/2} \to \alpha^{1/2}) \\
&= \beta * (\beta^{1/2} \to \alpha^{1/2}) \leq \alpha^{1/2} * \beta^{1/2} * (\beta^{1/2} \to \alpha^{1/2}) \leq \alpha .
\end{aligned}
$$

**Corollary 2.13** *Let $M = (L, \leq, *)$ be an integral, commutative Girard-monoid with square roots. Then the following relations are valid*

(xv)    $(\alpha * \beta)^{1/2} = ((\alpha^{1/2} * \beta^{1/2}) \to 0^{1/2}) \to 0^{1/2}$ .

(xxvi)    If M satsifies the algebraic strong de Morgan law, then
$(\alpha \vee \beta)^{1/2} = \alpha^{1/2} \vee \beta^{1/2}$ holds.

**Proof.** Referring to (iii), (xvi) and (xxi) we obtain

$$
\begin{aligned}
((\alpha^{1/2} * \beta^{1/2}) \to 0^{1/2}) \to 0^{1/2} &= (\alpha^{1/2} \to (\beta \to 0)^{1/2}) \to 0^{1/2} \\
&= ((\alpha \to (\beta \to 0)) \to 0)^{1/2} = (\alpha * \beta)^{1/2}.
\end{aligned}
$$

Thus (xxv) is established. Now we invoke (xxii) , the algebraic strong de Morgan law and use again (xxi)

$$
\begin{aligned}
(\alpha \vee \beta)^{1/2} &= (((\alpha \to 0) \wedge (\beta \to 0)) \to 0)^{1/2} \\
&= ((\alpha \to 0)^{1/2} \wedge (\beta \to 0)^{1/2}) \to 0^{1/2} \\
&= ((\alpha \to 0)^{1/2} \to 0^{1/2}) \vee ((\beta \to 0)^{1/2} \to 0^{1/2} \\
&= \alpha^{1/2} \vee \beta^{1/2} \quad ;
\end{aligned}
$$

hence the assertion (xxvi) is verified.
□

There exist different, but equivalent ways to define MV–algebras ([5]). Here we are looking from the point of view of integral, residuated l–monoids at MV–algebras. This leads to the following definition :

An integral, residuated,commutative l–monoid $(L, \leq, *)$ is said to be a MV–*algebra* (where MV stands for many valued logics) if and only if the important, additional condition

(MV) $\qquad (\alpha \to \beta) \to \beta \quad = \quad \alpha \vee \beta$

is satisfied. We can characterize MV–algebras as follows

**Lemma 2.14** *Let* $(L, \leq, *)$ *be an integral, residuated, commutative l–monoid. Then the following assertions are equivalent*

(1) $\qquad (L, \leq, *)$ is an MV–algebra

(2) $\qquad (L, \leq, *)$ is a divisible Girard–monoid

**Proof.** (cf. [17])

**Proposition 2.15** *In any MV–algebra the following relation holds*

(xxvii) $\qquad \alpha \to (\alpha * \beta) \quad = \quad (\alpha \to 0) \vee \beta \qquad .$

**Proof.** Referring to (iii) and (MV) we obtain

$$
\begin{aligned}
(\alpha \to 0) \vee \beta \quad &= \quad (\beta \to (\alpha \to 0)) \to (\alpha \to 0) \\
&= \quad ((\alpha * \beta) \to 0) \to (\alpha \to 0) \\
&= \quad \alpha \to (\alpha * \beta) \qquad .
\end{aligned}
$$

□

An immediate consequence of Proposition 2.6 , Lemma 2.9 and Lemma 2.14 is the following

**Corollary 2.16 (C.C. Chang [5])** *Every MV–algebra fulfills the algebraic strong de Morgan law.*

**Proposition 2.17** *Every MV-algebra with square roots has the properties*

(xxviii) $\qquad (\alpha * \beta)^{1/2} \quad = \quad (\alpha^{1/2} * \beta^{1/2}) \vee 0^{1/2}$

(xxix) $\qquad (0^{1/2} \to 0) * (0^{1/2} \to 0) \qquad$ is idempotent w.r.t. $\quad * \qquad .$

**Proof.** (xxviii) follows immediately from Corollary 2.13 and Axiom (MV) . In order to verify (xxix) we first use (xxi) and (xxvii)

$$
\begin{aligned}
(0^{1/2} \to 0)^{1/2} \quad &= \quad (0^{1/2})^{1/2} \to 0^{1/2} \\
&= \quad ((0^{1/2})^{1/2} \to 0) \vee (0^{1/2})^{1/2} \\
&\leq \quad (0^{1/2} \to 0) \vee (0^{1/2})^{1/2}.
\end{aligned}
$$

Now we take on both sides the fourth power and obtain

$$
(0^{1/2} \to 0) * (0^{1/2} \to 0) \quad \leq \quad [(0^{1/2} \to 0) * (0^{1/2} \to 0)]^2
$$

hence (xxix) is verified.
□

**Corollary 2.18** *In any MV-algebra* $(L, \leq, *)$ *with square roots there exists a unique element* $\iota_0 \in L$ *provided with the following properties*

(1)   $\iota_0$ is idempotent w.r.t. $*$ .

(2)   $\iota_0 \vee 0^{1/2} \quad = \quad 0^{1/2} \to 0$ .

**Proof.** In order to verify the existence of an element of L satisfying (1) and (2) we put   $\iota_0 \quad = \quad (0^{1/2} \to 0) * (0^{1/2} \to 0)$   . Because of Proposition 2.17(xxix) the element $\iota_0$ is idempotent. Further we obtain from Proposition 2.17(xxviii)

$$
\begin{aligned}
0^{1/2} \to 0 \quad = \quad (0^{1/2} \to 0) \vee 0^{1/2} \quad &= \quad (\iota_0)^{1/2} \\
&= \quad (\iota_0 * \iota_0)^{1/2} \quad = \quad \iota_0 \vee 0^{1/2} \quad ;
\end{aligned}
$$

i.e. $\iota_0$ fulfills also conditon (2) . Finally the uniqueness of $\iota_0$ follows from the properties (1),(2) and Proposition 2.11(xiii) .
□

**Proposition 2.19** *Let* $M = (L, \leq, *)$ *be an MV-algebra with square roots. Then the following assertions are equivalent*

(1)   M is a Boolean algebra (i.e. $* = \wedge$).

(2)   $0^{1/2} \quad = \quad 0$ .

**Proof.** The implication (1) $\Longrightarrow$ (2) is obvious. Now we assume that (2) holds. Then   $\alpha \wedge (\alpha \to 0) = 0$   follows from the inequality $\alpha \wedge (\alpha \to 0) \leq 0^{1/2}$ . We apply Proposition 2.8(xviii) and obtain

$$
\alpha \quad = \quad \alpha * ((\alpha \to 0) \vee \alpha) \quad = \quad \alpha * \alpha \quad ;
$$

i.e. every element of L is idempotent w.r.t. $*$ . Hence $(L, \leq, *)$ is a Boolean algebra.
□

**Remark 2.20** Let $M = (L, \leq, *)$ be an MV–algebra. By virtue of Corollary 2.7 and Corollary 2.16 the set $\mathbb{B}_M$ of all idempotent elements of L forms a *Boolean* algebra. Therefore every idempotent element of L determines a decomposition of M into a product of MV–algebras in the following way : Let $\iota \in L$ be idempotent w.r.t. $*$ and $[0, \iota] = \{\lambda \in L | \lambda \leq \iota\}$ be the order interval corresponding to $\iota$ . On $[0, \iota]$ we consider the MV–algebra structure $M_\iota$ inherited from M by restriction. In particular the "implication" $\overset{\iota}{\rightarrow}$ in $[0, \iota]$ is given by

$$\alpha \overset{\iota}{\rightarrow} \beta \quad = \quad \iota \wedge (\alpha \rightarrow \beta) \quad .$$

Now we use $\iota \vee (\iota \rightarrow 0) = 1$ and obtain after a moment's reflection that M isomorphic to the product $M_\iota \times M_{\iota \rightarrow 0}$ .

Finally we observe that the property of having square roots is also inherited from M onto $M_\iota$. In particular the square root $S_\iota(\alpha)$ w.r.t. $M_\iota$ has the following form $S_\iota(\alpha) \quad = \quad \iota \wedge (\alpha)^{1/2}$ .

□

In order to derive from Corollary 2.18 an important and far reaching consequence we need the following definition : An MV–algebra with square roots is called *strict* if and only if $\quad 0^{1/2}\quad$ is a fixpoint w.r.t. the "negation" – i.e. $0^{1/2} \rightarrow 0 = 0^{1/2}$ .

An immediate consequence from Proposition 2.19 is the statement that a *Boolean algebra* (with at least two elements) is *never* a *strict* MV–algebra. On the other hand, the real unit interval [0,1] propvided with Lukasiewicz's arithmetic conjunction $T_m$ (cf. [8] and section 3.3 infra) is an important example of a strict MV–algebra. Moreover we can classify MV–algebras with square roots as follows :

**Theorem 2.21 (Classification of MV-algebras with square roots)**
*Every MV–algebra $M = (L, \leq, *)$ with square roots is either a Boolean algebra or a strict MV–algebra or is isomorphic to a product of a Boolean algebra $M_b$ and a strict MV–algebra $M_s$ In the latter case $M_b$ and $M_s$ are uniquely determined by $M \cong M_b \times M_s$ up to an isomorphism.*

**Proof.** Without loss of generality we assume that M is neither a Boolean algebra nor a strict MV–algebra .

(a) (Existence) Let $\iota$ be an idempotent element of L satisfying the condition (2) in Corollary 2.18 . According to Remark 2.20 the MV–algebra M can be decomposed to the product $M_\iota \times M_{\iota \rightarrow 0}$ . Further we denote by $S_\iota$ (resp. $S_{\iota \rightarrow 0}$) the square root in $M_\iota$ (resp. $M_{\iota \rightarrow 0}$). Then the relations

$$S_\iota(0) \quad = \quad \iota \wedge (0)^{1/2} \quad = \quad \iota * 0^{1/2} \quad = \quad 0$$

$$S_{\iota \rightarrow 0}(0) \overset{\iota \rightarrow 0}{\longrightarrow} 0 \quad = \quad S_{\iota \rightarrow 0}(0)$$

follow from property (2) in 2.18 ; i.e. $M_\iota$ is a Boolean algebra (cf. Proposition 2.19) and $M_{\iota \rightarrow 0}$ is a strict MV–algebra. Hence the first part of Theorem 2.21 is verified.

(b) (Uniqueness) Let M be isomorphic to a product $M_b \times M_s$ of a Boolean algebra $M_b = (L_b, \leq, \wedge)$ and a strict MV–algebra $M_s = (L_s, \leq, *_s)$. If we denote by $1_b$ (resp. $0_b$) the universal upper (resp. lower) bound in $(L_b, \leq)$ and by $1_s$ (resp. $0_s$) the universal upper (resp. lower) bound in $(L_s, \leq)$ , then we obtain

$$
\begin{aligned}
(0_b, 0_s)^{1/2} \vee (1_b, 0_s) &= (0_b, (0_s)^{1/2}) \vee (((0_b)^{1/2} \to 0_b), 0_s) \\
&= (0_b, ((0_s)^{1/2} \to 0_s)) \vee (((0_b)^{1/2} \to 0_b), 0_s) \\
&= (((0_b)^{1/2} \to 0_b), ((0_s)^{1/2} \to 0_s)) \\
&= (0_b, 0_s)^{1/2} \to (0_b, 0_s)
\end{aligned}
$$

i.e. the element $(1_b, 0_s)$ of $L_b \times L_s$ satisfies the condition (2) in Corollary 2.18 . Therefore the uniqueness of the decomposition follows immediately from Corollary 2.18 .

□

**Proposition 2.22** *Let $M = (L, \leq, *)$ be a strict MV–algebra. Then the following relations are valid*

(1)    $\alpha^{1/2} \to (\alpha^{1/2} * \beta^{1/2}) = \beta^{1/2}$ .

(2)    If   $\gamma \leq \alpha \wedge \beta$   , then
$$(\alpha \to \gamma)^{1/2} * (\beta \to \gamma)^{1/2} = (\alpha^{1/2} * \beta^{1/2}) \to \gamma \ .$$

(3)    If $\iota$ is idempotent and satisfies the inequality $0^{1/2} \leq \iota$ , then $\iota = 1$ .

**Proof.** From    $0^{1/2} \to 0 = 0^{1/2}$    we infer    $\alpha^{1/2} \to 0 \leq 0^{1/2}$   ; hence the relation (1) follows from Proposition 2.15(xxvii) . Further we assume $\gamma \leq \alpha \wedge \beta$ and observe   $\gamma \leq \alpha^{1/2} * \gamma^{1/2} \leq \alpha^{1/2} * \beta^{1/2}$   . Now we apply Lemma 2.14 and Lemma 2.5(3) and obtain from (1)

$$
\begin{aligned}
(\alpha^{1/2} * \beta^{1/2}) \to \gamma &= ((\alpha^{1/2} * \beta^{1/2}) \to (\alpha^{1/2} * \gamma^{1/2})) * ((\alpha^{1/2} * \gamma^{1/2}) \to \gamma) \\
&= [\beta^{1/2} \to (\alpha^{1/2} \to (\alpha^{1/2} * \gamma^{1/2}))] * (\alpha^{1/2} \to (\gamma^{1/2} \to \gamma)) \\
&= (\beta^{1/2} \to \gamma^{1/2}) * (\alpha^{1/2} \to \gamma^{1/2}) \\
&= (\beta \to \gamma)^{1/2} * (\alpha \to \gamma)^{1/2} \ ;
\end{aligned}
$$

hence the relation (2) is verifed.Finally let us consider an idempotent element $\iota$ with $0^{1/2} \leq \iota$ ; then the strictness of M implies $0 = (\iota \to 0)^2 = \iota \to 0$ ; hence the relation (3) follows.

□

# 3   Examples

The aim of this section is to present some mainstream examples of integral, residuated,commutative l–monoids. As a by-product of these considerations we may obtain some indications of possible applications of the theory sketched in section 2 .

First of all there is the ring-theoretical aspect. Every ideal lattice of an associative, commutative ring with unity forms an integral, residuated, commutative l–monoid with respect to the ideal-multiplication ([4]) Since ideal lattices are in general *not* distributive (cf. [9]) , the distributivity of the underlying lattice of an integral, residuated, commutative l–monoid is always a special event.In this context the algebraic strong de Morgan law as well as the divisibility is a remarkable axiom (cf. Lemma 2.4 and Proposition 2.6) . Another significant example is the *Lindenbaum algebra* of the monoidal, propositional calculus. We present this link with *non-classical logics* in detail.

## 3.1   Lindenbaum algebra

Let $\mathcal{L}$ be a formalized language of order $0$ ([25]) and $\{\neg, \wedge, \vee, \rightarrow, \otimes\}$ be the set of logical symbols where $\neg$ is a unary and the remaining symbols are binary operations. The *logical axioms* of *monoidal logic* are the following axiom schemes

$(T_1)$ $\quad ((\alpha \rightarrow \beta) \rightarrow ((\beta \rightarrow \gamma) \rightarrow (\alpha \rightarrow \gamma)))$ $\hfill$ (Syllogism Law)

$(T_2)$ $\quad (\alpha \rightarrow (\alpha \vee \beta))$

$(T_3)$ $\quad (\beta \rightarrow (\alpha \vee \beta))$

$(T_4)$ $\quad ((\alpha \rightarrow \gamma) \rightarrow ((\beta \rightarrow \gamma) \rightarrow ((\alpha \vee \beta) \rightarrow \gamma)))$

$(T_5)$ $\quad ((\alpha \wedge \beta) \rightarrow \alpha)$

$(T_5')$ $\quad ((\alpha \otimes \beta) \rightarrow \alpha)$

$(T_6)$ $\quad ((\alpha \wedge \beta) \rightarrow \beta)$

$(T_6')$ $\quad ((\alpha \otimes \beta) \rightarrow (\beta \otimes \alpha))$

$(T_6'')$ $\quad ((\alpha \otimes (\beta \otimes \gamma)) \rightarrow ((\alpha \otimes \beta) \otimes \gamma))$

$(T_7)$ $\quad ((\gamma \rightarrow \alpha) \rightarrow ((\gamma \rightarrow \beta) \rightarrow (\gamma \rightarrow (\alpha \wedge \beta))))$

$(T_8)$ $\quad ((\alpha \rightarrow (\beta \rightarrow \gamma)) \rightarrow ((\alpha \otimes \beta) \rightarrow \gamma))$ $\hfill$ (Importation Law)

$(T_9)$ $\quad (((\alpha \otimes \beta) \rightarrow \gamma) \rightarrow (\alpha \rightarrow (\beta \rightarrow \gamma)))$ $\hfill$ (Exportation Law)

$(T_{10})$ $\quad ((\alpha \otimes \neg\alpha) \rightarrow \beta)$ $\hfill$ (Duns Scotus)

$(T_{11})$ $\quad ((\alpha \rightarrow (\alpha \otimes \neg\alpha)) \rightarrow \neg\alpha)$

As the only rule of inference we use *Modus Ponens* (MP) . The *monoidal, propositional calculus* SC is the usual propositional calculus based on the logical axioms $(T_1) - (T_5)$ , $(T_5')$, $(T_6)$, $(T_6')$, $(T_6'')$, $(T_7) - (T_{11})$ . Further we apply the usual notations – e.g. if $\alpha$ is provable, then this situation is denoted by $\vdash \alpha$ .

**Lemma 3.1.1** *Let SC be the monoidal, propositional calculus. Then for all* $\alpha, \beta, \gamma \in \mathcal{L}$ *the following relations hold*

(i)      $\vdash$   $(\alpha \rightarrow \alpha)$

(ii)     $\vdash$   $((\alpha \otimes (\alpha \rightarrow \beta)) \rightarrow \beta)$

(iii)    $\vdash$   $(\alpha \rightarrow (\beta \rightarrow (\beta \otimes \alpha)))$

(iv)     If $\vdash \alpha$ ,   then    $\vdash$   $(\beta \rightarrow (\beta \otimes \alpha))$

(v)      $\vdash$   $(\alpha \rightarrow (\beta \rightarrow \alpha))$

(vi)     $\vdash$   $((\beta \rightarrow \gamma) \rightarrow ((\alpha \otimes \beta) \rightarrow (\alpha \otimes \gamma)))$

**Proof.** Because of the exportation law $(T_9)$ the formula

$$(((((\alpha \otimes \alpha) \rightarrow \alpha) \otimes \alpha) \rightarrow \alpha) \rightarrow (((\alpha \otimes \alpha) \rightarrow \alpha) \rightarrow (\alpha \rightarrow \alpha)))$$

is provable. Then (i) follows from $(T_1), (T_5'), (T_6')$ and several applications of (MP) . The relation (iv) is a consequence of (iii) . Further (ii), (iii) and (v) follow immediately from (i), $(T_1); (T_5'), (T_6'), (T_8)$ and $(T_9)$ . In oder to verify (vi) we first infer from (ii), (iii), $(T_1)$ and $(T_6')$ that the formula

$$((\beta \otimes (\beta \rightarrow \gamma)) \rightarrow (\alpha \rightarrow (\alpha \otimes \gamma)))$$

is provable. Hence (vi) follows from $(T_1), (T_6'), (T_6''), (T_8)$ and $(T_9)$ .
□

An important consequence of the syllogism law and the assertion (i) is the fact that the relation $\triangleright$ defined by

$$\alpha \triangleright \beta \qquad \Longleftrightarrow \qquad \vdash \quad (\alpha \rightarrow \beta)$$

is a preorder on the set $\mathcal{L}$ of all well formed formulas. If $\sim$ is the equivalence relation associated with $\triangleright$ (i.e. $\alpha \sim \beta \iff \vdash (\alpha \rightarrow \beta)$ and $\vdash (\beta \rightarrow \alpha)$), then we can consider the quotient $L = \mathcal{L}/\sim$ of all equivalence classes of logically equivalent formulas. In particular $\triangleright$ induces a partial ordering $\leq$ on L . Referring to $(T_2) - (T_7), (T_{10})$ and (v) it is easy to see that $(L, \leq)$ is a lattice with the top elemnt $1 = \{\alpha \in L \mid \vdash \alpha\}$ .Because of $(T_1), (T_6')$ and (vi) the logical symbol $\otimes$ defines a binary operation $*$ on L as follows

$$[\alpha] * [\beta] \quad = \quad [\alpha' \otimes \beta'] \qquad \text{where} \quad \alpha' \in [\alpha], \quad \beta' \in [\beta] \qquad .$$

From the axioms $(T_5'), (T_6'), (T_6''), (T_8), (T_9)$ and (iv) we conclude that $(L, \leq, *)$ is an integral, residuated, commutative l–monoid. In particular $(L, \leq, *)$ is also called the *Lindenbaum algebra* of the monoidal, propositional calculus. We consider the following special cases.

**3.1.2** We add to the system of logical axioms of monoidal logic the *law of idempotency* – i.e. the axiom schema

$(T'_{12})$     $(\alpha \to (\alpha \otimes \alpha))$   .

Then the logical symbols $\otimes$ and $\wedge$ are logically equivalent and the Lindenbaum algebra is a Heyting algebra ([20]). In this context the extended axiom system reduces to the axioms of *intuitionistic logic*.

**3.1.3** If we adjoin to the axioms of monoidal logic the *law of double negation* – i.e. the axiom scheme

$(T''_{12})$     $(\neg\neg\alpha \to \alpha)$   ,

then the Lindenbaum algebra is a Girard–monoid, and we arrive at Girard's integral, commutative *linear logic* ([10]).

**3.1.4** If we add to the axiom system of monoidal logic the *law of divisibility* – i.e. the axiom schema

$(T'''_{12})$     $((\alpha \wedge \beta) \to (\alpha \otimes (\alpha \to \beta)))$

*and* the *law of double negation* $(T''_{12})$ , then the Lindenbaum algebra is a divisible Girard–monoid – i.e. an MV–algebra (cf. Lemma 2.14) . In this context the extended axiom system reduces to the *Wajsberg axioms* of *Łukasiewicz logic* (cf. 3.1.6 in [12]).

**3.1.5** Finally we adjoin to the axioms of monoidal logic the axiom $(T'_{12}), (T''_{12})$, $(T'''_{12})$ . Then we obtain the *classical* propositional calculus; and the Lindenbaum algebra is a Boolean algebra ([28]) .


Further examples of integral, residuated, commutative monoids are Heyting aslgebras ([20]) ; in this case the semigroup operation $*$ coincides with the meet operation of the underlying lattice. Typical examples of (complete) Heyting algebras are lattices of *open* subsets of *topological spaces*. Moreover there exists a simple extension procedure of integral, residuated, commutative monoids which we discuss now in detail.

## 3.2   Extension of integral, residuated, l–monoids

Let $(L, \leq *)$ be an integral, residuated, commutative l–monoid. On the set $R_L = \{(\alpha, \beta) \in L \times L \mid \alpha \leq \beta\}$   we consider a partial ordering $\preceq$ defined componentwise – i.e.

$(\alpha_1, \beta_1)$   $\preceq$   $(\alpha_2, \beta_2)$     $\Longleftrightarrow$     $\alpha_1 \leq \alpha_2$   and   $\beta_1 \leq \beta_2$   .

Then $(R_L, \preceq)$ is again a lattice. On $R_L$ we introduce a binary operation $\diamond$ as follows

$(\alpha_1, \beta_1) \diamond (\alpha_2, \beta_2)$     $=$     $((\alpha_1 * \alpha_2), ((\alpha_1 * \beta_2) \vee (\alpha_2 * \beta_1)))$   .

**Lemma 3.2.1** $(R_L, \preceq, \diamond)$ *is an integral, residuated, commutative l-monoid, and* $j : \quad L \longmapsto R_L \quad$ *defined by* $j(\alpha) = (\alpha, \alpha)$ *is an injective homomorphism from* $(L, \leq, *)$ *to* $(R_L, \preceq, \diamond)$ .

**Proof.** It is easy to verify that $(R_L, \diamond)$ is a commutative monoid with unit element $(1,1)$ . Further we observe

$$(\alpha_1, \beta_1) \diamond (\alpha_2, \beta_2) \quad \preceq \quad (\alpha_3, \beta_3) \qquad \Longleftrightarrow$$

$$(\alpha_1 * \alpha_2) \quad \leq \quad \alpha_3 \ , \qquad ((\alpha_1 * \beta_2) \vee (\alpha_2 * \beta_1)) \quad \leq \quad \beta_3 \qquad \Longleftrightarrow$$

$$(\alpha_2, \beta_2) \quad \preceq \quad (((\alpha_1 \to \alpha_3) \wedge (\beta_1 \to \beta_3)), (\alpha_1 \to \beta_3)) \qquad ;$$

hence $(R_L, \preceq, \diamond)$ is an integral, residuated l-monoid. Since j preserves the lattice operations as well as the algebraic operations $*$ and $\to$ , we obtain that j is an injective homomorphism.
□ .

Since the "negation" in $(R_L, \preceq, \diamond)$ is given by

$$(\alpha, \beta) \to (0, 0) \qquad = \qquad ((\beta \to 0), (\alpha \to 0))$$

we see immediately that $(R_L, \preceq, \diamond)$ is a Girard-monoid if and only if $(L, \leq, *)$ is a Girard-monoid. Moreover we can characterize the divisibility of $(R_L, \preceq, \diamond)$ as follows

**Corollary 3.2.2** *The following assertions are equivalent*

(i)     $(R_L, \preceq, \diamond)$ is divisible.

(ii)    $(L, \leq, *)$ is a Boolean algebra (i.e. $*$ and $\wedge$ coincide , and the "negation" is an involution).

(iii)   $(R_L, \preceq, \diamond)$ is a MV–algebra.

**Proof.** Let $\alpha$ be an element of L . If $(R_L, \preceq, \diamond)$ is divisible, then there exist elements $(\beta, \gamma)$ and $(\delta, \epsilon)$ of $R_L$ with the following properties

$$\begin{array}{llll} (\alpha, 1) \diamond (\beta, \gamma) & = & ((\alpha * \beta), ((\alpha * \gamma) \vee \beta)) & = & (\alpha, \alpha) \\ (\alpha, 1) \diamond (\delta, \epsilon) & = & ((\alpha * \delta), ((\alpha * \epsilon) \vee \delta)) & = & (0, 1). \end{array}$$

In particular we obtain

$$\alpha * \beta = \alpha \ , \quad \beta \leq \alpha \ , \quad \delta \leq \alpha \to 0 \ , \quad \alpha \vee \delta = 1 \ ;$$

i.e. $\alpha$ is idempotent w.r.t. $*$ , and the law of the excluded middle $\alpha \vee (\alpha \to 0) = 1$ holds. Thus $(L, \leq, *)$ is a Boolean algebra. Therewith the implication (i) $\Longrightarrow$ (ii) is established. In order to verify (ii) $\Longrightarrow$ (iii) we first observe that in view of Lemma 2.14 it is sufficient to show the divisibility of $(R_L, \preceq, \diamond)$. Therefore let us consider two elements $(\alpha_1, \beta_1)$ and $(\alpha_2, \beta_2)$ of $R_L$ . Since $(L, \leq, *)$ is a

Boolean algebra, we obtain

$$(\alpha_1 \wedge (\alpha_1 \to \beta_2)) \vee (\beta_1 \wedge (\alpha_1 \to \alpha_2) \wedge (\beta_1 \to \beta_2)) \quad =$$
$$(\alpha_1 \wedge \beta_2) \vee ((\beta_1 \wedge \beta_2) \wedge (\alpha_1 \to \alpha_2)) \quad =$$
$$\beta_1 \wedge \beta_2 \wedge (\alpha_1 \vee (\alpha_1 \to \alpha_2)) \quad = \quad \beta_1 \wedge \beta_2 \quad ;$$
$$\alpha_1 \wedge (\alpha_1 \to \alpha_2) \wedge (\beta_1 \to \beta_2) \quad = \quad \alpha_1 \wedge \alpha_2 \quad .$$

Hence the relation

$$\begin{aligned}
(\alpha_1, \beta_1) \diamond ((\alpha_1, \beta_1) \to (\alpha_2, \beta_2)) \quad &= \quad ((\alpha_1 \wedge \alpha_2), (\beta_1 \wedge \beta_2)) \\
&= \quad (\alpha_1, \beta_1) \wedge (\alpha_2, \beta_2)
\end{aligned}$$

follows. Because of Lemma 2.5 $(R_L, \preceq, \diamond)$ is divisible. The implication (iii) $\Longrightarrow$ (i) is obvious.
$\square$

**Remark 3.2.3 (Non validity of the algebraic strong de Morgan law)**
(a) If $(L, \leq, \wedge)$ is a linearly ordered Heyting algebra with at least three elements (e.g. a finite chain), then $(R_L, \preceq, \diamond)$ does not satisfy the algebraic strong de Morgan law.
(b) Let $L = \{0, 1/2, 1\}$ be the chain of three elements viewed as a MV–algebra – i.e. $1/2 * 1/2 = 0$. Then $(R_L, \preceq, \diamond)$ is a Girard–monoid. Since the "negation" has two fixpoints – namely $(0, 1)$ and $(1/2, 1/2)$, we conclude from Lemma 2.10 that $(R_L, \preceq, \diamond)$ cannot satisfy the algebraic strong de Morgan law.

In view of various applications to system theory monoidal structures on the real unit interval [0,1] play an important role. Details of this situation are presented in the next subsection.

## 3.3   Left–continuous t–norms

According to the terminology proposed by K. Menger, B. Schweizer and A. Sklar ([27]) a *t–norm* is a binary operation T on the real unit interval [0,1] provided with the following properties

(T1)     $T(x, 1) \quad = \quad x \quad , \quad T(x, 0) \quad = \quad 0$

(T2)     $T(x, y) \quad = \quad T(y, x)$     (Symmetry)

(T3)     $T(x, y) \quad \leq \quad T(\bar{x}, \bar{y})$   whenever   $x \leq \bar{x}, \quad y \leq \bar{y}$     (Isotonicity)

(T4)     $T(x, T(y, z)) \quad = \quad T(T(x, y), z)$     (Associativity)

It is easy to see that $([0, 1], \leq, T)$ is an integral, residuated, commutative l-monoid if and only if T is *left–continuous* t–norm . Moreover $([0, 1], \leq, T)$ is divisible if and only if T is a *continuous* t–norm. Referring to a result due to Mostert and Shields (cf. [23]) every continuous t–norm is the *ordinal* sum of the following t–norms

$$T_1(x,y) \quad = \quad Min(x,y)$$

$$T_2(x,y) \quad = \quad Prod(x,y) \quad = \quad x \cdot y$$

$$T_3(x,y) \quad = \quad T_m(x,y) \quad = \quad Max(x+y-1,0) \quad ;$$

hence the structure of integral,divisible, residuated, commutative l–monoids on [0,1] is well understood. In particular $T_m$ is also called *Lukasiewicz's arithmetic conjunction* (cf. [8]).

Contrary to the case of continuous t–norms, the structure of left–continuous t–norms is extremely complicated. As an example we quote the "Nilpotent Minimum" $T_0$ discovered by J. Fodor ([7])

$$T_0(x,y) \quad = \quad \left\{ \begin{array}{rcl} Min(x,y) & : & 1 < x+y \\ 0 & : & x+y \leq 1 \end{array} \right\} \quad .$$

Then $([0,1], \leq, T_0)$ is a *non* divisible Girard–monoid .

Since [0,1] is a chain, we finally remark that all integral, residuated,commutative, monoidal structures on [0,1] satisfy the algebraic strong de Morgan law.

# 4   Filter theory

Let $(L, \leq, *)$ be an integral, residuated,commutative l–monoid. $\alpha^n$ denotes always the n–th power of $\alpha$ w.r.t. the semigroup operation $*$.

A nonempty subset F of L is called a *filter* in L if and only if F satisfies the following conditions

(F1)        $\alpha \leq \beta, \quad \alpha \in F \quad \Longrightarrow \quad \beta \in F$

(F2)        $\alpha, \beta \in F \quad \Longrightarrow \quad \alpha * \beta \ \in F$

(F3)        $0 \ \notin F$   .

On the set $\mathcal{F}(L)$ of all filters F in L we use the partial ordering determined by the set inclusion $\subseteq$. A filter F is said to be *maximal* if and only if F is a maximal element of the partially ordered set $(\mathcal{F}(L), \subseteq)$ . A filter F is called *prime* if and only if F fulfills the additional property

(F4)        $\alpha \vee \beta \ \in \ F \quad \Longrightarrow \quad \alpha \in F \ \text{ or } \ \beta \in F$   .

**Proposition 4.1** *(a) Every filter is contained in a maximal filter. (b) Maximal filters are prime.*

**Proof.** The assertion (a) follows immediately from *Zorn's Lemma* . In order to verify (b) we proceed as follows : Let F be a maximal filter and $\alpha \vee \beta$ be an element of F . If $\alpha$ *and* $\beta$ are not elements of F , then the maximality of F implies that there exist $\gamma_1, \gamma_2 \in F$ and natural numbers $n_1, n_2$ with the properties $\alpha^{n_1} * \gamma_1 = 0$, $\beta^{n_2} * \gamma_2 = 0$ . Because of $(\alpha \vee \beta)^{n_1 + n_2} * \gamma_1 * \gamma_2 = 0$ we infer from (F3) that $\alpha \vee \beta$ is not an element of F . The last conclusion is an obvious contradiction to the choice of $\alpha \vee \beta$ ; hence F is prime.
□

**Lemma 4.2** *Let* $(L, \leq, *)$ *be an integral, residuated, commutative l–monoid and* F *be a filter in* L . *Then the binary relation* $\sim_F$ *defined by*

$$\alpha \sim_F \beta \qquad \Longleftrightarrow \qquad (\alpha \to \beta) \wedge (\beta \to \alpha) \quad \in \quad F$$

*is a non trivial congruence relation on* L *– i.e.*

(i) $\sim_F$ *is an equivalence relation on* L

(ii) $\alpha \sim_F \bar{\alpha}$, $\beta \sim_F \bar{\beta}$ $\implies$ $(\alpha \wedge \beta) \sim_F (\bar{\alpha} \wedge \bar{\beta})$, $(\alpha \vee \beta) \sim_F (\bar{\alpha} \vee \bar{\beta})$

(iii) $\alpha \sim_F \bar{\alpha}$, $\beta \sim_F \bar{\beta}$ $\implies$ $(\alpha * \beta) \sim_F (\bar{\alpha} * \bar{\beta})$, $(\alpha \to \beta) \sim_F (\bar{\alpha} \to \bar{\beta})$

(iv) $0 \not\sim_F 1$ .

**Proof.** With regard to (F1) , (F2) and Lemma 2.2(b2) it is easy to see that $\sim_F$ is an equivalence relation on L . Further it is not difficult to verify the following relations

$$(((\alpha \to \bar{\alpha}) \wedge (\beta \to \bar{\beta})) \quad \leq \quad (\alpha \wedge \beta)) \to (\bar{\alpha} \wedge \bar{\beta})$$
$$(((\alpha \to \beta) \wedge (\bar{\alpha} \to \bar{\beta})) \quad \leq \quad (\alpha \vee \beta)) \to (\bar{\alpha} \vee \bar{\beta})$$
$$((\alpha \to \bar{\alpha}) * (\beta \to \bar{\beta}) \quad \leq \quad \alpha * \beta) \to (\bar{\alpha} * \bar{\beta})$$
$$((\bar{\alpha} \to \alpha) * (\bar{\beta} \to \beta) \quad \leq \quad (\alpha \to \beta)) \to (\bar{\alpha} \to \bar{\beta}) \quad ;$$

hence the assertions (ii) and (iii) follow from (F1) and (F2) . Because of (F3) the assertion (iv) also holds.
□

If F is a filter in L, then we conclude from Lemma 4.2 and Proposition 2.1(iii) that $L / \sim_F$ can be provided with the structure of an integral, residuated, commutative l–monoid $(L / \sim_F, \preceq_F, *_F)$ such that the quotient map $q_F : \quad L \longmapsto L / \sim_F$ is a surjective homomorphism. Moreover we obtain

**Corollary 4.3** *Let* $(L_1, \leq_1, *_1)$ *be an integral, residuated, commutative l–mo-noid. Then there exists a bijective map between the set* $\mathcal{F}(L_1)$ *of all filters in* $L_1$ *and the set of all quotients of* $(L_1, \leq_1, *_1)$ *(in the sense of the category of (integral) residuated, commutative l–monoids with* $0 \neq 1$*).*

**Proof.** Let $h : (L_1, \leq_1, *_1) \longmapsto (L_2, \leq_2, *_2)$ be a surjective homomorphism. Then $F_h = \{\alpha \in L_1 |\ h(\alpha) = 1\}$ is a filter in $L_1$, because $L_2$ contains at least two elements. It is not difficult to show that there exists an isomorphism $\bar{h} : (L_1/\sim_{F_h}, \leq_{F_h}, *_{F_h}) \longmapsto (L_2, \leq_2, *_2)$ making the following diagram commutative

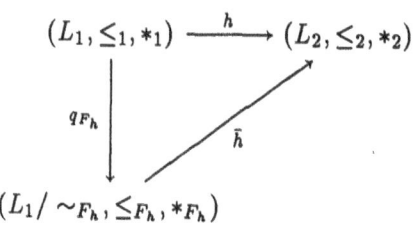

$\square$

According to the terminology introduced by C.C. Chang [5] an integral, residuated, commutative l-monoid $(L, \leq, *)$ is called *locally finite* if and only if for every elemnt $\alpha \in L$ with $\alpha \neq 1$ there exists a natural number n such that $\alpha^n = 0$ (i.e. all $\alpha \neq 1$ are *nilpotent*).

**Theorem 4.4 (C.C.Chang)** *Let $(L, \leq, *)$ be an integral, residuated, commutative l-monoid, and let F be a filter in L. Then the following assertions are equivalent*

(i)        F is a maximal filter.

(ii)       $(L/\sim_F, \leq_F, *_F)$ is locally finite.

**Proof.** Let us assume that F is maximal, and let us consider an element $[\alpha] \in L/\sim_F$ with $[\alpha] \neq [1]$ ; i.e. $\alpha \notin F$. Then the maximality of F forces the existence of an element $\beta \in F$ and a natural number n such that $\alpha^n * \beta = 0$ . Because of $\beta \leq (\alpha^n \to 0)$ we infer from (F1) $\alpha^n \sim_F 0$ –i.e. $[\alpha]^n = [0]$ .
(b) Let us assume that the assertion (ii) is valid. Further we choose a filter G containing F. If there would exist an elemnt $\alpha \in G$ with $\alpha \notin F$, then we invoke (ii) and obtain a natural number n with $\alpha^n \to 0 \in F$. Since F is contained in G, $\alpha^n \to 0$ is also an element of G. Now we apply (F2) and obtain $0 \in G$ which is an obvious contradiction to (F3) .Hence the assumption is false – i.e. F is maximal.
$\square$

**Remark 4.5** The structure of integral, divisible, residuated, commutative l-monoids as well as of integral,commutative Girard–monoids is preserved under the formation of quotients. In particular the quotient of an MV-algebra is again an MV–algebra.

**Lemma 4.6** *Let $(L, \leq, *)$ be an integral, residuated, commutative l-monoid.*
*(a) For every elemnt $\alpha \in L$ with $\alpha \neq 1$ there exists a prime filter P in L with*

$\alpha \notin P$ .

*(b) Let $\alpha$ be an elemnt of L with $\alpha \neq 1$ . Then the following assertions are equivalent*

(i)     There exists a maximal filter F with $\alpha \notin F$.

(ii)     There exists a natural number n s.t. $(\alpha^n \to 0)^m \neq 0$ for all $m \in \mathbb{N}$.

**Proof.** (a) Let $\alpha$ be an element of L with $\alpha \neq 1$ . We denote by $\mathcal{F}(\alpha)$ the set of all filters F in L with $\alpha \notin F$ . Because of $\alpha \notin \{1\}$ the set $\mathcal{F}(\alpha)$ is nonempty. Hence we conclude from *Zorn's Lemma* that $(\mathcal{F}(\alpha), \subseteq)$ contains maximal elements P . We show that P is prime. Therefore let us consider the following situation     $\beta_1 \vee \beta_2 \in P$,     $\beta_1 \notin P$,     $\beta_2 \notin P$  .   By virtue of the maximality of P viewed as an elment of $\mathcal{F}(\alpha)$ there exist elements $\gamma_1 \in P$ , $\gamma_2 \in P$ and natural numbers $n_1, n_2$ such that the following inequalities hold

$$\beta_1^{n_1} * \gamma_1 \;\; \leq \;\; \alpha \quad , \quad \beta_2^{n_2} * \gamma_2 \;\; \leq \;\; \alpha \quad .$$

Hence we obtain     $(\beta_1 \vee \beta_2)^{n_1+n_2} * \gamma_1 * \gamma_2 \;\; \leq \;\; \alpha$  . which is an obvious contradiciton to the fact that P is an element of $\mathcal{F}(\alpha)$ . Therefore P is prime.
(b) Let U be a maximal filter in L with $\alpha \notin U$ . Then the maximality of U forces the existence of an element $\beta \in U$ and a natural number n such that the equation $\alpha^n * \beta = 0$ holds . Because of (F1) the element $\alpha^n \to 0$ is contained in U. Now we invoke (F2) and (F3) and obtain that   $(\alpha^n \to 0)^m \;\; \neq \;\; 0$   for all   $m \in \mathbb{N}$  .  On the other hand, if $(\alpha^n \to 0)^m \neq 0$ holds for all $m \in \mathbb{N}$ , then $\alpha^n \to 0$ generates a filter F in L as follows

$$F \;\; = \;\; \{\beta \in L | \;\; \exists m \in \mathbb{N} : \;\; (\alpha^n \to 0)^m \;\; \leq \;\; \beta\} \quad .$$

By virtue of Zorn's Lemma there exists a maximal filter U containing F . Since $\alpha^n \to 0 \in F \subseteq U$ , we conclude from (F2) and (F3) that $\alpha$ is not an element of U .
□

**Corollary 4.7** *Let $(L, \leq, *)$ be an integral, divisible, residuated, commutative l-monoid satisfying the algebraic strong de Morgan law. Further let $\alpha$ be an element of L with $\alpha \neq 1$. Then the following assertions are equivalent*

(i)    There exists a maximal filter U with $\alpha \notin U$.

(ii)    There exists a natural number $n_0$ s.t. $(\alpha^{n_0} \to 0) \to \alpha \;\; \neq \;\; 1$.

**Proof.** In order to verify (i) $\Longrightarrow$ (ii) we show $\neg$ (ii) $\Longrightarrow \neg$ (i) . Let us assume that $(\alpha^n \to 0) \to \alpha = 1$ for all $n \in \mathbb{N}$ . Then

$$\alpha * (\alpha^{n+1} \to 0) \;\; = \;\; \alpha^n \to 0$$

follows from Lemma 2.2(b2) and Lemma 2.5(2) . Hence we obtain for all $n \in \mathbb{N}$:
$(\alpha^n \to 0)^{n+1} = 0$ ; i.e. the assertion (i) is false (cf. Lemma 4.6).

On the other hand, let us now assume that there exists a natural number $n_0 \in \mathbb{N}$ such that $(\alpha^{n_0} \to 0) \to \alpha \neq 1$ . Referring to Proposition 2.1(iii) and Proposition 2.6(xiv) we infer from the algebraic strong de Morgan law that the equation

$$((\alpha^{n_0} \to 0) \to \alpha)^m \vee (\alpha^{n_0+1} \to 0)^m = 1$$

holds for all $m \in \mathbb{N}$ . Because of $(\alpha^{n_0} \to 0) \to \alpha \neq 1$ we obtain $(\alpha^{n_0} \to 0)^m \neq 0$ for all $m \in \mathbb{N}$ ; hence the assertion (i) follows from Lemma 4.6 .

□

**Theorem 4.8 (Algebraic strong de Morgan law)** *Let $M = (L, \leq, *)$ be an integral, commutative, residuated l-monoid. Then the following asertions are equivalent*

(i)     M satisfies the algebraic strong de Morgan law.

(ii)    M is a l-submonoid of a product of integral, commutative, residuated *linearly ordered* l-monoids.

**Proof.** Since the linearity of the underlying order forces the algebraic strong de Morgan law, the implication (ii) $\Longrightarrow$ (i) is obvious. In oder to verify (i) $\Longrightarrow$ (ii) we proceed as follows : Let $\mathbf{P}(L)$ be the set of all prime filters in L (in the sense of M). Then the algebraic strong de Morgan law implies that for all $P \in \mathbf{P}(L)$ the quotient $M_P = (L/\sim_P, \leq_P, *_P)$ is an integral, commutative, residuated linearly ordered l-monoid. We infer from Lemma 4.6(a) that there exists a monomorphism from M into $\prod_{P \in \mathbf{P}(L)} M_P$ ; hence the assertion follows.

□

An integral, commutative, residuated l-monoid M is called *semi-simple* if and only if the intersection of all maximal filters coincides with the trivial filter – i.e. $\bigcap\{U \mid U \text{ maximal filter}\} = \{1\}$ .

It is easy to see that local finiteness implies semi-simplicity. Moreover we obtain immediately from Theorem 4.4 the following

**Proposition 4.9** *Let M be an integral, commutative, residuated l-monoid. Then the following assertions are equivalent*

(i)     M is semi-simple.

(ii)    M is a l-submonoid of a product of integral, commutative, residuated, locally finite l-monoids.

**Theorem 4.10** *Let $(L, \leq, *)$ be an integral, commutative, residuated semi-simple l-monoid . Then the following assertions are equivalent*

(i)     $(L, \leq, *)$ is divisible.

(ii)   $(L, \leq, *)$ is an MV–algebra.

The proof is based on the following

**Lemma 4.11** *Let $M = (L, \leq *)$ be an integral, commutative, residuated, divisible and locally finite l–monoid. Then $(L, \leq, *)$ is a MV–algebra.*

**Proof.** Referring to Lemma 2.14 it is sufficient to show that M is a Girard–monoid. Therefore we choose without loss of generality an element $\alpha \in L$ with $\alpha \neq 0$. It is easy to verify

$$\alpha \quad \leq \quad (\alpha \to 0) \to 0 \quad , \quad ((\alpha \to 0) \to 0) \to 0 \quad = \quad \alpha \to 0 \quad .$$

By virtue of the divisibility of M there exists an element $\gamma \in L$ such that $((\alpha \to 0) \to 0) * \gamma \quad = \quad \alpha$ holds. Taking on both sides the "negation" we obtain

$$\alpha \to 0 \quad = \quad \gamma \to (\alpha \to 0) \quad ;$$

hence we can establish by induction the following relation

$$\alpha \to 0 \quad = \quad \gamma^n \to (\alpha \to 0) \quad n \in \mathbb{N} \quad .$$

Now we use the local finiteness of M and obtain $\gamma = 1$ ; hence the "negation" is an involution.
□

**Proof of Theorem 4.10.** The assertion follows immediately from Proposition 4.9 and Lemma 4.11 .
□

**Proposition 4.12** *Let $(L, \leq, *)$ be an MV–algebra. Then the following assertions are equivalent*

(i)   $(L, \leq, *)$ is semi–simple .

(ii)   For every elemnt $\alpha \in L$ with $\alpha \neq 1$ there exists $n_0 \in \mathbb{N}$ such that
$(\alpha \to 0) \to \alpha^{n_0} \quad \neq \quad 1$ .

**Proof.** Since the "negation" is an involution, the assertion follows immediately from Corollary 2.16 and Corollary 4.7 .
□

Obviously every Boolean algebra is semi–simple (cf. also Proposition 4.12). Moreover we obtain from Theorem 4.10 the following

**Corollary 4.13** *Let $(L, \leq, \wedge)$ be a Heyting algebra. Then the following assertions are equivalent*

(i)    $(L, \leq, \wedge)$ is semi–simple.

(ii)   $(L, \leq, \wedge)$ is a Boolean algebra.


We finish this section with a short remark on strongly–atomless MV–algebras: According to the terminology proposed L.P. Belluce [3] an MV–algebra M = $(L, \leq, *)$ is called *strongly-atomless* if and only if for every element $\alpha \in L$ with $\alpha \neq 1$ there exists a prime filter P in M such that the equivalence classe $[\alpha] \in L/\sim_P$ is *not* a dual atom of the quotient MV–algebra $(L/\sim_P, \leq_P, *_P)$ – i.e. there exists $[\beta] \in L/\sim_P$ such that $[\alpha] \lneq_P [\beta] \lneq_P [1]$ .

**Lemma 4.14** *Let $M = (L, \leq, *)$ be an MV–algebra with square roots. If M is strongly-atomless, then M is a strict MV–algebra.*

**Proof.** Let $\iota_0$ be the idempotent element of L satisfying Condition (2) in Corollary 2.18 . Because of $\iota_0 = (0^{1/2} \to 0)^2$ it is sufficient to show $\iota_0 = 0$. Let us assume $\iota_0 \neq 0$. Since M is strongly atomless, there exists a prime filter P such that $\iota_0 \to 0 \notin P$ and $[\iota_0 \to 0]$ is *not* a dual atom of the quotient MV–algebra $(L/\sim_P, \leq_P, *_P)$. Referring to Condition(2) in 2.18 we obtain in the case of $[\alpha] \leq_P [\iota_0]$ :

$$[\alpha] \wedge [\alpha \to 0] \quad \leq_P \quad [0^{1/2}] \wedge [\iota_0] \quad = \quad [0^{1/2} * \iota_0] \quad = \quad [0] \quad .$$

Since $(L/\sim_P, \leq_P)$ is a chain (cf. Theorem 4.8), the alternative $[\alpha] = [0]$ or $[\alpha] = [\iota_0]$ follows; i.e. there does *not* exist $[\beta] \in L/\sim_P$ with $[\iota_0 \to 0] \lneq_P [\beta] \lneq_P [1]$. Hence $[\iota_0 \to 0]$ is a dual atom – an obvious contradiction to the assumption. Thus $\iota_0 = 0$ .
□


# 5   Integral commutative cl–monoids

A commutative, residuated l–monoid $(L, \leq, *)$ is said to be *complete* if and only if the underlying lattice $(L, \leq)$ is complete.

**Lemma 5.1** *Let $(L, \leq, *)$ be a commutative l–monoid([4]). If the underlying lattice is complete, then the following assertions are equivalent*

(1)    $(L, \leq, *)$ satisfies (AD) – i.e. $(L, \leq, *)$ is residuated.

(2)    The semigroup operation $*$ is distributive over arbitrary joins – i.e. the following infinite distributivity law holds

$$\alpha * (\bigvee_{i \in I} \beta_i) \quad = \quad \bigvee_{i \in I} (\alpha * \beta_i) \quad .$$

**Proof.** (a) ((1) $\implies$ (2)) Because of $0 \leq \alpha \to 0$ we infer from (AD) that the universal lower bound 0 acts as zero element w.r.t. $*$ . Hence the assertion (2) holds for the empty index set I . Now we assume that I is non empty, and we put $\gamma = \bigvee_{i \in I} (\alpha * \beta_i)$. Then we infer from (AD) that the inequality $\beta_i \leq \alpha \to \gamma$ holds for all $i \leq I$. Since the underlying lattice $(L, \leq)$ is complete, $\bigvee_{i \in I} \beta_i$ exists, and the inequality $\bigvee \beta_i \leq \alpha \to \gamma$ follows. Applying again (AD) we obtain $\alpha * (\bigvee_{i \in I} \beta_i) \leq \gamma$; hence the assertion (2) is verified.

(b) ((2) $\implies$ (1)) By virtue of the completeness of the lattice $(L, \leq)$ we are in the position to introduce a binary operation $\to$ on L in the following way

$$\alpha \to \beta \quad = \quad \bigvee \{\lambda \in L \ \mid \ \alpha * \lambda \leq \beta\} \tag{5.1}$$

Becasue of (2) the operation $\to$ fulfills the important property

$$\alpha * (\alpha \to \beta) \quad \leq \quad \beta \tag{5.2}$$

Since $(L, \leq, *)$ is in particular a partially ordered monoid ([4]), we conclude from the definition of $\to$ and the inequality (5.2) that $\to$ fulfills (AD) – i.e. $(L, \leq, *)$ is a residuated l-monoid.

$\square$

Referring to the previous lemma and the terminology introduced by G. Birkhoff [4] we call a commutative, residuated, complete l-monoid also a commutative *cl-monoid*.

In any commutative cl–monoid the following relations hold

(i) $\quad (\bigvee_{i \in I} \alpha_i) \to \beta \quad = \quad \bigwedge_{i \in I} (\alpha_i \to \beta)$

(ii) $\quad \alpha \to (\bigwedge_{i \in I} \beta_i) \quad = \quad \bigwedge_{i \in I} (\alpha \to \beta_i)$ .

An integral, divisible, commutative cl–monoid is said to be a *GL–monoid* (where GL stands for generalized logics ([16],[17]). In view of the terminology used in [26] an integral, commutative, residuated, complete Girard–monoid is also called an integral, commutative *Girard quantale*.

**Theorem 5.2** *Let $(L, \leq, *)$ be an integral, commutative, cl–monoid.*
*(a) If $(L, \leq, *)$ is a GL–monoid, then the underlying lattice $(L, \leq)$ is a complete Heyting algebra.*
*(b) If $(L, \leq, *)$ is an integral, commutative Girard quantale, then the de Morgan law*

(iii) $\quad (\bigwedge_{i \in I} \alpha_i) \to 0 \quad = \quad \bigvee_{i \in I} (\alpha_i \to 0)$

*holds.*
*(c) Let $(L, \leq, *)$ be a complete MV-algebra. Then the underlying lattice $(L, \leq)$*

is a complete Heyting algebra and simultaneously a complete Brouwerian lattice ([4]). In particular $*$ is distributive over arbitrary meets – i.e. the following infinite distributivity law holds

$$\alpha * (\bigwedge_{i \in I} \beta_i) \quad = \quad \bigwedge_{i \in I} (\alpha * \beta_i) \qquad .$$

**Proof.** If $(L, \leq, *)$ is divisible, we obtain from Lemma 2.5(2) and Lemma 5.1(2) the following estimation

$$
\begin{aligned}
\alpha \wedge (\bigvee_{i \in I} \beta_i) \quad &= \quad (\bigvee_{i \in I} \beta_i) * ((\bigvee_{i \in I} \beta_i) \to \alpha) \\
&= \quad (\bigvee_{i \in I} \beta_i) * (\bigwedge_{i \in I} (\beta_i \to \alpha)) \\
&\leq \quad \bigvee_{i \in I} (\beta_i * (\beta_i \to \alpha)) \quad = \quad \bigvee_{i \in I} (\alpha \wedge \beta_i) \qquad ;
\end{aligned}
$$

hence the infinite distributivity law

$$\alpha \wedge (\bigvee_{i \in I} \beta_i) \quad = \quad \bigvee_{i \in I} (\alpha \wedge \beta_i)$$

holds – i.e. $(L, \leq)$ is a complete Heyting algebra ([20]). Therewith the assertion (a) is established.

If the "negation" is an involution, then (iii) follows immeadiately from (i) . Finally the first assertion in (c) is a consequence from the assertions (a) and (b) and Lemma 2.14 . In order to verify the second assertion in (c) we refer to Proposition 2.15 and obtain

$$
\begin{aligned}
\alpha \to (\bigwedge_{i \in I} (\alpha * \beta_i)) \quad &= \quad \bigwedge_{i \in I} (\alpha \to (\alpha * \beta_i)) \\
&= \quad \bigwedge_{i \in I} ((\alpha \to 0) \vee \beta_i) \\
&= \quad (\alpha \to 0) \vee (\bigwedge_{i \in I} \beta_i) \qquad .
\end{aligned}
$$

If we multiply both sides with $\alpha$, then we can apply Lemma 2.5(2) , and the desired infinite distributivity law follows.

$\square$

**Theorem 5.3** Let $(L, \leq, *)$ be an integral, commutative cl–monoid. If $(L, \leq, *)$ is divisible or satisfies the algebraic strong de Morgan law, then the following assertions are equivalent

(1)    $(L, \leq, *)$ has square roots .

(2)    Every element $\alpha$ of L is a *square* w.r.t. $*$ (i.e. there exists $\beta \in L$ s.t. $\alpha = \beta * \beta$).

**Proof.** Obviously (1) implies (2) . In order to verify (2) $\Longrightarrow$ (1) we first infer from Lemma 2.4(1) and Proposition 2.6(xiv) that the following relation holds in $(L, \leq, *)$

$$\alpha * \beta \quad \leq \quad (\alpha * \alpha) \vee (\beta * \beta) \tag{5.3}$$

Since the underlying lattice $(L, \leq)$ is complete, we can define an unary operation $S : L \longmapsto L$ by

$$S(\alpha) \quad = \quad \bigvee \{\lambda \in L \mid \lambda * \lambda \leq \alpha\} \quad .$$

We conclude from Lemma 5.1(2) , formula (5.3) and assertion (2) that $S$ satisfies the axioms (S1) and (S2) ; hence $(L, \leq, *)$ has square roots.
$\square$

In the following considerations we study the *MacNeille completion* of commutative, residuated l-monoids. First we recall the MacNeille completion of partially ordered sets $(P, \leq)$ . For every subset A of P we denote by $\mathcal{U}(A)$ (resp. $\mathcal{L}(A)$) the set of all upper (resp. lower) bounds of A w.r.t. $\leq$ . Further we associate with every subset A of P a further subset $A^\sharp$ determined by $A^\sharp = \mathcal{L}(\mathcal{U}(A))$ . It is not difficult to verify the following relations

$$A \subseteq A^\sharp \quad , \quad \mathcal{U}(A^\sharp) = \mathcal{U}(A) \quad , \quad (A^\sharp)^\sharp = A^\sharp \quad ,$$
$$A \subseteq B \quad \text{implies} \quad A^\sharp \subseteq B^\sharp \quad , \quad (\bigcap_{i \in I} A_i^\sharp)^\sharp = \bigcap_{i \in I} A_i^\sharp \quad .$$

Further let $\mathcal{P}^\sharp$ be the set of all $\sharp$-closed subsets of P – i.e. $\mathcal{P}^\sharp = \{A^\sharp \mid A \subseteq P\}$, and let $\preceq^\sharp$ be the partial ordering on $\mathcal{P}^\sharp$ determined by the set inclusion – i.e. $\preceq^\sharp = \subseteq$ . Then the map $j^\sharp : P \longmapsto \mathcal{P}$ defined by

$$j^\sharp(\alpha) \quad = \quad \{\alpha\}^\sharp \quad = \quad \{\lambda \in P \mid \lambda \leq \alpha\}$$

is an embedding from $(P, \leq)$ into $(\mathcal{P}^\sharp, \preceq^\sharp)$ .

**Lemma 5.4** *Let* $(P, \leq)$ *be a partially ordered set. Then* $(\mathcal{P}^\sharp, \preceq^\sharp)$ *is a complete lattice and the embedding* $j^\sharp : (P, \leq) \longmapsto (\mathcal{P}^\sharp, \preceq^\sharp)$ *has the following properties*

(a)    $A^\sharp \quad = \quad \vee \{j^\sharp(\alpha) \mid \alpha \in A\}$ *for all* $A \subseteq P$ .

(b)    $\{\alpha \in P \mid j^\sharp(\alpha) \preceq^\sharp A^\sharp\} \quad = \quad A^\sharp$ .

**Proof.** The join and meet operations in $(\mathcal{P}^\sharp, \preceq^\sharp)$ are given by

$$\bigvee_{i \in I} A_i^\sharp \quad = \quad (\bigcup_{i \in I} A_i^\sharp)^\sharp \quad , \quad \bigwedge_{i \in I} A_i^\sharp \quad = \quad \bigcap_{i \in I} A_i^\sharp \quad ;$$

hence $(\mathcal{P}^{\sharp}, \preceq^{\sharp})$ is a complete lattice. The relation (b) is obvious and relation (a) follows from $\quad \mathcal{U}(\bigcup_{\alpha \in A} j^{\sharp}(\alpha)) \;=\; \mathcal{U}(A)$ .

□

**Theorem 5.5 (MacNeille Completion)** *Let $(P, \le)$ be a partially ordered set. Then there exists a further partially ordered set $(L, \le)$ and a map $j_L : P \mapsto L$ satisfying the following conditions*

1.  $j_L$ *is an embedding – i.e.* $j_L(\alpha) \le j_L(\beta) \quad \Longleftrightarrow \quad \alpha \le \beta$ .

2.  $(L, \le)$ *is a complete lattice.*

3.  *For every element* $\lambda \in L$ *the following relations are valid*

   (a)   $\{\alpha \in P \mid j_L(\alpha) \le \lambda\}^{\sharp} \;=\; \{\alpha \in P \mid j_L(\alpha) \le \lambda\}$
   (b)   $\lambda \;=\; \vee\{j_L(\alpha) \mid \alpha \in P, \; j_L(\alpha) \le \lambda\}$   .

*Moreover $(j_L, (L, \le))$ is uniquely determined by 1. – 3. up to an order isomorphism.*

**Proof.** The existence of $(j_L, (L, \le))$ follows from Lemma 5.4 . In order to verify the uniqueness we proceed as follows : If $(j_{L_1}, (L_1, \le))$ and $(j_{L_2}, (L_2, \le))$ are two pairs provided with the properties 1. – 3. , then the completeness of $(L_2, \le)$ permits the definition of a map $\phi : L_1 \mapsto L_2$ in the following way

$$\phi(\lambda_1) \;=\; \vee\{j_{L_2}(\alpha) \mid \alpha \in P, \; j_{L_1}(\alpha) \le \lambda_1\}$$   .

Obviously $\phi$ is isotone and makes the diagram

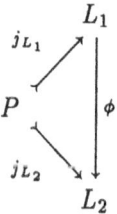

commutative. Interchanging the role of $L_1$ and $L_2$ we can also define an isotone map $\psi : L_2 \mapsto L_1$ by

$$\psi(\lambda_2) \;=\; \vee\{j_{L_1}(\alpha) \mid \alpha \in P, \; j_{L_2}(\alpha) \le \lambda_2\}$$   .

Further we use the following notations

$$A_{\lambda_1} \;=\; \{\alpha \in P \mid j_{L_1}(\alpha) \le \lambda_1\} \quad \text{for all} \quad \lambda_1 \in L_1$$
$$A_{\lambda_2} \;=\; \{\alpha \in P \mid j_{L_2}(\alpha) \le \lambda_2\} \quad \text{for all} \quad \lambda_2 \in L_2 \quad .$$

In order to verify the bijectivity of $\phi$ we first show

$$A_{\phi(\lambda_1)} = A_{\lambda_1} \quad , \quad A_{\lambda_2} = A_{\psi(\lambda_2)} \tag{5.4}$$

The inclusion $A_{\lambda_1} \subseteq A_{\phi(\lambda_1)}$ follows immediately from the definition of $\phi$ . Further let $\bar{\alpha}$ be an element of $A_{\phi(\lambda_1)}$ ; then for every upper bound $\beta$ of $A_{\lambda_1}$ we obtain

$$j_{L_2}(\bar{\alpha}) \quad \le \quad \bigvee_{\alpha \in A_{\lambda_1}} j_{L_2}(\alpha) \quad \le \quad j_{L_2}(\beta) \quad .$$

Because of Property 1. the element $\bar{\alpha}$ is a lower bound of $\mathcal{U}(A_{\lambda_1})$ – i.e. $\bar{\alpha} \in A_{\lambda_1}^{\sharp}$ . Since $A_{\lambda_1}$ is $\sharp$–closed (cf. 3.(a)) , the inclusion $A_{\phi(\lambda_1)} \subseteq A_{\lambda_1}$ follows – i.e. $A_{\phi(\lambda_1)} = A_{\lambda_1}$ . Analogously we verify $A_{\lambda_2} = A_{\psi(\lambda_2)}$ .
Now we apply formula (5.4) and obtain

$$A_{\lambda_1} \quad = \quad A_{\psi(\phi(\lambda_1))} \quad , \quad A_{\lambda_2} \quad = \quad A_{\phi(\psi(\lambda_2))} \quad ;$$

hence $\psi \circ \phi = id_{L_1}$ and $\phi \circ \psi = id_{L_2}$ follows from Property 3.(b) .
$\Box$

**Remark 5.6** The properties 1. and 3. in the previous theorem imply that the embedding $j_L$ preserves joins and meets provided they exist in $(P, \le)$ . In fact, let us consider the following situation

$$\gamma_0 \quad = \quad \bigvee_{i \in I} \gamma_i \quad , \quad j_L(\gamma_i) \le \lambda_1 \quad \text{where} \quad \gamma_0, \gamma_i \in P, \quad \lambda_1 \in L$$

$$\delta_0 \quad = \quad \bigwedge_{i \in I} \delta_i \quad , \quad \lambda_2 \le j_L(\delta_i) \quad \text{where} \quad \delta_0, \delta_i \in P, \quad \lambda_2 \in L$$

Then the relation

$$\{\gamma_i | i \in I\} \subseteq A_{\lambda_1}, \quad \gamma_0 \in A_{\lambda_1}^{\sharp}, \quad \{\delta_i | i \in I\} \subseteq \mathcal{U}(A_{\lambda_2}), \quad \delta_0 \in \mathcal{U}(A_{\lambda_2})$$

follows. Now we apply Property 3. and obtain

$$A_{\lambda_1}^{\sharp} = A_{\lambda_1} \quad , \quad \mathcal{U}(A_{\lambda_2}) = \{\beta \in P \mid \lambda_2 \le j_L(\beta)\} \quad ;$$

hence the inequalities $j_L(\gamma_0) \le \lambda_1$ and $\lambda_2 \le j_L(\delta_0)$ are valid.
$\Box$

**Theorem 5.7** Let $(P, \le *)$ be a residuated, commutative l-monoid. Then the MacNeille completion of $(P, \le, *)$ is a commutative cl-monoid, and the embedding from $(P, \le, *)$ into its MacNeille completion preserves the algebraic structure – i.e. is a homomorphism.

The proof is based on

**Lemma 5.8** Let $(P, \le, *)$ be a commutative, residuated l-monoid and $\star$ be the Minkowski multiplication on the power set of $P$ induced by $*$ – i.e.
$A \star B = \{\alpha * \beta \mid \alpha \in A, \quad \beta \in B\}$. Then the following relation holds

$$\mathcal{U}(A^{\sharp} \star B^{\sharp}) \quad = \quad \mathcal{U}(A^{\sharp} \star B) \quad = \quad \mathcal{U}(A \star B) \quad .$$

**Proof.** The inclusion $\mathcal{U}(A^\natural \star B^\natural) \subseteq \mathcal{U}(A^\natural \star B) \subseteq \mathcal{U}(A \star B)$ is obvious. On the other hand, let $\gamma$ be an upper bound of $A \star B$. Then the axiom (AD) implies

$$\alpha \ \leq \ \beta \to \gamma \quad \text{for all} \quad \alpha \in A, \quad \beta \in B \quad ;$$

hence $\beta \to \gamma$ is an upper bound of A for all $\beta \in B$. Applying again (AD) we infer from $\mathcal{U}(A) = \mathcal{U}(A^\natural)$ that $\bar{\alpha} \to \gamma$ is an upper bound of B for all $\bar{\alpha} \in A^\natural$. Because of $\mathcal{U}(B) = \mathcal{U}(B^\natural)$ the inequality $\bar{\beta} \leq \bar{\alpha} \to \gamma$ holds for all $\bar{\alpha} \in A^\natural$ and $\bar{\beta} \in B^\natural$. Therewith the inclusion $\mathcal{U}(A \star B) \subseteq \mathcal{U}(A^\natural \star B^\natural)$ is established. $\square$

**Proof of Theorem 5.7.** Let $(\mathcal{P}^\natural, \preceq^\natural)$ be the MacNeille completion of $(P, \leq)$ (cf. Lemma 5.4 and Theorem 5.5). On $\mathcal{P}^\natural$ we introduce a commutative binary operation $*^\natural$ as follows : Let $\star$ be the *Minkowski multiplication* corresponding to $*$ , then

$$A^\natural *^\natural B^\natural \ = \ (A^\natural \star B^\natural)^\natural \tag{5.5}$$

By virtue of Lemma 5.8 the operation $*^\natural$ is associative ,and the embedding $j^\natural : P \mapsto \mathcal{P}^\natural$ is a semigroup–homomorphism. Referring again to Lemma 5.8 and to the proof of lemma 5.4 we obtain

$$
\begin{aligned}
A^\natural *^\natural \left(\bigvee_{i \in I} B_i^\natural\right) &= \left(A^\natural \star \left(\bigcup_{i \in I} B_i^\natural\right)\right)^\natural \\
&= \left(\bigcup_{i \in I} (A^\natural \star B_i^\natural)\right)^\natural \\
&\preceq^\natural \bigvee_{i \in I} (A^\natural *^\natural B^\natural) \quad ;
\end{aligned}
$$

hence $(\mathcal{P}^\natural, \preceq^\natural, *^\natural)$ is a commutative cl–monoid. In particular $\{1\}^\natural$ acts as unit element in $\mathcal{P}^\natural$ w.r.t. $*^\natural$. Finally we show that $j^\natural$ is a homomorphism between commutative, residuated l–monoids. In view of the preceding considerations it is sufficient to verify that $j^\natural$ preserves the "implication" $\to$ . Let us consider the following situation

$$j^\natural(\alpha) *^\natural B^\natural \ \preceq^\natural \ j^\natural(\beta) \quad \text{,where} \quad \alpha, \beta \in P, \quad B \subseteq P \quad .$$

Then we infer from formula (5.5) and lemma 5.8

$$\{\gamma \in P \ | \ \beta \leq \gamma\} \ = \ \mathcal{U}(\{\beta\}^\natural) \ \subseteq \ \mathcal{U}(\{\alpha\} \star B) \quad ;$$

i.e. the following chain of equivalences holds true

$$
\begin{aligned}
j^\natural(\alpha) *^\natural B^\natural \ \preceq^\natural \ j^\natural(\beta) \quad &\Longleftrightarrow \quad \beta \in \mathcal{U}(\{\alpha\} \star B) \\
&\Longleftrightarrow \quad \delta \leq \alpha \to \beta \quad \forall \delta \in B \\
&\Longleftrightarrow \quad B^\natural \ \subseteq \ \{\alpha \to \beta\}^\natural.
\end{aligned}
$$

Referring to formula (5.1) we obtain

$$j^\sharp(\alpha) \to^\sharp j^\sharp(\beta) \overset{\text{def}}{=} \bigvee\{B^\sharp \in \mathcal{P}^\sharp \mid j^\sharp(\alpha) *^\sharp B^\sharp \preceq^\sharp j^\sharp(\beta)\}$$
$$= \{\alpha \to \beta\}^\sharp = j^\sharp(\alpha \to \beta) \quad .$$

Therewith the assertion of theorem 5.7 is verified.
□

It is easy to see that the MacNeille completion preserves the integrality of residuated, commutative l-monoids. Moreover we obtain

**Corollary 5.9** *Let* $(P, \leq, *)$ *be an integral, commutative Girard-monoid. Then the MacNeille completion of* $(P, \leq, *)$ *is an integral, commutative Girard quantale.*

**Proof.** Let $(\mathcal{P}^\sharp, \preceq^\sharp, *^\sharp)$ be the MacNeille completion of $(P, \leq, *)$ (cf. Theorem 5.7).
(a) Obviously $\mathcal{L}(\{\alpha \to 0 \mid \alpha \in A^\sharp\})$ is ♯-closed. Further we observe

$$\gamma \in \mathcal{L}(\{\alpha \to 0 \mid \alpha \in A^\sharp\}) \quad \Longleftrightarrow \quad \alpha \leq \gamma \to 0 \;\; \forall \alpha \in A^\sharp$$
$$\Longleftrightarrow \quad \gamma \to 0 \in \mathcal{U}(A) \quad .$$

Since the "negation" is an involution, we obtain

$$\mathcal{L}(\{\alpha \to 0 \mid \alpha \in A^\sharp\}) = \{\gamma \to 0 \mid \gamma \in \mathcal{U}(A)\} \tag{5.6}$$

(b) Referring to formula (5.5) it is not difficult to establish the following chain of equivalences

$$A^\sharp *^\sharp B^\sharp = \{0\}^\sharp = \{0\} \quad \Longleftrightarrow \quad \mathcal{U}(A^\sharp \star B^\sharp) = P$$
$$\Longleftrightarrow \quad A^\sharp \star B^\sharp = \{0\}$$
$$\Longleftrightarrow \quad \beta \leq \alpha \to 0 \;\; \forall \alpha \in A^\sharp \;\; \forall \beta \in B^\sharp$$
$$\Longleftrightarrow \quad B^\sharp \subseteq \mathcal{L}(\{\alpha \to 0 \mid \alpha \in A^\sharp\}) \;\; ;$$

hence the "negation" of $A^\sharp$ is given by

$$A^\sharp \to^\sharp \{0\}^\sharp = \mathcal{L}(\{\alpha \to 0 \mid \alpha \in A^\sharp\}) \quad .$$

Now we apply formula (5.6) and obtain

$$(A^\sharp \to^\sharp \{0\}^\sharp) \to^\sharp \{0\}^\sharp = \mathcal{L}(\{\beta \to 0 \mid \beta \in \{\gamma \to 0 \mid \gamma \in \mathcal{U}(A)\}\})$$
$$= \mathcal{L}(\mathcal{U}(A)) = A^\sharp \quad ;$$

hence the assertion is verified.
□

**Theorem 5.10** *Let* $(P, \leq, *)$ *be an integral, commutative, residuated l-monoid with square roots. Then the MacNeille completion of* $(P, \leq, *)$ *is an integral, commutative cl-monoid with square roots, and the embedding from* $(P, \leq, *)$ *into its MacNeille completion preserves the formation of square roots.*

**Proof.** Let $(\mathcal{P}^\natural, \preceq^\natural, *^\natural)$ be the MacNeille completion of $(P, \leq, *)$. We maintain the notations in the proof of Theorem 5.7. Since $(P, \leq, *)$ has square roots, the inclusions

$$\mathcal{U}(\{\beta * \beta \mid \beta \in B\} \subseteq \mathcal{U}(B^\natural \star B^\natural) = \mathcal{U}(B \star B)$$

$$\mathcal{U}(B_1^\natural *^\natural B_1^\natural \cup B_2^\natural *^\natural B_2^\natural) \subseteq \mathcal{U}(B_1^\natural \star B_2^\natural)$$

follow from Proposition 2.11(xxiii) and Lemma 5.8 . Hence we obtain

$$(B^\natural)^2 \overset{\text{def}}{=} B^\natural *^\natural B^\natural = \{\beta * \beta \mid \beta \in B\}^\natural \tag{5.7}$$

$$B_1^\natural *^\natural B_2^\natural \preceq^\natural (B_1^\natural)^2 \vee (B_2^\natural)^2 \tag{5.8}$$

Further we define a unary operation $\mathcal{S}$ on $\mathcal{P}^\natural$ as follows

$$\mathcal{S}(A^\natural) = \vee\{B^\natural \mid (B^\natural)^2 \preceq^\natural A^\natural\}$$

In order to verify that $\mathcal{S}$ is a square root it is sufficient to show $\mathcal{S}(A^\natural) *^\natural \mathcal{S}(A^\natural) = A^\natural$ . Obviously $\mathcal{S}(A^\natural) *^\natural \mathcal{S}(A^\natural) \preceq^\natural A^\natural$ follows from formula (5.8) . On the other hand we obtain from formula (5.7)

$$\{\alpha^{1/2} \mid \alpha \in A^\natural\}^\natural *^\natural \{\alpha^{1/2} \mid \alpha \in A^\natural\}^\natural = A^\natural \; ;$$

hence the equality $\mathcal{S}(A^\natural) *^\natural \mathcal{S}(A^\natural) = A^\natural$ is attained. Further let $j^\natural$ be the embedding from $(P, \leq, *)$ into $(\mathcal{P}^\natural, \preceq^\natural, *^\natural)$ . Because of (5.7) we infer from $(B^\natural)^2 \preceq^\natural j^\natural(\alpha)$

$$\{\lambda \in P \mid \alpha \leq \lambda\} \subseteq \mathcal{U}(\{\beta * \beta \mid \beta \in B\}) \quad ;$$

hence $\alpha^{1/2}$ is an upper bound of B , and the inclusion $B^\natural \subseteq \{\lambda \in P \mid \lambda \leq \alpha^{1/2}\}$ follows. Thus we obtain $j^\natural(\alpha^{1/2}) = \mathcal{S}(j^\natural(\alpha))$ ; i.e. $j^\natural$ preserves the formation of square roots.
□

**Proposition 5.11** *The MacNeille completion of a Heyting algebra is again a Heyting algebra.*

**Proof.** In the case of $* = \wedge$ we obtain from formula (5.5)

$$A^\natural \wedge^\natural A^\natural = \{\alpha \wedge \bar{\alpha} \mid \alpha, \bar{\alpha} \in A^\natural\}^\natural = A^\natural \; ;$$

i.e. $\wedge^\natural$ is idempotent. Hence the assertion follows from Theorem 5.7 .
□

**Theorem 5.12 ([16])** *Let $(P, \leq, *)$ be an integral, divisible, residuated commutative l-monoid. If $(P, \leq, *)$ satisfies the additional property*

(C) $\qquad \alpha \leq \beta^n \quad$ for all $\quad n \in \mathbb{N} \qquad \Longrightarrow \qquad \alpha = \alpha * \beta$ ,

*then the MacNeille completion of $(P, \leq, *)$ is a GL-monoid.*

**Proof.** In view of Theorem 5.7 it is sufficient to show that under the condition (C) the MacNeille completion preserves the divisibility. We maintain the notations in the proof of Theorem 5.7 . Let us consider a pair $(A^\sharp, B^\sharp) \in \mathcal{P}^\sharp \times \mathcal{P}^\sharp$ with $A^\sharp \preceq^\sharp B^\sharp$ – i.e. $A^\sharp$ is a subset of $B^\sharp$ . Then we introduce a subset C of P by

$$C \;=\; \{\epsilon \rightarrow \alpha \;\mid\; \epsilon \in \mathcal{U}(B^\sharp), \quad \alpha \in A^\sharp\} \qquad .$$

If $\gamma$ is an upper bound of $A^\sharp$ , then we obtain for all $\beta \in B^\sharp$, $\quad \epsilon \in \mathcal{U}(B^\sharp)$ and $\alpha \in A^\sharp$

$$(\epsilon \rightarrow \alpha) * \beta \quad \leq \quad \epsilon * (\epsilon \rightarrow \alpha) \quad \leq \quad \alpha \quad \leq \quad \gamma \;\; ;$$

i.e. $\gamma$ is also an upper bound of $C \star B^\sharp$ . Thus the inclusion $\mathcal{U}(A^\sharp) \subseteq \mathcal{U}(C \star B^\sharp)$ holds. In order to verify $\mathcal{U}(C \star B^\sharp) \subseteq \mathcal{U}(A^\sharp)$ we first fix $\gamma \in \mathcal{U}(C \star B^\sharp)$, $\quad \alpha \in A^\sharp$ and show

$$\epsilon \in \mathcal{U}(B^\sharp) \quad \text{implies} \quad \epsilon * (\alpha \rightarrow \gamma) \in \mathcal{U}(B^\sharp) \tag{5.9}$$

Since $(P, \leq, *)$ is divisible, we deduce from $A^\sharp \subseteq B^\sharp$

$$\epsilon * (\epsilon \rightarrow \alpha) \;=\; \alpha \qquad \text{for all} \quad \epsilon \in \mathcal{U}(B^\sharp) \qquad ;$$

hence we obtain for all $\beta \in B^\sharp$ and $\epsilon \in \mathcal{U}(B^\sharp)$

$$\begin{aligned} \beta \;&=\; \epsilon * (\epsilon \rightarrow \beta) \quad \leq \quad \epsilon \left( (\epsilon * (\epsilon \rightarrow \alpha)) \rightarrow (\beta * (\epsilon \rightarrow \alpha)) \right) \\ &\leq\; \epsilon * (\alpha \rightarrow \gamma) \qquad . \end{aligned}$$

Therewith the implication (5.9) is verified. In particular

$$(\alpha \rightarrow \gamma)^n \quad \in \mathcal{U}(B^\sharp) \qquad \text{for all} \quad n \in \mathbb{N} \tag{5.10}$$

follows by recursion from (5.9) . Since $A^\sharp$ is a subset of $B^\sharp$ , we obtain from (5.10) that the inequality $\alpha \leq (\alpha \rightarrow \gamma)^n$ holds for all $n \in \mathbb{N}$ ; hence $\alpha = \alpha * (\alpha \rightarrow \gamma)$ follows from Condition (C) . Thus $\gamma$ is an upper bound of $A^\sharp$, and the inculsion $\mathcal{U}(C \star B^\sharp) \subseteq \mathcal{U}(A^\sharp)$ is verified.

Now we infer from Lemma 5.8 and the preceding considerations that the equation $\mathcal{U}(A^\sharp) \;=\; \mathcal{U}(C^\sharp \star B^\sharp)$ holds ; hence $A^\sharp \;=\; C^\sharp *^\sharp B^\sharp$ follows from fromula (5.5) .
□

**Corollary 5.13** *The MacNeille completion of a locally finite MV-algebra is again a locally finite MV-algebra.*

**Proof.** Let $(P, \leq *)$ be a locally finite MV–algebra and $(\mathcal{P}^\natural, \preceq^\natural, *^\natural)$ its MacNeille completion. The local finiteness of $(P, \leq, *)$ implies the validity of Condition (C) . Therefore we can apply Corollary 5.9 and Theorem 5.12 together with Lemma 2.14 and obtain that $(\mathcal{P}^\natural, \preceq^\natural, *^\natural)$ is a MV–algebra. In order to verify the local finiteness of $(\mathcal{P}^\natural, \preceq^\natural, *^\natural)$ we choose an element $A^\natural \in \mathcal{P}^\natural$ with $A^\natural \neq \{1\}^\natural$; hence there exists $\alpha \in P$ provided with the following properties

$$\alpha \in \mathcal{U}(A^\natural) \quad , \quad \alpha \neq 1 \quad .$$

Since $(P, \leq, *)$ is locally finite, there exists a natural number $n_0 \in \mathbb{N}$ such that for all $n_0$-tuple $(\gamma_1, \ldots, \gamma_{n_0}) \in A^{n_0}$ the following relation holds

$$\gamma_1 * \cdots * \gamma_{n_0} \quad \leq \quad \alpha^{n_0} \quad = \quad 0$$

Referring to Lemma 5.8 and formula (5.5) we obtain $(A^\natural)^{n_0} = \{0\}^\natural$ ; hence the local finiteness of $(\mathcal{P}^\natural, \preceq^\natural, *^\natural)$ is established.
□

**Examples 5.14 (Integral, commutative cl–monoids)** Let $D$ be the set of all rational dyadic numbers of the real unit interval $[0,1]$ – i.e.
$D \quad = \quad \bigcup_{n \in \mathbb{N}} \{i/2^n \mid i \in \{0, 1, \ldots, 2^n\}\} \quad$ . On $D$ we consider a binary operation $\boxtimes$ defined by

$$\alpha \boxtimes \beta \quad = \quad Max(\alpha + \beta - 1, 0) \quad .$$

Then $\mathcal{D} = (D, \leq, \boxtimes)$ is an MV–algebra with square roots. In particular $\mathcal{D}$ is a MV–subalgebra of $([0,1], \leq, T_m)$ (cf. section 3.3) and the *Dedekind* (= MacNeille) completion of $\mathcal{D}$ coincides with $([0,1], \leq, T_m)$ .
Further let $(L, \leq, *)$ be an integral, commutative cl–monoid and $L([0,1])$ be the set of all maps $F : D \mapsto L$ satisfying the following conditions

(r1)  $F(1) \quad = \quad 0$                                      (Boundary Condition)

(r2)  $s, t \in D$ with $s \leq t \quad \Longrightarrow \quad F(t) \leq F(s)$           (Antitonicity)

(r3)  $\bigvee_{n \in \mathbb{N}} F(min(r + (1/2^n), 1)) \quad = \quad F(r)$           (Rightcontinuity)

$L([0,1])$ can be provided with the structure of an integral, commutative cl–monoid as follows

$$F_1 \preceq F_2 \quad \Longleftrightarrow \quad F_1(r) \leq F_2(r) \quad \text{for all} \quad r \in D$$
$$(F_1 \Diamond_* F_2)(r) \quad = \quad \bigvee \{F_1(u) * F_2(v) \mid r < u \boxtimes v\}$$

In particular $\epsilon_1$ defined by $\quad \epsilon_1(r) \quad = \quad \left\{ \begin{array}{ll} 1 & : \quad r \neq 1 \\ 0 & : \quad r = 1 \end{array} \right\}$ is the unit element in $L([0,1])$ w.r.t. $\Diamond_*$. Further it is not difficult to verify the subsequent equivalence

$$F \Diamond_* G \preceq H \quad \Longleftrightarrow \quad G(v) \quad \leq \quad \wedge\{F(u) \to H(r) \mid 1 - u + r \quad < \quad v\} \quad ;$$

hence the "implication" $\longrightarrow$ in $(L([0,1]), \preceq, \Diamond_*)$ is given by

$$\left.\begin{array}{rl}
(F \longrightarrow H)(r) & = \\
= & \bigvee_{n \in \mathbb{N}} (\bigwedge\{F(u) \to H(v) \mid 1 - u + v < Min(r + (1/2^n), 1)\}) \\
& \text{whenever} \quad r \neq 1 \quad .
\end{array}\right\} \quad (5.11)$$

Finally we notice that $(L, \leq, *)$ , respectivly $([0,1], \leq, T_m)$ can be embedded into $(L([0,1]), \preceq, \Diamond_*)$ in the following way

$$\phi : (L, \leq, *) \rightarrowtail (L([0,1], \preceq, \Diamond_*) \quad , \quad \psi : ([0,1], \leq, T_m) \rightarrowtail (L([0,1], \preceq, \Diamond_*)$$

$$\phi(\alpha) = F_\alpha \quad \text{where} \quad F_\alpha(r) = \left\{ \begin{array}{cc} \alpha & , \quad r \neq 1 \\ 0 & , \quad r = 1 \end{array} \right\} \quad \text{for all} \quad \alpha \in L$$

$$\psi(x) = \epsilon_x \quad \text{where} \quad \epsilon_x(r) = \left\{ \begin{array}{cc} 0 & , \quad x \leq r \\ 1 & , \quad r < x \end{array} \right\} \quad \text{for all} \quad x \in [0,1]$$

In this sense we can view $(L([0,1]), \preceq, \Diamond_*)$ as a common extension of the integral cl-monoid $(L, \leq, *)$ and the complete MV–algebra $([0,1], \leq, T_m)$ .
(a) Referring to formula (5.11) it is not difficult to show that the "negation" in $(L([0,1]), \preceq, \Diamond_*)$ has the following form

$$(F \longrightarrow \epsilon_0)(r) = \bigvee_{n \in \mathbb{N}} F(max(1 - r - (1/2^n), 0)) \to 0 \quad \text{where} \quad r \neq 1 \quad .$$

Hence $(L([0,1]), \preceq, \Diamond_*)$ is an integral commutative *Girard quantale* if and only if $(L, \leq, *)$ is an integral, commutative Girard quantale.
(b) Let $(L, \leq, *) = (\mathbb{B}, \leq, \wedge)$ be a complete Boolean algebra (i.e. $* = \wedge$ and the "negation" is an involution). Further let $(s, t)$ be a pair of rational dyadic numbers with $s \leq t$ . Then we derive from formula (5.11) the following relation

$$((F \longrightarrow H)(t) \to 0) \vee H(s) =$$
$$\bigwedge_{n \in \mathbb{N}} (\vee\{(F(u) \wedge (H(v) \to 0)) \vee H(s) \mid 1 - u + v < min(t + (1/2^n), 1)\}$$
$$\leq (\bigwedge_{n \in \mathbb{N}} F(max(1 - t - (1/2^n) + s, 0))) \vee H(s) \quad .$$

Hence we obtain again from (5.11)

$$((F \longrightarrow H) \longrightarrow H)(r) \leq$$
$$\bigvee_{m \in \mathbb{N}} (\wedge\{ \bigwedge_{n \in \mathbb{N}} F(max(1 - t - (1/2^n) + s, 0))) \vee H(s) \mid$$
$$1 - t + s < min(r + (1/2^m), 1)\})$$
$$\leq (\bigvee_{m \in \mathbb{N}} F(min(r + (1/2^{m+2}), 1))) \vee H(r) = F(r) \vee H(r) \quad ;$$

i.e. the axiom (MV) is established. Thus $(\mathbb{B}([0,1]), \preceq, \Diamond_\wedge)$ is a complete

*MV-algebra.*- Further we observe

$$F^{1/2} \Diamond_\wedge F^{1/2} \;=\; F \quad \text{where} \quad F^{1/2}(r) = \left\{ \begin{array}{lcl} 1 & , & 0 \leq r < 1/2 \\ F(2r-1) & , & 1/2 \leq r \leq 1 \end{array} \right\}$$

$F$ is idempotent w.r.t. $\Diamond_\wedge$ $\Longleftrightarrow$ $\exists\ \alpha \in \mathbb{B}$ with $F = F_\alpha$ ;

Hence $(\mathbb{B}([0,1]), \preceq, \Diamond_\wedge)$ is a complete MV-algebra with *square roots*, and the idempotent elements in $\mathbb{B}([0,1])$ can be identified with the elements of the underlying Boolean algebra $\mathbb{B}$ .

(c) Let $(L, \leq, *)$ be an integral, commutative cl-monoid and $(R_L, \preceq, \Diamond)$ be the example studied in section 3.2 . If we identify $(\alpha, \beta) \in R_L$ with an element $G_{(\alpha,\beta)} \in L([0,1])$ in the following way

$$G_{(\alpha,\beta)} \;=\; \left\{ \begin{array}{lcl} 0 & , & r = 1 \\ \alpha & , & 1/2 \leq r < 1 \\ \beta & , & 0 \leq r < 1/2 \end{array} \right\} \quad ,$$

then we obtain an embedding from $(R_L, \preceq, \Diamond)$ into $(L([0,1]), \preceq, \Diamond_*)$. Hence Corollary 3.2.2 implies that $(L([0,1]), \preceq, \Diamond_*)$ is *divisible* if and only if $(L, \leq, *)$ is a Boolean algebra.

(d) In the case of $(L, \leq *) = ([0,1], \leq, T_m)$ the elements of $L([0,1])$ are *probability distribuiton functions*, the support of which is contained in $[0,1]$ . With regard to possible applications in *probability theory and statistics* it is interesting to note that $([0,1]([0,1]), \preceq, \Diamond_*)$ is *not* an MV-algebra, but an *integral, commutative Girard quantale.*
□

# 6   $\sigma$-complete MV-algebras

An MV-algebra is called $\sigma$-complete if and only if the underlying lattice $(L, \leq)$ is $\sigma$-complete –i.e. countable joins and countable meets exist. If we replace in section 5 arbitrary joins (respectively, meets) by countable joins (respectively, meets), then the relations (i)-(iii) in section 5 and Theorem 5.2 (b) and (c) remain valid for $\sigma$-complete MV-algebras. In particular, we note that Examples 5.14(b) requires only $\sigma$-complete Boolean algebras.

**Lemma 6.1** *If $\alpha$ is an element of a $\sigma$-complete MV-algebra $(L, \leq, *)$ , then*

$$e_\alpha \;:=\; \bigwedge_{n \in \mathbb{N}} \alpha^n$$

*is idempotent w.r.t.* $*$ .

**Proof.** The assertion follows immediately from Theorem 5.2(c) .
□

**Lemma 6.2 ([2])** *Every $\sigma$-complete MV-algebra is semi-simple.*

**Proof.** Let $\alpha$ be an element of L with $\alpha \neq 1$ ; and let us assume that $(L, \leq, *)$ is not semi-simple. Then we conclude from Proposition 4.12 and Lemma 6.1 that $\alpha \to 0 \leq e_\alpha$ and $e_\alpha$ is idempotent. Referring to Proposition 2.6(xii) and Lemma 2.14 we obtain

$$0 \; = \; e_\alpha * (\alpha \to 0) \; = \; e_\alpha \wedge (\alpha \to 0) \; = \; \alpha \to 0 \quad ;$$

i.e. $\alpha = 1$ ; but his equality is a contradiction to the choice of $\alpha$ . Hence $(L, \leq, *)$ is semi-simple.
□

**Theorem 6.3 (Characterization of semi–simplicity of MV–algebras)**
Let $M = (L, \leq, *)$ be an MV-algebra. Then the following assertions are equivalent

   (i)     The MacNeille completion of M is an MV–algebra.

   (ii)    M is semi–semiple.

**Proof.** The implication (i) $\implies$ (ii) follows from Lemma 6.2 . In order to verify (ii) $\implies$ (i) we proceed as follows : In view of Proposition 4.12 , Corollary 5.9 and Theorem 5.12 it is sufficient to show that the assertion (ii) in 4.12 implies Condition (C) in 5.12 . Therefore we consider a pair $(\alpha, \beta) \in L \times L$ with $\alpha \leq \beta^n$ for all $n \in \mathbb{N}$ . Then we put $\gamma = \alpha \to (\alpha * \beta)$ and obtain

$$\gamma \to 0 \; \leq \; \alpha \; \leq \; \beta^n \; \leq \; \gamma^n \quad \forall n \in \mathbb{N} \quad .$$

Thus $\gamma = 1$ follows from Proposition 4.12 (ii) –i.e. $\alpha = \alpha * \beta$ holds.
□

**Theorem 6.4** *Let $M = (L, \leq, *)$ be a complete MV-algebra, and $\mathbb{B}_M$ be the set of all idempotent elements of L w.r.t. $*$. Then the following assertions are equivalent*

   1.  Every element $\alpha$ of L is a square (i.e. $\exists \beta \in L$ with $\alpha = \beta^2$).

   2.  Every element $\alpha \in L$ with the subsequent property

     (I)   $\{\lambda \in L | \lambda \leq \alpha\} \; = \; \{\iota \wedge \alpha | \; \iota \in \mathbb{B}_M\}$

    is idempotent (i.e. $\alpha \in \mathbb{B}_M$).

**Proof.** In order to verify the implication 1. $\implies$ 2. we consider an element $\alpha \in L$ satisfying Condition (I) . Since $\alpha \to 0$ is a square, there exists an element $\beta \in L$ with $\alpha \to 0 = \beta^2$ ; in particular $\beta \to 0 \leq \alpha$ . Now we use (I) and choose an idempotent element $\iota$ such that $\beta \to 0 = \iota \wedge \alpha$ . Because of

$$\alpha \; = \; \beta^2 \to 0 \; = \; ((\alpha \to 0)^2 \to 0) \wedge \iota$$

we obtain

$$\beta \to 0 \;\; = \;\; \alpha \;\; = \;\; (\beta^2) \to 0 \quad ;$$

i.e. $\beta$ is idempotent, hence $\alpha$ is also idempotent (cf. Corollary 2.7).
The verification of the implication 2. $\Longrightarrow$ 1. is based on the following lemmas

**Lemma 6.5** *Let* $M = (L, \leq, *)$ *be a $\sigma$-complete MV-algebra satisfying the assertion 2. in Theorem 6.4 . Then for all $\alpha \in L$ with $\alpha \neq 0$ there exists $\beta \in L$ such that $\beta \neq 0$ and* $(\beta \to 0)^2 \to 0 \;\; \leq \;\; \alpha$ .

**Proof.** Without loss of generality we consider a non idempotent element $\alpha \in L$. Then we infer from assertion 2. in Theorem 6.4 that there exists an element $\gamma \in L$ provided with the subsequent properties

(i)     $\gamma \neq 0$ ,   $\gamma \leq \alpha$    .

(ii)    For all  $e \in \mathbb{B}_M$  :   $\gamma \neq e \wedge \alpha$  .

Following an idea proposed by L.P. Belluce we put

$$\beta \;\; \overset{\text{def}}{=} \;\; \gamma \wedge (\alpha * (\gamma \to 0)) \quad .$$

(a) We show : $\beta \neq 0$ . Let us assume the contrary – i.e. $\beta = 0$ ; then we obtain
$1 \;\; = \;\; (\gamma \to 0) \vee (\alpha \to \gamma)$   which is equivalent to   $\gamma \;\; = \;\; \gamma * (\alpha \to \gamma)$ .
In particular the relation

$$\gamma \;\; = \;\; \gamma * (\alpha \to \gamma)^n$$

follows by recursion over $n \in \mathbb{N}$. Since the underlying lattice is $\sigma$-complete,

$$e_0 \;\; \overset{\text{def}}{=} \;\; \bigwedge_{n \in \mathbb{N}} (\alpha \to \gamma)^n$$

exists and is idempotent (cf. Lemma 6.1). Now we use the fact that $*$ is distributive over countable meets and obtain

$$\gamma = \gamma * e_0 = \gamma \wedge e_0 \quad , \quad e_0 \wedge \alpha = e_0 * \alpha \leq \gamma \quad ;$$

i.e. $\gamma = \alpha \wedge e_0$ which is a contradiction to (ii) . Hence $\beta \neq 0$ holds.

(b) We show : $(\beta \to 0)^2 \to 0 \leq \alpha$ . In order to simplify the presentation we introduce for all $\lambda \in L$ the following abbreviation

$$2\lambda \;\; = \;\; (\lambda \to 0)^2 \to 0 \quad .$$

The definition of $\beta$ implies

$$2\beta \;\; \leq \;\; (2\gamma) \wedge [2(\alpha * (\gamma \to 0))] \quad .$$

Now we embark on the algebraic strong de Morgan law (cf. Corollary 2.16) and obtain

$$
\begin{aligned}
2\beta \quad &= \quad (2\beta) * [((2\gamma) \to \alpha) \vee (\alpha \to (2\gamma))] \\
&\leq \quad \alpha \vee (2(\alpha * (\gamma \to 0))) * (\alpha \to (2\gamma))
\end{aligned}
\tag{6.1}
$$

Further we observe

$$
\begin{aligned}
2(\alpha * (\gamma \to 0)) \quad &= \quad ((\alpha \to \gamma)^2) \to 0 \quad = \quad (\alpha \to \gamma) \to (\alpha * (\gamma \to 0)) \quad , \\
\alpha \to (2\gamma) \quad &= \quad (\gamma \to 0)^2 \to (\alpha \to 0) \quad = \quad (\gamma \to 0) \to (\alpha \to \gamma) \quad ;
\end{aligned}
$$

i.e. the following relation holds

$$
\begin{aligned}
(2(\alpha * (\gamma \to 0))) * (\alpha \to (2\gamma)) \quad &= \\
((\gamma \to 0) \to (\alpha \to \gamma)) * ((\alpha \to \gamma) \to (\alpha * (\gamma \to 0))) \quad &\leq \\
(\gamma \to 0) \to (\alpha * (\gamma \to 0)) \quad = \quad \gamma \vee \alpha \quad \leq \quad &\alpha \quad .
\end{aligned}
\tag{6.2}
$$

Hence $2\beta \leq \alpha$ follows from (6.1) and (6.2) .

□

**Lemma 6.6** *Let $M = (L, \leq, *)$ be a $\sigma$-complete MV-algebra satisfying the assertion 2. in Theorem 6.4 . Further let $(\alpha, \beta)$ be an element of $L \times L$ with property $\beta^2 < \alpha$. Then there exists $\gamma \in L$ such that $\beta < \gamma$ and $\gamma^2 \leq \alpha$ .*

**Proof**. Because of $\epsilon \overset{\text{def}}{=} \alpha * (\beta^2 \to 0) \neq 0$ we can apply Lemma 6.5 and choose an element $\delta \in L$ provided with the following properties

$$
\delta \neq 0 \quad , \quad (\delta \to 0)^2 \to 0 \quad = \quad 2\delta \quad \leq \quad \epsilon
\tag{6.3}
$$

Now we put $\gamma = (\delta \to 0) \to \beta$ ; obviously $\beta \leq \gamma$.

(a) We show : $\beta \neq \gamma$ (i.e. $\beta < \gamma$ ). Let us assume the contrary – i.e. $\beta = \gamma$; hence the axiom (MV) implies $1 = \beta \vee (\delta \to 0)$ . Now we invoke Proposition 2.6(xiv) (resp. Lemma 2.4(1)) and obtain $1 = \beta^2 \vee (\delta \to 0)^2$ ; hence

$$
\delta \quad \leq \quad 2\delta \quad = \quad (\beta^2 \to 0) \wedge 2\delta \quad = \quad 0 \quad ;
$$

but this is a contradiction to (6.3)

(b) Because of $\gamma^2 \leq (\delta \to 0)^2 \to \beta^2$ we obtain

$$
\begin{aligned}
\gamma^2 \quad &\leq \quad ((2\delta) \to 0) \to \beta^2 \quad \leq \quad (\epsilon \to 0) \to \beta^2) \\
&= \quad (\alpha \to \beta^2) \to \beta^2 \quad = \quad \alpha \vee \beta^2 \quad = \quad \alpha
\end{aligned}
$$

hence $\gamma$ fulfills the desired properties.

□

**Continuation of the proof of Theorem 6.4** Let $\alpha$ be an element of L ; since the underlying lattice $(L, \leq)$ of M is *complete*, we can define an element $\beta_0$ of L as follows

$$
\beta_0 \quad = \quad \vee \{\lambda \in L \mid \lambda^2 \leq \alpha\} \quad .
$$

Obviously the inequality $\beta_0^2 \leq \alpha$ follows from Proposition 2.6(xiv). On the other hand, Lemma 6.6 implies $\beta_0^2 = \alpha$ ; hence the assertion 1. is verified.
□

**Corollary 6.7** *Every infinite, locally finite, complete MV–algebra M is a strict MV–algebra.*

**Proof.** Since M is locally finite, the algebraic strong de Morgan law (cf. Corollary 2.16) implies that the underlying lattice $(L, \leq)$ is a chain. Further the completeness of $(L, \leq)$ guarantees the existence of the following element of L

$$\lambda_0 \quad = \quad \vee \{\lambda \in L \mid \lambda \neq 1\} \quad .$$

If we assume $\lambda_0 \neq 1$ , then every element

$$\alpha \quad \in \quad \{\lambda \in L \mid \lambda_0^{n+1} \leq \lambda < \lambda_0^n\}$$

satisfies the equation $\lambda_0 = \lambda_0^n \to \alpha$ . By virtue of the divisibility and local finiteness of M we can find a natural number m such that
$L = \{1\} \vee \{\lambda_0^n \mid n \in \{1, \ldots, m\}\}$ . But the previous equality is an obvious contradiction to the hypothesis of the infiniteness of L. Therefore $\lambda_0$ coincides necessarily with 1 ; i.e. for each element $\alpha \in L$ with $\alpha \neq 0$ there exists $\beta \in L$ s.t. $\beta \neq 0$ and $\beta < \alpha$. Since the universal bounds in $(L, \leq)$ are the only idempotent elements in L , we conclude from Theorem 6.4 that every element of L is a square w.r.t. $*$. Thus M has *square roots* (cf. Theorem 5.3).
In order to verify the strictness of M it is now sufficient to establish the inequality $0^{1/2} \to 0 \leq 0^{1/2}$. First the local finiteness of $M = (L, \leq, *)$ and the infiniteness of L shows that M *cannot* be a Boolean algebra; hence $0^{1/2} \to 0 \neq 1$ follows from Proposition 2.19 . Applying again the local finiteness of M we conclude from Proposition 2.17(xxix) that $(0^{1/2} \to 0)^2 = 0$ –i.e. $0^{1/2} \to 0 \leq 0^{1/2}$ holds.
□

**Theorem 6.8** *Let $M = (L, \leq, *)$ be a $\sigma$–complete MV–algebra and $\mathbb{B}_M$ be the ($\sigma$–complete) Boolean algebra of all idempotent elemnts of L w.r.t. $*$ . Then the following assertions are equivalent*

1. M is isomorphic to $(\mathbb{B}_M([0,1]), \preceq, \Diamond_\wedge)$ (cf. Examples 5.14(b)).

2. There exists an injective MV–algebra–homomorphism $j$ form the MV–algebra $\mathcal{D}$ of all dyadic, rational numbers (cf. Examples 5.14) into M satisfying the contition

$$(\Gamma) \quad \bigvee_{n \in \mathbb{N}} j(1 - (1/2^n)) \quad = \quad 1 \quad .$$

**Proof.** (a) In order to verify the implication 1. $\Longrightarrow$ 2. we denote by $i : D \longmapsto \mathbb{B}_M([0,1])$ the restriction of $\psi : [0,1] \longmapsto \mathbb{B}_M([0,1])$ to D (cf. Examples 5.14). It is easy to see that $i$ is an MV–algebra-homomorphism satisfying

$$\bigvee_{n \in \mathbb{N}} i(1 - (1/2^n)) \quad = \quad \epsilon_1 \quad .$$ If $\theta$ is an isomorphism from $(\mathbb{B}_M([0,1]), \preceq, \Diamond_\wedge)$ to $M$, then $j = \theta \circ i$ fulfills the desired properties; hence the assertion 2. is verified.

(b) In order to prove 2. $\Longrightarrow$ 1. we first show that every $\alpha \in L$ induces an element $F_\alpha$ of $\mathbb{B}_M([0,1])$ in the following way

$$F_\alpha(r) \quad = \quad [\bigwedge_{n \in \mathbb{N}} (\alpha \to j(r))^{2^n}] \to 0 \qquad r \in D$$

(b1) We fix $r_0 \in D$ and consider a sequence $(r_n)_{n \in \mathbb{N}}$ in $D$ defined by $r_n = min(r_0 + (1/2^n), 1)$ . Because of $r_n \to r_0 = 1 - r_n + r_0 = 1 - (1/2^n)$ the condition ($\Gamma$) implies

$$j(r_0) \quad \leq \quad \bigwedge_{n \in \mathbb{N}} j(r_n) \quad = \quad \bigwedge_{n \in \mathbb{N}} j((r_n \to r_0) \to r_0)$$

$$= \quad \bigwedge_{n \in \mathbb{N}} (j(r_n \to r_0) \to j(r_0)) \quad = \quad j(r_0)$$

i.e. $j$ preserves all infima provided they exist in $D$ .

(b2) Because of Lemma 6.1 all elements $F_\alpha(r)$ $(r \in D)$ are idempotent. In particular $F_\alpha(1) = 0$ holds, and $F_\alpha$ is antitone. Referring to Lemma 2.4(1) , Corollary 2.16 and Theorem 5.2(c) we obtain by recursion over $n \in \mathbb{N}$

$$(\bigwedge_{i \in \mathbb{N}} \beta_i)^{2^n} \quad = \quad \bigwedge_{i \in \mathbb{N}} (\beta_i)^{2^n} \quad ;$$

hence the part (b1) implies that $F_\alpha$ is rightcontinuous. Therefore $F_\alpha$ is in fact an element of $\mathbb{B}_M([0,1])$.

(b3) In order to verify

$$\alpha \quad = \quad \bigvee_{r \in D} (j(r) * F_\alpha(r)) \qquad \alpha \in L \tag{6.4}$$

let us first consider finitely many elements $r_i \in D$ defined by $r_i = i/2^n$ $(i = 0, 1, \ldots, 2^n)$. We derive from the definition of $F_\alpha$ the subsequent relations

$$F_\alpha(0) \quad = \quad F_\alpha(r_0) \quad \geq \alpha \quad , \qquad F_\alpha(1) \quad = \quad F_\alpha(r_{2^n}) \quad = \quad 0 \quad ,$$

$$F_\alpha(r_0) \quad = \quad \bigvee_{i=1}^{2^n} (F_\alpha(r_{i-1}) \wedge (F_\alpha(r_i) \to 0)) \quad ,$$

$$F_\alpha(r_{i-1}) \wedge \alpha \wedge (F_\alpha(r_i) \to 0) \quad = \quad F_\alpha(r_{i-1}) * \alpha * (F_\alpha(r_i) \to 0) \quad ;$$

hence $\alpha$ can be decomposed as follows

$$\alpha \quad = \quad \bigvee_{i=1}^{2^n} F_\alpha(r_{i-1}) * \alpha * (F_\alpha(r_i) \to 0) \tag{6.5}$$

Because of

$$j(1 - (1/2^n)) * \alpha * (F_\alpha(r_i) \to 0) \quad \le \quad j(1 - (1/2^n)) * j(r_i) \quad = \quad j(r_{i-1})$$

the inequality

$$\alpha \quad \le \quad \bigvee_{r \in D} F_\alpha(r) * j(r) \tag{6.6}$$

follows from condition $(\Gamma)$ and formula (6.5) . On the other hand we conclude from Lemma 2.4(1) and Corollary 2.16

$$1 \quad = \quad (\alpha \to j(r))^{2^n} \vee (j(r) \to \alpha)^{2^n} \quad .$$

Referring to Theorem 5.2(c) we derive from the $\sigma$–completeness of the underlying lattice $(L, \le)$ the equation

$$1 \quad = \quad (\bigwedge_{n \in \mathbb{N}} (\alpha \to j(r))^{2^n}) \vee (\bigwedge_{n \in \mathbb{N}} (j(r) \to \alpha)^{2^n}) \quad ;$$

hence the relation

$$j(r) * F_\alpha(r) \quad \le \quad j(r) * (\bigwedge_{n \in \mathbb{N}} (j(r) \to \alpha)^{2^n}) \quad \le \quad \alpha \tag{6.7}$$

holds. Finally the relation (6.4) follows from (6.6) and (6.7)

(b4) We consider a map $\theta : L \longmapsto \mathbb{B}_M([0,1])$ defined by $\theta(\alpha) = F_\alpha$. Then the formula (6.4) guarantees the injectivity of $\theta$ . In order to verify the surjectivity of $\theta$ we define for each $F \in \mathbb{B}_M([0,1])$ an element $\alpha_F$ of L by

$$\alpha_F \quad = \quad \bigvee_{r \in D} j(r) * F(r) \quad .$$

Since $j$ is an MV–algebra–homomorphism and the range of F consists of idempotent elements, we derive immediately from Proposition 2.6(xii), Proposition 2.8(xvii) and Lemma 2.14 the following relation $(r_0 \in D)$

$$\bigwedge_{n \in \mathbb{N}} (\alpha_F \to j(r_0))^{2^n} \quad = \quad \bigwedge_{r \in D} (\bigwedge_{n \in \mathbb{N}} ((j(r) * F(r)) \to j(r_0))^{2^n})$$

$$= \quad \bigwedge_{\{r \in D | r_0 < r\}} (\bigwedge_{n \in \mathbb{N}} (F(r) \to j(r \to r_0))^{2^n}) \tag{6.8}$$

$$= \quad \bigwedge_{\{r \in D | r_0 < r\}} [(F(r) \to 0) \vee (\bigwedge_{n \in \mathbb{N}} (j(r \to r_0))^{2^n}))]$$

Now we observe that in the case of $r_0 < r$ the element $j(r \to r_0)$ is nilpotent; hence we obtain from (6.8)

$$\bigwedge_{n \in \mathbb{N}} (\alpha_F \to j(r_0))^{2^n} \quad = \quad \bigwedge_{\{r \in D | r_0 < r\}} F(r) \to 0 \quad = \quad F(r_0) \to 0 \quad ;$$

i.e. $\theta(\alpha_F) = F$ .

Further it is easy to see that $\theta$ is an order–isomorphism between $(L, \leq)$ and $(\mathbb{B}_M([0,1]), \preceq)$ . In order to show that $\theta$ is an MV–algebra-isomorphism it is therefore sufficient to verify that $\theta$ preserves the corresponding semigroup operations. Referring to formula (6.4) and Proposition 2.6(xii) we obtain

$$[\theta(\alpha * \beta)](r_0)$$
$$= \quad [\bigwedge_{n \in \mathbb{N}} [(\bigvee_{r_1, r_2 \in D} j(r_1) * j(r_2) * \theta(\alpha)(r_1) * \theta(\beta)(r_2)) \rightarrow j(r_0)]^{2^n}] \rightarrow 0$$
$$= \quad [\bigwedge_{n \in \mathbb{N}} [(\bigvee_{r \in D} j(r) * [\theta(\alpha)\lozenge_\wedge\theta(\beta)](r)) \rightarrow j(r_0)]^{2^n}] \rightarrow 0$$
$$= \quad [\bigwedge_{r \in D} \bigwedge_{n \in \mathbb{N}} ([\theta(\alpha)\lozenge_\wedge\theta(\beta)](r) \rightarrow j(r \rightarrow r_0))^{2^n}] \rightarrow 0$$
$$= \quad [\bigwedge_{\{r \in D | r_0 < r\}} (([\theta(\alpha)\lozenge_\wedge\theta(\beta)](r) \rightarrow 0) \vee \bigwedge_{n \in \mathbb{N}} (j(r \rightarrow r_0))^{2^n}] \rightarrow 0$$
$$= \quad [\bigwedge_{\{r \in D | r_0 < r\}} [\theta(\alpha)\lozenge_\wedge\theta(\beta)](r) \rightarrow 0] \rightarrow 0 \quad = \quad [\theta(\alpha)\lozenge_\wedge\theta(\beta)](r_0) \quad .$$

Hence the assertion 1. is verified.
□

**Theorem 6.9** *Let* $\mathcal{D} = (D, \leq, \boxtimes)$ *be the MV–algebra of dyadic, rational numbers and* $M = (L, \leq, *)$ *be a strict MV-algebra. Then there exists an injective MV–algebrahomomorphism* $j : \mathcal{D} \longmapsto M$ . *In particular, if M is* $\sigma$-*complete, then j satisfies the condition* $(\Gamma)$ *in Theorem 6.8 .*

**Proof.** We put $D_0 = \{0, 1\}$ , $D_n = \{i/2^n | i = 0, 1, \ldots, 2^n\}$ and notice $D_{n+1} = \{(r + s)/2 | r, s \in D_n\}$ . Further we observe that $D_n$ is a support set of a finite MV–subalgebra of $([0, 1], \leq, T_m)$ (cf. Section 3.3 and Examples 5.14). By recursion over $n \in \mathbb{N}$ we define a sequence $(j_n)_{n \in \mathbb{N}}$ of maps $j_n$ as follows

$$j_0 : D_0 \longmapsto L \quad \text{by} \quad j_0(0) = 0, \quad j_0(1) = 1$$
$$j_n : D_n \longmapsto L \quad \text{by} \quad j_n((r + s)/2) \quad = \quad (j_{n-1}(r))^{1/2} * (j_{n-1}(s))^{1/2}$$
$$\text{where} \quad r, s \in D_{n-1} \quad .$$

(a) We show that $j_n$ ($n \in \mathbb{N}_0$) is well defined and a MV–algebrahomomorphism. The assertion is obvious for $n=0$ . Let us assume that the assertion holds for $n-1$ ($n \in \mathbb{N}$). In the case of $(r + s)/2 = (\bar{r} + \bar{s})/2$ we assume without loss of generality $\bar{r} < r$ . Then we obtain $1 - r + \bar{r} = 1 - \bar{s} + s < 1$; hence

$$j_{n-1}(r) \rightarrow j_{n-1}(\bar{r}) \quad = \quad j_{n-1}(\bar{s}) \rightarrow j_{n-1}(s) \quad ,$$

because $j_{n-1}$ is a MV–algebrahomomorphism. Referring to Proposition 2.11 we obtain

$$(j_{n-1}(\bar{s}))^{1/2} * (j_{n-1}(r))^{1/2} * ((j_{n-1}(r))^{1/2} \rightarrow j_{n-1}(\bar{r}))^{1/2}) \quad =$$

$$(j_{n-1}(\bar{s}))^{1/2} * (j_{n-1}(\bar{r}))^{1/2} \quad = \quad (j_{n-1}(r))^{1/2} * (j_{n-1}(s))^{1/2} \quad ;$$

i.e. $j_n$ is well defined.

In order to show that $j_n$ is order preserving let us consider the situation $(r_1 + s_1)/2 \leq (r_2 + s_2)/2$ . We distinguish the following cases

*Case 1* : $r_1 \leq r_2$ and $s_1 \leq s_2$ ; then the isotonicity of $j_{n-1}$ implies $j_n((r_1 + s_1)/2) \leq j_n((r_2 + s_2)/2)$ .

*Case 2* : $r_1 \geq r_2$ and $s_1 \leq s_2$ ; then we obtain $1 - s_2 + s_1 \leq 1 - r_1 + r_2$ . Obviously $j_{n-1}(s_2) \to j_{n-1}(s_1) \leq j_{n-1}(r_1) \to j_{n-1}(r_2)$ follows from the isotonicity of $j_{n-1}$. Referring again to Proposition 2.11 we obtain

$$(j_{n-1}(r_1))^{1/2} * (j_{n-1}(s_2))^{1/2} * ((j_{n-1}(s_2))^{1/2} \to (j_{n-1}(s_1))^{1/2}) =$$
$$(j_{n-1}(r_1))^{1/2} * (j_{n-1}(s_1))^{1/2} \leq (j_{n-1}(s_2))^{1/2} * (j_{n-1}(r_2))^{1/2} \quad .$$

Further we derive from Proposition 2.22(2) the following relation

$$\begin{aligned}
j_n(1 - ((r + s)/2)) \quad &= \quad j_n(((1 - r) + (1 - s))/2) \\
&= \quad (j_{n-1}(1 - r))^{1/2} * (j_{n-1}(1 - s))^{1/2} \\
&= \quad (j_{n-1}(r) \to 0)^{1/2} * (j_{n-1}(s) \to 0)^{1/2} \\
&= \quad ((j_{n-1}(r))^{1/2} * (j_{n-1}(s))^{1/2}) \to 0 \\
&= \quad (j_n((r + s)/2)) \to 0 \quad ;
\end{aligned}$$

i.e. $j_n$ preserves the "negation". In order to show that $j_n$ preserves also the semigroup operations $\boxtimes$ and $*$ we proceed as follows : If $u \boxtimes v = 0$ , then $u \leq 1 - v$. With regard to the preceding considerations we obtain

$$j_n(u) * j_n(v) \quad \leq \quad j_n(1 - v) * j_n(v) \quad = \quad (j_n(v) \to 0) * j_n(v) \quad = 0 \quad ;$$

i.e. $j_n(u) * j_n(v) = 0$ . Therefore without loss of generality we assume

$$0 \quad < u \boxtimes v \quad = \quad u + v - 1 \quad < \quad 1 \quad \text{where} \quad u, v \in D_n \quad .$$

It is not difficult to show that there exist elements $r_1, r_2, s_1, s_2 \in D_{n-1}$ provided with the following properties

$$\frac{r_1 + r_2}{2} = u, \quad \frac{s_1 + s_2}{2} = v, \quad 1 - s_1 \leq r_1, \quad 1 - s_2 \leq r_2 \quad ,$$

$$\frac{(r_1 \boxtimes s_1) + (r_2 \boxtimes s_2)}{2} \quad = \quad u \boxtimes v \quad .$$

By virtue of the divisibility of M the relation

$$0^{1/2} \quad = \quad (j_{n-1}(s_i))^{1/2} * ((j_{n-1}(s_i) \to 0)^{1/2} \quad \leq \quad j_{n-1}(s_i))^{1/2} * (j_{n-1}(r_i))^{1/2}$$

follows for $i = 1, 2$ ; hence we obtain from Proposition 2.17(xxviii)

$$\begin{aligned}
j_n(u) * j_n(v) \quad &= \quad (j_{n-1}(r_1))^{1/2} * (j_{n-1}(r_2))^{1/2} * (j_{n-1}(s_1))^{1/2} * (j_{n-1}(s_2))^{1/2} \\
&= \quad ((j_{n-1}(r_1)) * (j_{n-1}(s_1))^{1/2} * ((j_{n-1}(r_2)) * (j_{n-1}(s_2))^{1/2} \\
&= \quad (j_{n-1}(r_1 \boxtimes s_1))^{1/2} * (j_{n-1}(r_2 \boxtimes s_2))^{1/2} \quad = \quad j_n(u \boxtimes v) \quad .
\end{aligned}$$

(b) We observe that the restriction of $j_n$ to $D_{n-1}$ coincides with $j_{n-1}$ ; hence the sequence $(j_n)_{n \in \mathbb{N}}$ determines an MV–algebrahomomorphism $j : D \to L$ by

$$j(r) \quad = \quad j_n(r) \qquad \text{whenever} \quad \exists n \in \mathbb{N} \quad \text{s.t.} \quad r \in D_n \quad .$$

Referring to Corollary 2.12 and Proposition 2.22(1) we obtain

$$\alpha < \beta \qquad \Longrightarrow \qquad \alpha \quad < \quad \alpha^{1/2} * \beta^{1/2} \quad < \quad \beta \qquad ;$$

hence $j$ is injective. Further, if M is $\sigma$–complete, we observe that the element $\iota_1 = \bigvee_{n \in \mathbb{N}} j(1 - (1/2^n))$ is idempotent. Hence $\iota_1 = 1$ follows from Proposition 2.22(3) ; i.e. $j$ satisfies Conditon($\Gamma$).
□

**Corollary 6.10** *Let $M = (L, \leq, *)$ be a $\sigma$-complete MV-algebra and $\mathbb{B}_M$ the ($\sigma$-complete) Boolean algebra of all idempotent elements of M. Then the following assertions are equivalent*

(i)    M is a strict MV–algebra.

(ii)   $([0,1], \leq, T_m)$   (cf. section 3.3) is a regular $\sigma$–complete MV–subalgebra of M (i.e. the embedding preserves countable joins and meets).

(iii)  M is isomorphic to $(\mathbb{B}_M([0,1]), \preceq, \Diamond_\wedge)$ .

**Proof.** The implications (i) $\Longrightarrow$ (ii) $\Longrightarrow$ (iii) follow from Examples 5.14 , Theorem 6.8 and Theorem 6.9 . On the other hand , it is easy to see that $(\mathbb{B}_M([0,1]), \preceq, \Diamond_\wedge)$ is a strict MV–algebra (cf. Examples 5.14(b)).
□

The equivalence of the assertions (i) and (iii) in Corollary 6.10 goes back to a result first established by F. Gößwald–Bernsau ([11]). Combining Theorem 2.21 with Corollary 6.10 we obtain a complete solution of the classification problem of complete MV–algebras with square roots :

**Remark 6.11 (Complete MV–algebras with square roots)** If M is a *complete*, strict MV–algebra, then Corollary 6.10 shows that M is isomorphic to a *Boolean valued model* of the MV–algebra $([0,1], \leq, T_m)$. Therefore, in view of Theorem 2.21, every complete MV–algebra with square roots is either a complete Boolean algebra or a Boolean valued model of the real unit interval viewed as a MV–algebra or a product of a complete Boolean algebra and some Boolean valued model of the real unit interval viewed as a MV–algebra ($* = \boxtimes = T_m$).
□

**Corollary 6.12** *Let $M = (L, \leq, *)$ be a complete MV-algebra containing only two idempotent elements – i.e. $\mathbb{B}_M = \{0,1\}$ . Then the following assertions are equivalent*

(i)     M is isomorphic to $([0,1], \leq, T_m)$ (cf. section 3.3).

(ii)    M is atomless – i.e. for every element $\alpha \in L$ with $\alpha \neq 0$ there exists $\beta \in L$ s.t. $\beta \neq 0$ and $\beta < \alpha$.

(iii)   M has square roots and L contains at least three elements.

(iv)    M is a strict MV–algebra.

**Proof.** The implication (i) $\Longrightarrow$ (ii) is obvious. If M is atomless, then L contains at least three elements, and Theorem 5.3 and Theorem 6.4 show that M has square roots; hence (ii) $\Longrightarrow$ (iii) is verified. The implication (iii) $\Longrightarrow$ (iv) follows from Corollary 2.18 and Proposition 2.19 . Since the universal bounds are the only idempotent elements in L, $(\mathbb{B}_M([0,1]), \preceq, \Diamond_\wedge)$ is isomorphic to $([0,1], \leq, T_m)$ (cf. Example 5.14); hence the implication (iv) $\Longrightarrow$ (i) follows from Theorem 6.8 and Theorem 6.9 .
□

**Corollary 6.13** *Every infinite, locally finite, complete MV–algebra is isomorphic to* $([0,1], \leq, T_m)$.

**Proof.** The assertion follows immediately from Corollary 6.7 and Corollary 6.12 .
□

**Corollary 6.14** *Every locally finite MV–algebra* $M = (L, \leq, *)$ *is isomorphic to a MV–subalgebra of* $([0,1], \leq, T_m)$. *In particular the embedding* $\xi : M \longmapsto ([0,1], \leq, T_m)$ *preserves meets and joins provided they exist in* $(L, \leq)$.

**Proof.** Referring to the proof of Corollary 6.7 it is easy to see that the assertion holds for finite lattices $(L, \leq)$. On the other hand, if $L$ is infinite, then the assertion follows from Corollary 5.13 and Corollary 6.13 .
□

By virtue of Corollary 6.14 we are now in the position to sharpen the statement of Proposition 4.9 in the case of MV–algebras as follows

**Corollary 6.15 (Representation of semi–simple MV–algebras ([1]))**
*Let* $\mathcal{M}$ *be the set of all maximal filters in an MV–algebras* $(L, \leq, *)$. *Then the following assertions are equivalent*

(i)     $(L, \leq, *)$ is semi–simple.

(ii)    $(L, \leq, *)$ is isomorphic to an MV–subalgebra of the product $\prod_{\mathcal{M}}([0,1], \leq, T_m)$ .

A further immediate consequence from Corollary 6.12 is the following theorem due to L.P. Belluce ([3]) :

**Theorem 6.16** *Let $M$ be a complete $MV$-algebra, and $\mathbb{B}_M$ be the (complete) Boolean algebra of all idempotent elements in $M$. Then the following assertions are equivalent*

(i)     M is atomless and $\mathbb{B}_M$ is atomic (or in the terminology of L.P. Belluce : M is subb–atomic).

(ii)    M is isomorphic to the product $\prod_{\mathcal{I}}([0,1], \leq, T_m)$ where $\mathcal{I}$ is the set of all atoms in $\mathbb{B}_M$.

**Proof.** The implication (ii) $\Longrightarrow$ (i) is obvious. Further let $\mathcal{I}$ be the set of all atoms of $\mathbb{B}_M$ . Then the MV-algebra $M_\iota$ induced by $\iota \in \mathcal{I}$ (in the sense of Remmark 2.20) contains exactly two idempotent elements; hence $M_\iota$ satisfies the assertion (ii) in Corollary 6.12 . If $\mathbb{B}_M$ is atomic, then $1 = \vee \mathcal{I}$, and M is isomorphic to $\prod_{\iota \in \mathcal{I}} M_\iota$. Thus the non trivial implication (i) $\Longrightarrow$ (ii) follows form Corollary 6.12.
$\square$

An interesting consequence of Lemma 4.14 and Corollary 6.10 is also the fact that L.P. Belluce's conjecture in [3] can be confirmed.

**Theorem 6.17** *Let $M$ be a complete $MV$-algebra. Then the following assertions are equivalent*

(i)     M is injective (in the category of MV–algebras).

(ii)    M is strongly–atomless.

(iii)   M is strict.

(iv)    M is divisible (in the sense of L.P. Belluce [3]).

**Proof.** (i) $\Longrightarrow$ (ii) is Corollary 2 in [3]. (ii) $\Longrightarrow$ (iii) follows from Theorem IV in [3] and Lemma 4.14. If M is a complete, strict MV–algebra, then M is isomorphic to a Boolean valued model of the MV–algebra $([0,1], \leq, T_m)$ (cf. Corollary 6.10 and Remark 6.11); hence M is divisible in the sense of the terminology used by L.P. Belluce in [3]. Finally (iv) $\Longrightarrow$ (i) has already been established in [21].
$\square$

We finish this section with the observation that a version of Tarsk's Lemma is valid in the case of $\sigma$–complete MV–algebras.

**Theorem 6.18 (Tarski's Lemma)** *Let $M = (L, \leq, *)$ be a $\sigma$–complete $MV$-algebra and $\alpha_0 \in L$ with $\alpha_0 \neq 1$ . Further let $\{A_n | n \in \mathbb{N}\}$ be a countable family of countable subsets $A_n$ of $L$. Then there exists an $MV$-algebrahomomorphism $h : M \longmapsto ([0,1], \leq, T_m)$ provided wioth the following properties*

(1)     $h(\alpha_0) \neq 1$   ,

(2)     $\displaystyle\inf_{\beta \in A_n} h(\beta)$   $=$   $h(\wedge A_n)$   for all   $n \in \mathbb{N}$

The proof is based on a sequence of lemmas.

**Lemma 6.19** *Let $M = (L, \leq, *)$ be a $\sigma$-complete MV-algebra. Then for every $\alpha \in L$ with $\alpha \neq 1$ there exists an idempotent element $\iota_0 \in L$ equipped with the properties*

(i)     $\iota_0 \neq 0$

(ii)    $\exists n_0 \in \mathbb{N}$   s.t.   $\iota_0$   $\leq$   $\alpha^{n_0+1} \to 0$   .

**Proof.** Since M is a $\sigma$-complete MV-algebra, M is semi-simple (cf. Lemma 6.1); hence for every $\alpha \in L$ with $\alpha \neq 1$ there exists a natural number $n_0$ such that   $(\alpha \to 0) \to \alpha^{n_0} \neq 1$   (cf. Proposition 4.12). Now we invoke the algebraic strong de Morgan law (cf. Corollary 2.16) and obtain

$$((\alpha \to 0) \to \alpha^{n_0})^m \vee (\alpha^{n_0+1} \to 0)^m   =   1   ;$$

hence

$$[\bigwedge_{m \in \mathbb{N}} ((\alpha \to 0) \to \alpha^{n_0})^m] \vee [\bigwedge_{m \in \mathbb{N}} (\alpha^{n_0+1} \to 0)^m]   =   1 \tag{6.9}$$

follows from the subsequent infinite distributivity law (cf. Theorem 5.2(c))

$$\bigwedge_{m \in \mathbb{N}} (\beta_m \vee \gamma)   =   (\bigwedge_{m \in \mathbb{N}} \beta_m) \vee \gamma   .$$

We put   $\iota_0$   $=$   $\displaystyle\bigwedge_{m \in \mathbb{N}} (\alpha^{n_0+1} \to 0)^m$   ; because of Lemma 6.1 the element $\iota_0$ is idempotent. In particular $\iota_0$ satisfies Property (ii) . Since $(\alpha \to 0) \to \alpha^{n_0} \neq 1$ , we infer from formula (6.9) that $\iota_0 \neq 0$ ; i.e. $\iota_0$ fulfills the desired properties.
$\square$

**Lemma 6.20** *Let $M = (L, \leq, *)$ be a $\sigma$-complete MV-algebra, $A = \{\alpha_k | k \in \mathbb{N}\}$ be a countable subset of $L$, and let $\beta$ be an element of $L$. Then the following implication holds*

$$\beta \wedge [\bigwedge_{m \in \mathbb{N}} ((\alpha_k)^n \to 0)^m]   =   0 \quad \forall (k, n) \in \mathbb{N} \times \mathbb{N}   \implies   \beta \leq \bigwedge_{k \in \mathbb{N}} \alpha_k   .$$

**Proof.** By virtue of the algebraic strong de Morgan law we are in the position to derive from the antecedent of the implication the following equation

$$\beta   =   \beta \wedge [\bigwedge_{m \in \mathbb{N}} ((\alpha_k \to 0) \to \alpha_k^{n-1})^m]$$

for all $k \in \mathbb{N}$ and $n \in \mathbb{N}$ ; hence the inequality $\beta \leq (\alpha_k \to 0) \to \alpha_k^{n-1}$ holds for all $n \in \mathbb{N}$ . Since $e_{\alpha_k} := \bigwedge_{n \in \mathbb{N}} \alpha_k^{n-1}$ is idempotent (cf. Lemma 6.1), we obtain

$$\beta \quad \leq \quad (\alpha_k \to 0) \to e_{\alpha_k} \quad = \quad \alpha_k \vee e_{\alpha_k} \quad = \quad \alpha_k \quad ; .$$

i.e. the inequality $\beta \leq \bigwedge_{k \in \mathbb{N}} \alpha_k$ follows.

□

**Lemma 6.21** ($\sigma$–stable maximal filters) *Let $M = (L, \leq, *)$ be a $\sigma$-complete MV-algebra and $\{A_n | n \in \mathbb{N}\}$ be a countable family of countable subsets $A_n$ of $L$. Further let $\alpha_0$ be an element of $L$ with $\alpha_0 \neq 1$ . Then there exists a maximal filter $\mathcal{U}$ in $M$ provided with the following properties*

(i)    $\alpha_0 \notin \mathcal{U}$   .

(ii)    The subsequent implication holds for all $n \in \mathbb{N}$ :
$(\forall \beta \in A_n : \beta \in \mathcal{U}) \quad \Longrightarrow \quad \wedge A_n \in \mathcal{U}$

**Proof.** According to Lemma 6.19 there exists a natural number $n_0$ and an idempotent element $\iota_0 \in L$ with $\iota_0 \neq 0$ and $\iota_0 \leq (\alpha_0)^{n_0+1} \to 0$ . Referring to Lemma 6.20 we define recursively a sequence $(\iota_n)_{n \in \mathbb{N}}$ of idempotent elements $\iota_n \in L$ satisfying the property

$$\left. \begin{array}{ll} \iota_n \neq 0 & , \qquad \iota_n \leq \iota_{n-1} \quad , \\[2mm] \iota_{n-1} \not\leq \wedge A_n \implies \exists \beta \in A_n, \exists k \in \mathbb{N} : \iota_n \leq \beta^k \to 0 \end{array} \right\} \quad \text{(P)}$$

for all $n \in \mathbb{N}$ . In fact; let us assume that we have already constructed $\iota_{n-1}$ ; we distinguish the following cases : *Case* 1 : $\iota_{n-1} \leq \wedge A_n$; then we put $\iota_n = \iota_{n-1}$. *Case* 2 : $\iota_{n-1} \not\leq \wedge A_n$ ; then we infer from Lemma 6.20 that there exists a natural number $l \in \mathbb{N}$ and $\beta \in A_n$ such that

$$0 \quad \neq \quad \iota_{n-1} \wedge (\bigwedge_{m \in \mathbb{N}} (\beta^l \to 0)^m) \quad \overset{\text{def}}{=} \quad \iota_n \quad .$$

In view of Lemma 6.1 the element $\iota_n$ is idempotent and satisfies by definition the desired properties. Let $\mathcal{F}$ be the filter generated by $\{\iota_n | n \in \mathbb{N}_0\}$ – i.e.

$$\mathcal{F} \quad = \quad \{\lambda \in L \mid \exists n \in \mathbb{N}_0 : \iota_n \leq \lambda\} \quad .$$

By virtue of Zorns's Lemma there exists a maximal filter $\mathcal{U}$ containg $\mathcal{F}$ . In particular the property (P) implies that $\mathcal{U}$ fulfills (i) and (ii).
□

**Proof of Theorem 6.18.** In order to simplify the notations we introduce the convention

$$m \cdot \alpha \quad = \quad ((\alpha \to 0)^m) \to 0 \quad , \quad m \in \mathbb{N} \quad .$$

Now we consider an element $\alpha_0 \in L$ with $\alpha_0 \neq 1$ and a countable family $\{A_n | n \in \mathbb{N}\}$ of countable subsets $A_n$ of L. Then we define a further, countable familiy $\{B_{mn} | (n, m) \in \mathbb{N} \times \mathbb{N}\}$ of countable subsets $B_{mn}$ of L as follows

$$B_{mn} \quad = \quad \{ \quad 2^m \cdot [\beta * ((\wedge A_n) \to 0)] \quad | \quad \beta \in A_n \quad \} \quad .$$

By virtue of Lemma 6.21 there exists a maximal filter $\mathcal{U}$ satisfying the conditions

(i)    $\alpha_0 \notin \mathcal{U}$    ,

(ii)   $\forall \quad \gamma \in B_{mn} : \gamma \in \mathcal{U} \quad \Longrightarrow \quad \wedge B_{mn} \in \mathcal{U}$  .

Referring to Lemma 2.4(1) and Theorem 5.2(b) we observe that for every sequence $(\delta_n)_{n \in \mathbb{N}}$ in L the relation $\bigwedge\limits_{n \in \mathbb{N}} 2^m \cdot \delta_n = 2^m \cdot (\bigwedge\limits_{n \in \mathbb{N}} \delta_n)$ holds; hence we obtain for all $(m, n) \in \mathbb{N} \times \mathbb{N}$ :

$$\wedge B_{mn} \quad = \quad 0 \quad .$$

Because of $0 \notin \mathcal{U}$ we conclude from Condition (ii) that for every pair $(m, n)$ there exists $\beta_{mn} \in A_n$ such that

(iii)   $2^m \cdot [\beta_{mn} * ((\wedge A_n) \to 0)] \quad \notin \quad \mathcal{U}$   .

Further let us denote by $\phi$ the quotient-homomorphism induced by $\mathcal{U}$ from $(L, \leq, *)$ onto $(L/ \sim_{\mathcal{U}}, \leq_{\mathcal{U}}, *_{\mathcal{U}})$. Since $(L/ \sim_{\mathcal{U}}, \leq_{\mathcal{U}}, *_{\mathcal{U}})$ is a locally finite MV–algebra (cf. Theorem 4.4), there exists an embedding $j_{\mathcal{U}} : (L/ \sim_{\mathcal{U}}, \leq_{\mathcal{U}}, *_{\mathcal{U}}) \longmapsto ([0,1], \leq, T_m)$    preserving joins and meets provided they exist in $(L/ \sim_{\mathcal{U}}, \leq_{\mathcal{U}}, *_{\mathcal{U}})$ (cf. Corollary 6.14). Thus $h = j_{\mathcal{U}} \circ \phi$ is an MV–algebrahomomorphism from M to $([0,1], \leq, T_m)$. Because of (i) h fulfills Property (1) . Let us now assume that h does not satisfy Property (2) – i.e. there exists a natural number n such that the strict inequality

$$h(\wedge A_n) \quad < \quad \inf_{\beta \in A_n} h(\beta)$$

is valid. Therefore we can choose a natural number $m_0$ with

$$1/2^{m_0} \quad \leq \quad (\inf_{\beta \in A_n} h(\beta)) - h(\wedge A_n) \quad = \quad \inf_{\beta \in A_n} T_m(h(\beta), 1 - h(\wedge A_n)) \quad ;$$

hence we obtain for all $\beta \in A_n$

$$2^{m_0} \cdot [\phi(\beta) *_{\mathcal{U}} (\phi(\wedge A_n) \to 0)] \quad = \quad 1 \quad ;$$

i.e. $2^{m_0} \cdot [\beta * ((\wedge A_n) \to 0)] \in \mathcal{U}$ for all $\beta \in A_n$. The last conclusion is an obvious contradiction to (iii). Thus h fulfills (2); and therewith the assertion is verified. $\square$

If we add to the predicate calculus w.r.t. Lukasiewicz logic (cf. 3.1.4) the infinitary inference rule

(R)    From $(\alpha \to 0) \to \alpha^n$ for all $n \in \mathbb{N}$, infer $\alpha$                    ,

then an important consequence from Theorem 6.18 is the fact that this modified non–classical predicate calculus is sound and complete – i.e. a well formed formula $\alpha$ is provable iff $\alpha$ attains for all [0,1]–valued interpretation and for all valuations the value 1 (cf. [13]). In this way H. Rasiowa and R. Sikorski's proof of Gödel's completeness theorem (cf. [24]) can be generalized to the case of Lukasiewicz logic ([18]).

# References

[1] L.P. Belluce . *Semi–simple algebras of infinite valued logic and bold fuzzy set theory*, Canad. J. Math. **38** (1986), 1356 – 1379.

[2] L.P. Belluce . *Semi–simple and complete MV–algebras*, Algebra Universalis **29** (1992), 1 – 9 .

[3] L.P. Belluce . *α–complete MV–algebras*, in "Non–Classical Logics and Their Applications to Fuzzy Subsets : A Handbook of the Mathematical Foundations of Fuzzy Set Theory", (Eds. U. Höhle and E.P Klement) (Kluwer, Boston 1994).

[4] G. Birkhoff . *Lattice Theory* , Amer. Math. Soc. Colloquium Publications, third edition (Amer. Math. Soc., RI, 1973).

[5] C.C. Chang . *Algebraic analysis of many valued logics*, Trans. Amer. Math. Soc. **88** (1958), 467 – 490 .

[6] C.C. Chang . *A new proof of the completeness of the Lukasiewicz axioms*, Trans. Amer. Math. Soc. **93** (1959), 74 – 80.

[7] J.C. Fodor . *Contrapositive symmetry of fuzzy implications*, Technical Report 1993/1, Eötvös Loránd University, Budapest 1993.

[8] O. Frink . *New algebras of logic*, Amer. Math. Monthly **45** (1938), 210 – 219 .

[9] L. Fuchs . *Über die Ideale arithmetischer Ringe*, Commentarii Math. Helv. **23** (1949),334 – 341.

[10] J.Y. Girard . *Linear logic*, Theor. Comp. Sci. **50** (1987), 1–102.

[11] F. Gößwald–Bernsau. *Kommutative, residuierte, verbandsgeordnete Monoide und Halbeinfachheit*, Diplomarbeit, Wuppertal 1994.

[12] S. Gottwald . *Mehrwertige Logik* (Akademie–Verlag, Berlin 1989).

[13] L.S. Hay . *Axiomatization of infinite-valued predicate calculus*, Journal of Symbolic Logic **28** (1963), 77 – 86.

[14] U. Höhle . *Editorial of the Special Issue : Mathematical Aspects of Fuzzy Set Theory* , Fuzzy Sets and Systems **40** (1991), 253 – 256 .

[15] U. Höhle and L.N. Stout . *Foundations of fuzzy sets*, Fuzzy Sets and Systems **40** (1991), 257-296.

[16] U. Höhle . *M-valued sets and sheaves over integral cl-monoids*, in "Applications of catgeory Theory to Fuzzy Subsets" , p. 33 – 72 , (Eds. S.E. Rodabaugh et al.) (Kluwer , Bosten 1992).

[17] U. Höhle . *Presheaves over GL-monmoids*, in "Non–Classical Logics and Their Applications to Fuzzy Subsets : A Handbook on the Mathematical Foundations of Fuzzy Set Theory," (Eds. U. Höhle and E.P. Klement) (Kluwer, Boston 1995).

[18] U. Höhle . *Monoidal Logic* in "Foundations in Fuzzy Systems" (Eds. R. Kruse et al.) (Vieweg 1995).

[19] H. MacNeille . *Partially ordered sets*, Trans. Amer. Math. Soc. **42** (1937), 416-460.

[20] P.T. Johnstone . *Stone Spaces* (Cambridge University Press, Cambridge 1982).

[21] F. Lacava . *Sulle L–algebre iniettive*, Bolletino UMI (7) 3-A (1989), 319 – 324.

[22] D. Mundici . *Interpretation of AF C\*-algebras in Lukasiewicz sentential logic*, Functional Analysis **65** (1986), 15–63.

[23] A.B. Paalman de Miranda . *Topological Semigroups* (Amsterdam : Math. Centrum 1964).

[24] H. Rasiowa and R. Sirkoski . *A proof of the completeness theorem of Gödel*, Fundamenta Math. **37** (1950), 193-200.

[25] H. Rasiowa and R. Sikorski . *The Mathematics of Metamathematics* , third edition (Polish Scientific Publishers , Warszawa 1970).

[26] K.I. Rosenthal . *Quantales and Their Applications* , Pitman Research Notes in Mathematics **234** (Longman, Burnt Mill, Harlow 1990).

[27] B. Schweizer and A. Sklar . *Probabilistic Metric Spaces* (North–Holland , Amsterdam 1983).

[28] R. Sikorski . *Booelan Algebras* (Springer–Verlag, Berlin 1964).

# V

# A proof of the completeness of the infinite–valued calculus of Łukasiewicz with one variable

D. Mundici and M. Pasquetto

## Prologue

In the literature one can find three quite different proofs of the completeness of the infinite–valued sentential calculus of Łukasiewicz [8]:

  (i) the syntactical proof of Rose and Rosser [7], using McNaughton's theorem,

 (ii) the algebraic proof of Chang [1, 2], using quantifier elimination in the first-order theory of divisible totally ordered abelian groups, and

(iii) the recent proof of Cignoli [3], using the representation of free lattice–ordered abelian groups.

In this paper we give a self–contained geometric proof of the completeness theorem for the one variable fragment.

## 1  Farey partitions and Schauder hats

For each $n = 0, 1, \ldots$ we define the $n$-th *Farey partition* $\Phi_n$ of the unit interval $[0, 1]$ as follows:

$$
\begin{aligned}
\Phi_0 &= \{0, 1\}, \\
\Phi_1 &= \{0, \tfrac{1}{2}, 1\}, \\
\Phi_2 &= \{0, \tfrac{1}{3}, \tfrac{1}{2}, \tfrac{2}{3}, 1\}, \\
\Phi_3 &= \{0, \tfrac{1}{4}, \tfrac{1}{3}, \tfrac{2}{5}, \tfrac{1}{2}, \tfrac{3}{5}, \tfrac{2}{3}, \tfrac{3}{4}, 1\}, \\
&\vdots \quad \vdots \quad \vdots
\end{aligned}
$$

and so on, where each $\Phi_{n+1}$ is obtained from $\Phi_n$ by inserting between any two consecutive elements $a/b$ and $c/d$ of $\Phi_n$ their *mediant* $(a+c)/(b+d)$, with $0 = 0/1$ and $1 = 1/1$. While being a mere variant of the familiar Farey sequence [4, § 3], the sequence of the $\Phi_n$'s will be more convenient to us for reasons of notational simplicity.

The following elementary properties of the Farey sequence, first proved by Cauchy, still hold, with the same proof, for our present $\Phi_n$'s:

(i) all fractions in $\Phi_{n+1}$ arising via the mediant operation from consecutive fractions in $\Phi_n$ are automatically in irreducible form,

(ii) for any two consecutive fractions $a/b < c/d$ in $\Phi_n$ we have the *unimodularity* law $cb - ad = 1$; moreover,

$$\frac{a}{b} < \frac{a+c}{b+d} < \frac{c}{d},$$

(iii) each irreducible fraction in $[0, 1]$ eventually occurs in some $\Phi_n$, see [4].

Note also that the number of elements of $\Phi_n$ is $2^n + 1$.

Let us display the elements of the Farey partition $\Phi_n$ in ascending order as follows:

$$0 < \alpha < \cdots < \gamma < \delta = \frac{c}{d} < \varepsilon < \cdots < \omega < 1.$$

A *(Schauder) hat* of $\Phi_n$ is a function of either form:

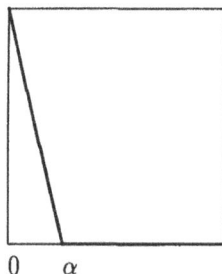 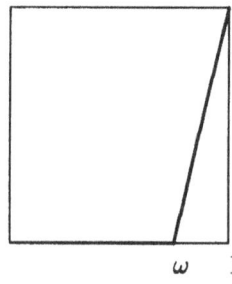 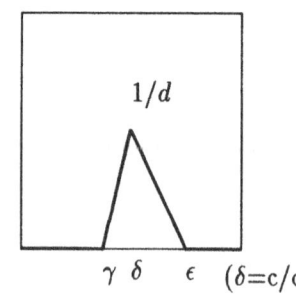

Thus each Schauder hat is a continuous piecewise linear function $h : [0, 1] \rightarrow [0, 1]$. The graph of $h$ consists of the four segments joining the points $(0, 0)$, $(\gamma, 0)$, $(\delta, 1/d)$, $(\varepsilon, 0)$ and $(1, 0)$. The graphs of the two extremal hats only consist of two segments.

For each $n = 0, 1, 2, \ldots$, we shall denote by $\Lambda_n = (h_1, \ldots, h_u)$ the naturally ordered sequence of all Schauder hats of $\Phi_n$, where $u = 2^n + 1$. The following is a picture of $\Lambda_0$, $\Lambda_1$ and $\Lambda_2$:

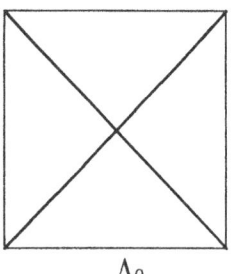 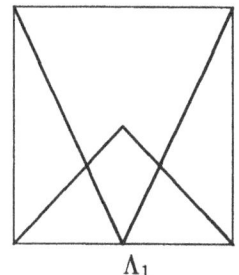 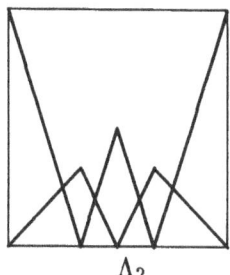

$\Lambda_0$ $\qquad\qquad\qquad\qquad$ $\Lambda_1$ $\qquad\qquad\qquad\qquad$ $\Lambda_2$

For each hat $h_i$ of $\Lambda_n$ the *multiplicity* $\mu_i$ of $h_i$ is defined as the inverse of the maximum value of $h_i$. By definition, $\mu_i$ coincides with the denominator $d_i$ of the rational point $c_i/d_i \in [0, 1]$ at which $h_i$ attains its maximum value. For any $n$, the Schauder hat $h_i$ precedes $h_j$ in $\Lambda_n$ iff $i < j$ iff $c_i/d_i < c_j/d_j$. From the sequence $(h_1, \ldots, h_u)$ one can unambiguously recover the multiplicity of each hat, as well as the number $n = \log_2(u - 1)$. It is easy to see that $\sum_i \mu_i h_i = 1$.

For all $x, y \in [0, 1]$ we write

$$\neg x = 1 - x, \qquad x \oplus y = \min(1, x + y), \qquad x \bullet y = \max(0, x + y - 1). \qquad (1)$$

For any two $[0, 1]$–valued functions $f$ and $g$ having the same domain, the functions $\neg f$, $f \oplus g$ and $f \bullet g$ are naturally defined by pointwise application of the corresponding operations. We sometimes write $f \vee g$ instead of $\max(f, g)$, and $f \wedge g$ instead of $\min(f, g)$. It is easy to see that

$$
\begin{aligned}
f \vee g &= \neg(\neg f \oplus g) \oplus g, \\
f \wedge g &= \neg(\neg f \bullet g) \bullet g, \\
f \bullet g &= \neg(\neg f \oplus \neg g).
\end{aligned}
\qquad (2)
$$

Throughout this paper, by a *node* of a piecewise linear function $f : [0, 1] \to [0, 1]$ we mean a nondifferentiability point in the domain of $f$. The extremes 0 and 1 are always included among the nodes of $f$. If $f$ arises from the identity function via a finite number of applications of the operations $\neg$, $\oplus$, $\bullet$, $\wedge$ and $\vee$, then the nodes of $f$ are rational numbers.

The following recurrence relation will be one of the key ingredients in the proof of the completeness theorem:

**Proposition 1.1** *Let $h_1, \ldots, h_u$ be the hats of $\Lambda_n$, in their natural order, and with their respective multiplicities $\mu_1, \ldots, \mu_u$. Let $k_1, \ldots, k_{2u-1}$ be the hats of*

$\Lambda_{n+1}$, with their respective multiplicities $\xi_1, \ldots, \xi_{2u-1}$. Then we have:

$$
\begin{aligned}
k_1 &= h_1 - (h_1 \wedge h_2) \\
&= h_1 \bullet \neg(h_1 \wedge h_2), \qquad && \text{with multiplicity } \xi_1 = \mu_1 = 1;
\end{aligned}
$$

$$
\begin{aligned}
k_{2u-1} &= h_u - (h_{u-1} \wedge h_u) \\
&= h_u \bullet \neg(h_{u-1} \wedge h_u), \qquad && \text{with multiplicity } \xi_{2u-1} = \mu_u = 1;
\end{aligned}
$$

$$
\begin{aligned}
k_{2i} &= h_i \wedge h_{i+1}, \qquad && \text{with multiplicity } \xi_{2i} = (\mu_i + \mu_{i+1}), \\
& && \text{for each } i = 1, 2, \ldots, u-1;
\end{aligned}
$$

$$
\begin{aligned}
k_{2i-1} &= h_i - (h_i \wedge (h_{i-1} \vee h_{i+1})) \\
&= h_i - (h_i \wedge (h_{i-1} + h_{i+1})) \\
&= h_i \bullet \neg(h_{i-1} \oplus h_{i+1}), \qquad && \text{with multiplicity } \xi_{2i-1} = \mu_i, \\
& && \text{for each } i = 2, 3, \ldots, u-1.
\end{aligned}
$$

*Proof.* Let $0 = c_1/d_1 < c_2/d_2 < \cdots < c_u/d_u = 1$ be the ascending sequence of elements of $\Phi_n$ . Then $d_i = \mu_i$. An elementary computation using the above mentioned unimodularity law (ii) shows that $h_i \wedge h_{i+1}$ attains its maximum value $1/(d_i + d_{i+1})$ at the mediant point $\rho = (c_i + c_{i+1})/(d_i + d_{i+1})$. Since the nodes of $h_i \wedge h_{i+1}$ are the five rational numbers $0, c_i/d_i$ , $\rho$ , $c_{i+1}/d_{i+1}$, 1 and the function coincides with $k_{2i}$ at each node, we conclude that $k_{2i}$ coincides with $h_i \wedge h_{i+1}$ over all of $[0,1]$. Similarly, the function $h_i - (h_i \wedge (h_{i-1} + h_{i+1}))$ is constantly equal to zero over both intervals $[0, (c_{i-1}+c_i)/(d_{i-1}+d_i)]$ and $[(c_{i+1}+ c_i)/(d_{i+1} + d_i), 1]$, and is linear over the intervals $[(c_{i-1} + c_i)/(d_{i-1} + d_i), c_i/d_i]$ and $[c_i/d_i, (c_{i+1}+c_i)/(d_{i+1}+d_i)]$. Since $(h_{i-1} \wedge h_{i+1}) = 0$, the function coincides with $k_{2i-1}$ at the point $c_i/d_i$ , where it attains its maximum value $1/d_i$. It is now easy to see that the identity $k_{2i-1} = h_i - (h_i \wedge (h_{i-1} + h_{i+1}))$ holds over all of $[0,1]$. One similarly proves the remaining identities. $\square$

Given the sequence $\Lambda_n = (h_1, \ldots, h_u)$ of Schauder hats of order $n$, with their respective multiplicities $\mu_1, \ldots, \mu_u$, by a *subsystem* $S$ of $\Lambda_n$ (in symbols, $S \subseteq \Lambda_n$) we mean a sequence of integers $\lambda_1, \ldots, \lambda_u$, satisfying the inequalities $0 \leq \lambda_i \leq \mu_i$ for each $i = 1, \ldots, u$. The *associated function* $f_S : [0,1] \to [0,1]$ is defined by $f_S = \sum_i \lambda_i h_i$. Thus, $f_S$ is a continuous piecewise linear function. By definition, the subsystem $S^\neg \subseteq \Lambda_n$ is given by $(\mu_1 - \lambda_1, \ldots, \mu_u - \lambda_u)$. We have the trivial identity $f_{S^\neg} = \neg f_S = 1 - f_S$.

Two subsystems $S \subseteq \Lambda_n$ and $R \subseteq \Lambda_m$ are said to be *equivalent* iff $f_R = f_S$. The following proposition is an immediate consequence of Proposition 1.1 and of the linear independence of the hats of $\Lambda_n$.

**Proposition 1.2** *If $S \subseteq \Lambda_n$ then for each integer $p \geq n$ there is precisely one subsystem $R \subseteq \Lambda_p$ equivalent to $S$.*

**Proposition 1.3** *Given subsystems $R \subseteq \Lambda_m$ and $S \subseteq \Lambda_n$ , for some $q \geq \max(m,n)$ and $T \subseteq \Lambda_q$ we have $f_T = f_R \oplus f_S$.*

*Proof.* Let $q$ be the smallest integer $\geq \max(m,n)$ such that all the nodes of the function $f_R \oplus f_S = \min(1, f_R + f_S)$ occur in $\Phi_q$. The existence of $q$ is

ensured by property (iii) above, together with Proposition 1.1 ensuring that all the nodes of $f_R \oplus f_S$ are rational numbers. Let $g = f_R + f_S$. For any two consecutive fractions $x$ and $y$ of $\Phi_q$ one cannot have $g(x) > 1$ and $g(y) < 1$. Let $R' = (\lambda_1, \ldots, \lambda_w)$ and $S' = (\nu_1, \ldots, \nu_w)$ be the equivalent subsystems of $S$ and of $R$ in $\Lambda_q$, as given by Proposition 1.2. Let $k_1$, ..., $k_w$ be the hats in $\Lambda_q$, with their multiplicities $\xi_1$, ..., $\xi_w$. Thus, for any two consecutive indexes $i, j \in \{1, \ldots, w\}$, if $\lambda_j + \nu_j < \xi_j$ then necessarily $\lambda_i + \nu_i \leq \xi_i$. Let $T \subseteq \Lambda_q$ be the subsystem $(\eta_1, \ldots, \eta_w)$, where $\eta_i = \min(\xi_i, \lambda_i + \nu_i)$. It follows that the function $f_T$ coincides with $f_R \oplus f_S$ at each point of $\Phi_q$, and both functions are linear on each interval between consecutive points of $\Phi_q$. Thus, $f_T = f_R \oplus f_S$ as required.
□

**Example 1.4** Let $m = n = 0$ and $R = (0,1) = S$. Then $\Lambda_0 = (1 - x, x)$, and $f_R = x = f_S$. Let $q = 1$. Then the nodes of the function $f_R \oplus f_S = \min(2x, 1)$ are 0, 1/2 and 1, which all occur in the Farey partition $\Phi_1$. Further, $\Lambda_1 = (\max(1 - 2x, 0), \min(x, 1 - x), \max(0, 2x - 1))$, with their respective multiplicities 1, 2, and 1. The equivalent subsystems $R' = S' \subseteq \Lambda_1$ are both given by the triplet of integers $(0, 1, 1)$. Taking $T = (0, 2, 1)$ we have $f_R \oplus f_S = f_T$, as required. For no subsystem $U$ of $\Lambda_0$ we would be able to write $f_R \oplus f_S = f_U$.

**Proposition 1.5** *For every function* $f : [0,1] \to [0,1]$ *the following are equivalent:*

(i) *$f$ arises from the identity function* $x : [0,1] \to [0,1]$ *via a finite number of applications of the operations* $\neg$ *and* $\oplus$;

(ii) *$f = f_T$ for some subsystem $T$.*

*Proof.* (ii) → (i): This follows from Proposition 1.1, since by (2) the operations • and ∧ are both definable in terms of $\neg$ and $\oplus$. (i) → (ii): As shown in Example 1.4, $x = f_T$ for some $T$. Since $f_{S\neg} = \neg f_S$, from Propositions 1.2 and 1.3 we have that the set of functions of the form $f_T$ is closed under pointwise applications of $\neg$ and $\oplus$. □

# 2  MV terms and equations

We assume the reader to be familiar with the rules of equational logic and with the rudiments of Chang's MV algebras [1]. Starting from the variables $x_1, \ldots,$ $x_n, \ldots$ and the operation symbols $0, 1, \neg, \oplus, \bullet$, one defines (MV-)*terms* in the usual way. For every such term $t = t(x_1, \ldots, x_n)$ we define its *associated* function $f_t : [0,1]^n \to [0,1]$ by induction on the *complexity* of (i.e., the number of occurrences of operation symbols in) $t$ as follows:

If $t = x_i$, then $f_t$ is the $i$-th canonical projection

$$f_t(x_1, \ldots, x_n) = x_i;$$

if $t = \neg r$, then

$$f_t = 1 - f_r;$$

if $t = (r \oplus s)$, then

$$f_t = \min(1, f_r + f_s) = f_r \oplus f_s;$$

if $t = (r \bullet s)$, then

$$f_t = \max(0, f_r + f_s - 1) = f_r \bullet f_s.$$

Arguing by induction on the complexity of $t$, one sees that $f_t$ is a continuous piecewise linear function, each piece being a linear polynomial with integral coefficients. An algebra $A = (A, 0, 1, \neg, \oplus, \bullet)$ is called an *MV algebra* iff it satisfies the following equations:

| | |
|---|---|
| MV1 | $(x \oplus y) \oplus z = x \oplus (y \oplus z)$ |
| MV2 | $x \oplus y = y \oplus x$ |
| MV3 | $x \oplus 0 = x$ |
| MV4 | $x \oplus 1 = 1$ |
| MV5 | $\neg 0 = 1$ |
| MV6 | $\neg 1 = 0$ |
| MV7 | $x \bullet y = \neg(\neg x \oplus \neg y)$ |
| MV8 | $\neg(\neg x \oplus y) \oplus y = \neg(\neg y \oplus x) \oplus x.$ |

Replacing $y$ by 0 in the last equation, we get $\neg\neg x = x$. Replacing $y$ by 1 we get $1 = \neg x \oplus x$. Then it is not hard to see [5, § 2.6] that these equations are equivalent to Chang's original equations.

When $A$ satisfies an equation $r = s$ we shall write $A \vDash r = s$. Of particular importance is the MV algebra $[0, 1]$ with natural operations, as defined in (1) above. We have $[0, 1] \vDash r = s$ iff $f_r = f_s$. When an equation $r = s$ follows from equations MV1–8 using the rules of equational logic, we say that the equation is *valid*. Trivially, each valid equation is satisfied in every MV algebra.

Following [1, 1.13] we write $r \leq s$ as an abbreviation of $\neg r \oplus s = 1$. Further, for every term $p$ and integer $n \geq 0$, we let $np$ be an abbreviation of $p \oplus \ldots \oplus p$ ($n$ times), and $0p = 0$.

As an elementary consequence of the above equations we have that $\leq$ is a partial order. As proved by Chang, upon defining $r \vee s = \neg(\neg r \oplus s) \oplus s$ and $r \wedge s = \neg(\neg r \bullet s) \bullet s$, it follows that the operations $\vee$ and $\wedge$ satisfy the axioms of distributive lattices with smallest element 0 and largest element 1. Further, the $\oplus$ operation distributes over $\wedge$, and $\bullet$ distributes over $\vee$. Also, both $\oplus$ and $\bullet$ are order preserving [1, 1.8], while $\neg$ is order reversing [1, 1.4(vi)]. We shall need the following elementary results:

**Proposition 2.1** *Let $r$ and $s$ be arbitrary terms. Then we have:*

(i) *Let $s_1, \ldots, s_n$ be terms. If for every $i = 1, \ldots, n$, the equation $s_i \wedge s = 0$ is valid, then so is the equation $(s_1 \oplus \cdots \oplus s_n) \wedge s = 0$.*

(ii) *$r = r \oplus s$ is valid iff so is $0 = \neg r \wedge s$;*

(iii) *$(\neg r \bullet s) \wedge (\neg s \bullet r) = 0$ is a valid equation;*

(iv) *if $r \wedge s = 0$ is valid then so is $r \vee s = r \oplus s$.*

*Proof.* (i) By induction on $n$. The basis is trivial. For the induction step, we have

$$
\begin{aligned}
0 &= ((s_1 \oplus \ldots \oplus s_k) \wedge s) \oplus (s_{k+1} \wedge s) \\
&= (((s_1 \oplus \ldots \oplus s_k) \wedge s) \oplus s_{k+1}) \wedge (((s_1 \oplus \ldots \oplus s_k) \wedge s) \oplus s) \\
&= (s_1 \oplus \ldots \oplus s_k \oplus s_{k+1}) \wedge (s \oplus s_{k+1}) \wedge (s_1 \oplus \ldots \oplus s_k \oplus s) \wedge (s \oplus s) \\
&= (s_1 \oplus \ldots \oplus s_{k+1}) \wedge (s \oplus (s_{k+1} \wedge (s_1 \oplus \ldots \oplus s_k))) \wedge (s \oplus s) \\
&= (s_1 \oplus \ldots \oplus s_{k+1}) \wedge (s \oplus (s_{k+1} \wedge (s_1 \oplus \ldots \oplus s_k) \wedge s)) \\
&= (s_1 \oplus \ldots \oplus s_{k+1}) \wedge s.
\end{aligned}
$$

(ii) See [1, 1.15].
(iii) See [1, 3.3].
(iv) Immediate from (ii). $\square$

Given a sequence of terms $(p_1, \ldots, p_u, q)$ we write $(p_1, \ldots, p_u) \sim q$ iff the following equations are valid :

$$
\begin{aligned}
\neg q \oplus p_2 \oplus \cdots \oplus p_u &= \neg p_1 \\
p_1 \oplus \neg q \oplus \cdots \oplus p_u &= \neg p_2 \\
&\vdots \\
p_1 \oplus p_2 \oplus \cdots \oplus \neg q &= \neg p_u \\
p_1 \oplus p_2 \oplus \cdots \oplus p_u &= q.
\end{aligned}
$$

As explained in [5, § 3], an MV algebra $A$ satisfies the system of equations $(p_1, \ldots, p_u) \sim q$ iff in the lattice-ordered abelian group $G$ corresponding to $A$ we have $p_1 + \cdots + p_u = q$. However, to make the present paper entirely independent of [5], we shall establish the basic properties of the relation $\sim$ directly from axioms MV1–8.

**Proposition 2.2** *Suppose $(p_1, \ldots, p_u) \sim q$. Let $\alpha_1, \ldots, \alpha_u \in \{0, 1\}$ be arbitrary integers. Then we have the following valid equations:*

(i) *$\alpha_1 p_1 \oplus \cdots \oplus \alpha_{k-1} p_{k-1} \oplus \alpha_{k+1} p_{k+1} \oplus \cdots \oplus \alpha_u p_u \oplus \neg q \leq \neg p_k$;*

(ii) *$(\alpha_1 p_1, \ldots, \alpha_u p_u) \sim \alpha_1 p_1 \oplus \cdots \oplus \alpha_u p_u$;*

*(iii)* $\alpha_1 p_1 \oplus \cdots \oplus \alpha_u p_u = \neg((1 - \alpha_1)p_1 \oplus \cdots \oplus (1 - \alpha_u)p_u \oplus \neg q)$.

*Proof.* (i) From the order preserving properties of the $\oplus$ operation it follows that

$$\alpha_1 p_1 \oplus \cdots \oplus \alpha_{k-1}p_{k-1} \oplus \alpha_{k+1}p_{k+1} \oplus \cdots \oplus \alpha_u p_u \oplus \neg q$$
$$\leq \ p_1 \oplus \cdots \oplus p_{k-1} \oplus p_{k+1} \oplus \cdots \oplus p_u \oplus \neg q$$
$$= \ \neg p_k.$$

(ii) Since the operation $\oplus$ is order preserving and $\neg$ is order reversing, we have

$$\alpha_2 p_2 \oplus \cdots \oplus \alpha_u p_u \leq \neg q \oplus p_2 \oplus \cdots \oplus p_u = \neg p_1 \leq \neg(\alpha_1 p_1).$$

By axiom MV8 we get

$$\neg(\alpha_1 p_1 \oplus \alpha_2 p_2 \oplus \cdots \oplus \alpha_u p_u) \oplus \alpha_2 p_2 \oplus \cdots \oplus \alpha_u p_u$$
$$= \ \neg(\alpha_1 p_1) \vee (\alpha_2 p_2 \oplus \cdots \oplus \alpha_u p_u)$$
$$= \ \neg(\alpha_1 p_1).$$

The rest of the proof is similar.

(iii) By induction on $\sigma = \alpha_1 + \alpha_2 + \cdots + \alpha_u$. The basis is trivial. For the induction step, without loss of generality we may assume $\alpha_1 = 1$. By (i) together with the induction hypothesis we have

$$\alpha_1 p_1 \oplus \cdots \oplus \alpha_u p_u = \ p_1 \oplus (0p_1 \oplus \alpha_2 p_2 \oplus \cdots \oplus \alpha_u p_u)$$
$$= \ p_1 \oplus \neg(p_1 \oplus (1 - \alpha_2)p_2 \oplus \cdots \oplus (1 - \alpha_u)p_u \oplus \neg q)$$
$$= \ p_1 \vee \neg((1 - \alpha_2)p_2 \oplus \cdots \oplus (1 - \alpha_u)p_u \oplus \neg q)$$
$$= \ \neg((1 - \alpha_2)p_2 \oplus \cdots \oplus (1 - \alpha_u)p_u \oplus \neg q)$$
$$= \ \neg((1 - \alpha_1)p_1 \oplus \cdots \oplus (1 - \alpha_u)p_u \oplus \neg q). \ \square$$

**Proposition 2.3** *Suppose* $(p_1, \ldots, p_u) \sim q$ *and, for suitable* $\alpha_1, \ldots, \alpha_u \in \{0, 1\}$, $(\alpha_1 p_1, \ldots, \alpha_u p_u) \sim p_0$. *Then* $(p_0, (1 - \alpha_1)p_1, \ldots, (1 - \alpha_u)p_u) \sim q$.

*Proof.* By Proposition 2.2(iii) we have

$$\neg q \oplus (1 - \alpha_1)p_1 \oplus \cdots \oplus (1 - \alpha_u)p_u = \neg(\alpha_1 p_1 \oplus \cdots \oplus \alpha_u p_u) = \neg p_0.$$

Further,

$$p_0 \oplus \neg q \oplus (1 - \alpha_2)p_2 \oplus \cdots \oplus (1 - \alpha_u)p_u$$
$$= \ p_0 \oplus \neg(p_1 \oplus \alpha_2 p_2 \oplus \cdots \oplus \alpha_u p_u)$$
$$= \ p_0 \oplus \neg((1 - \alpha_1)p_1 \oplus \alpha_1 p_1 \oplus \alpha_2 p_2 \oplus \cdots \oplus \alpha_u p_u)$$
$$= \ p_0 \oplus \neg((1 - \alpha_1)p_1 \oplus p_0)$$
$$= \ \neg((1 - \alpha_1)p_1) \vee p_0$$
$$= \ \neg((1 - \alpha_1)p_1).$$

The last equality follows from

$$
\begin{aligned}
p_0 &= \alpha_1 p_1 \oplus \cdots \oplus \alpha_u p_u \\
&= \neg((1-\alpha_1)p_1 \oplus \cdots \oplus (1-\alpha_u)p_u \oplus \neg q) \\
&\leq \neg((1-\alpha_1)p_1).
\end{aligned}
$$

We also have

$$
\begin{aligned}
p_0 \oplus (1-\alpha_1)p_1 &\oplus \cdots \oplus (1-\alpha_u)p_u \\
&= \alpha_1 p_1 \oplus \cdots \oplus \alpha_u p_u \oplus (1-\alpha_1)p_1 \oplus \cdots \oplus (1-\alpha_u)p_u \\
&= p_1 \oplus \cdots \oplus p_u \\
&= q.
\end{aligned}
$$

The rest of the proof is similar. □

**Proposition 2.4** *Assume* $(p_1,\ldots,p_u) \sim q$ *and* $(t_1,\ldots,t_w) \sim p_i$, *for some index* $i$ *with* $1 \leq i \leq u$. *Then* $(p_1,\ldots,p_{i-1},t_1,\ldots,t_w,p_{i+1},\ldots,p_u) \sim q$.

*Proof.* We may safely assume $i = 1$. We have

$$
\begin{aligned}
\neg q \oplus t_2 \oplus \cdots \oplus t_w \oplus p_2 \oplus \cdots \oplus p_u &= t_2 \oplus \cdots \oplus t_w \oplus \neg q \oplus p_2 \oplus \cdots \oplus p_u \\
&= t_2 \oplus \cdots \oplus t_w \oplus \neg p_1 \\
&= \neg t_1.
\end{aligned}
$$

Analogously,

$$
t_1 \oplus \cdots \oplus t_w \oplus \neg q \oplus p_3 \oplus \cdots \oplus p_u = p_1 \oplus \neg q \oplus p_3 \oplus \cdots \oplus p_u = \neg p_2
$$

and

$$
t_1 \oplus \cdots \oplus t_w \oplus p_2 \oplus \cdots \oplus p_u = p_1 \oplus p_2 \oplus \cdots \oplus p_u = q.
$$

The rest of the proof is similar. □

**Proposition 2.5** *For each* $n = 2,3,4,\ldots$, *the following are equivalent:*

(i) $\underbrace{(p,p,\ldots,p)}_{n \text{ times}} \sim np,$

(ii) *the equation* $(n-1)p \leq \neg p$ *is valid.*

*Proof.* First note that

$$
\neg(np) \oplus (n-1)p = \neg((n-1)p \oplus p) \oplus (n-1)p = \neg p \vee (n-1)p.
$$

Therefore, from (i) we have $\neg(np) \oplus (n-1)p = \neg p$, and (ii) holds. Conversely, from (ii) we get $\neg(np) \oplus (n-1)p = \neg p \vee (n-1)p = \neg p$, whence (i) holds. □

# 3   Schauder terms

As a syntactic counterpart of the Schauder hats of Section 1, we now introduce
Schauder terms and their multiplicities via the following inductive definition:

BASIS: The sequence $H_0$ of Schauder terms of order 0 is given by $H_0 = (\neg x, x)$, and the multiplicities of $x$ and $\neg x$ are both equal to 1.

INDUCTION STEP: Let $u = 2^n + 1$, and $(t_1, \ldots, t_u)$ be the sequence $H_n$ of
Schauder terms of order $n$, with their respective multiplicities $\mu_1, \ldots, \mu_u$. Then
the sequence $H_{n+1} = (p_1, \ldots, p_{2u-1})$ of Schauder terms of order $n+1$, with
their respective multiplicities $\xi_1, \ldots, \xi_{2u-1}$ is defined as follows:

$$p_1 = t_1 \bullet \neg(t_1 \wedge t_2) \qquad \text{with multiplicity 1,}$$

$$p_{2u-1} = t_u \bullet \neg(t_{u-1} \wedge t_u) \qquad \text{with multiplicity 1,}$$

$$p_{2i} = t_i \wedge t_{i+1} \qquad \begin{array}{l}\text{with multiplicity } \xi_{2i} = (\mu_i + \mu_{i+1}), \\ \text{for each } i = 1, 2, \ldots, u-1,\end{array}$$

$$p_{2i-1} = t_i \bullet \neg(t_{i-1} \oplus t_{i+1}) \qquad \begin{array}{l}\text{with multiplicity } \xi_{2i-1} = \mu_i, \\ \text{for each } i = 2, 3, \ldots, u-1.\end{array}$$

**Proposition 3.1** *For each $n = 0, 1, 2, \ldots$, the map $t \to f_t$ sending MV terms
into their associated $[0,1]$-valued functions induces a one-one correspondence
from $H_n$ onto $\Lambda_n$. Moreover, under this correspondence, multiplicities are preserved.*

*Proof.* By definition of the MV algebra $[0,1]$ and recalling Proposition 1.1. $\square$

The fact that two nonconsecutive hats in $\Lambda_n$ are disjoint has the following
equational generalization:

**Proposition 3.2** *Let $t_a$ and $t_b$ be Schauder terms in $H_n$. If $|a - b| > 1$, the
equation $t_a \wedge t_b = 0$ is valid.*

*Proof.* By induction on $n$. For $n = 0$ there is nothing to prove. For $n = 1$ the
validity of

$$\left(\neg x \bullet \neg(\neg x \wedge x)\right) \wedge \left(x \bullet \neg(\neg x \wedge x)\right) = \left(\neg x \bullet x \vee \neg x \bullet \neg x\right) \wedge \left(x \bullet x \vee x \bullet \neg x\right)$$
$$= \left(\neg x \bullet \neg x\right) \wedge \left(x \bullet x\right)$$
$$= 0$$

is an easy consequence of Proposition 2.1(iii). For the induction step, assume
the theorem holds for $H_n = (t_1, \ldots, t_u)$. Let $p_s$ and $p_r$ be terms of $H_{n+1}$, and
assume $|s - r| > 1$. There are four possible cases:

CASE 1: $s = 2i$ and $r = 2j$ for suitable $i, j = 1, \ldots, u-1$.
Then $|i - j| \geq 1$. If $|i - j| = 1$, say $i = j+1$, it follows by induction that $p_s \wedge p_r = (t_{j+1} \wedge t_{j+2}) \wedge (t_j \wedge t_{j+1}) = 0$. On the other hand, if $|i - j| > 1$, again by induction
hypothesis we have $t_i \wedge t_j = 0$, whence $p_s \wedge p_r = (t_i \wedge t_{i+1}) \wedge (t_j \wedge t_{j+1}) = 0$.

CASE 2: $s = 2i$ and $r = 2j - 1$ for some $i = 1, \ldots, u - 1$ and $j = 1, \ldots, u$. If $s > r$ then $i > j$ and, by induction hypothesis, $t_{i+1} \wedge t_j = 0$, whence $p_s \wedge p_r \leq (t_i \wedge t_{i+1}) \wedge t_j = 0$. If, on the other hand, $s < r$ then $j - i > 1$, whence the desired conclusion immediately follows from the induction hypothesis.

CASE 3: $s = 2i - 1$ and $r = 2j$ for some $i = 1, \ldots, u$ and $j = 1, \ldots, u - 1$. This is similar to Case 2.

CASE 4: $s = 2i - 1$ and $r = 2j - 1$ for $i, j = 1, \ldots, u$. We may safely assume $i - j \geq 1$. If $i - j > 1$ then by induction hypothesis $p_s \wedge p_r \leq t_i \wedge t_j = 0$. If $i = 1 + j$ then

$$
\begin{aligned}
p_s \wedge p_r &= (t_i \bullet \neg(t_{i-1} \oplus t_{i+1})) \wedge (t_j \bullet \neg(t_{j-1} \oplus t_{j+1})) \\
&= (t_{j+1} \bullet \neg(t_j \oplus t_{j+2})) \wedge (t_j \bullet \neg(t_{j-1} \oplus t_{j+1})) \\
&\leq (t_{j+1} \bullet \neg t_j) \wedge (t_j \bullet \neg t_{j+1}) \\
&= 0.
\end{aligned}
$$

This last equation follows from Proposition 2.1(iii). The modifications for either case $j = 1$ or $i = u$ are trivial. $\square$

The following proposition gives an equational generalization of the fact that each hat of $\Lambda_n$ is the sum of at most three consecutive hats of $\Lambda_{n+1}$'s:

**Proposition 3.3** *Let* $H_n = (t_1, \ldots, t_u)$ *and* $H_{n+1} = (p_1, \ldots, p_w)$ *, where* $w = 2u - 1$. *Then we have*

*(i)* $(p_1, p_2) \sim t_1$;

*(ii)* $(p_{2u-2}, p_{2u-1}) \sim t_u$;

*(iii) For each* $i = 2, 3, \ldots, u - 1$, $(p_{2i-2}, p_{2i-1}, p_{2i}) \sim t_i$.

*Proof.* (iii) From Propositions 3.2 and 2.1(iv) we get $(t_{i-1} \oplus t_{i+1}) = (t_{i-1} \vee t_{i+1})$. We then have the valid equations $(t_i \wedge t_{i+1}) \leq \neg(t_i \wedge t_{i-1})$, $(t_i \wedge t_{i-1}) \leq \neg(t_i \wedge t_{i+1})$ and $(t_i \wedge t_{i+1}) \oplus (t_i \wedge t_{i-1}) = t_i \wedge (t_{i-1} \oplus t_{i+1})$. Using these equations we get

$$
\begin{aligned}
\neg t_i \oplus p_{2i-1} \oplus p_{2i} &= \neg t_i \oplus (t_i \bullet \neg(t_{i-1} \oplus t_{i+1})) \oplus (t_i \wedge t_{i+1}) \\
&= (\neg t_i \vee \neg(t_{i-1} \oplus t_{i+1})) \oplus (t_i \wedge t_{i+1}) \\
&= \neg(t_i \wedge (t_{i-1} \oplus t_{i+1})) \oplus (t_i \wedge t_{i+1}) \\
&= \neg((t_i \wedge t_{i-1}) \oplus (t_i \wedge t_{i+1})) \oplus (t_i \wedge t_{i+1}) \\
&= \neg(t_i \wedge t_{i-1}) \vee (t_i \wedge t_{i+1}) \\
&= \neg(t_i \wedge t_{i-1}) \\
&= \neg p_{2i-2}.
\end{aligned}
$$

Further,

$$
p_{2i-2} \oplus \neg t_i \oplus p_{2i} = (t_i \wedge t_{i-1}) \oplus \neg t_i \oplus (t_i \wedge t_{i+1})
$$

$$\begin{aligned}
&= (t_i \wedge (t_{i-1} \oplus t_{i+1})) \oplus \neg t_i \\
&= (t_i \oplus \neg t_i) \wedge ((t_{i-1} \oplus t_{i+1}) \oplus \neg t_i) \\
&= \neg t_i \oplus (t_{i-1} \oplus t_{i+1}) \\
&= \neg p_{2i-1}.
\end{aligned}$$

Similarly,

$$\begin{aligned}
p_{2i-2} \oplus p_{2i-1} \oplus \neg t_i &= (t_i \wedge t_{i-1}) \oplus (t_i \bullet \neg(t_{i-1} \oplus t_{i+1})) \oplus \neg t_i \\
&= (t_i \wedge t_{i-1}) \oplus (\neg t_i \vee \neg(t_{i-1} \oplus t_{i+1})) \\
&= (t_i \wedge t_{i-1}) \oplus \neg(t_i \wedge (t_{i-1} \oplus t_{i+1})) \\
&= (t_i \wedge t_{i-1}) \oplus \neg((t_i \wedge t_{i-1}) \oplus (t_i \wedge t_{i+1})) \\
&= \neg(t_i \wedge t_{i+1}) \vee (t_i \wedge t_{i-1}) \\
&= \neg(t_i \wedge t_{i+1}) = \neg p_{2i}.
\end{aligned}$$

Using now the valid equation

$$(x \wedge y) \oplus (x \bullet \neg y) = (\neg(x \bullet \neg y) \bullet x) \oplus (x \bullet \neg y) = x \vee (x \bullet \neg y) = x,$$

we have

$$\begin{aligned}
p_{2i-2} \oplus p_{2i-1} \oplus p_{2i} &= (t_i \wedge t_{i-1}) \oplus (t_i \bullet \neg(t_{i-1} \vee t_{i+1})) \oplus (t_i \wedge t_{i+1}) \\
&= (t_i \wedge (t_{i-1} \vee t_{i+1})) \oplus (t_i \bullet \neg(t_{i-1} \vee t_{i+1})) \\
&= t_i.
\end{aligned}$$

The proofs of (i) and (ii) are similar. $\square$

**Proposition 3.4** *Let $t_i$ and $t_{i+1}$ be two Schauder terms in $\mathbf{H}_n$, with multiplicities $\mu_i$ and $\mu_{i+1}$ respectively. Then the following equations are valid:*

*(i)* $(\mu_i - 1)t_i \oplus \mu_{i+1}t_{i+1} \leq \neg t_i,$

*(ii)* $\mu_i t_i \oplus (\mu_{i+1} - 1)t_{i+1} \leq \neg t_{i+1}.$

*Proof.* (i) By induction on $n$. The basis is trivial. By induction hypothesis, the proposition holds for $\mathbf{H}_0$, $\mathbf{H}_1$, ..., $\mathbf{H}_n$. Let $p_s$ and $p_{s+1}$ be two Schauder terms in $\mathbf{H}_{n+1}$ with multiplicity $\xi_s$ and $\xi_{s+1}$ respectively. By Proposition 3.3 we have $p_{2i-1} \oplus p_{2i} \leq t_i$ and $p_{2i} \oplus p_{2i+1} \leq t_{i+1}$. If, on the one hand, $s = 2i - 1$ then

$$\begin{aligned}
(\xi_s - 1)p_s \oplus \xi_{s+1}p_{s+1} &= (\mu_i - 1)p_{2i-1} \oplus (\mu_i + \mu_{i+1})p_{2i} \\
&= (\mu_i - 1)(p_{2i-1} \oplus p_{2i}) \oplus p_{2i} \oplus \mu_{i+1}p_{2i} \\
&\leq (\mu_i - 1)t_i \oplus (t_i \wedge t_{i+1}) \oplus \mu_{i+1}(t_i \wedge t_{i+1}) \\
&\leq (\mu_i - 1)t_i \oplus \mu_{i+1}t_{i+1} \oplus t_{i+1} \\
&\leq \neg t_i \oplus t_{i+1}
\end{aligned}$$

$$\begin{aligned}
&\leq\ t_{i-1} \oplus \neg t_i \oplus t_{i+1} \\
&=\ \neg(t_i \bullet \neg(t_{i-1} \oplus t_{i+1})) \\
&=\ \neg p_{2i-1} \\
&=\ \neg p_s.
\end{aligned}$$

On the other hand, if $s = 2i$, then

$$\begin{aligned}
(\xi_s - 1)p_s \oplus \xi_{s+1}p_{s+1} &=\ (\mu_i + \mu_{i+1} - 1)p_{2i} \oplus \mu_{i+1}p_{2i+1} \\
&=\ (\mu_i - 1)p_{2i} \oplus \mu_{i+1}(p_{2i} \oplus p_{2i+1}) \\
&\leq\ (\mu_i - 1)(t_i \wedge t_{i+1}) \oplus \mu_{i+1}t_{i+1} \\
&\leq\ (\mu_i - 1)t_i \oplus \mu_{i+1}t_{i+1} \\
&\leq\ \neg t_i \\
&\leq\ \neg t_i \vee \neg t_{i+1} \\
&=\ \neg p_{2i} \\
&=\ \neg p_s.
\end{aligned}$$

The proof of (ii) is analogous. □

The fact that the sum of $\mu_i$ copies of a hat $h_i$ never exceeds the value 1, has the following equational counterpart:

**Proposition 3.5** *Let $t_i$ be a term in $\mathbf{H}_n = (t_1, \ldots, t_u)$ with multiplicity $\mu_i$. Then*

$$\underbrace{(t_i, t_i, \ldots, t_i)}_{\mu_i \text{ times}} \sim \mu_i t_i.$$

*Proof.* By induction on $n$. The basis is trivial. For the induction step, assume the theorem to hold for $\mathbf{H}_0$, $\mathbf{H}_1$, ..., $\mathbf{H}_n$. Let $p_s$ be a Schauder term in $\mathbf{H}_{n+1}$ with multiplicity $\xi_s$. The cases $s = 1$ and $s = 2u - 1$ are trivial. For the proof of the other cases, in the light of Proposition 2.5, it is sufficient to establish the validity of the equation $(\xi_s - 1)p_s \leq \neg p_s$. There are two cases:

CASE 1: $s = 2i$ for some $i = 1, 2, \ldots, u - 1$.

Then $p_s = t_i \wedge t_{i+1}$ and $\xi_s = \mu_i + \mu_{i+1}$. By Proposition 3.4(i) we have

$$\begin{aligned}
(\xi_s - 1)p_s &=\ (\mu_i + \mu_{i+1} - 1)(t_i \wedge t_{i+1}) \\
&=\ (\mu_i - 1)(t_i \wedge t_{i+1}) \oplus \mu_{i+1}(t_i \wedge t_{i+1}) \\
&\leq\ (\mu_i - 1)t_i \oplus \mu_{i+1}t_{i+1} \\
&\leq\ \neg t_i \\
&\leq\ \neg(t_i \wedge t_{i+1}) \\
&=\ \neg p_s.
\end{aligned}$$

CASE 2: $s = 2i - 1$ for some $i = 2, 3, \ldots, u - 1$.

Then $p_s = t_i \bullet \neg(t_{i-1} \oplus t_{i+1})$ and $\xi_s = \mu_i$. From the induction hypothesis, again together with Proposition 2.5, it follows that

$$
\begin{aligned}
(\xi_s - 1)p_s &= (\mu_i - 1)(t_i \bullet \neg(t_{i-1} \oplus t_{i+1})) \\
&\leq (\mu_i - 1)t_i \\
&\leq \neg t_i \\
&\leq \neg t_i \oplus t_{i-1} \oplus t_{i+1} \\
&= \neg(t_i \bullet \neg(t_{i-1} \oplus t_{i+1})) \\
&= \neg p_s. \quad \square
\end{aligned}
$$

The fact that $\sum_i \mu_i h_i = 1$ has the following equational generalization:

**Proposition 3.6** *Let* $H_n = (t_1, \dots, t_u)$ *with multiplicities* $\mu_1, \dots, \mu_u$ *respectively. Then* $(\mu_1 t_1, \dots, \mu_u t_u) \sim 1$.

*Proof.* By induction on $n$. The case $n = 0$ is trivial. Assume the theorem to hold for $H_0$, $H_1$, ..., $H_n$. Let $t_1$, ..., $t_u$ be the Schauder terms in $H_n$, with their multiplicities $\mu_1$, ..., $\mu_u$. Let $p_1$, ..., $p_{2u-1}$ be the Schauder terms in $H_{n+1}$, with multiplicities $\xi_1$, ..., $\xi_{2u-1}$. From the induction hypothesis together with Propositions 3.5 and 2.4, we obtain $(t_1, \dots, t_1, \dots, t_u, \dots, t_u) \sim 1$, where each $t_i$ appears $\mu_i$ times. ¿From Proposition 3.3 we get

$$
\begin{aligned}
(p_1, p_2, \dots, p_1, p_2, p_2, p_3, p_4, \dots, p_2, p_3, p_4, \dots, \\
p_{2u-4}, p_{2u-3}, p_{2u-2}, \dots, p_{2u-4}, p_{2u-3}, p_{2u-2}, \\
p_{2u-2}, p_{2u-1}, \dots, p_{2u-2}, p_{2u-1}) \quad \sim \quad 1.
\end{aligned}
$$

The pairs $(p_1, p_2)$ and $(p_{2u-2}, p_{2u-1})$ occur $\mu_1 = 1$, and $\mu_u = 1$ times respectively, while every triplet $(p_{2i-2}, p_{2i-1}, p_{2i})$ occurs precisely $\mu_i$ times. It follows that $p_{2i}$ occurs $(\mu_i + \mu_{i+1})$ times and $p_{2i-1}$ occurs $\mu_i$ times. So, each term $p_s$ occurs $\xi_s$ times, where $\xi_s$ is its multiplicity. From Proposition 3.5, together with Proposition 2.3, we obtain $(\xi_1 p_1, \dots, \xi_{2u-1} p_{2u-1}) \sim 1$ as required. $\square$

# 4   Normal forms

Given a term $t$ with one variable, we say that $t$ is *n-decomposable*   ($n = 0, 1, 2, \dots$) iff, letting $t_1, \dots, t_u$ be the Schauder terms of $H_n$ with their respective multiplicities $\mu_1, \dots, \mu_u$, there exist integers $0 \leq \lambda_i \leq \mu_i$ such that the equation

$$
t = \lambda_1 t_1 \oplus \dots \oplus \lambda_u t_u \tag{3}
$$

is valid. Following [6], we regard (3) as a normal form for $t$. The fact, proved in Section 1, that $f_{s\neg} = \neg f_s$ has the following equational generalization:

**Proposition 4.1** *If t is n-decomposable, then so is ¬t. Specifically, if equation (3) is valid, then so is* $\neg t = (\mu_1 - \lambda_1)t_1 \oplus \ldots \oplus (\mu_u - \lambda_u)t_u$.

*Proof.* By Propositions 3.6 and 2.4 we have

$$(t_1, \ldots, t_1, t_2, \ldots, t_2, \ldots, t_u, \ldots, t_u)$$
$$= (t_{11}, \ldots, t_{1\mu_1}, t_{21}, \ldots, t_{2\mu_2}, \ldots, t_{u1}, \ldots, t_{u\mu_u}) \sim 1,$$

where each $t_i$ occurs $\mu_i$ times, and $t_{i\theta}$ denotes the $\theta$-th occurrence of $t_i$, for $i = 1, \ldots, u$ and $\theta = 1, \ldots, \mu_i$. Let $\alpha_{i\theta} \in \{0, 1\}$ be defined by $\alpha_{i\theta} = 1$ iff $\theta \leq \lambda_i$, and $\alpha_{i\theta} = 0$ iff $\theta > \lambda_i$. The desired conclusion now follows by Propositions 2.2(iii) and 2.3. □

To describe the normal form of $t' \oplus t''$ we first establish the following equational generalization of Proposition 1.2:

**Proposition 4.2** *If t is n-decomposable, then t is* $(n + 1)$*-decomposable.*

*Proof.* Adopting the above notation, assume we have the valid equation (3). Let $p_1, \ldots, p_{2u-1}$ be the Schauder terms of $H_{n+1}$, with their respective multiplicities $\xi_1, \ldots, \xi_{2u-1}$. Then by Proposition 3.3 we have

$$t = \lambda_1 p_1 \oplus (\lambda_1 + \lambda_2)p_2 \oplus \lambda_2 p_3 \oplus (\lambda_2 + \lambda_3)p_4 \oplus \ldots \oplus \lambda_i p_{2i-1}$$
$$\oplus (\lambda_i + \lambda_{i+1})p_{2i} \oplus \ldots \oplus \lambda_u p_{2u-1}.$$

This yields the desired $(n + 1)$-decomposition of $t$, since by assumption, $\lambda_i \leq \mu_i = \xi_{2i-1}$ and $(\lambda_i + \lambda_{i+1}) \leq (\mu_i + \mu_{i+1}) = \xi_{2i}$. □

**Proposition 4.3** *Let* $t_1, \ldots, t_u$ *be the Schauder terms of* $H_n$ *with their multiplicities* $\mu_1, \ldots, \mu_u$. *Then we have the following valid equations:*

(i) $(\mu_1 t_1 \oplus \nu_2 t_2) = (\mu_1 t_1 \oplus \nu_2 t_2) \oplus t_1,$ *for each integer* $\nu_2 \geq \mu_2$;

(ii) $(\mu_{i-1} t_{i-1} \oplus \mu_i t_i \oplus \nu_{i+1} t_{i+1}) = (\mu_{i-1} t_{i-1} \oplus \mu_i t_i \oplus \nu_{i+1} t_{i+1}) \oplus t_i,$
*for each* $i = 2, \ldots, u - 1$, *and each integer* $\nu_{i+1} \geq \mu_{i+1}$;

(iii) $(\mu_{u-1} t_{u-1} \oplus \mu_u t_u) = (\mu_{u-1} t_{u-1} \oplus \mu_u t_u) \oplus t_u.$

*Proof.* (ii) By Propositions 3.2 and 2.1(i), we have $\mu_j t_j \wedge t_i = 0$ whenever $|i - j| > 1$. By Proposition 4.1 we have

$$0 = (\mu_1 t_1 \oplus \ldots \oplus \mu_{i-2} t_{i-2} \oplus \mu_{i+2} t_{i+2} \oplus \ldots \oplus \mu_u t_u) \wedge t_i$$
$$= \neg(\mu_{i-1} t_{i-1} \oplus \mu_i t_i \oplus \mu_{i+1} t_{i+1}) \wedge t_i$$
$$\geq \neg(\mu_{i-1} t_{i-1} \oplus \mu_i t_i \oplus \nu_{i+1} t_{i+1}) \wedge t_i.$$

The desired conclusion now follows from Proposition 2.1(ii). The proofs of (i) and (iii) are analogous. □

**Proposition 4.4** *Let $s'$ and $s''$ be terms with one variable. If $s'$ is $n'$- decomposable and $s''$ is $n''$-decomposable, then for some $n \geq \max(n', n'')$, the term $s' \oplus s''$ is $n$-decomposable.*

*Proof.* By repeatedly applying Proposition 4.2 we can safely assume $n' = n'' = n$. Recalling property (iii) in Section 1, we can also assume $n$ to be so large that all nodes of $f_{s'} \oplus f_{s''}$ occur in the Farey partition $\Phi_n$. Let $t_1$, ..., $t_u$ be the Schauder terms in $H_n$ with their multiplicities $\mu_1, \ldots, \mu_u$. Then there are integers $0 \leq \lambda_i \leq \mu_i$ and $0 \leq \nu_i \leq \mu_i$ such that $s' = \lambda_1 t_1 \oplus \ldots \oplus \lambda_u t_u$, and $s'' = \nu_1 t_1 \oplus \ldots \oplus \nu_u t_u$. Recalling now Proposition 3.1 and the proof of Proposition 1.3, our assumption about $n$ ensures that for any two consecutive indexes $i, j = 1, \ldots, u$ we cannot have $\lambda_i + \nu_i > \mu_i$ and $\lambda_j + \nu_j < \mu_j$. So, let $i$ be the smallest index such that $\lambda_i + \nu_i > \mu_i$ (if no such $i$ exists we are done). It follows that $\lambda_{i-1} + \nu_{i-1} = \mu_{i-1}$ and $\lambda_{i+1} + \nu_{i+1} \geq \mu_{i+1}$, with trivial modifications in case $i = 1$ or $i = u$. By repeatedly applying Proposition 4.3 we can safely replace $(\lambda_i + \nu_i) t_i$ by $\mu_i t_i$ in the valid equation $s' \oplus s'' = (\lambda_1 + \nu_1) t_1 \oplus \cdots \oplus (\lambda_u + \nu_u) t_u$. Proceeding in this way, we finally obtain the required $n$-decomposition $s' \oplus s'' = \tau_1 t_1 \oplus \cdots \oplus \tau_u t_u$, where for each $j = 1, \ldots, u$, $\quad \tau_j = \min(\mu_j, \lambda_j + \nu_j)$. $\square$

Proposition 1.5 has the following equational generalization, whose proof directly follows from Propositions 4.1 and 4.4, recalling from MV7 that the $\bullet$ operation is definable in terms of $\neg$ and $\oplus$:

**Corollary 4.5** *Every term $t$ with one variable is $n$-decomposable for some $n$.*

# Epilogue: The Completeness Theorem

**Theorem 4.6** *Let $t$ be a term with one variable. Suppose that $[0, 1] \models t = 1$, i.e., the associated function $f_t$ is constantly equal to 1. Then the equation $t = 1$ is valid.*

*Proof.* By Corollary 4.5, for suitably large $n$, letting $t_1$, ..., $t_u$ be the sequence of Schauder terms of $H_n$ with their multiplicities $\mu_1, \ldots, \mu_u$, there exist integers $0 \leq \gamma_i \leq \mu_i$ such that the following equation is valid:

$$\gamma_1 t_1 \oplus \ldots \oplus \gamma_u t_u = t. \tag{4}$$

Since this equation holds in every MV algebra, we have in particular

$$[0, 1] \models \gamma_1 t_1 \oplus \ldots \oplus \gamma_u t_u = t.$$

By Proposition 3.1, the functions $f_{t_i}$ are the Schauder hats of $\Lambda_n$. For each fraction $c_i/d_i$ in the Farey partition $\Phi_n$, the only hat of $\Lambda_n$ which is nonzero at $c_i/d_i$ is $f_{t_i}$. Our hypothesis on $f_t$ is to the effect that $1 = \gamma_i f_{t_i}(c_i/d_i) = \gamma_i/d_i$, whence $\gamma_i = d_i = \mu_i$. Therefore, from Proposition 3.6 we have the valid equation

$\gamma_1 t_1 \oplus \ldots \oplus \gamma_u t_u = 1$, which, together with (4) yields the valid equation $t = 1$.
□

**Corollary 4.7** *Let $t$ be a term with one variable. Then the following conditions are equivalent for the equation $t = 1$:*

*(i) The equation is valid;*

*(ii) The equation is satisfied by the MV algebra $[0, 1]$;*

*(iii) The equation is satisfied by every MV algebra.*

*Proof.* (ii) implies (i) by the completeness theorem. The implications (i) → (iii) and (iii) → (ii) are trivial.

**Remark.** Corollary 4.5 yields an effective procedure to decompose each term $t$ with one variable into a sum (3) of Schauder hats. Upon looking at the multiplicities of each summand $t_i$ of $t$ in this decomposition, one can effectively decide whether the equation $t = 1$ follows from the equations MV1–8.

**Problem.** Generalize the above proof to the case of $n$ variables, $n = 2, 3, \ldots$

# References

[1] C.C. CHANG, Algebraic analysis of many valued logics, *Trans. Amer. Math. Soc.* **88** (1958) 467–490.

[2] C.C. CHANG, A new proof of the completeness of the Lukasiewicz axioms, *Trans. Amer. Math. Soc.* **93** (1959) 74–80.

[3] R. CIGNOLI, Free lattice–ordered abelian groups and varieties of MV algebras, *Proc. Latin American Symp. Logic, Bahia Blanca 1992*, Notas de Logica Matematica, Univ. Nacional del Sur, Bahia Blanca, Argentina, 1994.

[4] G.H. HARDY, E.M. WRIGHT, *An Introduction to the theory of numbers*, Fifth Edition, Oxford University Press, 1979.

[5] D. MUNDICI, Interpretation of AF $C^*$-algebras in Lukasiewicz sentential logic, *J. Functional Analysis* **65** (1986) 15–63.

[6] D. MUNDICI, Normal forms in infinite–valued logic: the case of one variable, *Lecture Notes in Computer Science* **626** (1992) 272–277.

[7] A. ROSE, J.B. ROSSER, Fragments of many–valued statement calculi, *Trans. Amer. Math. Soc.* **87** (1958) 1–53.

[8] A. TARSKI, J. LUKASIEWICZ, Investigations into the Sentential Calculus, In: *Logic, Semantics, Metamathematics*, Oxford University Press, 1956, pp. 38–59. Reprinted by Hackett Publishing Company, 1981.

# Part   B

# Non–Classical Models And Topos–Like Categories

# VI

# Presheaves over GL–monoids

## U. Höhle

## 1 Introduction

The use of lattice–valued maps in logic and model theory goes back to A. Mostowski ([19]). The essence of his idea can be described as follows : Any realization of a non-classical, formalized theory is based on a suitable, complete lattice $(L, \leq)$, the structure of which corresponds with the underlying system of logical axioms. In this context predicate symbols are interpreted as L-valued maps. It is remarkable to see that neither A. Mostowski nor one of his successors (e.g. H. Rasiowa and R. Sirkorski [20]) attachs to lattice-valued maps any special meaning. Only L.A. Zadeh ([25]) motivated by problems from system theory performs this step and considers in the case of the real unit interval [0,1]-valued maps as *generalized* characteristic functions characterizing a new type of subsets – so-called *fuzzy sets*. Subsequently, J.A. Goguen ([8], [9]) surrounds Zadeh's idea with a lattice-theoretical frame. In the case of integral, commutative cl-monoids ([4]) $(L, \leq, *)$ he also views L-valued maps as generalized characteristic functions and coined for this type of functions the name L-*fuzzy sets*. Since ordinary characteristic functions on a given set X describe subsets of X and vice versa, J.A. Goguen's (respectively L.A. Zadeh's) understanding of lattice-valued maps provokes the following categorical question

- Which (new) type of subobjects are descibed by generalized characteristic functions?

Returning to Zadeh's paper from 1965 it is interesting to note that he does not make any attempt to give a mathematically explicit definition of a fuzzy set ([13])

The answer of the preceding question depends on the structure of the underlying lattice. In the case of complete Heyting algebras H it is well known that strict and extensional H-valued maps and subsheaves over H come to the same thing. In particular this relationsship is expressed by the subobject classifier diagram which relates characteristic morphims (which are precisely the internal version of strict and extensional H-valued maps) to subsheaves and vice-versa

127

([11]). In the general case of integral, commutative cl-monoids this question is still open. In order to illustrate this situation we consider as a typical example the real unit interval [0,1] viewed as a complete MV–algebra – i.e. [0,1] is provided with the algebraic and lattice-theoretical structure determined by Łukasiewicz's connectives (cf. Example 2.4(b) ). It is easy to see that extensional, [0,1]-valued maps as well as Łukasiewicz's connectives can be internalized as nonexpansive maps within the category MET1 of metric spaces $(X, \rho)$ with $\rho \leq 1$ ([2] , [14]).Moreover extremal subobjects in MET1 admit a classification by an extensional, [0,1]-valued map, but unfortunately this classification is not unique ([14]).

The aim of this paper is to overcome these difficulties and to give in the following sense a complete solution of the above problem: For any integral, divisible, commutative cl-monoid M = $(L, \leq, *)$ ([4]) there exists a complete and cocomplete category $\mathcal{C}_M$ satisfying the following conditions

(I) There exists a $\mathcal{C}$–object $\Omega$ and a $\mathcal{C}$–morphism $t : 1 \longmapsto \Omega$ such that

    (I.1) strict and extensional L-valued maps admit an internal version as $\mathcal{C}$–morphisms with codomain $\Omega$ .

    (I.2) every $(\Omega, t)$–classifiable subobject is unique $(\Omega, t)$–classifiable –i.e. if $\phi : U \longmapsto X$ is any $\mathcal{C}$–monomorphim and $(\chi_1, \chi_2)$ is a pair of $\mathcal{C}$–morphisms $\chi_i : X \longmapsto \Omega$    (i=1,2) such that the diagram

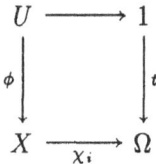

    is a pullback square for $i \in \{1, 2\}$ , then $\chi_1$ and $\chi_2$ coincide.

(II) If M = $(L, \leq, *)$ is a complete Heyting algebra (i.e. $\wedge = *$ ) , then $\mathcal{C}_M$ is the quasitopos Spsh(L) of (ordinary) separated presheaves over L .

(III) If $*$ is Łukasiewicz's arithmetic conjunction (cf. Example 2.4(b) ) and if M = $([0, 1], \leq *)$ , then MET1 is a full isomorphism-closed subcategory of $\mathcal{C}_M$ .

Since the unique classification of extremal subobjcets in Spsh(L) is essentially based on the existence of restriction maps, the properties (I) – (III) suggest themselves to impose on metric spaces also a concept of *restrictions*. Obviously restriction maps themselves require a concept of *local existence*. Hence the key idea consists in the introduction of a "localized version" of metric spaces (cf. Proposition 3.1) . This approach leads to the concept of M–*valued sets* and subsequently to *presheaves over* GL–*monoids* M . In section 6 we show that the category SPSH(M) of separated presheaves over M fulfills the desired properties

(I) –(III) . Further SPSH(M) is an (extremal mono)-topological category over Ssph(L) and is a monadic category over the category M-SET of separated M-valued sets. In particular, if the terminal object 1 in M-SET is provided with the monoidal structure (m,e) induced by the underlying GL-monoid M , then separated presheaves over M and (1,m,e)–*actions* come to the same thing.

In order to fix the necessary notations we start with a section recalling some basic facts about integral,divisible, residuated, commutative l-monoids.

## 2 Lattice-theoretical foundations

A triple $(L, \leq, *)$ is called a *residuated, commutative, l-monoid* ([4]) if and only if $(L, \leq)$ is a lattice, $(L, *)$ is a commutative monoid and there exists a further binary operation $\rightarrow$ on L such that the condition

$$(AD) \qquad \alpha * \beta \ \leq \ \gamma \qquad \Longleftrightarrow \qquad \alpha \ \leq \ \beta \rightarrow \gamma$$

holds for all $\alpha, \beta, \gamma \in L$ . It is easy to see that $\rightarrow$ is uniquely determined by (AD) .

**Proposition 2.1 (General properties)** *Let $(L, \leq, *)$ be a residuated, commutative l-monoid. Then the following assertions are valid*

(i)    $(L, \leq, *)$    is a partially ordered monoid.

(ii)    $\alpha * (\alpha \rightarrow \beta) \ \leq \ \beta.$

(iii)    $\alpha \rightarrow (\beta \rightarrow \gamma) \ = \ (\alpha * \beta) \rightarrow \gamma .$

(iv)    $\alpha * (\beta \vee \gamma) \ = \ (\alpha * \beta) \vee (\alpha * \gamma) .$

(v)    $\alpha \rightarrow (\beta \wedge \gamma) \ = \ (\alpha \rightarrow \beta) \wedge (\alpha \rightarrow \gamma) .$

        $(\alpha \vee \beta) \rightarrow \gamma \ = \ (\alpha \rightarrow \gamma) \wedge (\beta \rightarrow \gamma) .$

(vi)    $\alpha * (\alpha \rightarrow \beta) \ = \ \beta \ \Longleftrightarrow \ \exists \gamma \in L \ \ \text{s.t.} \ \ \alpha * \gamma = \beta.$

(vii)    $\alpha \rightarrow (\alpha * \beta) \ = \ \beta \ \Longleftrightarrow \ \exists \gamma \in L \ \ \text{s.t.} \ \ \alpha \rightarrow \gamma = \beta .$

$\square$

Let $(L, \leq)$ be a bounded lattice – i.e $L$ has universal upper and lower bounds. A residuated, commutative l-monoid $(L, \leq, *)$ is *integral* if and only if the universal upper bound 1 acts as unit element w.r.t. $*$ . The integrality of $(L, \leq, *)$ implies the validity of the following relations

(viii)    $\alpha * \beta \ \leq \ \alpha \wedge \beta .$

(ix)    $1 \ = \ \alpha \rightarrow \beta \ \Longleftrightarrow \ \alpha \leq \beta .$

(x)    The universal lower bound 0 is the zero element w.r.t. $*$ .

An integral, residuated, commutative l-monoid $(L, \leq, *)$ is called *divisible* if and only if for every pair $(\alpha, \beta) \in L \times L$ with $\beta \leq \alpha$ there exists $\gamma \in L$ s.t. $\beta = \alpha * \gamma$ .

**Lemma 2.2 (Consequences from the divisibility)** *If $(L, \leq, *)$ is an integral, divisible, residuated, commutative l-monoid, then the following relations are valid*

(xi)    $\alpha * (\alpha \to \beta) = \alpha \wedge \beta$ .

(xii)    $\alpha * \alpha = \alpha \implies \alpha \wedge \beta = \alpha * \beta$ for all $\beta \in L$.

(xiii)    $\alpha_1 \leq \alpha_2 \implies \alpha_1 * \beta = \alpha_1 * (\alpha_2 \to (\alpha_2 * \beta))$.

(xiv)    $\alpha * (\beta \wedge \gamma) = (\alpha * \beta) \wedge (\alpha * \gamma)$ .

(xv)    $\alpha \to (\beta \wedge \gamma) = (\alpha \to \beta) * ((\alpha \wedge \beta) \to \gamma)$.

**Proof.** Since $(L, \leq, *)$ is divisible, the relation (xi) follows immediately from (v),(vi) and (ix). Now we use (viii) , (xi) and the hypothesis of (xii)

$$\alpha * \beta \leq \alpha \wedge \beta = \alpha * (\alpha \to \beta) = \alpha * \alpha * (\alpha * \beta) \leq \alpha * \beta \;;$$

hence (xii) holds. In the case of $\alpha_1 \leq \alpha_2$ we obtain from (xi)

$$\begin{aligned}
\alpha_1 * (\alpha_2 \to (\alpha_2 * \beta)) &= \alpha_2 * (\alpha_2 \to \alpha_1) * (\alpha_2 \to (\alpha_2 * \beta)) \\
&= (\alpha_2 \to \alpha_1) * \alpha_2 * \beta = \alpha_1 * \beta \quad .
\end{aligned}$$

In order to verify (xiv) we proceed as follows : In view of (iii) and (xi) we obtain

$$\begin{aligned}
(\alpha * \beta) \wedge (\alpha * \gamma) &= \alpha * (\beta * [\beta \to (\alpha \to (\alpha * \gamma))]) \\
&= \alpha * (\beta \wedge (\alpha \to (\alpha * \gamma))) \\
&= \alpha * (\alpha \to (\alpha * \gamma)) * ((\alpha \to (\alpha * \gamma)) \to \beta) \\
&\leq \alpha * \gamma * (\gamma \to \beta) = \alpha * (\beta \wedge \gamma) \quad ;
\end{aligned}$$

hence relation (xiv) is established. Finally

$$\begin{aligned}
\alpha \to (\beta \wedge \gamma) &= (\alpha \to \beta) * ((\alpha \to \beta) \to (\alpha \to \gamma)) \\
&= (\alpha \to \beta) * ((\alpha \wedge \beta) \to \gamma)
\end{aligned}$$

follows from (iii), (v) and (xi) .
□

Let $(L, \leq)$ be a bounded lattice and $(L, \leq, *)$ be an integral, residuated, commutative l-monoid. $(L, \leq, *)$ is called an MV–*algebra* ([3],[5]) if and only if $(L, \leq *)$

satisfies the additional important property

(MV) $\qquad (\alpha \to \beta) \to \beta \quad = \quad \alpha \vee \beta$ .

**Lemma 2.3** *If $(L, \leq, *)$ is an integral, residuated, commutative l-monoid, then the following assertions are equivalent*

(i) $\quad (L, \leq, *)$ is a MV–algebra .

(ii) $\quad (L, \leq, *)$ is divisible and the "negation" is an involution –i.e.
$\quad (\alpha \to 0) \to 0) \quad = \quad \alpha$ for all $\alpha \in L$ .

**Proof.** The axiom (MV) implies immediately that the "negation" is an involution. Using again (MV) we obtain in the case of $\alpha \leq \beta$

$$
\begin{aligned}
\alpha \to 0 \quad &= \quad ((\alpha \to 0) \to (\beta \to 0)) \to (\beta \to 0) \\
&= \quad (\beta \to \alpha) \to (\beta \to 0) \quad = \quad (\beta * (\beta \to \alpha)) \to 0
\end{aligned}
$$

i.e. $\quad \alpha = \beta * (\beta \to \alpha)$ holds; hence the implication (i) $\Longrightarrow$ (ii) is verfied. In oder to show (ii) $\Longrightarrow$ (i) we proceed as follows : Referring to Proposition 2.1 (v) and Lemma 2.2 (xi) we obtain

$$
\begin{aligned}
(\alpha \vee \beta) \to 0 \quad &= \quad (\alpha \to 0) \wedge (\beta \to 0) \\
&= \quad (\beta \to 0) * ((\beta \to 0) \to (\alpha \to 0)) \quad .
\end{aligned}
$$

Since the "negation" is an involution , the relation

$$
\alpha \vee \beta \quad = \quad ((\beta \to 0) * (\alpha \to \beta)) \to 0 \quad = \quad (\alpha \to \beta) \to \beta
$$

follows.
□

**Examples 2.4** (a) Every Heyting algebra $(L, \leq)$ is an integral, divisible, residuated, commutative l-monoid with respect to $* = \wedge$ ([16]) .
(b) Let $\leq$ be the usual partial ordering on the real unit interval $[0,1]$ , and $*$ be Lukasiewicz's *arithmetic conjunction* ([7]) on $[0,1]$ . In particular $*$ is determined by

$$
\alpha * \beta \quad = \quad Max(\alpha + \beta - 1, 0) \qquad \alpha, \beta \in [0, 1] \quad .
$$

Then $([0, 1], \leq, *)$ is an MV–algebra.
(c) Any continuous t-norm T ([22]) induces on $[0,1]$ the structure of an integral, divisible, residuated, commutative l-monoid.
□

An integral, divisible, residuated, commutative l-monoid $(L, \leq, *)$ is called a *GL–monoid* (where GL stands for generalized logics) if and only if the underlying lattice $(L, \leq)$ is complete. In any GL–monoid the following infinite distributive laws are valid (cf. Lemma 1.2 in [15])

$$\alpha * \left( \bigvee_{i \in I} \beta_i \right) = \bigvee_{i \in I} (\alpha * \beta_i)$$

$$\alpha \wedge \left( \bigvee_{i \in I} \beta_i \right) = \bigvee_{i \in I} (\alpha \wedge \beta_i)$$

In particular the underlying lattice $(L, \leq)$ of any GL–monoid is a complete Heyting algebra.

The relationship between complete Boolean algebras, complete MV–algebras and GL–monoids can be described by the subsequent diagram

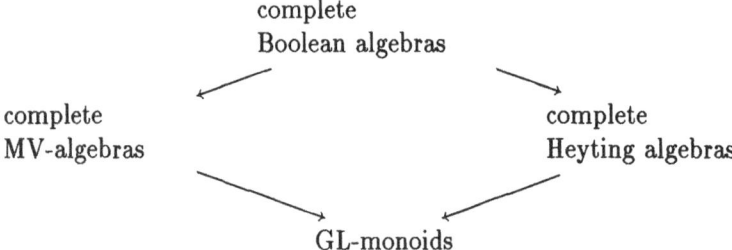

A GL–monoid M is a complete Boolean algebra if and only if M is simultaneously a complete Heyting algebra *and* a complete MV-algebra.

## 3  M-valued sets

In the following considerations we denote by $M = (L, \leq, *)$ always a GL–monoid. First let us recall some simple facts about M-valued sets ([15]). A M-*valued set* is a pair (X,E) where X is a set and E is a M-*valued equality* on X –i.e. $E : \; X \times X \longmapsto L$ is a map satisfying the axioms

(E1)    $E(x, y) \leq E(x, x) \wedge E(y, y)$                          (Strictness)

(E2)    $E(x, y) = E(y, x)$                                          (Symmetry)

(E3)    $E(x, y) * (E(y, y) \to E(y, z)) \leq E(x, z)$            (Transitivity)

We interpret the value E(x,y) as the degree of *overlap* of x and y and E(x,x) as the extent to which x *exists*. From a logical point of view we can understand M-valued sets as M-valued interpretations of the formalized theory of *identity* with *existence* predicate.

In the case of the real unit interval viewed as a complete MV–algebra (cf. Example 2.4 (b) ) we can give a characterization of [0,1]–valued sets as follows

**Proposition 3.1** *Let $M$ be a GL-monoid determined by $([0.1], \leq, *)$ (cf. 2.4(b)). Further let $X$ be a nonempty set, $\mathcal{E}(X)$ be the set of all $[0,1]$-valued equalities on $X$, and let $\mathcal{D}(X)$ be the set of all pairs $(d, \rho)$ where $d : X \longmapsto [0,1]$ is a map and $\rho$ a pseudometric on $X$ satisfying the following conditions*

(D1)  $1/2 \leq d(x), \qquad \rho(x, y) \leq 1 \qquad$ for all $x, y \in X$.

(D2)  $d$ is nonexpansive –i.e. $|d(x) - d(y)| \leq \rho(x, y)$ .

(D3)  $\rho(x, y) \leq d(x) + d(y) - 1$ .

*Then there exists a bijective map from $\mathcal{D}(X)$ onto $\mathcal{E}(X)$ .*

**Proof.** Referring to Exemple 2.2 in [15] we can consider two maps
$\Gamma : \quad \mathcal{D}(X) \longmapsto \mathcal{E}(X) \quad$ and $\quad \Theta : \quad \mathcal{E}(X) \longmapsto \mathcal{D}(X) \quad$ defined as follows

$$\Gamma(d, \rho) \quad = \quad E_{(d,\rho)} \quad , \quad \Theta(E) \quad = \quad (d_E, \rho_E) \qquad \text{where}$$

$$E_{(d,\rho)}(x, y) \quad = \quad d(x) + d(y) - \rho(x, y) - 1$$

$$d_E(x) \quad = \quad \frac{E(x, x) + 1}{2} \quad , \quad \rho_E(x, y) \quad = \quad \frac{E(x, x) + E(y, y)}{2} \; - \; E(x, y).$$

It is not difficult to show that $E_{(d,\rho)}$ is a $[0,1]$-valued equality and $(d_E, \rho_E)$ is an element of $\mathcal{D}(X)$ (cf. Example 2.2 in [15]) Moreover we obtain

$$\Theta \circ \Gamma \quad = \quad id_{\mathcal{D}(X)} \quad \text{and} \quad \Gamma \circ \Theta \quad = \quad id_{\mathcal{E}(X)} \quad ;$$

hence the assertion is verified.
□

**Examples 3.2** Let $M = (L, \leq, *)$ be a GL-monoid.
(a) Every map $\quad \mathbb{E} : X \longmapsto L \quad$ induces at least two different M-valued equalities $\quad E_{ind} \quad$ and $\quad E_{dis} \quad$ as follows

$$E_{ind}(x, y) \quad = \quad \mathbb{E}(x) \wedge \mathbb{E}(y)$$

$$E_{dis}(x, y) \quad = \quad \left\{ \begin{array}{cc} 0 & , \quad x \neq y \\ \mathbb{E}(x) & , \quad x = y \end{array} \right\}$$

Obviously $E_{ind}$ (resp. $E_{dis}$ ) is the coarsest (resp. finest) M-valued equality E on X statisfying

$$E(x, x) \quad = \quad \mathbb{E}(x) \qquad \text{for all} \quad x \in X.$$

(b) $(L, \wedge)$ is a M-valued set.
(c) Let $\quad R_L \quad$ be the set of all pairs $(\alpha, \lambda) \in L \times L \quad$ with $\quad \lambda \leq \alpha \quad$ . Then $E_L : \quad R_L \times R_L \longmapsto L \quad$ defined by

$$E_L((\alpha_1, \lambda_1), (\alpha_2, \lambda_2)) \quad = \quad ((\alpha_1 * (\lambda_1 \rightarrow \lambda_2) \wedge (\alpha_2 * (\lambda_2 \rightarrow \lambda_1))$$

is a M–valued equality on $R_L$ (cf. [15] ).

(d) Since the underlying lattice $(L, \leq)$ of any GL–monoid $M = (L, \leq, *)$ is a complete Heyting algebra, we can also consider the concept of L–valued sets. According to the terminologoy of M. Fourman and D.S. Scott [6] a pair (X,E) is called a *L–valued set* if and only if X is a set and $E : X \times X \longmapsto L$ is a map satisfying the following conditions

(E2)     $E(x, y) \;=\; E(y, x)$

(E3*)     $E(x, y) \wedge E(y, z) \;\leq\; E(x, z)$ .

The axioms (E2) and(E3*) imply immediately (E1) . Further

$$
\begin{aligned}
E(x, y) * (E(y, y) \to E(y, z)) \;&\leq\; E(x, y) * (E(x, y) \to E(y, z)) \\
&\leq\; E(x, y) \wedge E(y, z)
\end{aligned}
$$

follows from Lemma 2.2 (xi) . Hence every L–valued equality is also a M–valued equality.

$\square$

The category M*–SET of M–valued sets consists of the following data : *Objects* are M–valued sets and *morhisms* are structure preserving maps – i.e. $\phi : \;\; (X, E) \longmapsto (Y, F) \;\;$ is a morphism if and only if $\phi : X \longmapsto Y$ is a map satisfying the axioms

(m1)     $F(\phi(x), \phi(x)) \;\leq\; E(x, x)$                                              (Strictness)

(m2)     $E(x_1, x_2) \;\leq\; F(\phi(x_1), \phi(x_2))$                        (Preservation of equality)

The *composition* is the usual composition of maps and the *identity* of (X,E) is the identical map $id_X$ of X.

Further let $SET \downarrow L$ be the comma category of SET over the underlying lattice $L$ – i.e. *objects* are maps $f : X \longmapsto L$ and *morphisms* $\phi : f \longmapsto g$ are maps $\phi : X \longmapsto Y$ making the diagram

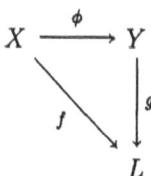

commutative. As a first consequence of the axioms (m1) and (m2) we obtain the existence of a functor     $\mathcal{E} : \;\; \text{M*–SET} \longmapsto SET \downarrow L \;\;$ defined by

$$\mathcal{E}(X, E) \;=\; (X, f_E) \quad \text{where} \quad f_E(x) \;=\; E(x, x) \quad \text{for all} \quad x \in X$$

Referring to Example 3.2(a) it is easy to see that each $\mathcal{E}$-structured source in $SET \downarrow L$ has a unique $\mathcal{E}$-initial lift. Since $Set \downarrow L$ is complete and cocomplete, we obtain immediately from Theorem 21.16 in [1]

**Theorem 3.3** *$M^*$-SET is complete and cocomplete.*

$\square$

A M-valued set $(X,E)$ is called *separated* if and only if E fulfills the additional property

(E4)     $E(x,x) \vee E(y,y) \;\leq\; E(x,y)$     implies   $x = y$     (Separation)

If $(X,E)$ is not necessarily separated, then we can replace $(X,E)$ by its *associated separated* M–valued set $(\hat{X}, \hat{E})$ . In fact, the relation

$$x \sim y \qquad \Longleftrightarrow \qquad E(x,x) \vee E(y,y) \;\leq\; E(x,y)$$

is an ordinary equivalence relationon X . We put $\hat{X} = X/\sim$ and denote by $\hat{x}$ the equivalence class induced by $x \in X$ . On $\hat{X}$ there exists a natural *separated* M–valued equality $\hat{E}$ making the quotient map   $q_{(X,E)} : \quad X \longmapsto \hat{X}$   into a $M^*$–SET–morphism. In particular $\hat{E}$ is determined by

$$\hat{E}(\hat{x}, \hat{y}) \;=\; E(x', y') \qquad \text{where} \qquad x' \in \hat{x}, \;\; y' \in \hat{y}.$$

Moreover for any morphism   $\phi : \quad (X, E) \longmapsto (Y, F)$   there exists a unique arrow   $\hat{\phi} : \; (\hat{X}, \hat{E}) \longmapsto (\hat{Y}, \hat{F})$   such that the diagram

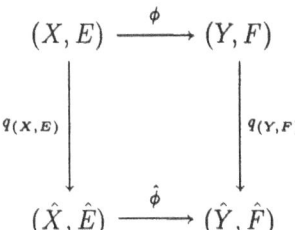

commutes.

In the following considerations we denote by M–SET the full subcategory of $M^*$–SET consisting of all *separated* M-valued sets. Obviously M–SET is isomorphism-closed. Further the preceding separation process leads to a functor   $\mathcal{S} :$ $M^*$–SET $\longmapsto$ M–SET   sending M–valued sets to their associated, separated, M–valued sets and morphism $\phi$ to $\hat{\phi}$ . It is not difficult to see that $\mathcal{S}$ is a reflector of the subcategory M–SET –i.e. is left adjoint to the embedding functor $\mathcal{I} :$   M–SET $\longmapsto M^*$–SET   . Summing up we obtain

**Theorem 3.4** *M–SET is a reflective, isomorphism-closed, full subcategory of $M^*$-SET .* $\square$

An immediate consequence of Theorem 3.3 , Theorem 3.4 and Corollary 36.14, Corollary 36.18 in [1] is the following

**Corollary 3.5** *M–SET is complete and cocomplete.*

□

We continue now the study of M\*–SET and note that M\*–SET carries a natural monoidal structure. With every pair $((X,E),(Y,F))$ of M–valued sets we associate a further M-valued set $(X\times Y, E\otimes F)$ where
$E\otimes F: \quad (X\times Y)\times(X\times Y)\longmapsto L \quad$ is defined by

$$E\otimes F((x_1,y_1),(x_2,y_2)) \quad = \quad E(x_1,x_2)*F(y_1,y_2)$$

The strictness and symmetry of $E\otimes F$ are evident. The verification of the transitivity of $E\otimes F$ is based on the subsequent relation

$$E\otimes F((x_1,y_1),(x_2,y_2))*((E(x_2,x_2)*F(y_2,y_2))\to E\otimes F((x_2,y_2),(x_3,y_3)))$$
$$= \quad E(x_1,x_2)*(E(x_2,x_2)\to E(x_2,x_3))*F(y_1,y_2)*(F(y_2,y_2)\to F(y_2,y_3)).$$

Further we consider the (separated) M-valued set I consisting of a one point set $\{\cdot\}$ provided with *total* existence – i.e. $I=(\{\cdot\},\approx)$ where $\quad\cdot\approx\cdot\quad=\quad 1\quad$. We introduce two bifunctors

$$\otimes: \quad \text{M\*–SET}\times\text{M\*–SET}\longmapsto\text{M\*–SET} \qquad \text{where}$$

$$(X,E)\otimes(Y,F) \quad = \quad (X\times Y, E\otimes F) \quad, \qquad \phi\otimes\psi \quad = \quad \phi\times\psi \quad,$$

$$\Pi: \quad \text{M\*–SET}\times\text{M\*–SET}\longmapsto\text{M\*–SET}\times\text{M\*–SET} \qquad \text{where}$$

$$\Pi((X,E),(Y,F)) \quad = \quad ((Y,F),(X,E)) \quad, \qquad \Pi(\phi,\psi) \quad = \quad (\psi,\phi) \quad,$$

and observe that the subsequent M\*–SET–morphisms

$$a_{(X,E)(Y,F)(Z,G)}: \quad (X,E)\otimes((Y,F)\otimes(Z,G))\longmapsto((X,E)\otimes(Y,F))\otimes(Z,G)$$
$$a_{(X,E)(Y,F)(Z,G)}(x,(y,z)) \quad = \quad ((x,y),z) \quad,$$
$$b_{(X,E)}: \quad I\otimes(X,E)\longmapsto(X,E) \quad, \qquad b_{X,E)}(\cdot,x) \quad = \quad x$$
$$c_{(X,E)}: \quad (X,E)\otimes I\longmapsto(X,E) \quad, \qquad c_{(X,E)}(x,\cdot) \quad = \quad x$$
$$d_{(X,E)(Y,F)}: \quad (X,E)\otimes(Y,F)\longmapsto(Y,F)\otimes(X,E) \quad,$$
$$d_{(X,E)(Y,F)}(x,y) \quad = \quad (y,x)$$

form the respective components of the following natural isomorphisms

$$a: \quad \otimes\circ(id\times\otimes) \quad\longmapsto\quad \otimes\circ(\otimes\times id)$$
$$b: \quad I\otimes - \longmapsto id$$
$$c: \quad -\otimes I\longmapsto id$$
$$d: \quad \otimes\longmapsto\otimes\circ\Pi$$

**Theorem 3.6** *(M\*-SET,⊗,a,b,c,d) is a symmetric, monoidal closed category.*

**Proof.** It is not difficult to show that (M\*-SET,⊗,a,b,c,d) satisfies the axioms of a symmetric monoidal category (cf. [18] , [17]). In order to verify the closedness of (M\*-SET,⊗,a,b,c,d) we first introduce a M-valued set $(Z_{(X,E)(Y,F)}, G_{(X,E)(Y,F)})$ which will turn out as a candidate for the cofree object hom((X,E),(Y,F)) over (Y,F) .
Let us denote by $Z_{(X,E)(Y,F)}$ the set of all pairs $(\alpha, f)$ such that $\alpha \in L$ and $f : X \longmapsto Y$ is a map satisfying the following conditions

$$\alpha * E(x,x) \quad = \quad F(f(x), f(x))$$

$$\alpha * E(x_1, x_2) \quad \leq \quad F(f(x_1), f(x_2))$$

On $Z_{X,E)(Y,F)}$ we consider a M-valued equality $G_{(X,E)(Y,F)}$ determined by

$$G_{(X,E)(Y,F)}((\alpha_1, f_1), (\alpha_2, f_2)) \quad = \quad \alpha_1 \wedge \alpha_2 \wedge (\bigwedge_{x \in X} (E(x,x) \rightarrow F(f_1(x), f_2(x))))$$

The strictness and symmetry of $G_{(X,E)(Y,F)}$ is obvious. The transitivity follows from

$$G((\alpha_1, f_1), (\alpha_2, f_2)) \quad * \quad (\alpha_2 \rightarrow G((\alpha_2, f_2), (\alpha_3, f_3))) \qquad \leq$$
$$\leq \quad ((\alpha_1 \wedge \alpha_2) * (\alpha_2 \rightarrow \alpha_3)) \quad \wedge$$
$$\wedge \quad \left( (\bigwedge_{x \in X} (E(x,x) \rightarrow G(f_1(x), f_2(x)))) \quad * \right.$$
$$\left. * \quad (\bigwedge_{x \in X} ((\alpha_2 * E(x,x)) \rightarrow G(f_2(x), f_3(x)))) \right) .$$

Now we are in the position to introduce a bifunctor

$$hom : \quad (M\text{*-SET})^{op} \times M\text{*-SET} \longmapsto M\text{*-SET}$$

$$hom((X, E), (Y, F)) \quad = \quad (Z_{(X,E)(Y,F)}, G_{(X,E)(Y,F)})$$

$$[hom(f, g)](h) \quad = \quad g \circ h \circ f \qquad \text{for all} \quad h \in Z_{(X,E)(Y,F)}$$

and the corresponding evaluation arrow

$$ev_{X,E)} \quad : \quad hom((X, E), (Y, F)) \quad \otimes \quad (X, E) \qquad \longmapsto \qquad (Y, F)$$

$$ev_{(X,E)}((\alpha, f), x) \quad = \quad f(x) \quad .$$

In order to verify the closedness of (M\*-SET,⊗,a,b,c,d) it is sufficient to show that for every morphism $\quad \phi : \quad (Z, H) \otimes (X, E) \longmapsto (Y, F) \quad$ there exists a unique arrow $\ulcorner\phi\urcorner: \quad (Z,H) \longmapsto hom((X,E),(Y,F))$ making the following diagram commutative

$$(Z,H) \otimes (X,E) \xrightarrow{\ulcorner\phi\urcorner \otimes id_{(X,E)}} hom((X,E),(Y,F)) \otimes (X,E)$$

with $\phi$ and $ev_{(X,E)}$ to $(Y,F)$

The uniqueness of $\ulcorner\phi\urcorner$ is evident. Further every element $z \in Z$ induces a map $f_z : \quad X \longmapsto Y$ by $f_z(x) = \phi(z,x)$. Since $\phi$ is a morphism, $(H(z,z),f_z)$ is an elemnt of $Z_{(X,E)(Y,F)}$ ; hence the correspondence $z \longmapsto (H(z,z),F_z)$ dertermines a map $\ulcorner\phi\urcorner : Z \longmapsto Z_{(X,E)(Y,F)}$ . Applying again the morphism properties of $\phi$ we easily verify the inequality

$$H(z_1,z_2) \leq \bigwedge_{x\in X} E(x,x) \to F(f_{z_1}(x), f_{z_2}(x)) \quad ;$$

hence $\ulcorner\phi\urcorner$ is a morphism from (Z,H) to hom((X,E),(Y,F)). Finally the definition of $\ulcorner\phi\urcorner$ shows that $\ulcorner\phi\urcorner$ fills in the above diagram.
□

Let us recall that $\mathcal{I}$ is the embedding of M–SET into M*–SET and $\mathcal{S}$ is the corresponding reflector from M*–SET to M–SET determined by the separation process. Then the tensorproduct $\otimes$ induces a bifunctor
$\hat{\otimes} :$ M–SET $\times$ M–SET $\longmapsto$ M–SET as follows : $\hat{\otimes} = \mathcal{S} \circ \otimes \circ (\mathcal{I} \times \mathcal{I})$ .
Since there exists a natural isomorphism $T : \hat{\otimes} \circ (\mathcal{S} \times \mathcal{S}) \longmapsto \mathcal{S} \circ \otimes$ , it is easy to see that the reflector $\mathcal{S}$ transfers the monoidal structure from M*–SET (cf. Theorem 3.6) onto M–SET . In this sense M–SET is also a symmetric, monoidal closed category.

We finish this section with the observation that strict and extensional L-valued maps can be internalized as M*–SET–morphims.

**Theorem 3.7 (Internalization of lattice-valued maps)** *Let $(R_L, E_L)$ be the (separated) M-valued set defined in Example 3.2(c) . Further let $(X,E)$ be an arbitrary M-valued set and $\chi : X \longmapsto R_L$ be a map . Then the following assertions are equivalent*

*(i) $\chi$ is a M*–SET–morphism from $(X,E)$ to $(R_L, E_L)$ .*

*(ii) There exists a map $\mu : X \longmapsto L$ provided with the subsequent properties*

$$1. \quad \mu(x) \leq E(x,x) \qquad \text{(Strictness)}$$
$$2. \quad \mu(x) * (E(x,x) \to E(x,y)) \leq \mu(y) \qquad \text{(Extensionality)}$$

3. $\quad \chi(x) \quad = \quad (E(x,x), \mu(x))$

**Proof.** (a) In order to prove the implication(i) $\Longrightarrow$ (ii) let us first denote by
$\Pi_i : \quad R_L \longmapsto L \quad$ the projection onto the $i^{\text{th}}$ coordinate (i=1,2). Since $\chi$
fulfills (m1) and (m2) , the equation $\quad \Pi_1 \circ \chi(x) \quad = \quad E(x,x) \quad$ holds for all
$x \in X$ . On the other hand we put $\quad \mu \quad = \quad \Pi_2 \circ \chi \quad$ and invoke again the
axiom of preservation of equality

$$
\begin{aligned}
\mu(x) * ((E(x,x) \to E(x,y)) \quad &= \quad (E(x,x) \to \mu(x)) * E(x,y) \\
&\leq \quad (E(x,x) \to \mu(x)) * (E(x,x) * (\mu(x) \to \mu(y)) \\
&\leq \quad \mu(y) \quad ;
\end{aligned}
$$

hence the properties 1. and 2. are verified. Property 3. holds by definition.
(b) Now we assume that the assertion (ii) holds. The strictness axiom (m1) is
an immediate consequence of 1. and 3. . Since $\mu$ is extensional (cf. Property
2. ) , we obtain

$$
E(x,y) \quad = \quad E(x,x) * (E(x,x) \to E(x,y)) \quad \leq \quad E(x,x) * (\mu(x) \to \mu(y)) \quad ;
$$

hence the axiom (m2) follows from the symmetry of E and the definition of $E_L$.
Therewith the implication (ii) $\Longrightarrow$ (i) has been established.
$\square$

# 4  Presheaves over GL-monoids

Since the underlying lattice $(L, \leq)$ of every GL-monoid M $= (L, \leq, *)$ is a complete Heyting algebra (cf. section 2 ) , we first recall the concept of ordinary
presheaves over complete Heyting algebras L . According to the terminology introduced by M. Fourman and D.S. Scott ([6]) a triple $(X, \mathbb{E}, \rho)$ is called a *presheaf*
*over* L if and only if X is a set, $\quad \mathbb{E}: \quad X \longmapsto L \quad$ and $\quad \rho: \quad X \times L \longmapsto X$
are maps satisfying the following axioms

(P1) $\quad \rho(x, \mathbb{E}(x)) \quad = \quad x \quad$ for all $\quad x \in X$ .

(P2) $\quad \rho(\rho(x, \alpha), \beta) \quad = \quad \rho(x, (\alpha \wedge \beta))$ .

(P3) $\quad \mathbb{E}(\rho(x, \alpha)) \quad = \quad \mathbb{E}(x) \wedge \alpha$ .

A *presheaf-morphism* $\quad \phi: \quad (X, \mathbb{E}, \rho) \longmapsto (Y, \mathbb{F}, \sigma) \quad$ is a map $\quad \phi: X \longmapsto Y$
provided with the following properties

(pm1) $\quad \mathbb{E} \quad = \quad \mathbb{F} \circ \phi$ .

(pm2) $\quad \phi(\rho(x, \alpha)) \quad = \sigma(\phi(x), \alpha)$ .

Presheaves over L and presheaf-morphisms form in an obvious way a category psh(L) . Further there exists a functor $\mathcal{F}$ : psh(L) $\longmapsto$ SET$\downarrow$L sending every presheaf $(X, \mathbb{E}, \rho)$ to its underlying *bundle* (X,$\mathbb{E}$). We show that every bundle $(X, \mathbb{E})$ generates a presheaf $(\tilde{X}, \tilde{\mathbb{E}}, \rho)$ in a natural way. Let us denote by $\tilde{X}$ the set of all pairs $(x, \alpha) \in X \times L$ with $\alpha \le E(x, x)$ and by $\tilde{\mathbb{E}}$ : $\tilde{X} \longmapsto L$ , $\rho$ : $\tilde{X} \times L \longmapsto \tilde{X}$ maps defined as follows

$$\tilde{\mathbb{E}}(x, \alpha) \;=\; \alpha \;, \qquad \rho((x, \alpha), \beta) \;= (x, \alpha \wedge \beta) \;.$$

It is easy to see that $(\tilde{X}, \tilde{\mathbb{E}}, \rho)$ is in fact a presheaf. Moreover $(\tilde{X}, \tilde{\mathbb{E}}, \rho)$ is the *free presheaf generated by* $(X, \mathbb{E})$ . For this purpose we denote by $\eta_{(X, \mathbb{E})}$ : $(X, \mathbb{E}) \longmapsto (\tilde{X}, \tilde{\mathbb{E}})$ the bundle-morphims determined by $\eta_{(X, \mathbb{E})}(x) \;=\; (x, \mathbb{E}(x))$ .

**Lemma 4.1 (Universal property of $(\tilde{X}, \tilde{\mathbb{E}}, \rho)$)** *Let* $(X, \mathbb{E})$ *be a bundle over* L *and* $(Y, \mathbb{F}, \sigma)$ *be a presheaf over* L *. Then for every bundle morphism* $\phi$ : $(X, \mathbb{E}) \longmapsto (Y, \mathbb{F})$ *there exists a unique presheaf-morphism* $\phi^{\sharp}$ : $(\tilde{X}, \tilde{\mathbb{E}}, \rho) \longmapsto (Y, \mathbb{F}, \sigma)$ *such that the diagram*

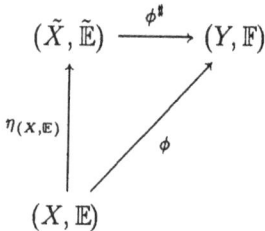

*commutes.*

**Proof**. It is not difficult to show that $\phi^{\sharp}$ : $\tilde{X} \longmapsto Y$ defined by $\phi^{\sharp}(x, \alpha) \;=\; \sigma(\phi(x), \alpha)$ fulfills the desired properties.
$\square$

We can reformulate the preceding lemma as follows :

**Corollary 4.2** *The functor* $\mathcal{F}$ : psh(L) $\longmapsto$ SET$\downarrow$L *has a left adjoint functor. In particular* $\eta$ *is the unit of this adjoint situation.*
$\square$

Finally we recall that every presheaf $(X, \mathbb{E}, \rho)$ over L there exists a *unique* L–*valued equality* E on X satisfying the condition (cf. Proposition 4.3 in [6])

(P4) $E(\rho(x, \alpha), y) \;=\; \alpha \wedge E(x, y)$ for all $\alpha \in L, \; x, y \in X$ .

In particular E is given by

$$E(x, y) \;=\; \vee\{\lambda \in L \mid \lambda \le \mathbb{E}(x) \wedge \mathbb{E}(y) \;, \; \rho(x, \lambda) = \rho(y, \lambda)\} \qquad (4.1)$$

In order to avoid any confusion with arbitrary M–valued equalities we call E the *canonical* L–valued equality associated with $(X, \mathbb{E}, \rho)$ , and we write instead of E also $E_{(\mathbb{E}, \rho)}$ .

Now we are in the position to refine the concept of ordinary presheaves over complete Heyting algebras. Let $M = (L, \leq, *)$ be an arbitrary GL-monoid. A quadruple $(X, E, \mathbb{E}, \rho)$ is said to be a *presheaf over* M if and only if (X,E) is a M–valued set and $(X, \mathbb{E}, \rho)$ is an ordinary presheaf over L such that the following compatibility conditions are satisfied

(Π1)  $\mathbb{E}(x) = E(x, x)$

(Π2)  $E(x, y) * ((E(x, x) \to \alpha) \wedge (E(y, y) \to \beta) \leq E(\rho(x, \alpha), \rho(y, \beta))$

**Lemma 4.3** *Let* $(X, E, \mathbb{E}, \rho)$ *be a presheaf over M and* $E_{(\mathbb{E}, \rho)}$ *be the canonical L–valued equality associated with* $(X, \mathbb{E}, \rho)$ *. Then the inequality*

$$E_{(\mathbb{E}, \rho)}(x, y) \leq E(x, y)$$

*holds for all* $x, y \in X$ .

**Proof.** Let us consider an element $\lambda \in L$ with the properties

$$\lambda \leq \mathbb{E}(x) \wedge \mathbb{E}(y) , \qquad \rho(x, \lambda) = \rho(y, \lambda) .$$

Then we infer from Lemma 2.2 (xi) , (E1), (P1), (P3), (Π1) and (Π2)

$$E(x, \rho(x, \lambda)) = \lambda , \qquad E(\rho(x, \lambda), \rho(x, \lambda)) = \lambda ,$$
$$E(y, \rho(x, \lambda)) = \lambda , \qquad E(\rho(y, \lambda), \rho(y, \lambda)) = \lambda .$$

Now we invoke the transitivity of E and obtain

$$\lambda = E(x, \rho(x, \lambda)) * (E(\rho(x, \lambda), \rho(x, \lambda)) \to E(\rho(x, \lambda), y)$$
$$\leq E(x, y) .$$

$\square$

**Examples 4.4** On every ordinary presheaf $(X, \mathbb{E}, \rho)$ over L there exists a "smallest" (respectively "largest") M–valued equality $E_0$ (respectively $E_1$ ) satisfying the axioms (Π1) and (Π2) . In particular $E_0$ and $E_1$ are given by

$$E_0 = E_{(\mathbb{E}, \rho)} (= \text{canonical,L–valued equality on X})$$
$$E_1(x, y) = \mathbb{E}(x) \wedge \mathbb{E}(y) .$$

Because of Lemma 4.3 , Axiom (Π1) and of the strictness axiom (E1) , it is sufficient to show that $E_0$ and $E_1$ fulfill the condition (Π2) . Using (P3) , (P4) and formula (4.1) we obtain the following estimations

$$E_{(\mathbb{E},\rho)}(x,y) * \left((E_{(\mathbb{E},\rho)}(x,x) \to \lambda) \wedge (E_{(\mathbb{E},\rho)}(y,y) \to \mu)\right) \quad \leq$$
$$E_{(\mathbb{E},\rho)}(x,y) * \left((E_{(\mathbb{E},\rho)}(x,y) \to (\lambda \wedge \mu)\right) \quad = \quad E_{(\mathbb{E},\rho)}(x,y) \wedge \lambda \wedge \mu \quad =$$
$$E_{(\mathbb{E},\rho)}(\rho(x,\lambda),\rho(y,\mu)) \quad ,$$

$$E_1(x,y) * \left((E_1(x,x) \to \lambda) \wedge (E_1(y,y) \to \mu)\right) \quad =$$
$$(\mathbb{E}(x) \wedge \mathbb{E}(y)) * \left((\mathbb{E}(x) \to \lambda) \wedge (\mathbb{E}(y) \to \mu)\right) \quad \leq$$
$$\mathbb{E}(x) \wedge \mathbb{E}(y) \wedge \lambda \wedge \mu \quad = \quad E_1(\rho(x,\lambda), \rho(y,\mu)) \quad ;$$

hence $(X, E_0, \mathbb{E}, \rho)$ and $(X, E_1, \mathbb{E}, \rho)$ are presheaves over M .
□

The category PSH(M) of presheaves over M consists of the following data :
*Objects* are presheaves over M and *morphism* are structure preserving maps –
i.e. $\quad \phi : \quad (X, E, \mathbb{E}, \rho) \longmapsto (Y, F, \mathbb{F}, \sigma) \quad$ is a *presheaf-morphism* if and only
if $\quad \phi \quad (X, E) \longmapsto (Y, F) \quad$ is a M*–SET-morphism and $\phi$ is compatible with
the *restriction maps* $\rho$ and $\sigma \quad$ –i.e. $\quad \phi(\rho(x,\lambda)) \quad = \quad \sigma(\phi(x),\lambda) \quad$ for all
$x \in X, \lambda \in L$ .
There exists an obvious functor $\quad \mathcal{P} : \quad PSH(M) \longmapsto psh(L) \quad$ sending
$(X, E, \mathbb{E}, \rho)$ to its underlying ordinary presheaf $(X, \mathbb{E}, \rho)$ . Since psh(L) is complete and cocomplete and each $\mathcal{P}$–structured source in psh(L) has a unique
$\mathcal{P}$–initial lift (cf. Examples 4.4 ) , we obtain immediately the following

**Theorem 4.5** *Let $(L, \leq)$ be the underlying complete Heyting algebra of a GL-
monoid M . Then $(PSH(M),\mathcal{P})$ is a topological category over psh(L) . In particular PSH(M) is complete and cocomplete.* □

Further there exists also an obvious functor $\quad \hat{\mathcal{F}} : \quad$ PSH(M) $\longmapsto$ M*–SET
assigning to each presheaf $(X, E, \mathbb{E}, \rho)$ its underlying M–valued set (X,E) . We
note that the diagram

$$
\begin{array}{ccc}
\text{PSH(M)} & \xrightarrow{\hat{\mathcal{F}}} & \text{M*–SET} \\
\mathcal{P} \downarrow & & \downarrow \varepsilon \\
\text{psh(L)} & \xrightarrow{\mathcal{F}} & SET \downarrow L
\end{array}
$$

commutes. Since $\mathcal{F}$ has a left adjoint functor (cf. Corollary 4.2) , we pose the
question whether $\hat{\mathcal{F}}$ has also a left adjoint functor. As a preparation for the
application of the Taut–Lift–Theorem we verify the following

**Lemma 4.6** $\hat{\mathcal{F}}$ *sends* $\mathcal{P}$*-initial sources into* $\mathcal{E}$*-initial sources.*

**Proof.** Let $\left( (X, E, \mathbb{E}, \rho) \xrightarrow{\phi_i} (Y_i, F_i, \mathbb{F}_i, \sigma_i) \right)_{i \in I}$ be a $\mathcal{P}$-initial source in PSH(M) . Then the M-valued equality E is given by

$$E(x_1, x_2) = \bigwedge_{i \in I} F_i(\phi_i(x_1), \phi_i(x_2)) \qquad \text{for all} \quad x_1, x_2 \in X \qquad (4.2)$$

We show that the source

$$\mathbf{S} = \left( \hat{\mathcal{F}}(X, E, \mathbb{E}, \rho) \xrightarrow{\hat{\mathcal{F}}(\phi_i)} \hat{\mathcal{F}}(Y_i, F_i, \mathbb{F}_i, \sigma_i) \right)_{i \in I} =$$

$$= \left( (X, E) \xrightarrow{\hat{\mathcal{F}}(\phi_i)} (Y_i, F_i) \right)_{i \in I}$$

in M*-SET is $\mathcal{E}$-initial. Therefore let us consider a further source **T** in M*-SET having the same codoamin as **S** (i.e. $\mathbf{T} = \left( (Z, G) \xrightarrow{\psi_i} (Y_i, F_i) \right)_{i \in I}$ ) and a bundle-morphism $\theta : \mathcal{E}(Z, G) \longmapsto \mathcal{E}(X, E)$ s.t. $\hat{\mathcal{F}}(\phi_i) \circ \theta = \psi_i$ holds on the level of $SET \downarrow L$ . From formula (4.2) we conclude that $\theta$ is a M*-SET-morphism ; hence **S** is $\mathcal{E}$-initial .
□

**Theorem 4.7** *The functor* $\hat{\mathcal{F}} : PSH(M) \longmapsto M^*\text{-}SET$ *has a left adjoint functor.*

**Proof.** By virtue of the Taut–Lift–Theorem 21.28 in [1] the assertion follows from Corollary 4.2, Lemma 4.6 and the preceding diagram.
□

An explicit description of the left adjoint functor to $\hat{\mathcal{F}}$ is also available For this purpose let us denote by (X,E) a M-valued set and by $\tilde{X}$ the set of all pairs $(x, \alpha)$ with $\alpha \le E(x, x)$ . On $\tilde{X}$ we define the following maps

$$\tilde{E}((x_1, \alpha_1), (x_2, \alpha_2)) = E(x_1, x_2) * ((E(x_1, x_1) \to \alpha_1) \wedge (E(x_2, x_2) \to \alpha_2))$$
$$\tilde{\mathbb{E}}(x, \alpha) = E(x, x) \wedge \alpha$$
$$\rho((x, \alpha), \lambda) = (x, (\lambda \wedge \alpha))$$

**Lemma 4.8** $(\tilde{X}, \tilde{E}, \tilde{\mathbb{E}}, \rho)$ *is a presheaf over* $M$ .

**Proof.** (a) First we show that $(\tilde{X}, \tilde{E})$ is a M-valued set. The strictness of $\tilde{E}$ follows from Lemma 2.2 (xi) . The symmetry of $\tilde{E}$ is evident . In order to verify the transitivity of $\tilde{E}$ we use the assertions (xi) ,(xiii) and (xiv) in Lemma 2.2 and obtain

$$\tilde{E}((x_2, \alpha_2), (x_3, \alpha_3)) =$$
$$= (E(x_2, x_2) \to E(x_2, x_3)) * ((E(x_2, x_2) \wedge \alpha_2) \wedge (E(x_2, x_2) * (E(x_3 x_3) \to \alpha_3)))$$

$$= \quad (E(x_2, x_2) \to E(x_2, x_3)) * (E(x_2, x_2) \land \alpha_2)*$$
$$*[(E(x_2, x_2) \land \alpha_2) \to (E(x_2, x_2) * (E(x_3, x_3) \to \alpha_3))] \quad ,$$

$$[(E(x_2, x_2) \land \alpha_2) \to (E(x_2, x_2) * (E(x_3, x_3) \to \alpha_3))]*$$
$$*[E(x_2, x_2) \land \alpha_2 \land (E(x_2, x_2) * (E(x_1, x_1) \to \alpha_1))] \quad \leq$$
$$\leq \quad (E(x_2, x_2) * (E(x_3, x_3) \to \alpha_3)) \land (E(x_2, x_2) * (E(x_1, x_1) \to \alpha_1)) \quad =$$
$$= \quad E(x_2, x_2) * ((E(x_1, x_1) \to \alpha_1) \land (E(x_3, x_3) \to \alpha_3)) \quad ,$$

$$\tilde{E}((x_1, \alpha_1), (x_2, \alpha_2)) * ((E(x_2, x_2) \land \alpha_2) \to \tilde{E}((x_2, \alpha_2), (x_3, \alpha_3))) \quad =$$
$$= (E(x_2, x_2) \to E(x_1, x_2)) * ((E(x_2, x_2) \land \alpha_2) \land (E(x_2, x_2) * (E(x_1, x_1) \to \alpha_1)))*$$
$$*(E(x_2, x_2) \to E(x_2, x_3))*$$
$$*((E(x_2, x_2) \land \alpha_2) \to (E(x_2, x_2) * (E(x_3, x_3) \to \alpha_3))) \quad \leq$$
$$\leq \quad (E(x_2, x_2) \to E(x_1, x_2)) * (E(x_2, x_2) \to E(x_2, x_3)) * (E(x_2, x_2)*$$
$$*((E(x_1, x_1) \to \alpha_1) \land (E(x_3, x_3) \to \alpha_3))) \quad \leq$$
$$\leq \quad \tilde{E}((x_1, \alpha_1), (x_3, \alpha_3)) \quad .$$

(b) Obviously $(\tilde{X}, \tilde{E}, \rho)$ is an ordinary presheaf over L satisfying (II1) . In order to verify (II2) we proceed as follows

$$\tilde{E}((x_1, \alpha_1), (x_2, \alpha_2)) * ((E(x_1, x_1) \land \alpha_1) \to \lambda_1) \land ((E(x_2, x_2) \land \alpha_2) \to \lambda_2) \quad =$$
$$= \quad E(x_1, x_2) * ((E(x_1, x_1) \to \alpha_1) \land (E(x_2, x_2) \to \alpha_2))*$$
$$*(((E(x_1, x_1) \land \alpha_1) \to \lambda_1) \land ((E(x_2, x_2) \land \alpha_2) \to \lambda_2)) \quad \leq$$
$$\leq \quad E(x_1, x_2) * ((E(x_1, x_1) \to (\alpha_1 \land \lambda_1)) \land (E(x_2, x_2) \to (\alpha_2 \land \lambda_2))) \quad =$$
$$= \quad \tilde{E}((x_1, (\alpha_1 \land \lambda_1)), (x_2, (\alpha_2 \land \lambda_2))) \quad ;$$

hence (II2) is verified . Q.E.D.

There exists an obvious M*–SET–morphism $\hat{\eta}_{(X,E)} : \quad (X, E) \longmapsto (\tilde{X}, \tilde{E})$ determined by $\quad \hat{\eta}_{(X,E)}(x) \quad = \quad (x, E(x, x)) \quad$ for all $\quad x \in X$ .

**Theorem 4.9 (Universal property of $(\tilde{X}, \tilde{E}, \tilde{\mathbb{E}}, \rho)$)** Let $(X, E)$ be a M-valued set and $(Y, F, \mathbb{F}, \sigma)$ be a presheaf over M . Then for every M*–SET-morphism $\phi : \quad (X, E) \longmapsto (Y, F) \quad$ there exists a unique PSH(M)–morphism

$\tilde{\phi}: \quad (\tilde{X}, \tilde{E}, \tilde{\mathbb{E}}, \rho) \longmapsto (Y, F, \mathbb{F}, \sigma) \quad$ *such that the following diagram commutes*

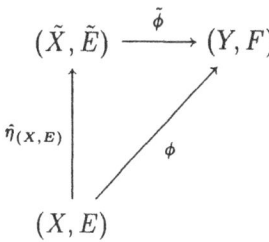

*In particular* $\tilde{\phi}$ *is given by* $\quad \tilde{\phi}(x, \alpha) \quad = \quad \sigma(\phi(x), \alpha) \quad$ *for all* $\quad (x, \alpha) \in X$ .

**Comment**. Because of Theorem 4.9 we call also $(\tilde{X}, \tilde{E}, \tilde{\mathbb{E}}, \rho)$ the presheaf over M *generated freely by* the M–valued set (X,E) .

## 5  Separated presheaves over GL–monoids

A presheaf $(X, E, \mathbb{E}, \rho)$ over a GL–monoid M is *separated* if and only if the underlying M–valued set (X,E) is separated.

With every presheaf $(X, E, \mathbb{E}, \rho)$ over M we can associate a separated presheaf $(\hat{X}, \hat{E}, \hat{\mathbb{E}}, \hat{\rho})$ as follows : Let $(\hat{X}, \hat{E})$ be the associated, separated, M–valued set corresponding to (X,E) , and let $\quad \hat{\mathbb{E}}: \quad \hat{X} \longmapsto L \quad$ be a map induced by $\mathbb{E}$ – i.e. $\hat{\mathbb{E}}(\hat{x}) \quad = \quad \mathbb{E}(x') \quad = \quad \hat{E}(\hat{x}, \hat{x})$ where $\quad x' \in \hat{x}$ . Referring to the compatibility conditions (II1) and (II2) we deduce from $E(x, x) \vee E(\bar{x}, \bar{x}) \quad \leq \quad E(x, \bar{x})$ the following relation

$$
\begin{aligned}
(E(x, x) \wedge \lambda) \vee (E(\bar{x}, \bar{x}) \wedge \lambda) \quad &= \quad (E(x, x) \vee E(\bar{x}, \bar{x})) \wedge \lambda \\
&\leq \quad E(x, \bar{x}) * ((E(x, x) \vee E(\bar{x}, \bar{x})) \to \lambda) \\
&\leq \quad E(\rho(x, \lambda), \rho(\bar{x}, \lambda)) \quad ;
\end{aligned}
$$

hence the map $\quad \hat{\rho}: \quad \hat{X} \times L \longmapsto \hat{X} \quad$ determined by $\quad \hat{\rho}(\hat{x}, \lambda) \quad = \quad \rho(x', \lambda)$ (where $\quad x' \in \hat{x}$) is well defined. Then the quadruple $(\hat{X}, \hat{E}, \hat{\mathbb{E}}, \hat{\rho})$ is a separated presheaf over M .

Let us denote by SPSH(M) the full subcategory of PSH(M) of all separated presheaves over M . The preceding separation process leads to a functor $\quad \mathcal{T}: PSH(M) \longmapsto SPSH(M) \quad$ which is left adjoint to the embedding functor from SPSH(M) to PSH(M) ; i.e. $\mathcal{T}$ is the reflector of SPSH(M) . Moreover SPSH(M) is isomorphism-closed subcategory of PSH(M) . Summing up we obtain from Theorem 4.5

**Theorem 5.1** *SPSH(M) is a reflective subcategory of PSH(M) . In particular SPSH(M) is complete and cocomplete.*

Referring to 4.1 and 4.7 in [6] it is easy to see that the underlying, ordinary presheaf $(X, \mathbb{E}, \rho)$ over L of any separated presheaf $(X, E, \mathbb{E}, \rho)$ over M is also separated. Hence SPSH(M) is an (extremal mono)-topological category over Spsh(L) of ordinary, separated presheaves over L. Moreover we show that SPSH(M) is isomorphic to the category of (1,m,e)-*actions* where (m,e) is a *natural monoidal* structure on the terminal object 1 of M–SET.The situation is as follows :

Let us consider the monoidal category (M–SET,$\hat{\otimes}$) introduced in section 3 . Then the semigroup operation $*$ on the underlying lattice $(L, \leq)$ induces a monoidal structure (m,e) on the terminal object $1 = (L, \wedge)$ of M–SET in the following way :

$$m : \quad 1 \hat{\otimes} 1 \; \longmapsto \; 1, \quad m([(\alpha, \beta)]) \;\; = \;\; \alpha * \beta$$

where $[(\alpha, \beta)]$ is an element of the support set of $\; 1 \hat{\otimes} 1$

$$e : \quad I \longmapsto 1 \;\;, \quad e\{\cdot\} \;\; = \;\; 1 \;\; (\in L)$$

Referring to Exercise 4 on page 214 in [18] we can describe (1,e,m)-*actions* as a triple $(X, E, \cdot)$ such that (X,E) is a separated M–valued set and
$\cdot : \quad (X, E) \hat{\otimes} 1 \; \longmapsto \; (X, E) \quad$ is a M–SET–morphism satisfying the following conditions

(A1) $\quad x \cdot 1 \;\; = \;\; x \quad$ for all $\;\; x \in X$

(A2) $\quad (x \cdot \alpha) \cdot \beta) \;\; = \;\; x \cdot (\alpha * \beta) \quad$ for all $\;\; x \in X; \alpha, \beta \in L.$

An $(1, m, e)$-*action-homomorphism* $\quad \phi : \quad (X, E, \cdot) \longmapsto (Y, F, \cdot) \quad$ is a M–SET–morphism $\quad \phi : \quad (X, E) \longmapsto (Y, F) \quad$ provided with the additional property

(m3) $\quad \phi(x \cdot \alpha) \;\; = \;\; (\phi(x)) \cdot \alpha.$

The category of (1,m,e)-actions over (M–SET,$\hat{\otimes}$) is denoted by $\mathcal{A}_{(1,*)}(M)$ .

**Theorem 5.2** *The categories SPSH(M) and $\mathcal{A}_{(1,*)}(M)$ are isomorphic.*

**Proof.** (a) Let $(X, E, \cdot)$ be an (1,m,e)-action . Since 1 is the terminal object in M–SET , the diagram

$$
\begin{array}{ccc}
(X, E) \hat{\otimes} 1 & \xrightarrow{\;\;\cdot\;\;} & (X, E) \\
\Big\downarrow {\scriptstyle !_{X,E} \hat{\otimes} id_1} & & \Big\downarrow {\scriptstyle !_{(X,E)}} \\
1 \hat{\otimes} 1 & \xrightarrow[\;\;m\;\;]{} & 1
\end{array}
$$

commutes; hence the relation

(A3) $\qquad E(x \cdot \alpha, x \cdot \alpha) \qquad = \qquad E(x, x) * \alpha$

holds for all $x \in X, \alpha \in L$. Using the separation property of (X,E) we obtain

(A4) $\qquad x \cdot \alpha \quad = \quad x \cdot (E(x, x) \to (E(x, x) * \alpha))$ .

(b) With any (1,m,e)–action $(X, E, \cdot)$ we associate an ordinary presheaf $(X, \mathbb{E}, \rho)$ over L as follows

$$\mathbb{E}(x) \quad = \quad E(x, x) \; , \qquad \rho(x, \alpha) \quad = \quad x \cdot (E(x, x) \to \alpha) \qquad (5.1)$$

The axioms (P1) and (P3) follow immediately from (A1) and (A3) and Lemma 2.2 (xi). The verification of (P2) is based on Lemma 2.2 (xv) and (A2)

$$
\begin{aligned}
\rho(\rho(x, \alpha), \beta)) \quad &= \quad (\rho(x, \alpha)) \cdot ((E(x, x) \wedge \alpha) \to \beta) \\
&= \quad (x \cdot (E(x, x) \to \alpha)) \cdot ((E(x, x) \wedge \alpha) \to \beta) \\
&= \quad x \cdot ((E(x, x) \to \alpha) * ((E(x, x) \wedge \alpha) \to \beta)) \\
&= \quad x \cdot (E(x, x) \to (\alpha \wedge \beta) \quad = \quad \rho(x, (\alpha \wedge \beta) \quad .
\end{aligned}
$$

Since $\quad \cdot : (X, E) \hat{\otimes} 1 \longmapsto (X, E) \quad$ is a M–SET–morphism, the definition of $\hat{\otimes}$ shows that the compatibility conditions (Π1) and (Π2) between (X,E) and $(X, \mathbb{E}, \rho)$ are satisfied. Thus $(X, E, \mathbb{E}, \rho)$ is a separated presheaf over M .

(c)Let $(X, E, \mathbb{E}, \rho)$ be a separated presheaf over M . Then we infer from the compatibility conditions (Π1) , (Π2) and Lemma 2.2 (xi)

$$E(x, x) \wedge \alpha \quad \leq \quad E(\rho(x, \alpha) , \; \rho(x, (E(x, x) \wedge \alpha))) \qquad ;$$

hence the separation property of (X,E) and again Lemma 2.2 (xi) implies

(P4') $\qquad \rho(x, \alpha) \quad = \quad \rho(x, (E(x, x) * (E(x, x) \to \alpha)))$

for all $x \in X$ and $\alpha \in L$ .

(d) With every separated presheaf $(X, E, \mathbb{E}, \rho)$ over M we associate an (1,m,e)–action $\quad \cdot : \quad (X, E) \hat{\otimes} 1 \longmapsto (X, E)$ as follows

$$\cdot([(x, \alpha)]) \quad = \quad \rho(x, (E(x, x) * \alpha)) \qquad (5.2)$$

First we show that $\cdot$ is well defined. Therefore we consider two pairs $(x, \alpha)$ and $(\bar{x}, \bar{\alpha})$ with

$$(E(x, x) * \alpha) \vee (E(\bar{x}, \bar{x}) * \bar{\alpha}) \qquad \leq \qquad E(x, \bar{x}) * (\alpha \wedge \bar{\alpha}) \quad .$$

From the compatibility condition (Π2) we infer

$$
\begin{aligned}
(E(x, x) * \alpha) \vee (E(\bar{x}, \bar{x}) * \bar{\alpha}) \quad &\leq \quad E(x, \bar{x}) * (\alpha \wedge \bar{\alpha}) \qquad &\leq \\
E(x, \bar{x}) * [(E(x, x) \to (\alpha * E(x, x))) \wedge (E(\bar{x}, \bar{x}) \to (\bar{\alpha} * E(\bar{x}, \bar{x})))] \quad &\leq \\
E(\rho(x, (E(x, x) * \alpha)), \rho(\bar{x}, (E(\bar{x}, \bar{x}) * \bar{\alpha}))) \quad & &;
\end{aligned}
$$

hence $\rho(x, (E(x, x) * \alpha)) = \rho(\bar{x}, (E(\bar{x}, \bar{x}) * \bar{\alpha}))$ follows from the separation property of (X,E) –i.e. · is well defined.

Referring again to the compatibility condition (II2) we obtain

$$E(x, \bar{x}) * (\alpha \wedge \bar{\alpha}) \leq$$

$$E(x, \bar{x}) * (E(x, x) \to (E(x, x) * \alpha)) \wedge (E(\bar{x}, \bar{x}) \to (E(\bar{x}, \bar{x}) * \bar{\alpha})) \leq$$

$$E(\cdot([(x, \alpha)]), \cdot([(\bar{x}, \bar{\alpha})]))$$

–i.e. · is a M–SET–morphism from $(X, E) \hat{\otimes} 1$ to (X,E) .Finally the axioms (A1) and (A2) follow immediately from (P1) and (P2) ; hence $(X, E, \cdot)$ is a (1,m,e)-action over (M–SET,$\hat{\otimes}$) .

(e) By virtue of (A4) ,(P4′) , (5.1) and (5.2) it is not difficult to show that SPSH(M) and $\mathcal{A}_{(1,*)}(M)$ are isomorphic. Q.E.D.

**Remark 5.3** (a) The preceding theorem shows that separated presheaves over a GL–monoid M $= (L, \leq, *)$ are uniquely determined by the monoidal structure on M–SET and, in particular, by the semigroup operation $*$ on the underlying lattice $(L, \leq)$ . This situation justifies the above chosen terminology , namely to talk of *presheaves over M* . Moreover we underline that the transfer between separated presheaves over M and (1,m,e)-actions over M–SET is based on the *divisibility* of M .

(b) The forgetful functor from SPSH(M) to M–SET is adjoint – i.e. has a left adjoint functor. If (X,E) is a separated,M–valued set, then the separated presheaf *generated freely by* (X,E) is given by the associated separated presheaf corresponding to $(\tilde{X}, \tilde{E}, \tilde{\mathbb{E}}, \rho)$ (cf. Theorem 4.9 ) .

(c) If $E_{(\mathbb{E}, \rho)}$ is the canonical L-valued equality corresponding to an ordinary, separated presheaf $(X, \mathbb{E}, \rho)$ over L , then $(X, E_{(\mathbb{E}, \rho)}, \mathbb{E}, \rho)$ is a discrete object of SPSH(M) (cf. Examples 4.4 and Lemma 4.3) . Hence the forgetful functor from SPSH(M) to Spsh(L) is adjoint.

(d) Let M $= (L, \leq, \wedge)$ be a complete Heyting algebra (i.e. $* = \wedge$) . Then (1,m,e)-actions over M–SET and separated presheaves over L (in the sense of M. Fourman and D.S. Scott ([6])) come to the same thing.
□

# 6   Classification of subobjects in SPSH(M)

In this section we discuss the classification of subobjects in SPSH(M) . As the reader will see this investigation leads implicitly to a version of a subobject classifier axiom in SPSH(M).

First we note that there exists at least two different ways to lift $(R_L, E_L)$ to a separated presheaf over M (cf. Example 3.2 (c) ). We maintain the notations of 3.2(c) and define two "restriction maps"

$$\rho_L: \quad R_L \times L \longmapsto R_L \quad , \qquad \rho_L^*: \quad R_L \times L \longmapsto R_L \qquad \text{by}$$

$$\rho_L((\alpha,\lambda),\beta) \;=\; (\alpha \wedge \beta, \lambda \wedge \beta), \quad \rho_L^*((\alpha,\lambda),\beta) \;=\; (\alpha \wedge \beta, (\alpha \wedge \beta) * (\alpha \to \lambda)).$$

**Lemma 6.1** *Let* $\Pi_1: \quad R_L \longmapsto L$ *be the projection onto the first coordinate. Then* $\Omega \;=\; (R_L, E_L, \Pi_1, \rho_L)$ *and* $\Omega^* \;=\; (R_L, E_L, \Pi_1, \rho_L^*)$ *are separated presheaves over* $M$.

**Proof.** Obviously $(R_L, \Pi_1, \rho_L)$ is an ordinary presheaf over L. Because of Lemma 2.2 (xiii) we obtain

$$\rho_L^*(\rho_L^*((\alpha,\lambda),\beta_1),\beta_2) \;=$$
$$\rho_L^*(((\alpha \wedge \beta_1),(\alpha \wedge \beta_1)*(\alpha \to \lambda)),\beta_2) \;=$$
$$(\alpha \wedge \beta_1 \wedge \beta_2, (\alpha \wedge \beta_1 \wedge \beta_2)*((\alpha \wedge \beta_1) \to ((\alpha \wedge \beta_1)*(\alpha \to \lambda)))) \;=$$
$$(\alpha \wedge \beta_1 \wedge \beta_2),(\alpha \wedge \beta_1 \wedge \beta_2)*(\alpha \to \lambda)) \;=\; \rho_L^*((\alpha,\lambda),(\beta_1 \wedge \beta_2)) \quad;$$

hence $(R_L, \Pi_1, \rho_L^*)$ is also an ordinary presheaf over L . In order to verify the compatibility conditions (II1) and (II2) we proceed as follows

$$((\alpha \to \beta) \wedge (\bar\alpha \to \bar\beta)) * E_L((\alpha,\lambda),(\bar\alpha,\bar\lambda)) * ((\alpha \wedge \beta) \to (\lambda \wedge \beta)) \quad \leq$$
$$[(\alpha \to \beta)*\alpha*(\lambda \to \bar\lambda)*((\alpha \wedge \beta) \to (\lambda \wedge \beta))] \wedge ((\bar\alpha \to \bar\beta)*\bar\alpha) \quad \leq$$
$$[((\lambda \wedge \beta) \to (\beta \wedge \bar\lambda))*(\lambda \wedge \beta)] \wedge \bar\beta \quad \leq \quad \bar\lambda \wedge \bar\beta \quad;$$

i.e.

$$E_L((\alpha,\lambda),(\bar\alpha,\bar\lambda)) * ((\alpha \to \beta) \wedge (\bar\alpha \to \bar\beta)) \quad \leq$$
$$(\alpha \wedge \beta) \wedge [((\alpha \wedge \beta) \to (\lambda \wedge \beta)) \to (\bar\lambda \wedge \bar\beta)] \;=\; (\alpha \wedge \beta)*((\lambda \wedge \beta) \to (\bar\lambda \wedge \bar\beta)) \quad.$$

Analogously we verify

$$E_L((\alpha,\lambda),(\bar\alpha,\bar\lambda)) \quad \leq \quad (\bar\alpha \wedge \bar\beta)*((\bar\lambda \wedge \bar\beta) \to (\lambda \wedge \beta)) \quad;$$

hence we obtain

$$E_L((\alpha,\lambda),(\bar\alpha,\bar\lambda))*((\alpha \to \beta) \wedge (\bar\alpha \to \bar\beta)) \quad \leq \quad E_L(\rho_L((\alpha,\lambda),\beta),\rho_L((\bar\alpha,\bar\lambda),\bar\beta)).$$

Further we observe

$$((\alpha \to \beta) \wedge (\bar\alpha \to \bar\beta))*E_L((\alpha,\lambda),(\bar\alpha,\bar\lambda))*[(\alpha \wedge \beta) \to ((\alpha \wedge \beta)*(\alpha \to \lambda))] \;=$$
$$((\alpha \to \beta) \wedge (\bar\alpha \to \bar\beta))*(\alpha \to \lambda)*E_L((\alpha,\lambda),(\bar\alpha,\bar\lambda)) \quad \leq$$
$$(\bar\alpha \to \bar\beta)*\alpha*(\lambda \to \bar\lambda)*(\alpha \to \lambda) \quad \leq \quad (\bar\alpha \to \bar\beta)*\bar\lambda \;=\; (\bar\alpha \wedge \bar\beta)*(\bar\alpha \to \bar\lambda) \quad;$$

i.e.

$$((\alpha \to \beta) \wedge (\bar{\alpha} \to \bar{\beta})) * E_L((\alpha, \lambda), (\bar{\alpha}, \bar{\lambda})) \quad \leq$$

$$(\alpha \wedge \beta) \wedge [((\alpha \wedge \beta) \to ((\alpha \wedge \beta) * (\alpha \to \lambda))) \to ((\bar{\alpha} \wedge \bar{\beta}) * (\bar{\alpha} \to \bar{\lambda}))] \quad =$$

$$(\alpha \wedge \beta) * [((\alpha \wedge \beta) * (\alpha \to \lambda)) \to ((\bar{\alpha} \wedge \bar{\beta}) * (\bar{\alpha} \to \bar{\lambda}))] \quad .$$

Analogously we verify

$$(\alpha \to \beta) \wedge (\bar{\alpha} \to \bar{\beta}) * E_L((\alpha, \lambda), (\bar{\alpha}, \bar{\lambda})) \quad \leq$$

$$(\bar{\alpha} \wedge \bar{\beta}) * (((\bar{\alpha} \wedge \bar{\beta}) * (\bar{\alpha} \to \bar{\lambda})) \to ((\alpha \wedge \beta) * (\alpha \to \lambda))) \quad ;$$

hence the inequality

$$E_L((\alpha, \lambda), (\bar{\alpha}, \bar{\lambda})) * ((\alpha \to \beta) \wedge (\bar{\alpha} \to \bar{\beta})) \quad \leq \quad E_L(\rho_L^*((\alpha, \lambda), \beta), \rho_L^*((\bar{\alpha}, \bar{\lambda}), \bar{\beta}))$$

holds.
□

**Lemma 6.2** *A SPSH(M)–morphism* $\phi : (X, E, \mathbb{E}, \rho) \longmapsto (Y, F, \mathbb{F}, \sigma)$ *is monic if and only if* $\phi : X \longmapsto Y$ *is an injective map.*

**Proof.** If we view SPSH(M)–morphisms as maps between the underlying support sets , then the injectivity of $\phi$ implies that $\phi$ is monic . On the other hand, let us assume that there exist elements $x_1$ and $x_2$ of X with $\phi(x_1) = \phi(x_2)$. Then we obtain $\mathbb{E}(x_1) = \mathbb{E}(x_2)$ , and we put $\alpha = \mathbb{E}(x_i)$ $(i \in \{1, 2\})$ . Further we consider on $U_\alpha = \{\lambda \in L \mid \lambda \leq \alpha\}$ the presheaf structure inherited from the terminal object $1_{PS} = (L, \wedge, id_L, \wedge)$ in SPSH(M) . By virtue of the compatibility conditions (II1) and (II2) we are in the position to define SPSH(M)–morphisms $\theta_i : (U_\alpha, \wedge, id_{U_\alpha}, \wedge) \longmapsto (X, E, \mathbb{E}, \rho)$ by $\theta_i(\lambda) = \rho(x_i, \lambda)$ for all $\lambda \in U_\alpha$, $i \in \{1, 2\}$ . Now we invoke the presheaf-morphism property of $\phi$ and obtain $\phi \circ \theta_1 = \phi \circ \theta_2$ . If we assume that $\phi$ is a monomorphism , then the morphisms $\theta_1$ and $\theta_2$ coincide ; hence $\phi : X \longmapsto Y$ is an injective map .
□

It is easy to see that the terminal object $1_{PS}$ in SPSH(M) is *freely generated by* the unit object $I = (\{\cdot\}, \approx)$ of the monoidal category $(\text{M–SET}, \hat{\otimes})$ . Hence the M–SET–morphism $s : (\{\cdot\}, \approx) \longmapsto (R_L, E_L)$ determined by $s(\cdot) = (1, 1)$ has a unique extension to a presheaf-morphims $s^{\natural} : 1_{PS} \longmapsto \Omega$ in the sense of SPSH(M) . $s^{\natural}$ is called the *arrow true* and will also be denoted by t . In particular t is given by $t(\lambda) = (\lambda, \lambda)$ for all $\lambda \in L$ .

**Definition 6.3** A monomorphism $\phi : (U, F, \mathbb{F}, \sigma) \longmapsto (X, E, \mathbb{E}, \rho)$ in the sense of SPSH(M) is said to be $(\Omega, t)$–*classifiable* if and only if there exists a

SPSH(M)-morphisms   $\chi : (X, E, \mathbb{E}, \rho) \longmapsto \Omega$   such that the diagram

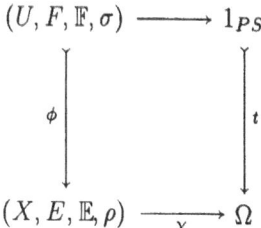

is a pullback square .
□


Since the underlying ordinary presheaf of $\Omega$ is the extremal subobject classifier in the *quasitopos* Spsh(L) ([24]) we obtain immediately

**Theorem 6.4** *$(\Omega, t)$-classifiable PSPH(M)-monomorphims are unique classifiable; i.e. , if*

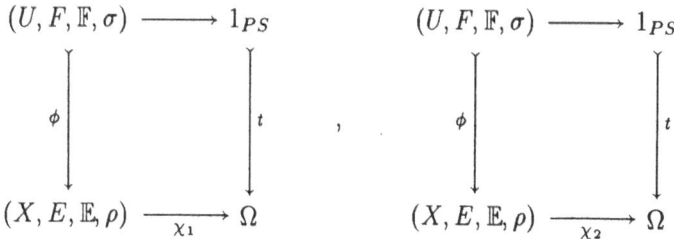

*are pullback squares, then $\chi_1$ and $\chi_2$ coincide.*

**Proof**. Since the forgetful functor from SPSH(M) to Spsh(L) has a left adjoint functor (cf. Remark 5.3 (c)) , the assertion follows.
□


**Theorem 6.5** *Let   $\phi :$   $(U, F, \mathbb{F}, \sigma) \longmapsto (X, E, \mathbb{E}, \rho)$   be a monomorphism in the sense of SPSH(M) , and let $E_{(\mathbb{E}, \rho)}$ , $F_{(\mathbb{F}, \sigma)}$ respectively, be the corresponding L-valued equality (cf. Lemma 4.3 ) . Then the following assertions are equivalent*

(i)    $\phi$ is $(\Omega, t)$-classifiable .

(ii)

    (1)    $\phi :$  $(U, \mathbb{F}, \sigma) \longmapsto (X, \mathbb{E}, \rho)$   is an extremal monomorphism in the sense of Spsh(L) – i.e.

        (1.1)   $F_{(\mathbb{F},\sigma)}(u, \bar{u}) = E_{(\mathbb{E},\rho)}(\phi(u), \phi(\bar{u}))$ .

        (1.2)   If $\bigvee\limits_{u \in U} E_{(\mathbb{E},\rho)}(x, \phi(u)) = \mathbb{E}(x)$ , then there exists $u_0 \in U$ s.t. $\phi(u_0) = x$ .

    (2)   $\bigvee\limits_{u \in U} E_{(\mathbb{E},\rho)}(x, \phi(u)) = \bigvee\limits_{u \in U} E(x, \phi(u))$  for all  $x \in X$ .

    (3)   $F(u, \bar{u}) = E(\phi(u), \phi(\bar{u}))$ .


**Proof.** (a)($(i) \Longrightarrow (ii)$) Let $\phi$ be a SPSH(M)–monomorphims such that there exists a SPSH(M)–morphims $\chi$ making the diagram

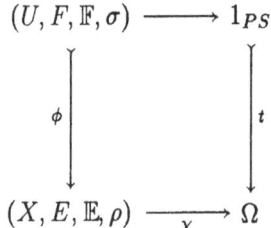

into a pullback square. Since the forgetful functors from SPSH(M) to Spsh(L) and SPSH(M) to M–SET, respectively , are adjoint (cf. Remark 5.3 (b) and (c)), we obtain that the subsequent diagrams

$$
\begin{array}{ccc}
(U, \mathbb{F}, \sigma) & \longrightarrow & (L, id_L, \wedge) \\
\phi \downarrow & & \downarrow t \\
(x, \mathbb{E}, \rho) & \xrightarrow{\ \chi\ } & (R_L, \Pi_1, \rho_L)
\end{array}
\qquad , \qquad
\begin{array}{ccc}
(U, F) & \longrightarrow & 1 \\
\phi \downarrow & & \downarrow t \\
(X, E) & \xrightarrow{\ \chi\ } & (R_L, E_L)
\end{array}
$$

are again pullback squares. Hence $\phi$ is an extremal monomorphims in Spsh(L), and $\chi : X \longmapsto R_L$  is given by  $\chi(x) = (\mathbb{E}(x), \mu(x))$   where

$$
\mu(x) = \bigvee\limits_{u \in U} E_{(\mathbb{E},\rho)}(x, \phi(u))
$$

From the commutativity of the right diagram we conclude

$$
\bigvee\limits_{u \in U} E(x, \phi(u)) \leq \mu(x)
$$

hence the property (2) follows from Lemma 4.3 . Finally the universal property of the right diagram implies (3) .

(b) $((ii) \implies (i))$ We construct a map $\quad \chi : X \longmapsto R_L \quad$ as follows

$$\chi(x) \quad = \quad (\mathbb{E}(x), \mu(x)) \quad \text{where} \quad \mu \quad = \quad \bigvee_{u \in U} E_{(\mathbb{I}, \rho)}(x, \phi(u)) \quad .$$

Then $\quad \chi : \quad (X, \mathbb{E}, \rho) \longmapsto (R_L, \Pi_1, \rho_L) \quad$ is an ordinary presheaf-morphism (in the sense of Spsh(L)) classifying the extremal monomorphism $\phi : \quad (U, \mathbb{F}, \sigma) \longmapsto (X, \mathbb{E}, \rho) \quad$ . Moreover we conclude from the properties (2) and (3) that $\chi$ is a M–SET-morphism (cf. Theorem 3.7) making the diagram

$$
\begin{array}{ccc}
(U, F) & \longrightarrow & 1 \\
\phi \downarrow & & \downarrow t \\
(X, E) & \underset{\chi}{\longrightarrow} & (R_L, E_L)
\end{array}
$$

into a pullback square. Referring to Lemma 6.2 we can put all things together and obtain that

$$
\begin{array}{ccc}
(U, F, \mathbb{F}, \sigma) & \longrightarrow & 1_{PS} \\
\phi \downarrow & & \downarrow t \\
(X, E, \mathbb{E}, \rho) & \underset{\chi}{\longrightarrow} & \Omega
\end{array}
$$

is a pullback square in SPSH(M) .
□

**Examples 6.6** Let (X,E) be a separated M–valued set and $(\tilde{X}, \tilde{E}, \tilde{\mathbb{E}}, \rho)$ be the presheaf over M generated freely by (X,E) (cf. Lemma 4.8 and Lemma 4.9) . Further let $\quad \mu : \quad X \longmapsto L \quad$ be a E–strict and E–extensional map –i.e.

$$\mu(x) \quad \leq \quad E(x, x) \quad , \quad \mu(x) * (E(x, x) \to E(x, \bar{x}) \quad \leq \quad \mu(\bar{x}) \quad .$$

Let us denote by $U_\mu$ the subset of all elements $(x, \alpha) \in \tilde{X}$ with $\alpha \leq \mu(x)$ . We note that $U_\mu$ is closed w.r.t. the separation process – i.e. , if $(x, \alpha) \in U_\mu$ and $\alpha \vee \bar{\alpha} \quad \leq \quad \tilde{E}((x, \alpha), (\bar{x}, \bar{\alpha})) \quad$ , then $\quad (\bar{x}, \bar{\alpha}) \in U_\mu$ . In fact the preceding hypothesis implies

$$
\begin{aligned}
\bar{\alpha} \quad & \leq \quad E(x, \bar{x}) * ((E(x, x) \to \alpha) \wedge (E(\bar{x}, \bar{x}) \to \bar{\alpha})) \\
& \leq \quad E(x, \bar{x}) * (E(x, x) \to \mu(x)) \quad \leq \quad \mu(\bar{x})
\end{aligned}
$$

Let us consider on $U_\mu$ the presheaf structure inherited from $(\tilde{X}, \tilde{E}, \tilde{\mathbb{E}}, \rho)$ ; then the inclusion map $i: U_\mu \longmapsto \tilde{X}$ is a presheaf-monomorphism in the sense of PSPH(M) . Taking on both sides the separated presheaves we obtain that the morphism $\hat{i}$ corresponding to $i$ is $(\Omega, t)$–classifiable. The conditions (1.1) and (3) are obvious . In order to verify (1.2) and (2) we proceed as follows : Let $E_{(\tilde{\mathbb{E}}, \rho)}$ be the canonical, L–valued equality associated with $(\tilde{X}, \tilde{E}, \tilde{\mathbb{E}}, \rho)$ ; then the equality $E_{(\tilde{\mathbb{E}}, \rho)}((x, \alpha), (x, \beta)) = \alpha \wedge \beta$ holds by definition. Further the strictness and extensionality of $\mu$ implies

$$E(x, y) * ((E(x, x) \to \alpha) \wedge (E(y, y) \to \mu(y))) \quad \le$$

$$(E(x, y) * (E(x, y) \to \alpha)) \wedge ((E(y, y) \to E(x, y)) * \mu(y)) \quad \le \quad \alpha \wedge \mu(x) \quad ,$$

$$E(x, x) * ((E(x, x) \to \alpha) \wedge (E(x, x) \to \mu(x))) \quad = \quad \alpha \wedge \mu(x) \quad ;$$

hence we obtain

$$\bigvee_{(y, \beta) \in U_\mu} \tilde{E}((x, \alpha), (y, \beta)) \quad =$$

$$\bigvee_{(y, \beta) \in U_\mu} E(x, y) * ((E(x, x) \to \alpha) \wedge (E(y, y) \to \beta)) \quad = \quad \alpha \wedge \mu(x) \quad =$$

$$\bigvee_{(y, \beta) \in U_\mu} E_{(\tilde{\mathbb{E}}, \rho)}((x, \alpha), (y, \beta)) \quad ;$$

i.e. properties (1.2) and (2) are established. According to Theorem 6.5 the monomorphism $\hat{i}$ is $(\Omega, t)$–classifiable.
□

**Remark 6.7 (Solution of the classification problem)** We maintain the notations in Examples 6.6 . Further let $\chi: (X, E) \longmapsto (R_L, E_L)$ be the M–SET–morphism corresponding to an strict and extensional map $\mu: X \longmapsto L$ (cf. Theorem 3.7) . Then the last equation in Examples 6.6 shows that the unique *characteristic* morphism classifying $\hat{i}$ coincides with the extension $\chi^\sharp$ of $\chi$ to a presheaf-morphism (cf. Theorem 4.9 ) . In this sense Examples 6.6 solve the classification problem mentioned in the introduction of this paper :
Let (X,E) be a separated, M–valued set and $(\tilde{X}, \tilde{E}, \tilde{\mathbb{E}}, \rho)\hat{\ }$ be the separated presheaf generated freely by (X,E) . Then the subobject corresponding to a strict and extensional map $\mu: X \longmapsto L$ is a separated, subpresheaf of $(\tilde{X}, \tilde{E}, \tilde{\mathbb{E}}, \rho)\hat{\ }$, the support set of which is given by $U_\mu$ .
□

We finish this paper with a brief discussion of *truth arrows* in SPSH(M) . The *conjunction* $\chi_\wedge: \Omega \times \Omega \longmapsto \Omega$ is obviously the characteristic morphism (i.e. classification) of the $(\Omega, t)$–classifiable subobject

$< t, t >: \quad 1_{PS} \longmapsto \Omega \times \Omega \quad$ . In particular $\chi_\wedge$ is given by

$$\chi_\wedge((\alpha, \lambda_1), (\alpha, \lambda_2)) \quad = \quad (\alpha, (\lambda_1 \wedge \lambda_2)) \quad .$$

Also the *disjunction* can be internalized as a presheaf-morphism
$\chi_\vee : \quad \Omega \times \Omega \longmapsto \Omega \quad$ within SPSH(M) and is given by

$$\chi_\vee((\alpha, \lambda_1), (\alpha, \lambda_2)) \quad = \quad (\alpha, (\lambda_1 \vee \lambda_2)) \quad .$$

What is more surprising is the fact that also the *negation* admits an internal-ization in SPSH(M) . For this purpose we consider the following map
$\chi_\neg : \quad R_L \longmapsto R_L \quad$ defined by

$$\chi_\neg(\alpha, \lambda) \quad = \quad (\alpha, (\alpha * (\lambda \to 0))) \quad .$$

**Proposition 6.8** $\quad \chi_\neg : \quad \Omega^* \longmapsto \Omega \quad$ *is a presheaf-morphism in the sense of SPSH(M) .*

**Proof.** (a) We show that $\chi_\neg$ is an ordinary presheaf-morphism. From the definition of $\rho_L^*$ (cf. Lemma 6.1) we derive the following relation

$$\chi_\neg(\rho_L^*((\alpha, \lambda), \beta)) \quad = \quad ((\alpha \wedge \beta), ((\alpha \wedge \beta) * [(\alpha \wedge \beta) \to ((\alpha \to \lambda) \to 0)]))$$
$$((\alpha \wedge \beta), (\alpha \wedge \beta \wedge ((\alpha \to \lambda) \to 0))) \quad .$$

Because of

$$\alpha \wedge ((\alpha \to \lambda) \to 0) \quad = \quad \alpha * (\alpha \to ((\alpha \to \lambda) \to 0)) \quad = \quad \alpha * (\lambda \to 0)$$

we obtain $\quad \chi_\neg(\rho_L^*((\alpha, \lambda), \beta)) \quad = \quad \rho_L(\chi_\neg(\alpha, \lambda), \beta) \quad .$
(b) We show that $\chi_\neg$ is a M–SET–morphism from $(R_L, E_L)$ to $(R_L, E_L)$ . Re-ferring to Example 3.2 (c) it is easy to verify the following relation

$$(\bar{\alpha} \to E_L((\alpha, \lambda), (\bar{\alpha}, \bar{\lambda}))) * \bar{\alpha} * (\bar{\lambda} \to 0) \quad \leq$$
$$\alpha * (\lambda \to \bar{\lambda}) * (\bar{\lambda} \to 0) \quad \leq \quad \alpha * (\lambda \to 0) \quad ;$$

hence the inequality

$$E_L((\alpha, \lambda), (\bar{\alpha}, (\bar{\lambda})) \quad \leq \quad \bar{\alpha} * [(\bar{\alpha} * (\bar{\lambda} \to 0)) \to (\alpha * (\lambda \to 0))]$$

holds. Analogously we verify

$$E_L((\alpha, \lambda), (\bar{\alpha}, \bar{\lambda})) \quad \leq \quad \alpha * [(\alpha * (\lambda \to 0)) \to (\bar{\alpha} * (\bar{\lambda} \to 0))].$$

Summing up we obtain $\quad E_L((\alpha, \lambda), (\bar{\alpha}, \bar{\lambda})) \quad \leq \quad E_L(\chi_\neg(\alpha, \lambda), \chi_\neg(\bar{\alpha}, \bar{\lambda})) \quad .$
Q.E.D.

**Corollary 6.9** *If the underlying GL–monoid $M = (L, \leq, *)$ is a complete MV–algebra (cf. section 2) , then $\quad \chi_\neg : \quad \Omega^* \longmapsto \Omega \quad$ is an SPSH(M)-isomorphism.*

**Concluding Remark.** The preceding considerations indicate that an internalization of intuitionistic as well as Lukasiewicz's connectives seems to be possible in SPSH(M) . In this sense the categories of separated presheaves over GL–monoids can be viewed as a sort of *"weak-quasitoposes"* – an idea arising also in L.N.Stout's survey article on relations between fuzzy set and topos theory ([23]) . Even though the axiomatization of this kind of structures is beyond the scope of this paper, we already would like to point out here that one of the starting points of this investigation will be the fact that the category SPSH(M) is (*extremal mono*)–*topological* over the *quasitopos* Spsh(L) and *monadic* over the *symmetric, monoidal closed* category M–SET.

# References

[1] J. Adámek, H. Herrlich and G.E.Strecker . *Abstract and Concrete Categories* (John Wiley & Sons , New York 1990).

[2] M.A. Arbib and E.G. Manes . *Arrows, Structures and Functors. The Categorical Imperative* (Academic Press , NewYork 1975).

[3] L.P. Belluce . *Semi-simple and complete MV-algebras*, Algebra Universalis **29** (1992), 1 – 9 .

[4] G. Birkhoff . *Lattice Theory* , Amer. Math. Soc. Colloquium Publications, third edition (Amer. Math. Soc., RI, 1973).

[5] C.C. Chang . *Algebraic analysis of many valued logics*, Trans. Amer. Math. Soc. **88** (1958), 467 – 490 .

[6] M. Fourman and D.S. Scott . *Sheaves and logic*, in : "Applications of Sheaves" , Lecture Notes in Mathematics **753**, 302 – 401 (Springer–Verlag 1979).

[7] O. Frink . *New algebras of logic*, Amer. Math. Monthly **45** (1938), 210 – 219 .

[8] J.A. Goguen . *L-Fuzzy sets* , J. Math. Anal. Appl. **18** (1967), 145 – 174 .

[9] J.A. Goguen . *The Tychonoff theorem* , J. Math. Anal. Appl. **43** (1973), 734 – 742 .

[10] J.A. Goguen . *Concept representation in natural and artificial languages* , Internat. J. Man-Machine Stud. **6** (1974), 513 – 561 .

[11] R. Goldblatt . *Topoi : The Categorial Analysis of Logic* (North–Holland , Amsterdam 1979).

[12] H. Herrlich and G.E. Strecker . *Category Theory* (Allyn and Bacon, Boston 1973).

[13] U. Höhle . *Editorial of the Special Issue : Mathematical Aspects of Fuzzy Set Theory* , Fuzzy Sets and Systems **40** (1991), 253 – 256 .

[14] U. Höhle . *Monoidal closed categories, weak topoi and generalized logics* , Fuzzy Sets and Systems **42** (1991), 15 – 35 .

[15] U. Höhle . *M-valued sets and sheaves over integral cl-monoids*, in "Applications of catgeory Theory to Fuzzy Subsets" , p. 33 – 72 , (Eds. S.E. Rodabaugh et al.) (Kluwer , Bosten 1992).

[16] P.T. Johnstone . *Stone Spaces* (Cambridge University Press, Cambridge 1982).

[17] S. Maclane . *Categories for the Working Mathematician* (Springer–Verlag 1971).

[18] E.G. Manes . *Algebraic Theories*, Graduate Texts in Mathematics **26** (Springer–Verlag, Berlin, New York 1976).

[19] A. Mostowski . *Proofs of non-deducibility in intuistionistic functional calculus* , J. Symbolic Logic **13** (1948), 204 – 207 .

[20] H. Rasiowa and R. Sikorski . *The Mathematics of Metamathematics* , third edition (Polish Scientific Publishers , Warszawa 1970).

[21] K.I. Rosenthal . *Quantales and Their Applications* , Pitman Research Notes in mathematics **234** (Longman, Burnt Mill, Harlow 1990).

[22] B. Schweizer and A. Sklar . *Probabilistic Metric Spaces* (North–Holland , Amsterdam 1983).

[23] L.N. Stout . *A survey of fuzzy set and topos theory*, Fuzzy Sets and Sytsems **42** (1991), 3 – 14 .

[24] O. Wyler . *Lecture Notes on Topoi and Quasitopoi* (World Scientific, Singapore 1991).

[25] L.A. Zadeh . *Fuzzy sets*, Information and Control **8** (1965), 338 – 353 .

# VII

# Quantales: Quantal sets

C.J. Mulvey and M. Nawaz

## Introduction

In this paper, we investigate the ideas involved in extending to the non-commutative case the concepts of presheaf and sheaf on a locale. The context within which we shall work will be that of quantales, introduced by the first author [10,11] to provide a framework for the development of the Gelfand representation of not necessarily commutative C*-algebras, extending the lattice theoretic ideas of Dilworth and Ward [3], and the spectral insights of Giles and Kummer [7] and of Akemann [1]. The ideas which we present are largely contained in the thesis of the second author [15], although their development here will be from a very slightly different perspective. That this has taken a little time to bring to print is due in part to the publication of a summary of the thesis elsewhere [2], developing the ideas further in another direction. An alternative approach to the problem discussed there will be undertaken in a sequel to the present paper [14].

The particular quantales that will be considered here arise, amongst other places, in the context of considering the spectrum Max $A$ of a C*-algebra $A$. At the time that the work was done, the quantales which will be considered here, which we now refer to as *right Gelfand quantales*, were being actively investigated in this context, following the work of Giles and Kummer [7] on spectral representations of C*-algebras. Subsequently, due to the insightful contributions of Rosicky [17], a broader context, namely that of Gelfand quantales, was found for these ideas [12]. These right Gelfand quantales were then seen to arise typically as the right sides of Gelfand quantales, with this same right side carrying the topological structure of the spectrum within the ambient structure of the Gelfand quantale [13].

Although the concept of sheaf which one might now wish to define would be one that is defined on the Gelfand quantale itself, the present study of a concept defined on a right Gelfand quantale would be expected to form the right reflection, in some sense, of that more general concept yet to be studied in any depth. The beginnings of approaches to this may be found in the papers of Nawaz [16] and of French [5]. A definitive attack on these ideas, with the aim of obtaining a sheaf theoretic context for the Gelfand representation of non-commutative C*-algebras, remains to be mounted. Our belief, however, is that the concepts of quantal set and of sheaf arising in that more general context will inevitably admit a reflection to the right (and left) sides of the Gelfand quantale Max $A$ concerned. For the moment, therefore, we investigate only the case of a right Gelfand quantale, both for its own interest, and to provide a point of reference for the development of the more general theory.

Within this context, in which many of the difficulties involved in working over a quantale are overcome due to the particular properties encountered, it proves possible to arrive at appropriate definitions of the concepts of presheaf and of sheaf, and to show the existence of a reflection of the category of presheaves into the category of sheaves on the quantale, corresponding to the process of sheafification. The approach taken to achieving this, however, is from the concept of a quantal set, generalising that of Heyting-valued, or as we would prefer to call it, local, set, introduced initially by Higgs [8] as a generalisation, and extension, of the notion of a Boolean-valued set [18], and further developed by Fourman and Scott [6].

The concept of quantal set is fundamental to the discussion, the principle being that a definition of the category of quantal sets over a right Gelfand quantale $Q$ is already an exacting task within this non-commutative context. To show that the concept of completeness may be extended to this situation itself provides some indication that the definitions taken are in some measure correct. While to find that these ideas may be tied consistently in to concepts of presheaf, and of sheaf, in a way that directly extends the localic case, while introducing aspects that reflect the non-commutative context within which one is working, gives an impression that the concepts introduced are, in some sense or other, quite natural.

The agenda, in some sense, is that one knows in advance the theorems that are to be proved. The question is whether one can find the definitions that allow their proof. The constraints are that, certainly, the concepts should become those with which one is familiar in the case that the quantale concerned is actually a locale, while, in the case that it is not, the quantale $Q$ itself should still be able to be recovered from the quantal sets defined over it, by considering the subsets of the quantal set $1_Q$. Indeed, these ideas should extend to the interpretation of a non-commutative predicate calculus in which the structure of quantal sets is reflected in the existence of quantal sets of subsets based on exponentiation of a quantal subset classifier [14]. Furthermore, as a subtext to this agenda, one has the requirement that the concepts should appear natural within, indeed should reflect, the non-commutative context within which the theory is developed.

The development of the logical aspects of these ideas gives further indication that the structures concerned are of intrinsic interest, although these aspects are those left to a later paper. The ideas involved will be alluded to in the final section of the present paper, in which a particularly murky aspect of the theory developed in the earlier sections is finally revealed. In those earlier sections, it has been our decision to include more detail in the proofs than might normally be considered necessary. This is in part due to our own experience of finding that the manipulations needed to establish a particular inequality were often more easily lost than found. Having found them, it seemed wise to write them down.

# 1 Quantales

The concept of a quantale was introduced [10,11] to provide a non-commutative generalisation of that of a locale [9]. On the one hand, this generalisation was intended to allow the consideration of non-commutative spaces, in particular that of the spectrum of a C*-algebra. On the other, the concept was expected to relate to the semantics of non-commutative logics, in particular that of quantum mechanics. The formalisation of these ideas leads to the following:

**Definition 1** *By a quantale $Q$ will be meant a lattice having arbitrary joins $\bigvee$, together with an associative product & satisfying*

$$a \ \& \ \bigvee b_i \ = \ \bigvee a \& b_i$$
$$and \quad \bigvee a_i \ \& \ b \ = \ \bigvee a_i \& b$$

*for all $a, b, a_i, b_i \in Q$.*

It may be remarked that any quantale $Q$ has a greatest element, $1_Q$, and a least element, $0_Q$, which will be referred to respectively as the *identity* and the *zero* elements of the quantale. As one consequence of the distributivity of the product & over any arbitrary join $\bigvee$, it may be remarked that in any quantale $Q$ one has that

$$p \leq q \quad \text{implies that} \quad p \& r \leq q \& r$$

for any $p, q, r \in Q$, with a similar result holding when taking the product on the left. One also has that

$$0_Q \& q = 0_Q = q \& 0_Q$$

for any $q \in Q$.

Any locale $L$ is a quantale, by taking the product & to be the operation of meet $\wedge$ in the lattice $L$. Since the identity $1_L$ in this case satisfies the condition that

$$1_L \wedge q = q = q \wedge 1_L$$

for any $q \in L$, the identity of a locale is actually a unit for the quantale, in the following sense:

**Definition 2** *By a* unit *$e \in Q$ for a quantale $Q$ will be meant an element of $Q$ satisfying*

$$e \,\&\, q = q = q \,\&\, e$$

*for all $q \in Q$. A quantale $Q$ will be said to be* unital *provided that it admits a unit, necessarily unique.*

Although in any locale $L$ the identity $1_L$ is a unit for the locale, in a quantale that is unital this is generally far from being the case. Indeed, a more typical example of a quantale that is unital is given by the following [13]:

**Definition 3** *By the* spectrum Max $A$ *of a $C^*$-algebra $A$ will be meant the quantale of closed linear subspaces of $A$, together with the join $\bigvee$ given by*

$$\bigvee_i M_i = \overline{\sum_i M_i}$$

*for any $M_i \in \mathrm{Max}\, A$, and the product $\&$ defined by*

$$M \,\&\, N = \overline{M \cdot N}$$

*for any $M, N \in \mathrm{Max}\, A$.*

The product $\&$ is given by taking the closure of the algebraic product

$$M \cdot N = \{ \sum_i a_i b_i \mid a_i \in M, \, b_i \in N \}$$

of the closed linear subspaces of $A$, while the join $\bigvee$ is given by taking the closure of the algebraic sum

$$\sum_i M_i = \{ \sum_j a_{i_j} \in A \mid a_{i_1} \in M_{i_1}, \ldots, a_{i_n} \in M_{i_n} \}.$$

It is evident that the closed linear subspace $e \in \mathrm{Max}\, A$ generated by the unit $1 \in A$ of the $C^*$-algebra $A$, always assumed to be unital, is a unit for the quantale Max $A$. It may also be remarked that this is certainly not the identity of the quantale, which is the $C^*$-algebra $A$ itself.

Although the identity of a quantale $Q$ is generally not a unit for the quantale, it is of interest to consider those elements on which it acts on one side or another as a unit for the product, in the sense of the following:

**Definition 4** *An element $q \in Q$ of a quantale $Q$ will be said to be* right-sided *(respectively* left-sided*) in the case that*

$$q \,\&\, 1_Q \leq q \ (\text{respectively } 1_Q \,\&\, q \leq q \ ).$$

*An element $q \in Q$ that is both right-sided and left-sided will be said to be* two-sided.

It may be remarked immediately that in any quantale $Q$ that is unital, the inequalities may be replaced by equalities, since

$$q = q \& e \leq q \& 1_Q \quad \text{(respectively} \quad q = e \& q \leq 1_Q \& q \text{)}$$

for any $q \in Q$.

We have already noted that in a locale $L$, the identity is a unit for the quantale, hence every element of the locale is two-sided. However, in the quantale Max $A$ that is the spectrum of a C*-algebra $A$, the right-sided (respectively left-sided) elements are exactly the closed right (respectively left) ideals of $A$, since, for any $M \in$ Max $A$, the product

$$M \& A \text{ (respectively } A \& M \text{ )}$$

is exactly the closure of the right (respectively left) ideal generated by the closed linear subspace $M$. In particular, the two-sided elements of Max $A$ are the closed ideals of the C*-algebra $A$.

For any quantale $Q$, the right-sided (respectively left-sided) elements again form a quantale with respect to the operations of the quantale $Q$, which we shall denote by

$$R(Q) \text{ (respectively } L(Q) \text{ )} ,$$

referred to as the *right* (respectively the *left*) *side* of the quantale $Q$. Similarly, the two-sided elements are closed under product and arbitrary joins, hence form a quantale

$$I(Q) ,$$

the *two-sided part* of the quantale $Q$. It should, however, be noted that the right sided part $R(Q)$ of a unital quantale $Q$ may in general no longer be unital. Indeed, if it has a unit, then this is necessarily the identity of the quantale $Q$, from which it follows that

$$R(Q) = I(Q) .$$

It may be remarked that for the spectrum Max $A$ of a C*-algebra $A$, the quantale

$$I(\text{Max } A)$$

is actually a locale. Moreover, for a commutative C*-algebra $A$, this locale is the spectrum of $A$ in the conventional sense that the locale of closed ideals of $A$ is that of the weak* topology on the space of multiplicative linear functionals on $A$. Concerning the quantale

$$R(\text{Max } A)$$

(with corresponding assertions about the quantale $L(\text{Max } A)$), it may be shown that

$$I \& I = I$$

for every closed right ideal $I \in R(\mathrm{Max}\,A)$, a result which is a consequence [15] of the existence of approximate units in arbitrary C*-algebras [4], and which leads us to the following:

**Definition 5** *A quantale $Q$ will be said to be* idempotent *provided that*

$$p \,\&\, p = p$$

*for each $p \in Q$, and to be* right-sided *provided that*

$$p \,\&\, 1_Q \leq p$$

*for each $p \in Q$.*

Any locale $L$ is an idempotent right-sided quantale that is also unital. Moreover, conversely, any idempotent, right-sided quantale $Q$ that is unital has its product given by the meet of the lattice $Q$, hence is actually a locale. For, one observes firstly that the unit $e \in Q$ of such a quantale $Q$ must actually be the identity of $Q$, since $e \,\&\, 1_Q = e$ by the right-sidedness of the quantale, while $e \,\&\, 1_Q = 1_Q$ since $e \in Q$ is a unit. Hence, any element $p \in Q$ is necessarily two-sided. Thus, given any elements $p, q \in Q$ of the quantale $Q$, we have, on the one hand, that

$$p \,\&\, q \leq p \,\&\, 1_Q = p \quad \text{and} \quad p \,\&\, q \leq 1_Q \,\&\, q = q \, ,$$

which implies that $p \,\&\, q \leq p \wedge q$, while, on the other,

$$p \wedge q \leq p \quad \text{and} \quad p \wedge q \leq q \, ,$$

which implies that $p \wedge q = (p \wedge q) \,\&\, (p \wedge q) \leq p \,\&\, q$, yielding the required equality. Since the product, now known to be the meet, distributes over arbitrary joins, the quantale $Q$ is a locale.

For an idempotent, right-sided quantale $Q$ which is not necessarily unital, there is still a significant way that one may go along the path towards locales, in the following [15]:

**Theorem 6** *For any idempotent, right-sided quantale $Q$ , one has that*

$$p \,\&\, q \,\&\, r = p \,\&\, r \,\&\, q$$

*for any elements $p, q, r \in Q$. In particular, for any element $p \in Q$, the quantale*

$$Q_p = \{ p \,\&\, q \in Q \mid q \in Q \}$$

*obtained by localising at the element $p \in Q$ is a locale.*

Although its consequences are extensive, the condition asserted, which we shall refer to as the *right-symmetricity* of the quantale $Q$, is almost trivially proved. For any $p, q, r \in Q$, one has that

$$p \,\&\, q \,\&\, r = p \,\&\, q \,\&\, r \,\&\, p \,\&\, q \,\&\, r$$

$$\leq p \,\&\, 1_Q \,\&\, r \,\&\, 1_Q \,\&\, q \,\&\, 1_Q$$

$$= p \,\&\, r \,\&\, q \,,$$

by respectively the idempotency of $Q$, the monotonicity of the product of $Q$, and the right-sidedness of the elements of $Q$. Similarly,

$$p \,\&\, r \,\&\, q \leq p \,\&\, q \,\&\, r \,,$$

giving the required result.

Considering now the subset $Q_p$, we observe that, by the distributivity of the product over the arbitrary joins in the quantale, the subset is closed under arbitrary joins; while, by the idempotency and the right-symmetricity of the quantale, one has that

$$(p \,\&\, q) \,\&\, (p \,\&\, r) = p \,\&\, p \,\&\, q \,\&\, r = p \,\&\, (q \,\&\, r) \,,$$

giving that it is closed under the product. The quantale is idempotent, since by the above identity, one has that

$$(p \,\&\, q) \,\&\, (p \,\&\, q) = p \,\&\, (q \,\&\, q) = p \,\&\, q \,.$$

It is also right-sided: for $p = p \,\&\, 1_Q \in Q_p$, by the right-sidedness of $Q$. Hence, $p \in Q_p$ is the identity of the quantale $Q_p$, since $p \,\&\, q \leq p \,\&\, 1_Q = p$ for any $q \in Q$. But then,

$$(p \,\&\, q) \,\&\, p = p \,\&\, p \,\&\, q = p \,\&\, q \,.$$

But, by a similar argument on the left side, we have that

$$p \,\&\, (p \,\&\, q) = (p \,\&\, p) \,\&\, q = p \,\&\, q \,,$$

giving that $p \in Q_p$ is actually a unit for the quantale $Q_p$. Thus, by our earlier remarks, the quantale $Q_p$ is actually a locale. $\square$

There is therefore a sense in which an idempotent, right-sided quantale $Q$ may be considered as being constructed from the locales $Q_p$ obtained by localising at each $p \in Q$, with the quantale $Q$ itself binding these locales together in a non-commutative manner. An immediate corollary to this result is that obtained by localising at the identity element of the quantale, extending the observation already made in the case of the spectrum Max $A$ of a C*-algebra $A$ :

**Corollary 7** *For any idempotent, right-sided quantale* $Q$ *, the quantale*

$$I(Q)$$

*of two-sided elements of* $Q$ *is necessarily a locale. Moreover, the idempotent, right-sided quantale* $Q$ *is a locale exactly if* $Q$ *is unital.*   $\square$

It may be observed in passing that in any idempotent quantale $Q$, one again has that an element is right-sided (respectively left-sided) exactly if

$$q \,\&\, 1_Q = q \quad (\text{respectively} \quad 1_Q \,\&\, q = q)$$

for any $q \in Q$, since already we have that

$$q = q \,\&\, q \leq q \,\&\, 1_Q \quad (\text{respectively} \quad q = q \,\&\, q \leq 1_Q \,\&\, q).$$

In the case of the spectrum Max $A$ of a C*-algebra $A$, it is the idempotent, right-sided quantale

$$R(\text{Max } A) \ ,$$

or equivalently the involutively isomorphic idempotent, left-sided quantale $L(\text{Max } A)$, that appears to carry the topological structure of the spectrum [13]. Indeed, it was these lattices of closed right and left ideals that were identified by Giles and Kummer [7], and by Akemann [1], as providing the non-commutative topology in terms of which the Gelfand representation of the C*-algebra might be expressed.

The development of these ideas into the context of quantales, and particularly the introduction of a concept of complete regularity within this context, and the observation that this could be defined intrinsically within the framework of Gelfand quantales, provide incentive for investigating the concepts of quantal set, and of presheaf and sheaf, over quantales that are idempotent and right-sided.

In view of the connection with Gelfand quantales [12], on the one hand, and with the Gelfand representation of C*-algebras [13], on the other, we shall make the following:

**Definition 8** *By a* right Gelfand *quantale* $Q$ *will be meant a quantale that is idempotent, and right-sided.*

It is with these quantales that we shall be concerned for the rest of this paper, and the quantales considered should throughout be assumed to be right Gelfand quantales unless otherwise explicitly stated.

## 2   Quantal Sets

Consider now a quantale $Q$, henceforth always assumed to be idempotent and right-sided, that is to say, a right Gelfand quantale. It will be recalled that such a quantale $Q$ is a locale exactly if it is unital. Our concern now is to describe a concept of quantal set over the quantale, in the sense of a set together with extent and equality relations taking values in $Q$, in such a way that this coincides with that of a Heyting-valued set over $Q$ in the case that the quantale is actually a locale [8], yet which reflects its non-commutativity in the case that it is not.

**Definition 9** *Given a right Gelfand quantale $Q$, by a* quantal set *over $Q$ will be meant a set $A$, together with mappings:*

$$E : A \to Q$$

$$and \quad [\cdot = \cdot] : A \times A \to Q ,$$

*which together satisfy the following conditions:*

*i)*   $Ea \& Ea \leq [a = a]$ ;

*ii)*   $[a = b] \leq Ea \& Eb$ ;

*iii)*   $Ea \& [b = a] \leq [a = b]$ ;

*iv)*   $[a = b] \& [b = c]$ ,

*for all $a, b, c \in A$.*

*The elements $Ea$ and $[a = b]$ of the quantale $Q$ will be called respectively the* extent *and the* equality *of the quantal set $A$.*

It may be observed immediately that:

$$Ea = [a = a]$$

for each $a \in A$, because of the idempotency of the quantale $Q$. It will be noted that the equality of a quantal set is not required to be symmetric, but rather to satisfy a condition akin to that of right symmetricity satisfied by the quantale. It will be seen that this condition implies that the equality is actually symmetric in the case that the quantale is indeed a locale.

For any quantale $Q$, it may be deduced from its right symmetricity that the equality of a quantal set satisfies the condition that:

$$Ea \& [a = b] = [a = b] = [a = b] \& Eb$$

for all $a, b \in A$. From this observation it follows that the equality of the quantal set is right symmetric, in the sense that:

$$p \& [a = b] = p \& [b = a] ,$$

for any $p \in Q$ and for all $a, b \in A$. That the equality is not generally symmetric can be seen from the following:

EXAMPLE.    Given a quantale $Q$, consider the quantal set

$$1_Q$$

obtained by taking the set $Q$, together with the extent and equality relations defined by:

$$Eq = q \, ,$$

$$\text{and} \quad [p = q] = p \& q \, ,$$

for all $p, q \in Q$.

It may be observed that in showing that this is a quantal set only the assumption that the quantale is right-sided is needed. The equality evidently is symmetric exactly if the quantale is commutative. In the presence of the assumption of idempotency also, this is again equivalent to the quantale being a locale, since the identity is then a unit for the quantale.

Now the idea of a map of quantal sets over the quantale $Q$ may be introduced by the following:

**Definition 10** *Given a right Gelfand quantale $Q$, by a* map of quantal sets

$$f : A \to B$$

*over the quantale $Q$ will be meant a mapping*

$$f : A \times B \to Q$$

*which satisfies the following conditions:*

> *i)*   $f(a, b) \leq Ea \& Eb$ ;
> *ii)*  $[a = a'] \& f(a', b) \leq f(a, b)$ ;
> *iii)* $f(a, b) \& [b = b'] \leq f(a, b')$ ;
> *iv)*  $Eb \& Eb' \& f(a, b) \& f(a, b') \leq [b = b']$ ;
> *v)*   $Ea \leq \bigvee_b f(a, b)$ ,

*for all $a, a' \in A$ and $b, b' \in B$.*

One is therefore describing a relation defined on $A \times B$ over the quantale $Q$, which is required to be functional with respect to the quantale $Q$. Again, the existence of a right identity element in the quantale allows it to be shown that:

$$Ea = \bigvee_b f(a, b) \, ,$$

for each $a \in A$. Moreover, from the right symmetricity of the quantale, one deduces also that:

$$Ea \& f(a, b) = f(a, b) = f(a, b) \& Eb ,$$

for all $a \in A$ and $b \in B$.

EXAMPLE. For any quantal set $A$ over $Q$, there is a map

$$\tau_A : A \rightarrow 1_Q$$

of quantal sets, defined by assigning to each $a \in A$ and each $q \in Q$ the element

$$\tau_A(a, q) = Ea \& q$$

of the quantale $Q$.

It may be seen straightforwardly that this is a map of quantal sets. For instance, to verify that

$$[\![a = a']\!] \& \tau_A(a', q) \leq \tau_A(a, q) ,$$

we note that $[\![a = a']\!] \leq Ea \& Ea'$, hence, $[\![a = a']\!] \& Ea' \leq Ea \& Ea' \leq Ea$, and so, $[\![a = a']\!] \leq Ea' \& q \leq Ea \& q$. It will be seen later that this map is the unique map of quantal sets from $A$ to the quantal set $1_Q$.

With these preliminaries, we may now state the following:

**Theorem 11** *For any right Gelfand quantale $Q$ , the quantal sets and the maps of quantal sets over $Q$ determine a category*

## $Q$-Sets

*of quantal sets over the quantale $Q$ .*

Evidently, it remains to define the notions of the identity map on any quantal set, and of the composition of maps of quantal sets: firstly, for any quantal set $A$, define

$$id_A : A \rightarrow A$$

to be given by the equality relation

$$[\![ \cdot = \cdot ]\!] : A \times A \rightarrow Q ;$$

next, for any maps $f : A \rightarrow B$ , $g : B \rightarrow C$ of quantal sets, define

$$gf : A \rightarrow C$$

to be the relation given by setting

$$(gf)(a, c) = \bigvee_b f(a, b) \& g(b, c)$$

for each $a \in A$ and $c \in C$.

Then it may be verified straightforwardly that these yield a category of quantal sets and maps of quantal sets over $Q$, although it may be remarked that the functionality of the composite of maps depends critically on the right symmetricity of the quantale. Explicitly, to verify that

$$Ec \& Ec' \& h(a, c) \& h(a, c') \leq [\![c = c']\!]$$

for the composite $h$ of $f$ and $g$, one observes that the left hand side is given by:

$$Ec \& Ec' \& \bigvee_b f(a, b) \& g(b, c) \& \bigvee_{b'} f(a, b') \& g(b', c') \ ,$$

which by distributivity may be written

$$\bigvee_{b, b'} Ec \& Ec' \& f(a, b) \& g(b, c) \& f(a, b') \& g(b', c') \ .$$

Applying the right symmetricity of the quantale, and the functionality of $f$ and of $g$, one obtains successively:

$$= \bigvee_{b, b'} Ec \& Ec' \& f(a, b) \& f(a, b') \& g(b, c) \& g(b', c')$$

$$= \bigvee_{b, b'} Ec \& Ec' \& Eb \& Eb' \& f(a, b) \& f(a, b') \& g(b, c) \& g(b', c')$$

$$\leq \bigvee_{b, b'} Ec \& Ec' \& [\![b = b']\!] \& g(b, c) \& g(b', c')$$

$$= \bigvee_{b, b'} Ec \& Ec' \& g(b, c) \& [\![b = b']\!] \& g(b', c')$$

$$\leq [\![c = c']\!],$$

which establishes the functionality of the composite. That the identity maps of the category compose identically follows from the extensionality of any map of quantal sets, together with the observation made earlier that:

$$Ea \& f(a, b) = f(a, b) = f(a, b) \& Eb \ ,$$

for any $a \in A$ and $b \in B$. All other details of the proof will be omitted.  $\square$

The immediate observation that one may make is then:

**Corollary 12** *For any locale $Q$ , the category*

$$Q\text{-Sets}$$

*is exactly the category of local sets and of maps of local sets over the locale $Q$ .*

It has already been remarked that in this case the commutativity of the multiplication of the quantale yields that equality is necessarily symmetric on any quantal set: for,

$$[\![a = b]\!] = [\![a = b]\!] \& Eb = Eb \& [\![a = b]\!]$$

implies that $[\![a = b]\!] \leq [\![b = a]\!]$, by the condition which gives the right symmetricity of equality, from which it follows that:

$$[\![a = b]\!] = [\![b = a]\!]$$

for all $a, b \in A$, by interchanging the elements considered. But, conversely, any local set over the locale has an equality which *a fortiori* is right symmetric. Hence, the concept of quantal set coincides with that of local set, in the case of a locale $Q$.

Equally, the conditions which define a map of quantal sets are equivalent to those for a map of local sets over the locale $Q$. Only that expressing functionality needs the observation that:

$$Eb \& Eb' \& f(a, b) \& f(a, b') \;=\; f(a, b) \& Eb \& f(a, b') \& Eb' \;=\; f(a, b) \& f(a, b') \,,$$

in this case, which yields that expressing the functionality of a map of local sets over the locale $Q$.  □

More generally, one may obtain the following characterisation of isomorphisms in the category of quantal sets over any quantale:

**Corollary 13** *For any right Gelfand quantale $Q$ , a map*

$$f : A \rightarrow B$$

*of quantal sets over $Q$ is an isomorphism if, and only if, the following conditions are satisfied:*

$$i) \quad f(a, b) \& f(a', b) \leq [\![a = a']\!]$$

*for all $a, a' \in A$ and any $b \in B$; and,*

$$ii) \quad Eb \leq \bigvee_a Eb \& f(a, b),$$

*for all $b \in B$.*

For suppose that

$$f : A \rightarrow B$$

is an isomorphism of quantal sets over $Q$, of which the inverse is

$$g : B \rightarrow A \,.$$

Then, firstly, $\bigvee_b f(a,b)\&g(b,a') = [\![a = a']\!]$ for all $a, a' \in A$, because this composite is the identity map on the quantal set $A$. Hence,

$$f(a,b)\&g(b,a') \leq [\![a = a']\!]$$

for each $b \in B$, and

$$Ea' = \bigvee_{b'} f(a',b')\&g(b',a')$$

for all $a' \in A$. Then,

$$f(a,b)\&f(a',b) = f(a,b)\&Ea'\&f(a',b) = f(a,b)\&f(a',b)\&Ea'$$

$$= \bigvee_{b'} f(a,b)\&f(a',b)\&f(a',b')\&g(b',a')$$

$$\leq \bigvee_{b'} f(a,b)\&[\![b = b']\!]\&g(b',a') \leq f(a,b)\&g(b,a')$$

$$\leq [\![a = a']\!] \ .$$

Now, secondly, the fact that composition in the reverse order yields the identity map on the quantal set $B$ gives that

$$[\![b = b']\!] = \bigvee_a g(b,a)\&f(a,b')$$

for each $b, b' \in B$, hence, in particular, that

$$Eb = \bigvee_a g(b,a)\&f(a,b) = \bigvee_a Eb\&g(b,a)\&f(a,b) \leq \bigvee_a Eb\&f(a,b) \ .$$

So, both conditions are therefore satisfied.

Conversely, given a map of quantal sets

$$f : A \to B$$

satisfying these conditions, consider the mapping

$$g : B \times A \to Q$$

defined by

$$g(b,a) = Eb\&f(a,b)$$

for each $a \in A$ and $b \in B$. It is asserted, firstly, that this determines a map of quantal sets

$$g : B \to A \ ,$$

and, moreover, that this provides an inverse for the given map. For, the extensionality of the relation defined may be straightforwardly verified. That it is functional and total is seen by:

$$Ea\&Ea'\&g(b,a)\&g(b,a') = Ea\&Ea'\&Eb\&f(a,b)\&Eb\&f(a',b)$$

$$\leq Ea\&Ea'\&f(a,b)\&f(a,'b) \leq Ea\&Ea'\&[\![a=a']\!]$$
$$= Ea\&[\![a=a']\!]\&Ea'$$
$$= [\![a=a']\!]\ ,$$

by applying the first condition; while, by the second, one has that

$$[\![b=b']\!] = Eb\&[\![b=b']\!] \leq \bigvee_a Eb\&f(a,b)\&[\![b=b']\!]$$

$$\leq \bigvee_a Eb\&f(a,b)\&f(a,b)\&[\![b=b']\!] = \bigvee_a g(b,a)\&f(a,b')\ ,$$

while, conversely ,

$$\bigvee_a g(b,a)\&f(a,b') = \bigvee_a Eb\&f(a,b)\&f(a,b') \leq Eb\&[\![b=b']\!]$$

$$= [\![b=b']\!]\ .$$

Hence, the reverse composite yields the identity map on $B$. The map of quantal sets

$$f:A \to B$$

is therefore an isomorphism, of which the inverse has been explicitly constructed by the assignment defined above.  □

Finally, we note the following:

**Corollary 14** *For any right Gelfand quantale $Q$ , the quantal set*

$$1_Q$$

*is the terminal object in the category of quantal sets over $Q$ .*

For, it has already been remarked that, for any quantal set $A$, there exists a map

$$\tau_A : A \to 1_Q$$

in the category of quantal sets over $Q$. That this is the unique such map for any quantal set $A$ may be seen by noting that any map

$$t:A \to 1_Q$$

has the property that $t(a,q) = t(a,q)\&Eq = t(a,q)\&q$ for each $a \in A$ and each $q \in Q$. Then, $t(a,q) = t(a,q)\&q\&1 \leq t(a,1)$ by extensionality in $1_Q$ ; hence, $t(a,q) = t(a,q)\&q \leq t(a,1)\&q$. But, $t(a,1)\&q = t(a,1)\&1\&q \leq t(a,q)$, by extensionality again; hence, $t(a,q) = t(a,1)\&q$ for each $a \in A$ and each $q \in Q$. Now, $\bigvee_q t(a,q) = Ea$, by the totality of the map; hence $t(a,1) = Ea$, because one has that $t(a,q) \leq t(a,1)$ for all $q \in Q$. Thus,

$$t(a,q) = Ea\&q$$

for each $a \in A$ and each $q \in Q$. So the map considered is exactly the map

$$\tau_A : A \to 1_Q ,$$

which is therefore unique.   □

## 3   Complete $Q$-Sets

Once a concept of quantal set is given, one approach to the idea of a sheaf on the quantale, extending that developed in the context of locales [6], lies in determining an appropriate notion of completeness for a quantal set. Explicitly, one wishes to describe firstly the notion of a subset of a quantal set, of which certain subsets will then be defined to be the singletons. A quantal set will then be considered complete provided that each singleton subset is determined by a unique element of the set. Towards this end we make first the following:

**Definition 15** *By a subset of a quantal set $A$ over a right Gelfand quantale $Q$ will be meant a mapping*

$$s : A \to Q$$

*which satisfies the following conditions:*

$$i) \quad s(a)\&[\![a = a']\!] \le s(a') ;$$
$$ii) \quad s(a) \le s(a)\&Ea ,$$

*for each $a, a' \in A$.*

EXAMPLE.   Consider the quantal set

$$1_Q$$

which is the terminal object in the category of quantal sets over $Q$. For each $p \in Q$, consider the assignment

$$\tilde{p} : Q \to Q$$

defined by:

$$\tilde{p}(q) = p\&q$$

for each $q \in Q$. Then this yields a subset of $1_Q$. For, the first condition is satisfied since:

$$p\&q\&q\&q' \le p\&q' ,$$

by the right identity of the quantale, while the second:

$$p\&q \le p\&q\&q ,$$

is satisfied by the idempotency of the quantale. So, each element of the quantale $Q$ determines a subset of the quantal set $\mathbf{1}_Q$. In fact, it will be shown later that each subset of $\mathbf{1}_Q$ is obtained in this way for a unique $p \in Q$. The subsets of the quantal set $\mathbf{1}_Q$ therefore correspond precisely to the elements of the quantale $Q$.

The subsets of $\mathbf{1}_Q$ just described are particular instances of the following concept:

**Definition 16** *By a* singleton *of a quantal set $A$ over a quantale $Q$ will be meant a subset*

$$s : A \to Q$$

*of $A$ which satisfies the condition that:*

$$Ea\&s(a)\&s(a') \le [\![a = a']\!] \,,$$

*for each $a, a' \in A$.*

EXAMPLE.  For any quantal set $A$ over $Q$, each element $a \in A$ determines a subset

$$\tilde{a} : A \to Q$$

by the assignment:

$$\tilde{a}(b) = [\![a = b]\!] \,,$$

for each $b \in A$. The subset of $A$ obtained is a singleton of the quantal set $A$. It is a subset, because $[\![a = b]\!]\&[\![b = b']\!] \le [\![a = b']\!]$ and $[\![a = b]\!] \le [\![a = b]\!]\&Eb$ for all $b, b' \in A$. Moreover, it is a singleton, because

$$Eb\&[\![a = b]\!]\&[\![a = b']\!] \ \le [\![b = a]\!]\&[\![a = b']\!]$$

$$\le [\![b = b']\!] \,,$$

for all $b, b' \in A$. It may be remarked that the right symmetricity,

$$Eb\&[\![a = b]\!] \le [\![b = a]\!] \,,$$

of equality in a quantal set is exactly what is required in showing that this subset is indeed a singleton of $A$.

It may be noted generally that any subset

$$s : A \to Q$$

of a quantal set $A$ necessarily satisfies the condition:

$$s(a)\&Ea = s(a)$$

for each $a \in A$. In the case of a singleton subset of a quantal set $A$, one may see moreover that:

$$s(a)\&s(b) = s(b)\&s(a)$$

for each $a, b \in A$. For,

$$s(a)\&s(b) = s(a)\&s(b)\&Eb$$

$$= s(a)\&Eb\&s(b)$$

$$= s(a)\&Eb\&s(b)\&s(a)\&s(a)$$

$$\leq s(a)\&[b = a]\&s(a)$$

$$= s(a)\&[a = b]\&s(a)$$

$$\leq s(b)\&s(a) ,$$

from which the conclusion follows on interchanging $a, b \in A$.

With these preliminaries we may make the following:

**Definition 17** *A quantal set $A$ over a right Gelfand quantale $Q$ will be said to be complete provided that each singleton of $A$ is of the form*

$$\tilde{a} : A \to Q$$

*for some unique $a \in A$.*

EXAMPLE.   For any quantale $Q$, the quantal set $1_Q$ is complete. Indeed, it will be shown that any subset of $1_Q$ is determined by a unique element of $Q$. For, given any subset

$$s : Q \to Q$$

of the quantal set $1_Q$, consider the element

$$p = \bigvee_q s(q)$$

of the quantale $Q$. It is asserted that

$$s(q) = p\&q$$

for each $q \in Q$. For certainly one has that $s(q) \leq p$, by the construction of $p \in Q$ ; hence that $s(q) = s(q)\&q \leq p\&q$. But, equally,

$$p\&q = \bigvee_{q'} s(q')\&q = \bigvee_{q'} s(q')\&q'\&q = \bigvee_{q'} s(q')\&[q' = q] \leq s(q) .$$

Hence, $s(q) = p\&q = [p = q] = \tilde{p}(q)$ for each $q \in Q$. The uniqueness of the element $p \in Q$ having this property is given by the observation that

$$s(1) = [p = 1] = p\&1 = p .$$

The quantal set $\mathbf{1}_Q$ is therefore complete.

Now, for any quantal set $A$ over the quantale $Q$, consider the set

$$S(A)$$

of all singletons of $A$. Define the extent and the equality of singletons of $A$ in the following manner:

$$Es = \bigvee_{a \in A} s(a) \,,$$

$$\text{and} \quad [\![s = t]\!] = \bigvee_{a \in A} s(a)\&t(a) \,,$$

for each $s, t \in S(A)$. Then $S(A)$ is a quantal set. For, given $s \in S(A)$ one has that:

$$Es \& Es = \bigvee_{a,a'} s(a)\&s(a') \leq \bigvee_{a} s(a) = \bigvee_{a} s(a)\&s(a) = [\![s = s]\!] \,.$$

And, for any $s, t \in S(A)$, one has that:

$$[\![s = t]\!] = \bigvee_{a} s(a)\&t(a) \leq \bigvee_{a,a'} s(a)\&t(a') = \bigvee_{a} s(a) \& \bigvee_{a'} t(a') = Es \& Et.$$

Moreover, given $s, t \in S(A)$, one has that:

$$Et \& [\![s = t]\!] = \bigvee_{a,a'} t(a)\&s(a')\&t(a')$$

$$= \bigvee_{a,a'} t(a)\&t(a)\&s(a')\&t(a')$$

$$= \bigvee_{a,a'} t(a)\&s(a')\&t(a')\&t(a)$$

$$= \bigvee_{a,a'} t(a)\&s(a')\&Ea'\&t(a')\&t(a)$$

$$\leq \bigvee_{a,a'} t(a)\&s(a')\&[\![a' = a]\!]$$

$$\leq \bigvee_{a} t(a)\&s(a)$$

$$= [\![t = s]\!] \,.$$

Finally, given $s, t, u \in S(A)$, one argues similarly that:

$$[\![s = t]\!] \& [\![t = u]\!] = \bigvee_{a,a'} s(a)\&t(a)\&t(a')\&u(a')$$

$$= \bigvee_{a,a'} s(a)\&Ea\&t(a)\&t(a')\&u(a')$$

$$\leq \bigvee_{a,a'} s(a)\&[\![a = a']\!]\&u(a')$$

$$\leq \bigvee_{a'} s(a')\&u(a')$$

$$= [\![ s = u ]\!] \ ,$$

which shows that $S(A)$ is a quantal set.

REMARK. For any element $b \in A$ of the quantal set $A$, the singleton $\tilde{b} \in S(A)$ which it determines has equality with any other singleton $s \in S(A)$ which satisfies:

$$[\![ s = \tilde{b} ]\!] = s(b) \ .$$

For, in one direction, one has that:

$$[\![ s = \tilde{b} ]\!] = \bigvee_{a} s(a)\&[\![ b = a ]\!] = \bigvee_{a} s(a)\&[\![ a = b ]\!]$$

$$\leq s(b) \ ,$$

while, conversely, one sees that:

$$s(b) = s(b)\&Eb = s(b)\&[\![ b = b ]\!] \leq \bigvee_{a} s(a)\&[\![ b = a ]\!]$$

$$= [\![ s = \tilde{b} ]\!] \ .$$

In particular, it may be remarked that:

$$[\![ \tilde{a} = \tilde{b} ]\!] = [\![ a = b ]\!]$$

for any $a, b \in A$. Hence,

$$E\tilde{a} = Ea$$

for any $a \in A$.

With these preliminaries, we may now state the following:

**Theorem 18** *For any right Gelfand quantale $Q$ , there are adjoint functors*

$$Q\text{-Sets} \ \rightleftarrows \ \textbf{Complete } Q\text{-Sets} \ ,$$

*of which the adjoint to the inclusion of the category of complete quantal sets in that of quantal sets assigns to each quantal set $A$ the quantal set $S(A)$ of singletons of the quantal set $A$ .*

To prove the assertion of the theorem, it will be shown firstly that the quantal set $S(A)$ is indeed complete. Moreover, that there exists a map of quantal sets over $Q$ :

$$\eta_A : A \rightarrow S(A)$$

which is universal amongst maps from $A$ to complete quantal sets. From this, the assertion of the theorem follows.

So, consider firstly a singleton

$$\sigma : S(A) \to Q$$

of the quantal set $S(A)$. It is asserted that the mapping

$$s : A \to Q ,$$

defined by

$$s(a) = \sigma(\tilde{a})$$

for each $a \in A$, yields uniquely an element $s \in S(A)$ for which $\tilde{s} = \sigma$.

That $s$ is a singleton of the quantal set $A$ follows trivially from the conditions satisfied because $\sigma$ is a singleton of the quantal set $S(A)$, by recalling that $[\![a = a']\!] = [\![\tilde{a} = \tilde{a}']\!]$ and $Ea = E\tilde{a}$ for each $a, a' \in A$. But then, one sees that

$$\sigma(t) = \sigma(t) \& Et$$

$$= \sigma(t) \& \bigvee_a t(a)$$

$$= \bigvee_a \sigma(t) \& [\![t = \tilde{a}]\!] \& t(a)$$

$$\leq \bigvee_a \sigma(\tilde{a}) \& t(a)$$

$$= \bigvee_a s(a) \& t(a)$$

$$= [\![s = t]\!] ,$$

while, conversely,

$$[\![s = t]\!] = \bigvee_a s(a) \& t(a)$$

$$= \bigvee_a \sigma(\tilde{a}) \& t(a)$$

$$= \bigvee_a \sigma(\tilde{a}) \& [\![t = \tilde{a}]\!]$$

$$= \bigvee_a \sigma(\tilde{a}) \& [\![\tilde{a} = t]\!]$$

$$\leq \sigma(t) ,$$

from which the rest of the required result follows. The uniqueness of the element $s \in S(A)$ is immediate, because

$$s(a) = [\![s = \tilde{a}]\!] = \sigma(\tilde{a}) = [\![s' = \tilde{a}]\!] = s'(a)$$

for each $a \in A$, for any singleton $s' \in S(A)$ satisfying the required condition. Hence, the quantale set $S(A)$ is complete.

To obtain a map of quantal sets

$$\eta_A : A \to S(A) ,$$

consider the mapping from $A \times S(A)$ to the quantale $Q$, given by defining

$$\eta_A(a, s) = [\tilde{a} = s]$$

for each $a \in A$ and $s \in S(A)$. It may be remarked that one has, firstly, that:

$$\eta_A(a, s) = [\tilde{a} = s] \leq E\tilde{a}\&Es = Ea\&Es .$$

Then, that:

$$[a = a']\&\eta_A(a', s) = [\tilde{a} = \tilde{a}']\&[\tilde{a}' = s]$$
$$\leq [\tilde{a} = s] = \eta_A(a, s) ,$$
$$\text{and} \quad \eta_A(a, s)\&[s = s'] = [\tilde{a} = s]\&[s = s']$$
$$\leq [\tilde{a} = s'] = \eta_A(a, s') .$$

Finally, that

$$Es\&Es'\&\eta_A(a, s)\&\eta_A(a, s') = Es\&Es'\&[\tilde{a} = s]\&[\tilde{a} = s']$$
$$= Es\&[s = \tilde{a}]\&[\tilde{a} = s']\&Es'$$
$$= [s = \tilde{a}]\&[\tilde{a} = s']$$
$$\leq [s = s']$$
$$\text{and} \quad Ea = E\tilde{a} = [\tilde{a} = \tilde{a}] \leq \bigvee_s [\tilde{a} = s] = \bigvee_s \eta_A(a, s) .$$

Hence, one has indeed a map of quantal sets from $A$ to $S(A)$.

Now, suppose given a map

$$\varphi : A \to B$$

from the quantal set $A$ to a complete quantal set $B$. Consider the mapping

$$\psi : S(A) \times B \to Q$$

defined by setting $\psi(s, b) = \bigvee_a s(a)\&\varphi(a, b)$ for each $s \in S(A)$ and $b \in B$. It may be proved straightforwardly that this determines a map of quantal sets

$$\psi : S(A) \to B .$$

Moreover, this makes the diagram:

commute, and is unique in this respect.

To see that the diagram is commutative, observe firstly that:

$$[\![a = a']\!] \;=\; \bigvee_s [\![\tilde{a} = s]\!] \& [\![s = \tilde{a}']\!]$$

for each $a, a' \in A$, by the transitivity of equality on $S(A)$ and the observation that

$$[\![\tilde{a} = \tilde{a}']\!] \;=\; [\![a = a']\!] \;.$$

Then

$$\bigvee_s \eta_A(a, s) \& \psi(s, b) \;=\; \bigvee_s [\![\tilde{a} = s]\!] \& s(a) \& \varphi(a, b)$$

$$=\; \bigvee_s [\![\tilde{a} = s]\!] \& [\![s = \tilde{a}]\!] \& \varphi(a, b)$$

$$=\; Ea \& \varphi(a, b)$$

$$=\; \varphi(a, b)$$

for each $a \in A$ and $b \in B$. While, to see the uniqueness, note that:

$$[\![s = t]\!] \;=\; \bigvee_a [\![s = \tilde{a}]\!] \& [\![\tilde{a} = t]\!] \;.$$

Then, for any map $\psi \,:\, S(A) \to B$ for which $\varphi(a, b) = \bigvee_t \eta_A(a, t) \& \psi(t, b)$ for each $a \in A$ and $b \in B$, one has that:

$$s(a) \& \varphi(a, b) \;=\; \bigvee_a \bigvee_t s(a) \& \eta_A(a, t) \& \psi(t, b)$$

$$=\; \bigvee_t \bigvee_a [\![s = \tilde{a}]\!] \& [\![\tilde{a} = t]\!] \& \psi(t, b)$$

$$=\; \bigvee_t [\![s = t]\!] \& \psi(t, b)$$

$$=\; \psi(s, b) \;,$$

for each $s \in S(A)$ and $b \in B$.

The map considered is therefore unique, from which the existence of the adjointness and the naturality of the maps concerned then follows. $\square$

The functor that takes any quantal set $A$ to its completion $S(A)$ will be denoted by

$$S \,:\, Q\text{-Sets} \;\to\; \textbf{Complete } Q\text{-Sets} \;,$$

whilst the embedding of the full subcategory of complete quantal sets in the category of quantal sets will be denoted by

$$I \,:\, \textbf{Complete } Q\text{-Sets} \;\to\; Q\text{-Sets} \;.$$

In particular, the adjunction and coadjunction maps will henceforth be considered to be written as

$$\eta_A \,:\, A \to IS(A)$$

and
$$\varepsilon_B : SI(B) \to B$$

for any quantal set $A$ and any complete quantal set $B$.

It may be remarked immediately that we have proved more than simply the existence of adjoint functors. It will be recalled that the adjunction

$$\eta_A : A \to S(A)$$

was defined by the assignment

$$\eta_A(a, s) = [\![\tilde{a} = s]\!]$$

for each $a \in A$ and $s \in S(A)$. In the course of the proof, it has been noted that:

$$\bigvee_s [\![\tilde{a} = s]\!] \& [\![s = \tilde{a}']\!] = [\![a = a']\!]$$

for each $a, a' \in A$ ; and that

$$\bigvee_a [\![s = \tilde{a}]\!] \& [\![\tilde{a} = t]\!] = [\![s = t]\!]$$

for each $s, t \in S(A)$. These observe that the map of quantal sets

$$\varepsilon_A : S(A) \to A$$

defined by the assignment

$$\varepsilon_A(s, a) = [\![s = \tilde{a}]\!] ,$$

for each $s \in S(A)$ and $a \in A$, is indeed an inverse to the adjunction. Hence, one has the following:

**Corollary 19** *For any right Gelfand quantale $Q$ , the adjoint functors*

$$\begin{array}{c} S \\ \textbf{Q-Sets} \ \rightleftharpoons \ \textbf{Complete Q-Sets} \\ T \end{array}$$

*establish an equivalence of categories.*  □

Finally, it may be remarked that the definition of the map

$$S(f) : S(A) \to S(B)$$

of complete quantal sets determined by any map $f : A \to B$ of quantal sets may be reformulated slightly in the light of these results. The map was defined to be that given by the relation

$$S(f) : S(A) \times S(B) \to Q$$

that assigns to each pair $(s,t) \in S(A) \times S(B)$ of singletons the element

$$S(f)(s,t) = \bigvee_{a,b} s(a) \,\&\, f(a,b) \,\&\, \eta_B(b,t)$$

of the quantale $Q$ . By the remarks following the definition of the concept of a map of quantal sets, one has that $f(a,b) = f(a,b) \,\&\, Eb$ . Moreover, $Eb \,\&\, \eta_B(b,t) = Eb \,\&\, t(b)$ , since, on the one hand,

$$Eb \,\&\, \eta_B(b,t) = Eb \,\&\, [\![\tilde{b} = t]\!]$$

$$= \bigvee_c Ec \,\&\, [\![b = c]\!] \,\&\, t(c)$$

$$= \bigvee_c Ec \,\&\, t(c) \,\&\, [\![b = c]\!]$$

$$\leq Eb \,\&\, t(b) \ ,$$

while, on the other,

$$Eb \,\&\, t(b) = Eb \,\&\, Eb \,\&\, Eb \,\&\, t(b)$$

$$\leq Eb \,\&\, [\![b = b]\!] \,\&\, t(b)$$

$$\leq \bigvee_c Eb \,\&\, [\![b = c]\!] \,\&\, t(c)$$

$$= Eb \,\&\, \eta_B(b,t) \ .$$

Hence, $S(f)(s,t) = \bigvee_{a,b} s(a) \,\&\, f(a,b) \,\&\, t(b)$ , as asserted.

In conclusion, it may be remarked that this equivalence exactly extends to the context of a quantale the equivalence which exists in the case of a locale. The concept of completeness of a quantal set evidently reduces in the case of a locale to that already known, giving the equivalence between local sets and complete local sets which is known, if a little unexpected when first encountered, in that situation. Hence, one has in particular that:

**Corollary 20** *For any locale $Q$ , the adjoint functors*

$$\text{$Q$-Sets} \quad \overset{S}{\underset{T}{\rightleftarrows}} \quad \text{Complete $Q$-Sets}$$

*are exactly the canonical adjoint equivalencebetween the category of local sets and the category of complete local sets over the locale $Q$ .*   □

# 4   Presheaves on $Q$

The concept of completeness for a quantal set was introduced to motivate the definition of the notion of a sheaf on a quantale. In the context of locales, these concepts are known to coincide in a very strict sense: any complete local set over a locale is canonically a sheaf, and every sheaf is canonically a complete local set over the locale. The expectation, therefore, is that this coincidence can be maintained on extending these ideas to the case of right Gelfand quantales. The first step towards establishing such a correspondence is to use the notion of complete quantal set already established to motivate the concept of presheaf on a quantale $Q$, before considering the condition for a presheaf on the quantale to be a sheaf.

Recalling that any complete quantal set $A$ is isomorphic to its completion $S(A)$ by the map of quantal sets induced by the mapping

$$A \to S(A)$$

that assigns to each $a \in A$ the singleton $\tilde{a} \in S(A)$ that it represents uniquely, we begin by examining the case of a quantal set of that form:

EXAMPLE. For any quantal set $A$, given any singleton $s \in S(A)$ on $A$, and any $p \in Q$, the mappings

$$p \mid s : A \to Q \quad \text{and} \quad s \mid p : A \to Q$$

defined by

$$(p \mid s)(a) = p \& s(a) \quad \text{and} \quad (s \mid p)(a) = s(a) \& p$$

for each $a \in A$, respectively, again define singleton subsets of $A$. Moreover, these restrictions of a singleton $s \in S(A)$ by any $p \in Q$ satisfy the conditions that:

$$p \mid q \mid s = p \& q \mid s \qquad s \mid p \mid q = s \mid p \& q$$
$$Es \mid s = s \qquad s \mid Es = s$$
$$E(p \mid s) = p \& Es \qquad E(s \mid p) = Es \& p$$
$$\text{and} \qquad (p \mid s) \mid q = p \mid (s \mid q) \ ,$$

for any $s \in S(A)$ and any $p, q \in Q$.

That the mappings $p \mid a$ and $s \mid p$ define subsets of the quantal set $A$ may be verified straightforwardly from the definitions by applying the corresponding properties of the subset $s \in S(A)$, together with, in the case of $s \mid p$, the right symmetricity of the quantale $Q$. Thus, for instance:

$$(s \mid p)(a) \& [\![a = a']\!] = s(a) \& p \& [\![a = a']\!] = s(a) \& [\![a = a']\!] \& p$$
$$\leq s(a') \& p = (s \mid p))(a') \ .$$

That the subsets are again singletons follows similarly by invoking the corresponding property of the subset $s \in S(A)$ and applying the rightsidedness of the elements of the quantale $Q$. Thus, for instance:

$$Ea\& (s \mathbin{\uparrow} p)(a) \& (s \mathbin{\uparrow} p)(a') = Ea\&s(a) \&p\&s(a') \&p$$
$$\leq Ea\&s(a) \&s(a') \leq [\![a = a']\!] \,.$$

Hence, both the left and the right restrictions of any singleton subset $s \in S(A)$ are again singletons on the quantal set $A$.

The main point that emerges from considering this example is that the completion $S(A)$ of a quantal set $A$ admits both a left restriction and a right restriction for any element $p \in Q$, with the expectation that these restrictions will be distinct unless the quantale $Q$ is actually commutative. With this in mind, we arrive at the following:

**Definition 21** *By a presheaf $A$ on a right Gelfand quantale $Q$ will be meant a set $A$ together with mappings*

$$\mathbin{\downarrow} : Q \times A \to A \quad and \quad \mathbin{\uparrow} : A \times Q \to A \,,$$

*which will be referred to respectively as* left restriction *and* right restriction *over $Q$, and a mapping*

$$E : A \to Q$$

*which will be referred to as the* extent *of $A$, together satisfying the conditions that:*

$$p \mathbin{\downarrow} q \mathbin{\downarrow} a = p\&q \mathbin{\downarrow} a \qquad a \mathbin{\uparrow} p \mathbin{\uparrow} q = a \mathbin{\uparrow} p\&q$$

$$Ea \mathbin{\downarrow} a = a \qquad a \mathbin{\uparrow} Ea = a$$

$$E(p \mathbin{\downarrow} a) = p\&Ea \qquad E(a \mathbin{\uparrow} p) = Ea\&p$$

$$and \qquad (p \mathbin{\downarrow} a) \mathbin{\uparrow} q = p \mathbin{\downarrow} (a \mathbin{\uparrow} q) \,,$$

*for any $a \in A$ and any $p, q \in Q$.*

Now consider any complete quantal set $A$ over the quantale $Q$, and observe that we have the following:

CONSTRUCTION. For each element $a \in A$ of the complete quantal set $A$, consider the singleton subset $\tilde{a} \in S(A)$ defined by

$$\tilde{a}(b) = [\![a = b]\!]$$

for each $b \in A$. Note that, by the completeness of $A$, each singleton subset of $A$ is of this form for a unique element of $A$. For any element $p \in Q$, denote by

$$p \mathbin{\downarrow} a \qquad and \qquad a \mathbin{\uparrow} p$$

respectively the unique elements of $A$ that represent the singleton subsets

$$p \downarrow \tilde{a} \qquad \text{and} \qquad \tilde{a} \uparrow p$$

defined above. It is asserted that, with respect to these operations of left and right restriction, together with the extent operator of the quantal set $A$ , the underlying set of any complete quantal set $A$ is indeed a presheaf on the quantale $Q$ , which we shall denote by $\Gamma(A)$ .

For, observe firstly that, by the definition of the left and the right restrictions of a complete quantal set, for any element $a \in A$ and for any $p \in Q$ the element $p \downarrow a \in A$ is uniquely determined by the property that:

$$[\![ p \downarrow a = b ]\!] = p \& [\![ a = b ]\!]$$

for all $b \in A$ . Similarly, the element $a \uparrow p \in A$ uniquely satisfies the condition that:

$$[\![ a \uparrow p = b ]\!] = [\![ a = b ]\!] \& p$$

for all $b \in A$ . From this observation, the required conditions on the left restriction may be proved by noting respectively that
$[\![ q \downarrow p \downarrow a = b ]\!] = q \& [\![ p \downarrow a = b ]\!] = q \& p \& [\![ a = b ]\!] = [\![ q \& p \downarrow a = b ]\!]$ , and that
$E(p \downarrow a) = [\![ p \downarrow a = p \downarrow a ]\!] = p \& [\![ a = p \downarrow a ]\!] = p \& [\![ p \downarrow a = a ]\!] =$
$p \& p \& [\![ a = a ]\!] = p \& Ea$ for all $b \in A$ , by the right symmetricity of equality in a quantal set, with similar remarks in the case of the right restriction.

With respect to the condition concerning the extent, it may be noted that in the case of left restriction, one has that $[\![ Ea \downarrow a = b ]\!] = Ea \& [\![ a = b ]\!] = [\![ a = b ]\!]$ for each $b \in A$ , giving the required equality, with a similar argument on the right. Finally, the condition that implies the associativity of left and right restrictions follows from that of the product of the quantale by noting that
$[\![ p \downarrow (a \uparrow q) = b ]\!] = p \& [\![ a \uparrow q = b ]\!] = p \& [\![ a = b ]\!] \& q = [\![ p \downarrow a = b ]\!] \& q = [\![ (p \downarrow a) \uparrow q = b ]\!]$ for all $b \in A$ , establishing that the complete quantal set $A$ is indeed canonically a presheaf on $Q$.

Recalling the characterisation of the restrictions given above, we have the following:

**Definition 22** *By the* canonical presheaf

$$\Gamma(A)$$

*determined by a complete quantal set $A$ over the right Gelfand quantale $Q$ will be meant the presheaf of which the underlying set and the extent operator are those of the quantal set $A$ , and of which the left and right restrictions are uniquely determined by requiring that*

$$[\![ p \downarrow a = b ]\!] = p \& [\![ a = b ]\!]$$

$$\text{and} \quad [\![ a \uparrow p = b ]\!] = [\![ a = b ]\!] \& p$$

*for any $a, b \in A$ and $p \in Q$ .*

It may be remarked in passing that, although we had already shown that the completion $S(A)$ of any quantal set $A$ was canonically a presheaf, the existence of the canonical isomorphism

$$\eta_A : A \to S(A)$$

is not sufficient to imply that $A$ is again a presheaf. Just as completeness for a quantal set $A$ requires the existence of certain elements in its underlying set $A$, so the existence of the restrictions of an element requires the presence of the elements concerned. Indeed, the completeness of the quantal set $A$ is exactly what is needed in order to ensure that the restrictions existing in the completion are represented within the quantal set $A$.

It may also be remarked that although the description of these restriction mappings is straightforward to define, it may still be difficult in any particular case to obtain an intrinsic characterisation of the restrictions of an element, since this involves passing through the quantal set $S(A)$ of singletons on the quantal set $A$.

There is, however, one immediate case in which this can be computed directly and naturally, yielding in particular an instance of a presheaf on which the left and right restrictions are distinct unless the quantale $Q$ is commutative, namely the following:

EXAMPLE. The quantal set $\mathbf{1}_Q$, which is the terminal object in the category of quantal sets over $Q$, obtained by taking the set $Q$, together with the equality and extent given by:

$$[\![p = q]\!] = p\&q \quad \text{and} \quad Ep = p$$

for each $p, q \in Q$, is a presheaf with respect to the restriction mappings given by

$$p \mid q = p\&q \quad \text{and} \quad q \mid p = q\&p$$

for each $p, q \in Q$.

To see this, it is enough to check that the element $p\&q \in Q$ indeed represents each of the singleton subsets $p \mid \tilde{q}$ and $\tilde{p} \mid q$ of the quantal set $\mathbf{1}_Q$. However, in establishing the completeness of the quantal set $\mathbf{1}_Q$, we have shown that each singleton subset

$$s : Q \to Q$$

of this particular quantal set is represented uniquely by the element $s(1) \in Q$ which is the image of the top element $1 \in Q$. Hence, from the definition of the singleton subset $\tilde{q} \in S(Q)$ determined by an element $q \in Q$, and by the rightsidedness of the quantale $Q$, we have that

$$p \mid q = (p \mid \tilde{q})(1) = p\&\tilde{q}(1) = p\&[\![q = 1]\!] = p\&q\&1 = p\&q \text{ , and that}$$

$$q \mid p = (\tilde{q} \mid p)(1) = \tilde{q}(1)\&p = [\![q = 1]\!]\&p = q\&1\&p = q\&p$$

for any $p \in Q$ , yielding the required assertion.

Recalling that, by the equivalence between the category of quantal sets over a quantale $Q$ and that of complete quantal sets over $Q$ , any map

$$f : A \to B$$

of quantal sets over $Q$ is equivalent to the map

$$S(f) : S(A) \to S(B)$$

of completions that it determines, in the sense that the diagram

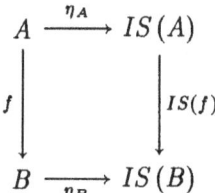

in which the maps $\eta_A$ and $\eta_B$ are the canonical isomorphisms from a quantal set to its completion, is commutative, we may begin to consider the concept of a map of presheaves on $Q$ by examining the case of a map of complete quantal sets determined in this way.

Observing firstly that, by the remarks made in the preceding section concerning the completion functor, the map

$$S(f) : S(A) \to S(B)$$

considered is that given by the relation

$$S(f) : S(A) \times S(B) \to Q$$

defined by

$$S(f)(s,t) = \bigvee_{a,b} s(a) \, \& f(a,b) \, \& t(b)$$

for any $s \in S(A)$ and $t \in S(B)$ , we consider the following:

EXAMPLE. For any map

$$f : A \to B$$

of quantal sets over $Q$ , the assignment to each singleton $s \in S(A)$ of the mapping

$$\varphi(s) : B \to Q$$

defined at each $b \in B$ by taking the element

$$\varphi(s)(b) = \bigvee_a s(a) \, \& f(a,b)$$

of the quantale $Q$ yields a mapping

$$\varphi : S(A) \rightarrow S(B)$$

from the set $S(A)$ of singletons on $A$ to the set $S(B)$ of singletons on $B$. Moreover, the mapping thereby obtained satisfies the conditions that:

$$p \mid \varphi(s) = \varphi(p \mid s) \quad \varphi(s) \uparrow p = \varphi(s \uparrow p)$$

$$\text{and} \quad E\varphi(s) = Es$$

for each singleton $s \in S(A)$ and any $p \in Q$.

That the mapping

$$\varphi(s) : B \rightarrow Q$$

defines a singleton subset $\varphi(s) \in S(B)$ may be verified in the following manner: it is a subset of $B$, since

$$\varphi(s)(b) \,\&\, [\![ b = b' ]\!] = \bigvee_a s(a) \,\&\, f(a,b) \,\&\, [\![ b = b' ]\!]$$

$$\leq \bigvee_a s(a) \,\&\, f(a,b')$$

$$= \varphi(s)(b') \ ,$$

and $\quad \varphi(s)(b) = \bigvee_a s(a) \,\&\, f(a,b) \leq \bigvee_a s(a) \,\&\, f(a,b) \,\&\, Eb = \varphi(s)(b) \,\&\, Eb;$

and it is a singleton of $A$, since

$$Eb \,\&\, \varphi(s)(b) \,\&\, \varphi(s)(b') = \bigvee_{a,a'} Eb \,\&\, s(a) \,\&\, f(a,b) \,\&\, s(a') \,\&\, f(a',b')$$

$$= \bigvee_{a,a'} Eb \,\&\, f(a,b) \,\&\, Ea \,\&\, s(a) \,\&\, s(a') \,\&\, f(a',b')$$

$$\leq \bigvee_{a,a'} Eb \,\&\, f(a,b) \,\&\, [\![ a = a' ]\!] \,\&\, f(a',b')$$

$$\leq \bigvee_a Eb \,\&\, f(a,b) \,\&\, f(a,b')$$

$$\leq [\![ b = b' ]\!] \ .$$

Thus, assigning to each element $s \in S(A)$ the element $\varphi(s) \in S(B)$ indeed yields a mapping

$$\varphi : S(A) \rightarrow S(B)$$

from the set of singletons on $A$ to the set of singletons on $B$.

That this mapping satisfies the conditions asserted with respect to restriction and extent may be verified by observing that

$$(p \mid \varphi(s))(b) = p \,\&\, \varphi(s)(b) = p \,\&\, \bigvee_a s(a) \,\&\, f(a,b) = \bigvee_a p \,\&\, s(a) \,\&\, f(a,b)$$

$$= \bigvee_a (p \mid s)(a) \ \& f(a,b) = \varphi(p \mid s)(b)$$

for each $b \in B$, giving the condition for left restriction, with a similar argument in the case of right restriction, and that

$$E\varphi(s) = \bigvee_b \varphi(s)(b) = \bigvee_{a,b} s(a) \ \& f(a,b) = \bigvee_a s(a) \ \& \left(\bigvee_b f(a,b)\right)$$

$$= \bigvee_a s(a) \ \& \ Ea = \bigvee_a s(a) = Es$$

in the case of extent.

With these observations in mind, we make the following:

**Definition 23** *By a map of presheaves*

$$\varphi : A \to B$$

*on a right Gelfand quantale $Q$ will be meant a mapping from the set $A$ to the set $B$ satisfying the following conditions with respect to the respective restriction and extent mappings:*

$$\varphi(p \mid a) = p \mid \varphi(a) \, \varphi(a \mid p) = \varphi(a) \mid p$$

$$\text{and} \quad E\varphi(a) = Ea$$

*for each $a \in A$ and for each $p \in Q$.*

Now consider any map $f : A \to B$ of complete quantal sets over the quantale $Q$, and observe that we have the following:

CONSTRUCTION. For each element $a \in A$ consider the singleton subset $\tilde{a} \in S(A)$ of $A$, which is therefore mapped by the map of presheaves

$$\varphi : S(A) \to S(B)$$

defined above to a singleton subset $\varphi(\tilde{a}) \in S(B)$ of $B$. Denoting by

$$f(a) \in B$$

the unique element of the complete quantal set $B$ that represents this singleton subset of $B$, it is asserted that the mapping

$$f : A \to B$$

from the underlying set of $A$ to the underlying set of $B$ thereby determined yields a map of presheaves from the canonical presheaf $\Gamma(A)$ to the canonical presheaf $\Gamma(B)$, which we shall denote by $\Gamma(f)$.

For, observe firstly that the element $f(a) \in B$ described above may equally be characterised by the condition that

$$[\![ f(a) = b ]\!] = f(a, b)$$

for each $b \in B$ . For, on the one hand,

$$[\![ f(a) = b ]\!] = \varphi(\tilde{a})(b) = \bigvee_{a'} [\![ a = a' ]\!] \& f(a', b) \leq f(a, b) \; ,$$

while, on the other, we have that

$$f(a, b) = Ea \& f(a, b) = [\![ a = a ]\!] \& f(a, b)$$

$$\leq \bigvee_{a'} [\![ a = a' ]\!] \& f(a', b) = \varphi(\tilde{a})(b) = [\![ f(a) = b ]\!] \; ,$$

which together yield the equality asserted. Extending to this context the convention that a mapping and the functional relation that defines it are denoted by the same symbol, we write:

$$f : A \rightarrow B$$

for the mapping from the set $A$ to the set $B$ thereby obtained.

It is asserted that this mapping indeed determines a map

$$\Gamma(f) : \Gamma(A) \rightarrow \Gamma(B)$$

of presheaves on $Q$ . For, recalling that for any $a \in A$ and any $p \in Q$ , the element $p \mid a$ is uniquely determined by the property that

$$[\![ p \mid a = a' ]\!] = p \& [\![ a = a' ]\!]$$

for each $a' \in A$ , it follows that $f(p \mid a) = p \mid f(a)$ for any $a \in A$ and $p \in Q$ . For, by the above remarks, it suffices to show that $f(p \mid a, b) = p \& f(a, b)$ for any $b \in B$ . But, on the one hand,

$$f(p \mid a) = E(p \mid a) \& f(p \mid a, b) = p \& Ea \& f(p \mid a, b)$$

$$= p \& p \& [\![ a = a ]\!] \& f(p \mid a, b)$$

$$= p \& [\![ p \mid a = a ]\!] \& f(p \mid a, b)$$

$$= p \& [\![ a = p \mid a ]\!] \& f(p \mid a, b)$$

$$\leq p \& f(a, b) \; ;$$

while conversely:

$$p \& f(a, b) = p \& Ea \& f(a, b) = p \& [\![ a = a ]\!] \& f(a, b)$$

$$= [\![p \downarrow a = a]\!] \& f(a, b) \leq f(p \downarrow a, b) \ ,$$

giving the required equality. The condition for right restriction may be established similarly, while, in the case of the extent, we observe that for any $b \in B$ we have that:

$$f(a, b) = f(a, b) \& f(a, b) = f(a, b) \& [\![f(a) = b]\!]$$
$$= f(a, b) \& [\![b = f(a)]\!] \leq f(a, f(a)) \ ,$$

since $f(a) \in B$ evidently implies that $f(a, f(a)) \leq \bigvee_b f(a, b)$ , establishing that this determines a canonical map $\Gamma(f) : \Gamma(A) \to \Gamma(B)$ of presheaves on $Q$.

Recalling the characterisation of the underlying mapping of the map of presheaves obtained, we have the following:

**Definition 24** *By the* canonical map of presheaves

$$\Gamma(f) : \Gamma(A) \to \Gamma(B)$$

*determined by a map $f : A \to B$ of complete quantal sets over the right Gelfand quantale $Q$ will be meant the map of which the underlying mapping is uniquely determined by requiring that*

$$[\![f(a) = b]\!] \ = \ f(a, b)$$

*for each $a \in A$ and $b \in B$ .*

It is now evident that, for any right Gelfand quantale $Q$, the presheaves on $Q$, together with the maps of presheaves on $Q$, form a category

**Presheaves $/Q$ .**

Moreover, from the observations that to each complete quantal set $A$ over the quantale $Q$ there corresponds a canonical presheaf

$$\Gamma(A)$$

on $Q$ , and to each map $f : A \to B$ of complete quantal sets over $Q$ a canonical map

$$\Gamma(f) : \Gamma(A) \to \Gamma(B)$$

of presheaves on $Q$ , it may be shown straightforwardly that one has a functor

$$\Gamma : \textbf{Complete } Q\textbf{-Sets} \ \to \textbf{Presheaves } /Q \ ,$$

which establishes the first part of the following:

**Theorem 25** *For any right Gelfand quantale $Q$ , there are adjoint functors*

$$
\textbf{Presheaves } /Q \overset{U}{\underset{\Gamma}{\rightleftarrows}} \textbf{Complete } Q\textbf{-Sets}
$$

*of which the coadjoint assigns to each complete quantal set over $Q$ its canonical presheaf on $Q$ .*

To establish the existence of the adjoint to the underlying presheaf functor, we shall first construct a functor

$$U : \textbf{Presheaves } /Q \rightarrow Q\textbf{-Sets}$$

by showing that each presheaf $A$ on the quantale $Q$ admits an underlying structure as a quantal set over $Q$ , and that each map of presheaves on $Q$ determines a map of the corresponding underlying quantal sets over $Q$ . Taking inspiration from the case of a presheaf on a locale, we define the equality of elements $a, b \in A$ of a presheaf $A$ to be the element of the quantale $Q$ given by:

$$[\![a = b]\!] = Ea \,\& \bigvee_{a|p=p|b} p \,\& \,Eb \,.$$

Intuitively, this is the join of those elements of the quantale on which the elements $a, b \in A$ are both defined and have identical restriction.

It is asserted first that, with respect to the equality

$$[\![\cdot = \cdot]\!] : A \times A \rightarrow Q$$

thereby defined, together with the extent mapping

$$E : A \rightarrow Q$$

of the presheaf, the underlying set $A$ of the presheaf is indeed a quantal set over $Q$. Verifying each of the axioms in turn, we argue as follows:

- $Ea\&Ea \leq [\![a = a]\!]$ since $Ea\&Ea \leq Ea\&Ea\&Ea \leq Ea\& \bigvee_{a|p=p|a} p\,\&Ea$ because $a \mid Ea = a = Ea \mid a$ ;

- $[\![a = b]\!] \leq Ea\&Eb$ since $[\![a = b]\!] = Ea\& \bigvee_{a|p=p|b} p\,\&\,Eb \leq Ea\&Eb$ by the right-sidedness of the quantale $Q$ ;

- $Ea\,\&[\![b = a]\!] \leq [\![a = b]\!]$ since

$$Ea\,\&[\![b = a]\!] = \bigvee_{b|p=p|a} Ea\&\,Eb\&\,p\&\,Ea$$

$$= \bigvee_{b|p=p|a} Ea\&(Eb\&\,p)\&Eb$$

$$\leq \bigvee_{a \mid q = q \mid b} Ea \, \& \, q \, \& \, Eb$$

since given any $p \in Q$ for which $b \restriction p = p \mid a$ we have that $p \mid a = p \mid p \mid a = p \mid b \restriction p$, by the idempotency and the choice of $p \in Q$, hence $Eb\&p \mid a = Eb\&p \mid b \restriction p = Eb\&p \mid b \restriction Eb\&p = E(b \restriction Eb\&p) \mid (b \restriction Eb\&p) = b \restriction Eb\&p$; and finally

- $[a = b]\&[b = c] \leq [a = c]$ since

$$[a = b]\&[b = c] = Ea\& \bigvee_{a \mid p = p \mid b} p \, \& \, Eb \, \& \, Eb\& \bigvee_{b \mid q = q \mid c} q \, \& \, Ec$$

$$\leq Ea\& \bigvee_{a \mid p = p \mid b} \bigvee_{b \mid q = q \mid c} p\&q \, \& \, Ec$$

$$\leq Ea\& \bigvee_{a \mid r = r \mid c} r \, \& \, Ec$$

since $a \restriction p\&q = p\&q \mid c$ because $a \restriction p\&q = (a \restriction p) \restriction q = (p \mid b) \restriction q = p \mid (b \restriction q) = p \mid (q \mid c) = p\&q \mid c$, as required.

The underlying set of the presheaf $A$ is therefore a quantal set, as asserted.

It may be remarked in passing that it is the particular form of this proof of the transitivity of equality that motivates the symmetry of the condition

$$a \restriction p = p \mid b$$

appearing in its definition.

Consider now a map

$$f : A \to B$$

of presheaves on $Q$, and define a mapping

$$f : A \times B \to Q$$

by writing

$$f(a, b) = [f(a) = b]$$

for each $a \in A$ and $b \in B$, where the equality relation is that of the quantal set canonically determined by the presheaf $B$. It is asserted that this defines a map

$$f : A \to B$$

of the quantal sets determined by the presheaves concerned.

Once again, we verify the axioms one by one, working principally from the properties of equality on the quantal set $A$:

- $f(a, b) \leq Ea\&Eb$ since $[f(a) = b] \leq Ef(a)\&Eb = Ea\&Eb$;

- $[a = a'] \& f(a', b) \leq f(a, b)$
  by $[a = a'] \& [f(a') = b] \leq [f(a) = f(a')] \& [f(a') = b] \leq [f(a) = b]$,
  since

$$[a = a'] = Ea \& \bigvee_{a|p=p|a'} p \& Ea' \leq Ef(a) \& \bigvee_{f(a|p)=f(p|a')} p \& Ef(b)$$

$$= Ef(a) \& \bigvee_{f(a)||p=p||f(a')} p \& Ef(b) = [f(a) = f(a')]$$

  by the properties of a map of presheaves on $Q$ ;

- $f(a, b) \& [b = b'] \leq f(a, b')$ since $[f(a) = b] \& [b = b'] \leq [f(a) = b']$ ;

- $Eb \& f(a, b) \& f(a, b') \& Eb' \leq [b = b']$ since

$$Eb \& [f(a) = b] \& [f(a) = b'] \& Eb' \leq [b = f(a)] \& [f(a) = b'] \& Eb'$$
$$\leq [b = b'] \& Eb' = [b = b'];$$

- $Ea \leq \bigvee_b f(a, b)$ by $Ea = Ea \& Ea = Ef(a) \& Ef(a) \leq$
  $[f(a) = f(a)] \leq \bigvee_b [f(a) = b] = \bigvee_b f(a, b)$ ,

which completes the verification that this yields a map of quantal sets.

It may be proved straightforwardly that this determines a functor, which we shall denote by

$$U : \textbf{Presheaves } /Q \rightarrow Q\textbf{-Sets}$$

from the category of presheaves on the quantale $Q$ to the category of quantal sets over $Q$ , of which the composite with the functor

$$\Gamma : \textbf{Complete } Q\textbf{-Sets} \rightarrow \textbf{Presheaves } /Q$$

that assigns to each complete quantal set over $Q$ its canonical presheaf is exactly the inclusion functor

$$I : \textbf{Complete } Q\textbf{-Sets} \rightarrow Q\textbf{-Sets}$$

of the category of complete quantal sets over $Q$ in that of quantal sets over $Q$ .

From this latter assertion, to the proof of which we now turn, the existence of the required adjoint may then be deduced from that of the completion of quantal sets over $Q$ . It must be proved, therefore, firstly, that, for any complete quantal set $A$ , the quantal set

$$U(\Gamma(A))$$

determined by the canonical presheaf $\Gamma(A)$ of $A$ is exactly the quantal set $A$ . Since the sets and the extent operators concerned are identical throughout, it is necessary only to show that the equality relations on the sets are the same.

Recalling the definition of the equality relation on the quantal set underlying a presheaf, we have to show that

$$\bigvee_{a\restriction p=p\restriction b} Ea\,\&\,p\,\&\,Eb \;=\; [\![a=b]\!]$$

for any elements $a, b \in A$ of the complete quantal set $A$. To prove this, observe firstly that

$$a\restriction [\![a=b]\!] = [\![a=b]\!]\restriction b \; ,$$

since, by the definition of the restrictions on the canonical presheaf of $A$, this is just the statement that $[\![a=c]\!]\&[\![a=b]\!] = [\![a=b]\!]\&[\![b=c]\!]$ for any $c \in A$. Certainly,

$$[\![a=b]\!]\&[\![b=c]\!] = [\![a=b]\!]\&[\![b=c]\!]\&[\![a=b]\!] \le [\![a=c]\!]\&[\![a=b]\!] \; ,$$

while conversely,

$$[\![a=c]\!]\&[\![a=b]\!] = Ea\,\&\,[\![a=c]\!]\&[\![a=b]\!]$$

$$= Ea\,\&\,[\![a=b]\!]\&[\![a=c]\!]$$

$$= Ea\,\&\,[\![a=b]\!]\&[\![b=a]\!]\&[\![a=c]\!]$$

$$\le \; Ea\,\&\,[\![a=b]\!]\&[\![b=c]\!]$$

$$= [\![a=b]\!]\&[\![b=c]\!] \; ,$$

establishing the required identity. Hence, observing that $[\![a=b]\!] = Ea\,\&\,[\![a=b]\!]\&\,Eb$, one has that

$$[\![a=b]\!] \le \bigvee_{a\restriction p=p\restriction b} Ea\,\&\,p\,\&\,Eb \; .$$

To establish the converse, it must be shown that

$$Ea\,\&\,p\,\&\,Eb \le [\![a=b]\!]$$

for each $p \in Q$ satisfying $a\restriction p = p\restriction b$. Observe firstly that $a\restriction p = p\restriction b$ means that for all $c \in A$ one has that $[\![a=c]\!]\&p = p\&[\![b=c]\!]$. Hence, in particular, $Ea\&p = p\&[\![b=a]\!] = p\&[\![a=b]\!]$, giving that $Ea\&p\&Eb = Ea\&Ea\&p\&Eb = Ea\&p\&[\![a=b]\!]\&Eb \le Ea\&[\![a=b]\!]\&Eb = [\![a=b]\!]$, as required. Thus,

$$\bigvee_{a\restriction p=p\restriction b} Ea\,\&\,p\,\&\,Eb \;=\; [\![a=b]\!] \; ,$$

as asserted.

Consider now a map $f : A \to B$ of complete quantal sets over $Q$, given by a functional relation

$$f : A \times B \to Q \; .$$

Observe that the map $\Gamma(f) : \Gamma(A) \to \Gamma(B)$ of canonical presheaves assigns to each $a \in A$ the element $f(a) \in \Gamma(B)$ uniquely satisfying

$$[\![ f(a) = b ]\!] = f(a, b)$$

for each $b \in B$. Applying the functor $U$ which gives each presheaf its underlying structure as a quantal set, we have that

$$U\Gamma(f) : U\Gamma(A) \to U\Gamma(B)$$

is given by the functional relation that assigns to each pair $(a, b) \in U\Gamma(A) \times U\Gamma(B)$ the element $[\![ f(a) = b ]\!] \in Q$ given by the equality relation on the quantal set underlying the presheaf $\Gamma(B)$. By the above remarks, this is exactly the element $f(a, b) \in Q$, yielding that map of canonical quantal sets thereby determined is exactly $I(f) : I(A) \to I(B)$. Hence, one has that the functors $U\Gamma$ and $I$ are identical, as asserted.

We now apply this to show that the functors

$$\textbf{Presheaves } /Q \underset{\Gamma}{\overset{\Sigma}{\rightleftarrows}} \textbf{Complete } Q\textbf{-Sets}$$

given respectively by $SU$ and $\Gamma$ are adjoint. This will be done by describing the adjunction and coadjunction maps

$$\tau : \textbf{Presheaves } /Q \to \Gamma\Sigma \quad \text{and} \quad \sigma : \Sigma\Gamma \to \textbf{Complete } Q\textbf{-Sets}$$

and verifying that the required adjunction identities are satisfied. Observe firstly that, by the above remarks, the functor $\Sigma\Gamma$ is exactly the functor $SI$ assigning to each complete quantal set the completion of its underlying quantal set. The coadjunction $\sigma$ may therefore be defined to be the coadjunction

$$\varepsilon : SI \to \textbf{Complete } Q\textbf{-Sets}$$

of the adjointness given by completion of quantal sets over $Q$. For the adjunction, we define $\tau$ by taking for each presheaf $A$ on $Q$ the map

$$\tau_A : A \to \Gamma SU(A)$$

that assigns to each $a \in A$ the element $\tilde{a} \in SU(A)$ of the completion of the quantal set canonically determined by the presheaf $A$. It may be verified straightforwardly that this is indeed a map of presheaves on $Q$ and that this yields a natural map of the required functors.

It remains to show that the maps

$$\Sigma(A) \xrightarrow{\Sigma(\tau_A)} \Sigma\Gamma\Sigma(A) \xrightarrow{\sigma_{\Sigma(A)}} \Sigma(A) \text{ and } \Gamma(B) \xrightarrow{\tau_{\Gamma(B)}} \Gamma\Sigma\Gamma(B) \xrightarrow{\Gamma(\sigma_B)} \Gamma(B)$$

are the respective identities for each presheaf $A$ on $Q$ and for each complete quantal set $B$ over $Q$ . Concerning the natural map $\tau$, it may be remarked firstly that

$$U\left(\tau_A\right) \;=\; \eta_{U(A)}$$

for any presheaf $A$ , since $\tau_A\left(a\right)$ is defined to be the singleton $\tilde{a}$ , and hence

$$U\left(\tau_A\right)\left(a,s\right) \;=\; [\![\tau_A\left(a\right)=s]\!]\;,$$

for each $a \in A$ and each $s \in \Gamma\Sigma\left(A\right)$ , where the equality relation is that of the quantal set underlying the presheaf $\Gamma\Sigma\left(A\right)$ . But this quantal set $U\Gamma\Sigma\left(A\right)$ is exactly $I\Sigma\left(A\right)$ , by the remarks made earlier, and hence its equality relation is exactly that of the completion of the quantal set $U(A)$ canonically determined by the presheaf $A$ . Hence,

$$U\left(\tau_A\right)\left(a,s\right) \;=\; [\![\tilde{a} = s]\!] \;=\; \eta_{U(A)}\left(a,s\right)$$

for each $a \in A$ and each $s \in \Gamma\Sigma\left(A\right)$ , by the definition of the natural map $\eta : Q\text{-Sets} \to IS$ determined by completion of quantal sets over $Q$ .

Recalling that the functor $\Sigma$ is defined to be the composite $SU$ , and that the natural map $\sigma$ is exactly the coadjunction:

$$\varepsilon : SI \to \textbf{Complete } Q\textbf{-Sets} \;,$$

we observe that $\Sigma\left(\tau_A\right) = SU\left(\tau_A\right) = S\eta_{U(A)}$ , and $\sigma_{\Sigma(A)} = \eta_{SU(A)}$ , from which it follows that

$$\Sigma(A) \xrightarrow{\;\Sigma(\tau_A)\;} \Sigma\Gamma\Sigma(A) \xrightarrow{\;\sigma_{\Sigma(A)}\;} \Sigma(A)$$

is the identity on $\Sigma\left(A\right)$ , by the corresponding identity between $\eta$ and $\varepsilon$ at the quantal set $U(A)$ . Now consider the other required identity, namely that

$$\Gamma(B) \xrightarrow{\;\tau_{\Gamma(B)}\;} \Gamma\Sigma\Gamma \xrightarrow{\;\Gamma(\sigma_B)\;} \Gamma(B)$$

is the identity on $\Gamma\left(B\right)$ for any complete quantal set $B$ . Observe that $\tau_{\Gamma(B)}$ assigns to each element $a \in \Gamma\left(B\right)$ of the canonical presheaf of $B$ the element $\tilde{a} \in \Gamma SU\,\Gamma\left(B\right)$ of the canonical presheaf of the completion of the quantal set $U\,\Gamma\left(B\right)$ . Applying again the fact that $U\,\Gamma$ is exactly $I$ , and observing that $\varepsilon_B : SI\left(B\right) \to B$ is given by assigning to each $s \in SI\left(B\right)$ and $b \in B$ the element $s\left(b\right) \in Q$ , we have that

$$\sigma_B\left(\tilde{a}, b\right) = [\![a = b]\!]\;.$$

Hence, $\Gamma\left(\sigma_B\right)\left(\tilde{a}\right)$ is exactly $a \in B$ , since this is the unique element of $B$ satisfying the above condition. The composite is therefore the identity on the underlying presheaf of the complete quantal set $B$ , as required, from which it follows that the functors

$$\textbf{Presheaves }/Q \;\underset{\Gamma}{\overset{\Sigma}{\rightleftarrows}}\; \textbf{Complete } Q\textbf{-Sets}$$

are indeed adjoint, as asserted. ☐

We have in fact proved more than simply that these functors are adjoint. For, recalling that the coadjunction is exactly the coadjunction

$$\varepsilon : SI \rightarrow \textbf{Complete } Q\textbf{-Sets}$$

determined by the completion of quantal sets over $Q$ , which has already been shown to be a natural isomorphism, we obtain the following:

**Corollary 26** *For any right Gelfand quantale $Q$ , the adjoint functors*

$$\textbf{Presheaves } /Q \stackrel{\Sigma}{\underset{\Gamma}{\leftrightarrows}} \textbf{Complete } Q\textbf{-Sets}$$

*of which the adjoint assigns to each presheaf its completion in the category of quantal sets provide a reflection of the category of presheaves on $Q$ into the category of complete quantal sets over $Q$ .* ☐

With this adjointness indicating firmly that the canonical presheaves of complete quantal sets are exactly the sheaves on the quantale $Q$ , we may now proceed to define the concept of a sheaf on the quantale $Q$. Before doing so, however, there is an important observation that may be made concerning the left and right restrictions on a presheaf $A$ , namely, that, by the idempotency and right-sidedness of the quantale $Q$ , one has that:

$$a \upharpoonright p = Ea \& p \downharpoonright a$$

for any $a \in A$ and any $p \in Q$ . In particular, the right restriction on any presheaf $A$ may be defined in terms of the left restriction.

To see this, we remark that

$$a \upharpoonright p = E(a \upharpoonright p) \downharpoonright (a \upharpoonright p) ,$$

since left (or right) restriction to the extent of an element leaves it unchanged. Then,

$$E\,(a \upharpoonright p) \downharpoonright (a \upharpoonright p) = Ea \& p \downharpoonright (a \upharpoonright Ea \& p \& Ea) ,$$

since $a \upharpoonright p = (a \upharpoonright Ea) \upharpoonright p = a \upharpoonright Ea \& p = a \upharpoonright Ea \& Ea \& p = a \upharpoonright Ea \& p \& Ea$ by the idempotency and right-sidedness of the quantale. But,

$$Ea \& p \downharpoonright (a \upharpoonright Ea \& p \& Ea) = (Ea \& p \downharpoonright a) \upharpoonright E\,(Ea \& p \downharpoonright a) ,$$

by the distributivity of left and right restriction, and

$$(Ea \& p \downharpoonright a) \upharpoonright E\,(Ea \& p \downharpoonright a) = Ea \& p \downharpoonright a ,$$

by right restriction to the extent being the identity. Hence,

$$a \upharpoonright p = Ea \,\&\, p \downharpoonleft a \,,$$

as asserted.

Although it may straightforwardly be shown that the concept of presheaf already introduced is indeed equivalent to one expressed simply in terms of left restriction, we have chosen to retain both left and right restriction because the concept of sheaf may be defined most naturally in that context, to which we now proceed.

# 5   Sheaves on $Q$

With the notion of presheaf established, together with its connection to that of quantal set, the identification of a concept of sheaf on a quantale comes down to describing a sheaf condition on a presheaf, corresponding to that of the completeness of a quantal set. The expectation is that this will extend that known in the case of local sets, and that the existing embedding of the category of complete quantal sets in the category of presheaves on the quantale will provide an isomomorphism with the category of sheaves on the quantale. Further, it may be expected that the idea of a compatible family of elements of a presheaf will be identifiable in some sense with that of a singleton subset of a quantal set. The existence of a unique element obtained by patching a given compatible family will then correspond to the completeness of the quantal set.

Consider first a complete quantal set $A$ , together with its canonical presheaf $\Gamma(A)$ . For any singleton subset $s : A \to Q$ , consider the family

$$(s(a) \downharpoonleft a)_{a \in A}$$

of elements of the presheaf $\Gamma(A)$ indexed by the quantal set $A$ . Then we assert that the elements of this family satisfy the condition that

$$(s(a) \downharpoonleft a) \upharpoonright s(b) = s(a) \downharpoonleft (s(b) \downharpoonleft b)$$

for any $a, b \in A$ .

For, to show these restrictions are equal, it is sufficient to show that, for any $c \in A$, one has

$$[\![ (s(a) \downharpoonleft a) \upharpoonright s(b) = c ]\!] = [\![ s(a) \downharpoonleft (s(b) \downharpoonleft a) = c ]\!]$$

in the quantal set $A$ . But, by the definitions of left and right restriction in a complete quantal set, one has that

$$[\![ (s(a) \downharpoonleft a) \upharpoonright s(b) = c ]\!] = s(a) \,\&\, [\![ a = c ]\!] \,\&\, s(b) \,,$$

and $\quad [\![ s(a) \mathbin{\downharpoonleft} (s(b) \mathbin{\downharpoonleft} b) = c ]\!] = s(a) \mathbin{\&} s(b) \mathbin{\&} [\![ b = c ]\!] = s(a) \mathbin{\&} [\![ b = c ]\!] \mathbin{\&} s(b) \; ,$

by the right symmetricity of the quantale. The required equality follows by remarking that, on the one hand,

$$s(a) \mathbin{\&} [\![ a = c ]\!] \mathbin{\&} s(b) = s(a) \mathbin{\&} s(a) \mathbin{\&} [\![ a = c ]\!] \mathbin{\&} s(b)$$

$$\leq s(a) \mathbin{\&} s(c) \mathbin{\&} s(b)$$

$$= s(a) \mathbin{\&} s(c) \mathbin{\&} s(b) \mathbin{\&} s(b) \mathbin{\&} Eb$$

$$= s(a) \mathbin{\&} Eb \mathbin{\&} s(b) \mathbin{\&} s(c) \mathbin{\&} s(b)$$

$$\leq s(a) \mathbin{\&} [\![ b = c ]\!] \mathbin{\&} s(b) \; ,$$

by applying the properties of a singleton subset, together with the right symmetricity and idempotency of the quantale; while, conversely, we have that

$$s(a) \mathbin{\&} [\![ b = c ]\!] \mathbin{\&} s(b) = s(a) \mathbin{\&} s(a) \mathbin{\&} [\![ b = c ]\!] \mathbin{\&} s(b) \mathbin{\&} s(b)$$

$$= s(a) \mathbin{\&} Ea \mathbin{\&} s(a) \mathbin{\&} s(b) \mathbin{\&} [\![ b = c ]\!] \mathbin{\&} s(b)$$

$$\leq s(a) \mathbin{\&} [\![ a = b ]\!] \mathbin{\&} [\![ b = c ]\!] \mathbin{\&} s(b)$$

$$\leq s(a) \mathbin{\&} [\![ a = c ]\!] \mathbin{\&} s(b) \; ,$$

by similar arguments.

Denoting by

$$a_s \in A$$

the element $s(a) \mathbin{\downharpoonleft} a$ obtained by localising $a \in A$ to the extent to which it lies in the subset $s \in S(A)$ , and observing that

$$E(s(a) \mathbin{\downharpoonleft} a) = s(a) \mathbin{\&} Ea = s(a) \; ,$$

we may express the condition satisfied by the family $(a_s)_{a \in A}$ determined by the singleton subset $s \in S(A)$ by writing that:

$$a_s \mathbin{\downharpoonleft} Eb_s = Ea_s \mathbin{\downharpoonleft} b_s$$

for each $a, b \in A$ . The fact that the family of elements is derived from a singleton subset of the complete quantal set therefore implies a certain compatibility between the elements chosen in the underlying presheaf.

It may be remarked that these observations in fact have a converse. One may, of course, obtain such a family

$$(a_s)_{a \in A}$$

of elements of the underlying presheaf $\Gamma(A)$ of a complete quantal set $A$ by localising at any subset

$$s : A \to Q$$

of the quantal set $A$ . The converse assertion is that, if the family so obtained satisfies the compatibility condition expressed above, then the subset is indeed a singleton subset of $A$ . For, given any $a, b \in A$, one has that:

$$Ea \,\&\, s\,(a) \,\&\, s\,(b) \;=\; Ea \,\&\, s\,(a) \,\&\, s\,(b) \,\&\, [\![b = b]\!]$$

$$=\; Ea \,\&\, [\![s\,(a) \mathbin{\downarrow} (s\,(b) \mathbin{\downarrow} b) = b]\!]$$

$$=\; Ea \,\&\, [\![(s\,(a) \mathbin{\downarrow} a) \mathbin{\uparrow} s\,(b) = b]\!]$$

$$=\; Ea \,\&\, s\,(a) \,\&\, [\![a = b]\!] \,\&\, s\,(b)$$

$$\leq\; Ea \,\&\, [\![a = b]\!]$$

$$=\; [\![a = b]\!] \;.$$

With these observations, we may now make the following:

**Definition 27** *For any presheaf $A$ on the right Gelfand quantale $Q$ , a family*

$$(a_i)_{i \in I}$$

*of elements of $A$ will be said to be* compatible *provided that*

$$a_i \mathbin{\uparrow} Ea_j \;=\; Ea_i \mathbin{\downarrow} a_j$$

*for each $i, j \in I$ .*

In these terms, we have shown above that in the underlying presheaf $\Gamma\,(A)$ of a complete quantal set $A$ , the family $(s\,(a) \mathbin{\downarrow} a)_{a \in A}$ obtained by localising the elements of $\Gamma\,(A)$ at a singleton subset $s \in S\,(A)$ is compatible. Moreover, we have shown that this compatibility characterises the property of a subset being a singleton. It may further be remarked that the completeness of the quantal set implies the existence of a unique element $b \in \Gamma\,(A)$ for which

$$Ea_s \mathbin{\downarrow} b = a_s$$

for each $a \in A$ , and

$$Eb \;=\; \bigvee_{a \in A} Ea_s \;.$$

For, taking for $b \in \Gamma\,(A)$ the unique element representing the singleton subset $s \in S\,(A)$ , in the sense that $s\,(a) = [\![b = a]\!]$ for each $a \in A$ , to show that $Ea_s \mathbin{\downarrow} b = a_s$ for any $a \in A$ , it suffices to show that

$$[\![Ea_s \mathbin{\downarrow} b = c]\!] = [\![a_s = c]\!]$$

for each $c \in A$ . But, on the one hand,

$$[\![Ea_s \mathbin{\downarrow} b = c]\!] \;=\; Ea_s \,\&\, [\![b = c]\!] \;=\; s\,(a) \,\&\, [\![b = c]\!] \;=\; [\![b = a]\!] \,\&\, [\![b = c]\!] \;,$$

while, on the other,

$$[\![a_s = c]\!] = [\![s(a) \downarrow a = c]\!] = s(a) \,\&\, [\![a = c]\!] = [\![b = a]\!] \,\&\, [\![a = c]\!] \,.$$

However,

$$[\![b = a]\!] \,\&\, [\![b = c]\!] = [\![b = a]\!] \,\&\, [\![b = a]\!] \,\&\, [\![b = c]\!]$$
$$= [\![b = a]\!] \,\&\, [\![a = b]\!] \,\&\, [\![b = c]\!]$$
$$\leq [\![b = a]\!] \,\&\, [\![a = c]\!] \,,$$

while, conversely,

$$[\![b = a]\!] \,\&\, [\![a = c]\!] = [\![b = a]\!] \,\&\, [\![b = a]\!] \,\&\, [\![a = c]\!]$$
$$\leq [\![b = a]\!] \,\&\, [\![b = c]\!] \,,$$

giving the required equality. Moreover, observing that $Ea_s = s(a)$ and that $b_s = s(b) \downarrow b$, so that $Eb_s = s(b) = [\![b = b]\!] = Eb$, we have that $Ea_s = s(a) = [\![b = a]\!] \leq Eb \,\&\, Ea \leq Eb$ for each $a \in A$, and hence that

$$Eb = \bigvee_{a \in A} Ea_s \,,$$

as asserted. Hence, the element $b \in A$ has the required properties. Moreover, it is unique: since if $b' \in A$ also has these properties, then necessarily

$$Eb = \bigvee_{a \in A} Ea_s = Eb' \,,$$

from which it follows that

$$b = Eb \downarrow b = s(b) \downarrow b = b_s = Eb_s \downarrow b' = Eb \downarrow b' = Eb' \downarrow b' = b' \,,$$

as required.

Again, with these observations in mind, we make the following:

**Definition 28** *By a sheaf on a right Gelfand quantale $Q$ will be meant a presheaf $A$ on $Q$ for which given any family*

$$(a_i)_{i \in I}$$

*of elements of $A$ that is compatible there exists a unique element $a \in A$ for which*

$$Ea_i \downarrow a = a_i \quad \text{for each} \ i \in I \,,$$

*and of which the extent $Ea$ is equal to $\bigvee_{i \in I} Ea_i$.*

*The element $a \in A$ obtained in this way will be said to be the* join *of the compatible family.*

At this point, we have shown only that the compatible family

$$(a_s)_{a \in A}$$

determined by localisation at any singleton subset $s \in S(A)$ of a complete quantal set $A$ has a unique join in its canonical presheaf $\Gamma(A)$. However, this may be extended to the case of an arbitrary compatible family

$$(a_i)_{i \in I}$$

of elements of the underlying presheaf $\Gamma(A)$, by considering the singleton subset $s : A \to Q$ of $A$ defined by writing

$$s(a) = \bigvee_{i \in I} [\![a_i = a]\!]$$

for each $a \in A$. That this defines a subset of $A$ holds irrespective of the compatibility of the family: for

$$s(a) \,\&\, [\![a = a']\!] = \bigvee_{i \in I} [\![a_i = a]\!] \,\&\, [\![a = a']\!]$$

$$\leq \bigvee_{i \in I} [\![a_i = a']\!]$$

$$= s(a') \ ,$$

for any $a, a' \in A$, and

$$s(a) = \bigvee_{i \in I} [\![a_i = a]\!] \leq \bigvee_{i \in I} [\![a_i = a]\!] \,\&\, Ea = s(a) \,\&\, Ea$$

for any $a \in A$. In the case of compatibility, the subset generated by the family is a singleton: since, for any $a, b \in A$ and $i, j \in I$, we have that

$$Ea \,\&\, [\![a_i = a]\!] \,\&\, [\![a_j = b]\!] = Ea \,\&\, Ea_i \,\&\, [\![a_i = a]\!] \,\&\, Ea_j \,\&\, [\![a_j = b]\!]$$

$$= Ea \,\&\, [\![a_i = a]\!] \,\&\, Ea_j \,\&\, Ea_i \,\&\, [\![a_j = b]\!]$$

$$= Ea \,\&\, [\![a_i \mid Ea_j = a]\!] \,\&\, [\![Ea_i \mid a_j = b]\!]$$

$$= Ea \,\&\, [\![a = a_i \mid Ea_j]\!] \,\&\, [\![Ea_i \mid a_j = b]\!]$$

$$\leq Ea \,\&\, [\![a = b]\!]$$

$$= [\![a = b]\!] \ ,$$

since $a_i \mid Ea_j = Ea_i \mid a_j$. From this, it follows that

$$Ea \,\&\, s(a) \,\&\, s(b) \leq Ea \,\&\, \bigvee_{i \in I} [\![a_i = a]\!] \,\&\, \bigvee_{j \in I} [\![a_j = b]\!] \leq [\![a = b]\!] \ ,$$

as asserted.

By the completeness of the quantal set $A$ , let $a \in A$ denote the unique element for which

$$s(b) = [\![a = b]\!]$$

for each $b \in A$ . It is asserted that the element $a \in \Gamma(A)$ is the unique join of the compatible family $(a_i)_{i \in I}$ of elements of the underlying presheaf $\Gamma(A)$ . For, firstly,

$$Ea = \bigvee_{i \in I} Ea_i \ ,$$

since, on the one hand,

$$Ea_i \leq \bigvee_{j \in I} [\![a_j = a_i]\!] = s(a_i) = [\![a = a_i]\!] \leq Ea$$

for each $i \in I$ , while, on the other,

$$Ea = [\![a = a]\!] = s(a) = \bigvee_{i \in I} [\![a_i = a]\!] \leq \bigvee_{i \in I} Ea_i \ .$$

Moreover, for any $c \in A$ , we have that:

$$[\![Ea_i \mid a = c]\!] = Ea_i \,\&\, [\![a = c]\!] = Ea_i \,\&\, s(c) = Ea_i \,\&\, \bigvee_{j \in I} [\![a_j = c]\!]$$

$$= \bigvee_{j \in I} Ea_i \,\&\, [\![a_j = c]\!] = \bigvee_{j \in I} [\![Ea_i \mid a_j = c]\!]$$

$$= \bigvee_{j \in I} [\![a_i \mid Ea_j = c]\!] = \bigvee_{j \in I} [\![a_i = c]\!] \,\&\, Ea_j$$

$$\leq [\![a_i = c]\!] \ ,$$

while conversely,

$$[\![a_i = c]\!] = Ea_i \,\&\, [\![a_i = c]\!] \leq Ea_i \,\&\, \bigvee_{j \in I} [\![a_j = c]\!] = Ea_i \,\&\, s(c)$$

$$= Ea_i \,\&\, [\![a = c]\!] = [\![Ea_i \mid a = c]\!] \ .$$

Hence,

$$[\![Ea_i \mid a = c]\!] = [\![a_i = c]\!]$$

for any $c \in A$ , from which it follows that:

$$Ea_i \mid a = a_i$$

for each $i \in I$ .

Finally, the uniqueness of this join follows by observing that any $a \in \Gamma(A)$ satisfying these conditions necessarily represents the singleton subset generated by the compatible family $(a_i)_{i \in I}$ , since, for any $c \in A$, we have that:

$$[\![a = c]\!] = Ea \,\&\, [\![a = c]\!]$$

$$= \bigvee_{i \in I} Ea_i \,\&\, [\![a = c]\!]$$

$$= \bigvee_{i \in I} [\![Ea_i \downarrow a = c]\!]$$

$$= \bigvee_{i \in I} [\![a_i = c]\!]$$

$$= s(c) \ .$$

Hence, by the uniqueness of the element $a \in A$ representing the singleton subset $s \in S(A)$ determined by the compatible family $(a_i)_{i \in I}$ in the presheaf $\Gamma(A)$, the join of the family is unique. In particular, the canonical presheaf $\Gamma(A)$ of any complete quantal set $A$ is necessarily a sheaf on the quantale $Q$.

With these remarks, denoting for any right Gelfand quantale $Q$ by

## Sheaves $/Q$

the full subcategory of the category **Presheaves** $/Q$ of presheaves on $Q$ determined by the sheaves, we see that the functor

$$\Gamma : \textbf{Complete } Q\textbf{-Sets} \ \rightarrow \ \textbf{Presheaves } /Q$$

that assigns to each complete quantal set over $Q$ its canonical presheaf may in fact be considered as a functor into the category of sheaves on $Q$, concerning which we have the following:

**Theorem 29** *For any right Gelfand quantale $Q$, the categories of complete quantal sets over $Q$ and of sheaves on $Q$ are isomorphic by the functors*

$$\textbf{Complete } Q\textbf{-Sets} \ \overset{\Gamma}{\underset{U}{\rightleftarrows}} \ \textbf{Sheaves } /Q$$

*that assign respectively to each complete quantal set its canonical presheaf, and to each sheaf its underlying quantal set.*

The existence of the functor from **Complete $Q$-Sets** to **Sheaves** $/Q$ has already been remarked. The inverse functor from **Sheaves** $/Q$ to **Complete $Q$-Sets** is obtained by establishing that the quantal set $U(A)$ canonically obtained from a sheaf $A$ by taking the extent of any element to be that in the sheaf $A$, and by considering the equality relation given by writing

$$[\![a = b]\!] = \bigvee_{a \downarrow p = p \downarrow b} Ea \,\&\, p \,\&\, Eb$$

for any elements $a, b \in A$, is actually complete.

Before proving this, we make an observation relating this equality relation defined on the underlying set of elements of the sheaf $A$ to its structure as a presheaf on the quantale $Q$, namely that

$$a \upharpoonright [\![ a = b ]\!] = [\![ a = b ]\!] \downharpoonright b,$$

for any elements $a, b \in A$ of the presheaf $A$. It is in this sense that the equality of elements of a sheaf behaves like that familiar in the case of a locale, thereby justifying the definition of the equality relation considered. To verify this assertion in the case of a sheaf $A$, observe firstly that

$$a \upharpoonright p = p \downharpoonright b$$

implies that $a \upharpoonright Ea \,\&\, p \,\&\, Eb = Ea \,\&\, p \,\&\, Eb \downharpoonright b$, since $a \upharpoonright Ea \,\&\, p \,\&\, Eb = a \upharpoonright p \,\&\, Eb = p \downharpoonright b \upharpoonright Eb = p \downharpoonright b$, and $Ea \,\&\, p \,\&\, Eb \downharpoonright b = Ea \,\&\, p \downharpoonright b = Ea \downharpoonright a \upharpoonright p = a \upharpoonright p$, by the remarks concerning the relationship between left and right restriction. Hence, the families

$$( a \upharpoonright Ea \,\&\, p \,\&\, Eb )_{a \upharpoonright p = p \downharpoonright b} \quad \text{and} \quad ( Ea \,\&\, p \,\&\, Eb \downharpoonright b )_{a \upharpoonright p = p \downharpoonright b}$$

of elements of the sheaf $A$ are identical. However, $E (a \upharpoonright Ea \,\&\, p \,\&\, Eb)$ $= Ea \,\&\, p \,\&\, Eb$ implies that

$$\bigvee_{a \upharpoonright p = p \downharpoonright b} E (a \upharpoonright Ea \,\&\, p \,\&\, Eb) = [\![ a = b ]\!] = E (a \upharpoonright [\![ a = b ]\!]), \quad \text{while}$$

$Ea \,\&\, p \,\&\, Eb \downharpoonright (a \upharpoonright [\![ a = b ]\!]) = a \upharpoonright Ea \,\&\, p \,\&\, Eb \,\&\, [\![ a = b ]\!] = a \upharpoonright Ea \,\&\, p \,\&\, Eb$, again by the relationship between left and right restrictions, and the observation that $Ea \,\&\, p \,\&\, Eb \leq [\![ a = b ]\!]$, implies that $E (a \upharpoonright Ea \,\&\, p \,\&\, Eb) \downharpoonright (a \upharpoonright [\![ a = b ]\!]) = a \upharpoonright Ea \,\&\, p \,\&\, Eb$ for any $p \in Q$ for which $a \upharpoonright p = p \downharpoonright b$. Hence,

$$a \upharpoonright [\![ a = b ]\!]$$

is the unique element of the sheaf $A$ which is the join of the compatible family

$$(a \upharpoonright Ea \,\&\, p \,\&\, Eb)_{a \upharpoonright p = p \downharpoonright b}$$

in the sheaf $A$. By similar remarks, we have that

$$[\![ a = b ]\!] \downharpoonright b$$

is the unique join of the compatible family

$$(Ea \,\&\, p \,\&\, Eb \downharpoonright b)_{a \upharpoonright p = p \downharpoonright b}$$

in the sheaf $A$. Since the families have already been remarked to be identical, it follows that

$$a \upharpoonright [\![ a = b ]\!] = [\![ a = b ]\!] \downharpoonright b$$

for any $a, b \in A$, as asserted.

Now consider this quantal set $U(A)$ underlying the sheaf $A$, and let $s \in S(U(A))$ be any singleton subset of $U(A)$. Then the elements $(a_s)_{a \in A}$ of the sheaf $A$, obtained by localising the elements of $A$ at the singleton subset by defining

$$a_s = s(a) \downharpoonleft a$$

for each $a \in A$, form a compatible family of elements of the sheaf $A$. For, given any elements $a, b \in A$, we observe first that

$$s(a) \,\&\, s(b) = s(a) \,\&\, s(b) \,\&\, [a = b],$$

since $Ea \,\&\, s(a) \,\&\, s(b) \leq [a = b]$ by the singleton axiom, hence $Ea \,\&\, s(a) \,\&\, s(b) = Ea \,\&\, s(a) \,\&\, s(b) \,\&\, [a = b]$ which gives the required equality on multiplying on the left by $s(a)$. Then,

$$(s(a) \downharpoonleft a) \upharpoonright s(b) = s(a) \,\&\, Ea \,\&\, s(b) \downharpoonleft a = s(a) \,\&\, s(b) \downharpoonleft a$$

$$= s(a) \,\&\, s(b) \,\&\, [a = b] \downharpoonleft = s(a) \,\&\, s(b) \downharpoonleft (a \upharpoonright [a = b])$$

$$= s(a) \,\&\, s(b) \downharpoonleft ([a = b] \downharpoonleft b) = s(a) \,\&\, s(b) \,\&\, [a = b] \downharpoonleft b$$

$$= s(a) \,\&\, s(b) \downharpoonleft b = s(a) \downharpoonleft (s(b) \downharpoonleft b),$$

hence,

$$a_s \upharpoonright Eb_s = Ea_s \downharpoonleft b_s.$$

Letting $a \in A$ denote the join of this family in the sheaf $A$, we assert that the corresponding element $a \in U(A)$ of the quantal set determined by $A$ represents the singleton subset $s \in S(U(A))$. Recalling that $Eb_s = s(b)$ for each $b \in A$, the fact that $a \in A$ is the join of the family implies that

$$Ea = \bigvee_{b \in A} s(b),$$

and that

$$s(b) \downharpoonleft a = s(b) \downharpoonleft b$$

for each $b \in A$. From these, we must deduce that

$$s(b) = [a = b]$$

for each $b \in U(A)$.

In one direction, consider the element $Ea \,\&\, s(b) \downharpoonleft a$, and observe that

$$Ea \,\&\, s(b) \downharpoonleft a = (Ea \,\&\, s(b) \downharpoonleft a) \upharpoonright Ea \,\&\, s(b) = Ea \,\&\, s(b) \downharpoonleft (a \upharpoonright Ea \,\&\, s(b))$$

$$= a \upharpoonright Ea \,\&\, s(b),$$

respectively by right restriction by the extent, the associativity of left and right restrictions, and by left restriction by the extent, since $E(Ea \,\&\, s(b) \downharpoonleft a)$

$= Ea \,\&\, s\,(b) \,\&\, Ea = Ea \,\&\, s\,(b)$, and $E\,(a \mid Ea \,\&\, s\,(b)) = Ea \,\&\, Ea \,\&\, s\,(b)$
$= Ea \,\&\, s\,(b)$. Hence, since $s\,(b) \downarrow a = s\,(b) \downarrow b$ and $s\,(b) = s\,(b) \,\&\, Eb$, we have
that $a \mid Ea \,\&\, s\,(b) \,\&\, Eb = Ea \,\&\, s\,(b) \,\&\, Eb \downarrow b$. Hence,

$$Ea \,\&\, s\,(b) \,\&\, Eb \leq [\![a = b]\!],$$

by the definition of equality in the quantal set $U(A)$. However, since $s\,(b) \leq Ea$
and $s\,(b) = s\,(b) \,\&\, Eb$, we have that $s\,(b) \leq Ea \,\&\, s\,(b) \,\&\, Eb$, from which it
follows that

$$s\,(b) \leq [\![a = b]\!].$$

In the other direction, we observe firstly that $s\,(b) \leq s\,(a)$ for each $b \in U\,(A)$.
For, by the inequality already shown, we have that $s\,(b) = s\,(b) \,\&\, s\,(b) \leq$
$s\,(b) \,\&\, [\![a = b]\!] \leq s\,(a)$. Hence, $Ea \leq s\,(a)$, since $Ea = \bigvee_{b \in A} s\,(b) \leq s\,(a)$.
Hence,

$$[\![a = b]\!] = Ea \,\&\, [\![a = b]\!] \leq s\,(a) \,\&\, [\![a = b]\!] \leq s\,(b),$$

from which it follows that

$$s\,(b) = [\![a = b]\!]$$

for each $b \in U\,(A)$, as required. Hence, the singleton subset $s \in S\,(U\,(A))$ is
representable by $a \in U\,(A)$, as asserted.

The uniqueness of the element $a \in U\,(A)$ follows from that of the join of any
compatible family in the sheaf $A$ by observing that if $a \in U\,(A)$ satisfies the
condition that

$$s\,(b) = [\![a = b]\!]$$

for all $b \in U\,(A)$, then necessarily

$$\bigvee_{b \in A} s\,(b) = Ea \quad \text{and} \quad s\,(b) \downarrow a = s\,(b) \downarrow b$$

for all $b \in A$. For firstly for any $b \in A$ we have that $s\,(b) = s\,(b) \,\&\, s\,(b) =$
$s\,(b) \,\&\, [\![a = b]\!] = s\,(b) \,\&\, [\![b = a]\!] \leq s\,(a) = [\![a = a]\!] = Ea$. Hence,

$$\bigvee_{b \in A} s\,(b) \leq Ea.$$

While conversely, $Ea = [\![a = a]\!] = s\,(a)$; hence,

$$Ea \leq \bigvee_{b \in A} s\,(b),$$

giving the first equality. While, for the second, since we have remarked already
that, for any elements $a, b \in A$ of the sheaf, it is the case that $a \mid [\![a = b]\!] =$
$[\![a = b]\!] \downarrow b$, it follows that for any $b \in A$ we have that
$[\![a = b]\!] \downarrow a = [\![a = b]\!] \downarrow b$, since

$$a \mid [\![a = b]\!] = E\,(a \mid [\![a = b]\!]) \downarrow (a \mid [\![a = b]\!])$$

$$= Ea \,\&\, [\![ a = b ]\!] \mid (a \uparrow [\![ a = b ]\!])$$

$$= (Ea \,\&\, [\![ a = b ]\!] \mid a) \uparrow [\![ a = b ]\!]$$

$$= (Ea \,\&\, [\![ a = b ]\!] \mid a) \uparrow Ea \,\&\, [\![ a = b ]\!] \,\&\, Ea$$

$$= Ea \,\&\, [\![ a = b ]\!] \mid a$$

$$= [\![ a = b ]\!] \mid a.$$

It follows that

$$s(b) \mid a = s(b) \mid b$$

for any $b \in A$. Hence, the element $a \in A$ is the unique join of the family $(b_s)_{b \in A}$ obtained by localising at the singleton subset $s \in S(U(A))$. Hence, the singleton subset is thus uniquely representable by the element $a \in A$, establishing the completeness of the quantal set $U(A)$.

Having established the existence of these functors

$$\textbf{Complete } Q\textbf{-Sets} \quad \overset{\Gamma}{\underset{U}{\rightleftarrows}} \quad \textbf{Sheaves } /Q,$$

it will now be shown that they are mutually inverse. In one direction, the work has already been done. Within the broader context of the category **Presheaves**$/Q$, it was shown that the composite

$$\textbf{Complete } Q\textbf{-Sets} \;\to\; \textbf{Presheaves } /Q \to Q\textbf{-Sets}$$

was indeed the inclusion functor of the full subcategory of complete quantal sets in the category of quantal sets over $Q$. In the present context, we therefore have that $U\,\Gamma$ is the identity functor on the category

$$\textbf{Complete } Q\textbf{-Sets}.$$

Consider then the composite

$$\textbf{Sheaves } /Q \to \textbf{Complete } Q\textbf{-Sets} \;\to\; \textbf{Sheaves } / Q$$

which assigns to each sheaf $A$ the presheaf $\Gamma U(A)$ canonically determined by its underlying quantal set $U(A)$. Since the underlying sets and the extent operators of each of these structures are identical, to show that $\Gamma U(A)$ is exactly $A$ is equivalent to showing that its restrictions are exactly those of the sheaf $A$. By the concluding observations of the preceding section, it suffices to show that the left restrictions coincide.

Suppose then that $A$ is a sheaf on the quantale $Q$. For any element $a \in A$, it is asserted that the left restriction $q \mid a \in A$ at any element $q \in Q$ is equal to that of the corresponding element of the sheaf $\Gamma U(A)$. By the construction of the left restriction on the canonical presheaf of a complete quantal set, this assertion is equivalent to showing that $q \mid a \in A$ represents the singleton subset of $U(A)$ obtained by assigning to each $c \in U(A)$ the element $q \& [\![ a = c ]\!] \in Q$. Hence, it must be shown that

$$[\![ q \mid a = c ]\!] = q \& [\![ a = c ]\!]$$

for each $c \in A$, in which the equality relation is that of the quantal set $U(A)$ underlying the sheaf $A$.

In one direction, it must be shown that

$$E(q \mid a) \& p \& Ec \leq q \& [\![ a = c ]\!]$$

for any $p \in Q$ for which $(q \mid a) \uparrow p = p \mid c$ in the sheaf $A$. Given that

$$q \& [\![ a = c ]\!] = q \& \bigvee_{a \uparrow p' = p' \mid c} Ea \& p' \& Ec,$$

it suffices, for any such $p \in Q$, to find $p' \in Q$ satisfying $a \uparrow p' = p' \mid c$ in the sheaf $A$, for which $E(q \mid a) \& p \& Ec \leq q \& Ea \& p' \& Ec$. Taking $p' = Ea \& q \& p \in Q$ and observing that $E(q \mid a) = q \& Ea$, we note that $(q \mid a) \uparrow p = q \mid (a \uparrow p) = q \& Ea \& p \mid a = q \& p \& Ea \mid a = q \& p \mid a$, which implies that $q \& p \mid a = p \mid c$. Hence, on the one hand, one has that

$$a \uparrow Ea \& q \& p = Ea \& q \& p \mid a = Ea \& q \& q \& p \mid a = Ea \& q \& p \mid c.$$

While, on the other, evidently $q \& Ea \& p \& Ec \leq q \& Ea \& (Ea \& q \& p) \& Ec$, by the right symmetricity of the quantale $Q$. Hence,

$$[\![ q \mid a = c ]\!] \leq q \& [\![ a = c ]\!].$$

For the converse, it must be shown that, for any $p \in Q$ for which $a \uparrow p = p \mid c$, one has that

$$q \& Ea \& p \& Ec \leq [\![ q \mid a = c ]\!].$$

Given that $[\![ q \mid a = c ]\!] = \bigvee_{(q \mid a) \uparrow p' = p' \mid c} E(q \mid a) \& p' \& Ec$, it suffices, for any such $p \in Q$, to find $p' \in Q$ satisfying $(q \mid a) \uparrow p' = p' \mid c$ in the sheaf $A$, for which $q \& Ea \& p \& Ec \leq E(q \mid a) \& p' \& Ec$. Taking $p' = q \& p \in Q$, we note that indeed, on the one hand, one has that

$$(q \mid a) \uparrow q \& p = q \mid (a \uparrow q \& p) = q \mid ((a \uparrow q) \uparrow p) = ((q \mid a) \uparrow q) \uparrow p$$

$$= (q \& Ea \& q \mid a) \uparrow p = q \mid (a \uparrow p) = q \mid (p \mid c) = q \& p \mid c.$$

While, on the other, evidently $q \& Ea \& p \& Ec \leq q \& Ea \& (q \& p) \& Ec$, once again by the right symmetricity of the quantale $Q$. Hence,

$$q \& [\![ a = c ]\!] \leq [\![ q \mid a = c ]\!],$$

giving the required equality for any $c \in A$. Hence, the left restriction (and so the right restriction) on the sheaf $\Gamma U (A)$ is exactly that of the sheaf $A$. The functor $\Gamma U$ is therefore the identity on objects.

Suppose finally that $f : A \rightarrow B$ is a map of sheaves on the quantale $Q$. Recall that the map

$$U (f) : U (A) \rightarrow U (B)$$

of quantal sets that it determines is given by the functional relation $U (f) :$ $A \times B \rightarrow Q$ defined by

$$U (f) (a, b) = [\![ f (a) = b ]\!]$$

for each $(a, b) \in A \times B$. By the remarks made in the preceding section concerning the underlying map of presheaves determined by any map of complete quantal sets, the map

$$\Gamma U (f) : \Gamma U (A) \rightarrow \Gamma U (B)$$

then assigns to each $a \in A$ the element $b \in B$ uniquely defined by the property that

$$[\![ b = c ]\!] = U (f) (a, c)$$

for each $c \in B$. However, by the above remarks, $U (f) (a, c) = [\![ f (a) = c ]\!]$ for each $c \in B$. Hence, $f (a) \in B$ is indeed the required element of the sheaf $B$, giving that $\Gamma U (f)$ is exactly $f$. The functor $\Gamma U$ is therefore the identity on maps, which completes the proof. □

This isomorphism, rather than equivalence, of categories states that sheaves on a quantale $Q$ and complete quantal sets over $Q$ are equivalent mathematical structures, interrelated by such identities as that

$$a \restriction [\![ a = b ]\!] = [\![ a = b ]\!] \restriction b$$

for all $a, b \in A$, observed earlier. The significance of this for the category of sheaves on $Q$ will be examined further later. For the moment, we note that the existence of the isomorphism allows us to express the adjointness derived in the preceding section is a more familiar form:

**Corollary 30** *For any right Gelfand quantale $Q$, there are adjoint functors*

$$\textbf{Presheaves } /Q \rightleftarrows \textbf{Sheaves } /Q,$$

*from the category of presheaves on $Q$ to the category of sheaves on $Q$, of which the coadjoint is the canonical embedding.* □

The adjoint is, of course, the functor which assigns to each presheaf $A$ on $Q$ the canonical sheaf of the completion of the quantal set underlying the presheaf $A$. This functor, the *sheafification functor* on the category of presheaves on $Q$ applies the identification of compatible families of elements of a presheaf with singleton subsets of the quantal set that it canonically determines to allow the completion of the presheaf to a sheaf, exactly as in the localic case [6].

# 6 Postlude

Over the preceding sections, we have described the way in which concepts of quantal set and of presheaf can be defined with respect to a right Gelfand quantale $Q$. The ideas have been intertwined, both in terms of motivation, and in terms of development of the notions of complete quantal set and of sheaf, in exactly the way that one observes classically in the case of a locale $L$. The adjoint functors

**Presheaves** $/Q \rightleftarrows$ **Sheaves** $/Q$    and    $Q$-**Sets** $\rightleftarrows$ **Complete** $Q$-**Sets**

have been arrived at not in isolation, but interlinked through what has become the isomorphism

**Sheaves** $/Q \rightleftarrows$ **Complete** $Q$-**Sets**

between the categories of sheaves on $Q$, and of complete quantal sets over $Q$.

One aspect of this that has, as yet, received little mention in its own right is that of the concept of a subset of a quantal set over $Q$. Through the idea of a singleton subset, it has played a critical role in developing the theory, leading explicitly to the concept of completeness of a quantal set, and, through that, motivating the definition of a sheaf on the quantale $Q$. Equally importantly, it has shown the quantale itself to be recoverable from its quantal sets, since the subsets of the complete quantal set $1_Q$ to the elements of the quantale $Q$, just as in the case of a locale.

What has been left unsaid about the concept of a subset of a quantal set is that it does not coincide with that of subobject in the category of quantal sets. It is at this point that we depart from that with which one is familiar in the case of local sets. Althought this will be explored in a later paper, it is perhaps important to note the extent to which these changes one's perceptions of the categories involved, since these changes are extensive. An interesting case to bear in mind in what follows is that of the right Gelfand quantale obtained from the spectrum

$$\operatorname{Max} M_n(\mathbb{C})$$

of a finite-dimensional C\*-algebra.

The subsets of a quantal set $A$ have a natural structure as a quantale, and with this comes the notion of a subset $s : A \to Q$ being *two-sided* in the event that

$$s(a) \in I(Q)$$

for each $a \in A$. It may also be remarked that the subsets of $A$ also have a natural structure as a quantal set over $Q$, in terms of which two-sidedness may also be defined. The subobjects of the quantal set $A$ in the category of quantal sets over $Q$ correspond bijectively with the two-sided subsets of $A$, with each subset having a two-sided closure that defines, and is defined by, a subobject in the category.

Having in mind the quantale already mentioned, we may note that in that case, the right side

$$R\left(\operatorname{Max} \operatorname{M}_n(\mathbb{C})\right)$$

is the quantale whose elements correspond to the linear subspaces of $\mathbb{C}^n$. The locale

$$I\left(\operatorname{Max} \operatorname{M}_n(\mathbb{C})\right)$$

however, has exactly two elements, corresponding to the closed ideals of $\operatorname{M}_n(\mathbb{C})$, namely the zero ideal, and the matrix ring itself. In particular, the quantal set

$$^1\operatorname{Max} \operatorname{M}_n(\mathbb{C})$$

has subsets corresponding bijectively with the linear subspaces of $\operatorname{M}_n(\mathbb{C})$, but exactly two subobjects in the category of quantal sets, namely the empty subset and the quantal set itself.

The concept of subset of a quantal set $A$ over a quantale $Q$ is not therefore intrinsic to the category of quantal sets over $Q$. It may be defined externally, which is the sense in which it has been dealt with in this paper. It may, however, equally be considered to be internal to the category, provided that the category is endowed with structure in the form of a subset classifier determined by the quantale $Q$. The details of this, directly generalising the situation in the case of a locale, will be found in a later paper [14]. The existence of this structure on the category of quantal sets, or equivalently the category of sheaves on the quantale $Q$, is fortuitous, in the face of another realisation concerning this category.

It will have been evident, throughout, that extent, whether of an element of a quantal set, or of an element of a sheaf, plays an important role in allowing, by the right symmetricity of the quantale, all manner of interchange to take place to its right. At the most basic level, it is its presence, on the left of expressions in certain of the axioms, that allows the composite of functional relations between quantal sets to be proved again functional. In terms of sheaves, the extent enters naturally in the condition that describes the extent of a restriction: namely, that

$$E(p \mid a) = p \mid Ea.$$

This axiom was not arbitrarily chosen, however reasonable it might be: it is present in the motivating instance of the completion of any quantal set, waiting to be noted down and axiomatised.

When working with quantal sets over a locale, or sheaves on a locale, conditions of this kind are considered to be part of an accounting process that keeps track of extent while restricting an element. Once the quantale concerned is not a locale, the condition has a more sinister, in the technical sense, contribution to make. Consider, for instance, the act of taking an element $a \in A$ of a sheaf on the quantale $Q$, and restricting it on the left by the identity element $1_Q \in Q$

of the quantale. One pauses to remember, of course, that this possibility of restricting an element of a sheaf to *any* element of the quantale, rather than those lying below its extent, is inherited from the logical, rather than the geometrical, way of looking at sheaves.

Now, look at the effect of this restriction on the extent: the element $a \in A$ of extent $Ea \in Q$ under restriction becomes the element $1_Q \mid a \in A$ with extent

$$E(1_Q \mid a) = 1_Q \, \& \, Ea \geq Ea$$

which is the two-sided closure of $Ea \in Q$, in the sense of the smallest two-sided element of the quantale containing the element $Ea \in Q$. Observe that by *restricting* the element $a \in A$, we have *extended* its extent. In the case of the quantale

$$R(\operatorname{Max} M_n(\mathbb{C})),$$

an element whose extent corresponds to a one-dimensional subspace of $\mathbb{C}^n$ is extended to an element whose extent corresponds to $\mathbb{C}^n$ itself.

The conclusion to which this leads is expressed in the following:

**Theorem 31** *For any right Gelfand quantale $Q$, the category of sheaves on $Q$ is equivalent to the category of sheaves on the locale $I(Q)$ of two-sided elements of $Q$, by the functors*

$$\textbf{Sheaves } /Q \rightleftarrows \textbf{Sheaves } / I(Q)$$

*that assign to each sheaf on $Q$ the sheaf on $I(Q)$ consisting of those elements having two-sided extent, and to any sheaf on $I(Q)$ its canonical extension to the quantale $Q$.*

The proof will be omitted, being relatively straightforward once one has realised this state of affairs. One literally takes, for any sheaf $A$ on the quantale $Q$, the set of elements with extent in $I(Q)$, together with its natural restrictions over $I(Q)$. Conversely, given any sheaf $A$ over the locale $I(Q)$, one associates with it the sheaf of which elements are those pairs $(q, a) \in Q \times A$ for which $1_Q \& q = Ea \in I(Q)$, with the natural definitions of extent and restrictions. The equivalence established by these functors then follows from the observation that left restriction by the identity of $Q$ provides for each $q \in Q$ an invertible mapping from the set $A(q)$ of elements of $A$ of extent $q \in Q$ and the set $A(1_Q \& q)$ of elements of extent $1_Q \& q \in I(Q)$.  □

The case of the quantale

$$R(\operatorname{Max} M_n(\mathbb{C}))$$

again provides an insight into what is happening. The category of sheaves on the quantale is now seen to be equivalent to the category of sheaves on the locale $I(\operatorname{Max} M_n(\mathbb{C}))$, which is the locale **1** corresponding to the topological space with a single point. The category of sheaves on the quantale is therefore equivalent to the category of sets. It is, however, the category of sets together with the structure afforded by the subset classifier determined by the right Gelfand quantale $R(\operatorname{Max} M_n(\mathbb{C}))$. In particular, we see that the category of sets may be endowed with a multiplicity of independent structures of this kind, from each of which the quantale concerned may be recovered by taking the quantale of subsets, in the sense determined by that structure, of the one-element set.

More generally, we have the following:

**Corollary 32** *For any right Gelfand quantale $Q$, the category of sheaves*

$$\text{Sheaves } /Q$$

*on the quantale $Q$ is a topos.*

One remarks that the quantale $I(Q)$ is actually a locale, from which the result follows.   □

The observation that the category

$$\text{Sheaves } /Q$$

is a topos is, however, one that is, in some sense, inappropriate, in that it introduces only one aspect of the structure that is canonically present. For the category of sheaves on $Q$ also admits a subset classifier, distinct from the subobject classifier which the category happens to have through being a topos. The subset classifier characterises the characteristic maps of subsets, while the subobject classifier performs the same function for two-sided subsets. It is the subset classifier which is the primary structure of interest in the category, carrying with it the logical structure of the category of sheaves on $Q$. It is to this that we return in the sequel [14] to this paper.

# References

[1] Akemann, C.A., Left ideal structure of C*-algebras, J. Functional Analysis 6 (1970), 305-317.

[2] Berni-Canani, U., Borceux, F. and Succi-Cruciani, R., A theory of quantal sets, J. Pure and Applied Algebra 62 (1989), 123-136.

[3] Dilworth, R., Ward, M., Residuated lattices, Trans. Amer. Math. Soc. 45 (1939), 335-354.

[4] Dixmier, J., C*-Algebras, North Holland, 1977.

[5] French, G.M., Syntactic and Semantic Aspects of Locales and Quantales, Thesis, University of Sussex, 1992.

[6] Fourman, M.P., Scott, D.S., Sheaves and logic, *in* Applications of Sheaves, Springer Lecture Notes 753 (1979), 302-401.

[7] Giles, R. and Kummer, H., A non-commutative generalization of topology, Indiana Univ. Math. J. 21 (1971), 91-102.

[8] Higgs, D., A category approach to boolean-valued set theory, Lecture Notes, University of Waterloo, 1973.

[9] Johnstone, P.T., Stone Spaces, Cambridge University Press, Cambridge, 1982.

[10] Mulvey, C.J., &, Tagungsbericht, Category Theory Meeting, Oberwolfach, 1983.

[11] Mulvey, C.J., &, Rendiconti Circ. Mat. Palermo 12 (1986), 99-104.

[12] Mulvey, C.J., Gelfand quantales, *to appear*.

[13] Mulvey, C.J., The spectrum of a $C^*$-algebra, *to appear*.

[14] Mulvey, C.J. and Nawaz, M., Quantales: Quantal logic, *to appear*.

[15] Nawaz, M., Quantal: Quantal Sets, Thesis, University of Sussex, 1985.

[16] Nawaz, M., Quantal sets and Cauchy-complete bicategories, *to appear*.

[17] Rosicky, J., Multiplicative lattices and $C^*$-algebras, Cahiers de Top. et Géom. Diff. Cat. XXX-2 (1989), 95-110.

[18] Scott, D.S., A proof of the independence of the continuum hypothesis, Math. Systems Theory 1 (1967), 89-111.

# VIII

# Categories of fuzzy sets with values in a quantale or projectale

## L.N. Stout

Properties of the lattice **L** are reflected in the properties of the categories **Set(L)**, **Set(L)**/$(A, \alpha)$, and the lattice $\mathcal{U}(A, \alpha)$. The lattices $\mathcal{U}(A, \alpha)$ best reflect the structures on the lattice if the structure is inherited by closed down segments and direct products. Operations at the level of the slice categories require distributivity too. The first object of this paper is to see how the additional structures given by an associative operation & which distributes over sups and hence has a right adjoint (a quantale in the sense of Rosenthal [7]) shows up at the three different levels for fuzzy sets. This structure generalizes the Heyting operation and the t-norms which give Lukaciewcz logic by allowing non-commutative connectives. Since a quantale structure does not always restrict to a quantale structure on down segments, we do not always get a quantale structure on $\mathcal{U}(A, \alpha)$. The second object is to see what a projectale structure on **L** as in Khatcherian ([3]), given as a consistent family of projection functors with right adjoints, gives in each of the levels. Here we no longer have a binary operation, though we do have notions of projection onto an element of **L** .

The category of **L**-fuzzy sets **Set(L)** was introduced by Goguen in [1] and studied further by Pultr [5,6,4] , Stout [8] and Höhle and Stout [2] . It has as objects pairs $(A, \alpha)$ where $A$ is a set and $\alpha : A \rightarrow \mathbf{L}$. The map $\alpha$ is thought of as giving a degree of membership. The morphisms are ordinary functions which do not decrease the degree of membership. If **L** is a completely distributive lattice then **Set(L)** is a quasitopos so, in particular, the slice categories **Set(L)**/$(A, \alpha)$ are cartesian closed. Goguen characterized these categories in terms of the structure of the lattice of all subobjects. Stout considered the structure of the lattice of unbalanced subobjects (those whose underlying set map is the identity), $\mathcal{U}(A, \alpha)$. All of the papers cited above considered how the additional structure given by a semigroup operation * on the lattice is reflected in a monoidal structure on **SET(L)** and as a Lukasiewcz logic on the subobject lattices.

In this paper we will consider two different kinds of structures on **L** given by more general kinds of operations. We will start by making explicit what properties of the categories **Set(L)**, **Set(L)**/$(A, \alpha)$, and the lattice $\mathcal{U}(A, \alpha)$ arise from weaker conditions on **L**. Then we will look at the structures given by the & of a quantale and the family of functors $a \sqcap -$ of a projectale.

# 1   Inheritance of Order Properties

The weakest kind of structure we can ask for **L** to have and still be able to define the category **Set(L)** is that of a partially ordered set. This is what is needed in order to be able to define the mappings in **Set(L)**. The reflexive law is needed in order to have identities and the transitive law is needed for a composition of morphisms of fuzzy sets to be a morphism.

The category **Set(L)** has rather few nice properties if all that **L** has is a partial order structure. The forgetful functor $U : \textbf{Set(L)} \to \textbf{Sets}$ preserves and reflects both monos and epis and is faithful, so we can always think of **Set(L)** as a concrete category; the fibers will be ordered sets. It does not have products or a terminal. It does, however, have equalizers constructed as in **Sets** and given the membership inherited from the codomain of the inclusion map. Colimits can be found using the colimit of the set part with the induced map to **L**.

If **L** has a partial order structure we will be able to find initial structures induced by single morphisms: If $f : A \to B$ and we are given a structure $(B, \beta)$, then the structure $(A, \beta(f(-)))$ is the largest such that $f$ is a **Set(L)** map. It also has the property that any map $g : U(C, \gamma) \to A$ gives a **Set(L)** map to $(A, \beta(f(-)))$ if and only if the composition $f \circ g : (C, \gamma) \to (B, \beta)$ is a **Set(L)** map. To get full initial structures we will need arbitrary sups.

Since **Set(L)** does not have nice properties, it is not reasonable to ask for the slice categories to have reasonable properties. However, we can define partially ordered sets of the form $\mathcal{U}(A, \alpha)$, though they will not be lattices unless **L** is. The elements of $\mathcal{U}(A, \alpha)$ are pairs $(A, \alpha')$ with $\alpha'(a) \le \alpha(a)$ for all $a \in A$.

**Proposition 1.1** $\mathcal{U}(A, \alpha)$ *can be thought of as a direct product of the partially ordered sets* $\downarrow \alpha(a) = \{l' \le \alpha(a)\}$ *for* $a \in A$.

PROOF:

> If $A$ has only a single element then $\downarrow \alpha(a)$ is $\mathcal{U}(A, \alpha)$. In general, giving an unbalanced subobject of $\mathcal{U}(A, \alpha)$ is the same as giving a value $\alpha'(a) \le \alpha(a)$ for each $a \in A$. Thus the underlying set of $\mathcal{U}(A, \alpha)$ is $\prod_{a \in A} \downarrow \alpha(a)$. The order relation is given componentwise: $(A, \alpha') \le (A, \alpha'')$ if and only if $\alpha'(a) \le \alpha(a)$ for all $a \in A$, exactly in the direct product order. ∎

This gives us a way of identifying properties of **L** which will be inherited by $\mathcal{U}(\mathcal{A}, \alpha)$: they should be properties inherited by closed downsegments and direct products of partial orders. In general, $\downarrow h$ has a terminal object $h$ and the direct product of partial orders with a terminal has a terminal, so the partial orders $\mathcal{U}(\mathcal{A}, \alpha)$ have terminal elements, even if there is no largest element of **L**.

For any kind of structure on **L** we get a functor transporting unbalanced subobjects back along a map $f : (A, \alpha) \to (B, \beta)$:

$$\mathcal{U}(B, \beta) \xrightarrow{f^{-1}} \mathcal{U}(A, \alpha)$$
$$(B, \beta') \quad \mapsto \quad (A, a \mapsto \alpha(a) \wedge \beta'(f(a))).$$

Here we need the $\alpha(a) \wedge$ in order to insure that we land in $\mathcal{U}(A, \alpha)$. When we look at projectales we will see another functor which achieves this objective in a different way.

If **L** has a terminal, then **Set(L)** will too, namely $(\{*\}, \top)$.

Strong subobjects are those of the form

$$\iota(A', \alpha|_{A'}) \hookrightarrow (A, \alpha).$$

They are characterized by maps

$$\chi_{A'} : (A, \alpha) \to (\{0, 1\}, \top)$$

exactly as in **Sets**. All we need here is the terminal, since the classifier uses the map $tr : (\{*\}, \top) \to (\{0, 1\}, \top)$, so the pullback in the usual subobject representation diagram is actually an equalizer (no products are needed).

If **L** has pairwise infs then **Set(L)** will have pairwise products:

$$(A, \alpha) \times (B, \beta) = (A \times B, (a, b) \mapsto \alpha(a) \wedge \beta(b)).$$

Pairwise infs and pairwise sups are inherited by downsegments and direct products of posets, so they will give pairwise infs and sups in $\mathcal{U}(\mathcal{A}, \alpha)$.

As soon as we have a lattice structure on **L** we will get pullbacks in **Set(L)**, so reasonable properties for the slice categories can be expected. In particular the slices will have products. We also get the pullback functor

$$f^* : \mathbf{Set(L)}/(B, \beta) \to \mathbf{Set(L)}/(A, \alpha)$$

for each morphism $f : (A, \alpha) \to (B, \beta)$, and we get the left adjoint $\Sigma_f$, given by composition with f, since it is available in any category with pullbacks. Furthermore $f^*$ lifts $f^{-1}$ in the sense that $\iota f^* = f^{-1}\iota$. Here $\iota$ is the inclusion functor.

Pullbacks in **Set(L)** will also be enough to give weak representation of unbalanced subobjects. First let us recall the definition from [8] :

**Definition 1.1** *A weak representer for unbalanced subobjects is an object* $\Lambda$ *equipped with an unbalanced mono* $\lambda : \Lambda' \to \Lambda$ *such that, given any unbalanced mono* $m : A' \to A$ *there is a (not necessarily unique) characteristic morphism* $\chi_m : A \to \Lambda$ *such that the square*

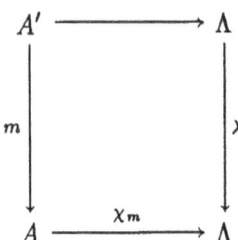

*is a pullback.*

For **Set(L)** we use

$$\Lambda = (L, \top)$$

and

$$\Lambda' = (L, \mathrm{id}_L)$$

with the identity used as the map $\lambda$. A characteristic map for $(A, \alpha') \to (A, \alpha)$ is then given by $\alpha'$.

If we let **L** be a complete lattice, then **Set(L)** will have all limits. Each of the lattices of unbalanced subobjects will be a complete lattice. The forgetful functor $U : \mathbf{Set(L)} \to \mathbf{Sets}$ becomes an initial structure functor, and hence has both right and left adjoints. The obvious inclusion functor $\iota : \mathcal{U}(A, \alpha) \to \mathbf{Set(L)}/(A, \alpha)$ has a left adjoint left inverse given by the factorization of a map into an epi followed by a strong mono, followed by an unbalanced mono:

$$(f(b), a \mapsto \bigvee_{\{b|f(b)=a\}} \beta(b)) \xrightarrow{\ inclusion\ } (A, a \mapsto \bigvee_{\{b|f(b)=a\}} \beta(b))$$

$$f \uparrow \qquad\qquad\qquad\qquad\qquad \downarrow \mathrm{id}_B$$

$$(B, \beta) \xrightarrow{\qquad\qquad f \qquad\qquad} (A, \alpha)$$

The adjoint is given by the functor $\sigma : \mathbf{Set(L)}/(A, \alpha) \to \mathcal{U}(A, \alpha)$ taking $f$ to $\mathrm{id}_A : (A, a \mapsto \bigvee_{\{b|f(b)=a\}} \beta(b)) \to (A, \alpha)$.

At the level of unbalanced subobject lattices we get the left adjoint to $f^{-1}$ as

$$\exists_f(\alpha') : b \mapsto \bigvee_{\{a|f(a)=b\}} \alpha'(a).$$

Notice that $\Sigma_f$ lifts $\exists_f$.

In order for a right adjoint $\Pi_f$ to exist for $f^*$ we need distributivity in the lattice. See [8] for an explicit formula. the quantifier at the level of unbalanced subobject lattices is obtained as follows:

$$\gamma(a) \;=\; \bigvee\{l \le \beta(f(a))|l \wedge \alpha(a) \le \alpha'(a)\}$$

$$\forall_f(\alpha')(b) \;=\; \bigwedge_{\{a|f(a)=b\}} \gamma(a)$$

The distributivity is needed to show that $\gamma(a) \wedge \alpha(a) \le \alpha'(a)$.

In that case we get the hyperdoctrinal diagram

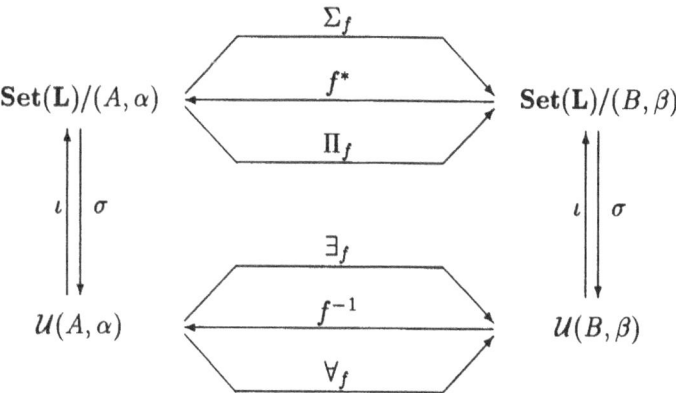

where

$$\Sigma_f \dashv f^* \dashv \Pi_f$$
$$\exists_f \dashv f^{-1} \dashv \forall_f$$

and the following equations hold:

$$\sigma\Sigma_f \;=\; \exists_f\sigma$$
$$\iota\forall_f \;=\; \Pi_f\iota$$
$$\iota f^{-1} \;=\; f^*\iota$$
$$\sigma f^* \;=\; f^{-1}\sigma.$$

Combining the existence of the right adjoint to $f^*$ with the representation of strong subobjects given above we see that if $\mathbf{L}$ is a complete distributive lattice, then $\mathbf{Set(L)}$ is a quasitopos.

$\mathbf{Set(L)}$ is unusual among topological categories over $\mathbf{Sets}$ in that there are lots of structures on the one point set. Indeed, we can recover the lattice $\mathbf{L}$ as the lattice of unbalanced subobjects of the crisp structure on the one point set.

This suggests our main theorem for this section:

**Theorem 1.2** *If $T \to$ Sets is a topological category in which the unbalanced subobject of crisp objects have unique characteristic maps, then $T$ is* Set($Hom(1, \Lambda)$).

PROOF:

   The first question to ask is exactly what we mean by "$T$ is Set($Hom(1, \Lambda)$)". Since both of these are concrete categories over **Sets**, one possible meaning would be that the two categories are concretely isomorphic. This will be true if $T$ has the property that any isomorphism which has underlying set map an identity *is* an identity, then we will get a concrete isomorphism. In general we will show that there are functors $F : T \to$ **Set**($Hom(1, \Lambda)$) and $G :$ **Set**($Hom(1, \Lambda)$) $\to T$ such that $FG = $id$_{\mathbf{Set}(Hom(1,\Lambda))}$ and $GF \approx$ id$_T$ with the isomorphism sitting over the identity in **Sets**. We start by defining the functor $F$: given an object $A$ in $T$ we can think of $A$ as an unbalanced subobject of the crisp structure on $U(A)$. Because unbalanced subobjects of crisp objects have unique characteristic morphisms this gives us a morphism $\alpha : (U(A), \text{crisp}) \to \Lambda$. We can then obtain a fuzzy set on $U(A)$ taking an element $a$ to the composition

$$1 \xrightarrow{\ulcorner a \urcorner} (A, \text{ crisp }) \xrightarrow{\alpha} \Lambda.$$

   Since this is to be a concrete functor we know what the action should be on morphisms: Given a map $f : A \to B$ in $T$, $F(f)$ should be $U(f) : U(A) \to U(B)$, so we need to check that the underlying map of a $T$-map does not decrease the degree of membership. Suppose $a \in U(A)$, then $\alpha \ulcorner a \urcorner$ characterizes an unbalanced subobject of the crisp structure on 1. This unbalanced subobject of 1 maps into $A$, so composing with $f$ gives a structure on 1 such that $\ulcorner f(a) \urcorner$ maps through $B$. Now if $\beta$ is the characteristic map of $B$ as an unbalanced subobject of $(U(B), \text{crisp})$, this tells us that $\beta \ulcorner f(a) \urcorner \geq \alpha \ulcorner a \urcorner$. This tells us that we have a map in **Set**($Hom(1, \Lambda)$). Thus we have a functor. Next we construct the functor $G :$ **Set**($Hom(1, \Lambda)$) $\to T$. Given an object $(A, \alpha)$ in **Set**($Hom(1, \Lambda)$) we need a $T$-structure on $A$. Now $(A, \alpha)$ is completely determined by the value of $\alpha$ at the elements of $A$. Let $a \in A$. Then $\alpha(a) : 1 \to \Lambda$ corresponds to a structure in $T$ on 1; let us call this $T$ object $(\{a\}, \alpha(a))$. We obtain $G(A, \alpha)$ as the final $T$ structure on $A$ determined by the family of maps $\{\ulcorner a \urcorner : U((\{a\}, \alpha(a))) \to A\}$. Now if $f : (A, \alpha) \to (B\beta)$ is a map in **Set**($Hom(1, \Lambda)$) then each of the maps $(\{a\}, \alpha(a)) \mapsto \{f(a)\}, \beta(f(a)))$ is a $T$ map, so all of the compositions $(\{a\}, \alpha(a)) \to G(B, \beta)$ are $T$ maps. Thus so is $f : G(A, \alpha) \to (B, \beta)$. Thus we have a functor. Much of what remains in the proof depends on the fact that the characteristic morphism of the structure on $A$ obtained as a final structure induced by structures on the elements of $A$ of the form $\alpha^\ulcorner a^\urcorner$ is $\alpha$. If we take that final structure we will certainly get $\alpha$ as a map in $\mathcal{T}$ from $(A, final)$ to $\Lambda'$. This tells us that the structure characterized by $\alpha$, $(A, \alpha)$, maps into $(A, final)$. On the other hand, all of the maps $^\ulcorner a^\urcorner : (1, \alpha^\ulcorner a^\urcorner) \to (A, \text{crisp})$ factor through $(A, \alpha)$, so $(A, final)$ also factors through $(A, \alpha)$. Next we show that the composition $FG = \text{id}_{\mathbf{Set}}(\text{Hom}(1, \Lambda))$. First we take the final structure determined by the structures on elements induced by the fuzzy set map $\alpha$. Then we take the characteristic map of the unbalanced subobject this produces. We have just shown that this is again $\alpha$ because of the uniqueness of characteristic morphisms of unbalanced subobjects of crisp objects. The other composition is equally easy: if we start with a structure $(A, \alpha)$, we get the characteristic morphism of $GF(A, \alpha)$ is also $\alpha$. This means that $(A, \alpha)$ and $GF(A, \alpha)$ are in the same unbalanced subobject of $(A, crisp)$, hence they can differ by at most an isomorphism sitting over the identity on $A$.

∎

In the remaining sections of the paper we will see what happens when we add in other operations on the lattice **L**.

# 2   Projectale Valued Fuzzy Sets

In [3], Khatcherian defines a *projectorial lattice* as a lattice **A** such that each element $a \in \mathbf{A}$ is equipped with a functor $a \sqcap (-) : \mathbf{A} \to \mathbf{A}$ called $a$-projection, such that

1. For all $x \in \mathbf{A}$, $a \sqcap x \leq a$.

2. For all $x \in \mathbf{A}$, if $x \leq a$ then $a \sqcap x = x$.

3. For all $x, a \in \mathbf{A}$, $x \sqcap a = \bot$ if and only if $a \sqcap x = \bot$.

If in addition we are given a right adjoint $a \Downarrow (-)$ to $a \sqcap (-)$, then the structure is called a *projectale*.

**Proposition 2.1** *If* **L** *is a projectorial lattice, then so are all of the lattices of unbalanced subobjects* $\mathcal{U}(A, \alpha)$.

PROOF:

We need only check that if **L** is a projectorial lattice then so are all of the closed downsegments $\downarrow l$ for $l \in \mathbf{L}$ and that the direct product of projectorial lattices is a projectorial lattice.

Axiom 1 tells us that when we restrict our attention to objects below $l$ we will always get $a \sqcap x \leq l$, so the operation will be well defined on $\downarrow l$. Axiom 1 will hold in the downsegment by inheritance. Since axioms 2 and 3 are equations they will also be inherited on downsegments.

The result for products is also easy:

$$(a_i)_{i \in I} \sqcap (x_i)_{i \in I} = (a_i \sqcap x_i)_{i \in I}$$

gives an assignment with all of the desired properties. ∎

The right adjoints which give a projectale structure also restrict and respect products, so we get

**Proposition 2.2** *If* **L** *is a projectale so are all of the lattices of unbalanced subobjects* $\mathcal{U}(A, \alpha)$.

PROOF:

We need to modify the adjoint $a \Downarrow (-)$ to get a map $\alpha' \Downarrow_\alpha (-) :$ $\mathcal{U}(A, \alpha) \to (A, \alpha)$. This is done by taking $\alpha''$ to

$$\alpha' \Downarrow_\alpha \alpha'' : x \mapsto (\alpha'(x) \Downarrow \alpha''(x)) \wedge \alpha(x).$$

The modification with $\wedge \alpha(x)$ does not change the adjointness since in all cases we are working with unbalanced subobjects of $(A, \alpha)$, so we always have all relevant values $\alpha'''(x)$ less than or equal to $\alpha(x)$. ∎

We can also use the projection $\alpha \sqcap (-)$ to get a functor transporting unbalanced subobjects back along a map $f : (A, \alpha) \to (B, \beta)$:

$$\begin{array}{ccc} \mathcal{U}(B, \beta) & \xrightarrow{f^\sqcap} & \mathcal{U}(A, \alpha) \\ (B, \beta') & \mapsto & (A, a \mapsto \alpha(a) \sqcap \beta'(f(a))). \end{array}$$

Since in general a projectale is neither distributive nor complete, we will not get a right adjoint for $f^{-1}$ to use for universal quantification. We will instead search for conditions giving adjoints to $f^\sqcap$ to use as quantifiers. Completeness will be a necessary starting point. We will return to this matter when we study the properties of the slice categories.

First, however, let us turn our attention to the properties of **Set(L)** as a concrete category.

The bottom structure on the terminal object plays a special role rather akin to that of the bottom of the lattice, so we will give it a special name: $0 = (\{*\}, \perp)$. Notice that this is not the initial object of **Set(L)**.

**Proposition 2.3** *If* **L** *is a projectorial lattice then each object* $(A, \alpha)$ *in* **Set(L)** *gives rise to an endofunctor*

$$(A, \alpha) \, \textcircled{n} \, (-) : \mathbf{Set(L)} \to \mathbf{Set(L)}$$

*with the following properties:*

1. *The underlying set of* $(A, \alpha) \, \textcircled{n} \, (B, \beta)$ *is* $A \times B$

2. *There is a natural map* $\pi : (A, \alpha) \, \textcircled{n} \, (B, \beta) \to (A, \alpha)$ *with underlying map given by projection*

3. *If* $(A, \alpha')$ *is an unbalanced subobject of* $(A, \alpha)$ *then the diagonal lifts to a map* $\Delta : (A, \alpha') \to (A, \alpha) \, \textcircled{n} \, (A, \alpha')$

4. *The structure on* $A$ *induced by the structure* $(A, \alpha) \, \textcircled{n} \, (A, \alpha')$ *using the diagonal map factors through* **0** *if and only if the structure induced by* $(A, \alpha') \, \textcircled{n} \, (A, \alpha)$ *does.*

PROOF:

   The definition of $(A, \alpha) \, \textcircled{n} \, (B, \beta)$ is $(A \times B), (a, b) \mapsto \alpha(a) \sqcap \beta(b)$. The properties then follow immediately from those for a projectorial lattice. ∎

In order to get further properties of **Set(L)** we will need to impose a completeness condition. Specifically we ask that **L** be a complete lattice which is also a projectale.

**Proposition 2.4** *If* **L** *is a complete projectale, then each of the functors* $(A, \alpha) \, \textcircled{n} \, (-)$ *has a right adjoint* $(A, \alpha) \, \textcircled{w} \, (-)$.

PROOF:

   We define $(A, \alpha) \, \textcircled{w} \, (B, \beta)$ as $(\{f | f : A \to B\}, \zeta)$ where

$$\zeta(f) = \bigwedge_{a \in A} (\alpha(a) \Downarrow \beta(f(a))).$$

A map $h : (C, \gamma) \to (A, \alpha) \, \textcircled{w} \, (B, \beta)$ induces a map $(A, \alpha) \, \textcircled{n} \, (C, \gamma) \to (B, \beta)$ taking $(a, c)$ to $(h(c))(a)$. Since for every $c \in C$ we will have

$$\gamma(c) \leq \zeta(h(c)) = \bigwedge_{a \in A} (\alpha(a) \Downarrow \beta(h(c)(a))),$$

we will also get

$$\gamma(c) \leq \alpha(a) \Downarrow \beta((h(c))(a))$$

for every $a$ and $c$. Thus

$$\alpha(a) \sqcap \gamma(c) \leq \beta((h(c))(a))$$

for all $a$ and $c$, making the induced map from the product a **Set(L)** map.

Similarly, given a map $k : (A, \alpha) \bigcirc (C, \gamma) \to (B, \beta)$ we can define the adjoint by $\hat{k} : (C, \gamma) \to (A, \alpha) \textcircled{\Downarrow} (B, \beta)$ has $\hat{k}(c) : a \mapsto k(a, c)$. It is easy to see that $\hat{k}$ is a **Set(L)** map if and only if $k$ is. ∎

The structure that this gives on **Set(L)** is weaker than a monoidal closed structure, since in general we have no guarantee that any form of associative law will hold for $\bigcirc$. If **L** is not a complete Heyting algebra, we will be working with a noncommutative operation. The terminal object $(\{*\}, \top)$ is a unit for $\bigcirc$.

We can next ask what structure the slice categories **Set(L)**$/(A, \alpha)$ have. First we note that there is an induced functor $I_{(A,\alpha)}$ from **Set(L)** to **Set(L)**$/(A, \alpha)$ taking $(B, \beta)$ to $\pi : (A, \alpha) \bigcirc (B, \beta) \to (A, \alpha)$. We can hope that this functor will preserve all of the structure of **Set(L)**, but first we need to describe the structures on the slice categories given by $\sqcap$ and $\Downarrow$.

**Definition 2.1** *For each object* $(B, \beta) \xrightarrow{f} (A, \alpha)$ *in* **Set(L)**$/(A, \alpha)$, *we get a functor*

$$(B, \beta) \xrightarrow{f} (A, \alpha) \bigcirc_{(A,\alpha)}(-) : \textbf{Set(L)}/(A, \alpha) \to \textbf{Set(L)}/(A, \alpha)$$

*by*

$$\begin{pmatrix} (B, \beta) \\ f \downarrow \\ (A, \alpha) \end{pmatrix} \bigcirc_{(A,\alpha)} \begin{pmatrix} (C, \gamma) \\ \downarrow g \\ (A, \alpha) \end{pmatrix} = \begin{pmatrix} (B_f \times_g C, \beta \sqcap \gamma) \\ \pi_1 \downarrow \\ (B, \beta) \\ f \downarrow \\ (A, \alpha) \end{pmatrix}$$

This functor has all of the properties listed in Proposition 2.3 when they are properly reinterpreted in the context of concrete categories over slice categories in **Sets**. In particular, there is a right adjoint.

**Proposition 2.5** *Each of the functors*

$$(B, \beta) \xrightarrow{f} (A, \alpha) \bigcirc_{(A,\alpha)}(-)$$

*has a right adjoint*

$$(B, \beta) \xrightarrow{f} (A, \alpha) \textcircled{\Downarrow}_{(A,\alpha)}(-).$$

PROOF:

We we make the following definition

$$
\begin{pmatrix} (B,\beta) \\ f \downarrow \\ (A,\alpha) \end{pmatrix} \Downarrow_{(A,\alpha)} \begin{pmatrix} (C,\gamma) \\ \downarrow g \\ (A,\alpha) \end{pmatrix} = \begin{pmatrix} (\{(a,h)|h:f^{-1}(a)\to g^{-1}(a)\},\zeta) \\ \pi_1 \downarrow \\ (A,\alpha) \end{pmatrix}
$$

where
$$
\zeta(a,h) = \alpha(a) \wedge \bigwedge_{b\in f^{-1}(a)} (\beta(b) \Downarrow \gamma(h(b))).
$$

Notice that this is the exponential in **Sets**$/A$ equipped with a degree of membership, so all we will need to do is show that the maps given by the adjointness in **Sets**$/A$ preserve membership if and only if their adjoints do.

In defining the membership the $\alpha(a)\wedge$ part is included so that the projection map to $(A,\alpha)$ is membership preserving. Since all of the objects under discussion have the map to $(A,\alpha)$ we can concentrate our attention on the other part of the membership function.

Suppose we have $k : (D,\delta) \to (B,\beta) \Downarrow_{(A,\alpha)} (C,\gamma)$ over $(A,\alpha)$, then we get $\hat{k} : B \times_A D \to C$ taking $(b,d)$ to $(k(d))(b)$. Now since $(D,\delta)$ sits over $(A,\alpha)$ and we work in fibers, we will know that $\delta(d) \leq \alpha(a)$ so it will suffice to show

$$
\delta(d) \leq \bigwedge_{b\in f^{-1}(a)} (\beta(b) \Downarrow \gamma((k(d))(b)))
$$

if and only if $\beta(b) \sqcap \delta(d) \leq \gamma((k(d))(b))$ for every $(b,d)$ over a given $a$. This follows from the equivalence of

$$
\beta(b) \sqcap \delta(d) \leq \gamma((k(d))(b)) \text{ and } \delta(d) \leq (\beta(b) \Downarrow \gamma((k(d))(b)))
$$

for each $b$. ∎

As in the case of local cartesian closed categories we can use similar constructions for transferring between slices: given a morphism $f : (A,\alpha) \to (B,\beta)$ we get functor

$$
f^{\mho} : \quad \mathbf{Set(L)}/(B,\beta) \quad \to \quad \mathbf{Set(L)}/(A,\alpha)
$$
$$
((C,\gamma) \xrightarrow{g} (B,\beta)) \quad \mapsto \quad ((A_f \times_g C, \alpha(a) \sqcap \gamma(c))) \to (A,\alpha)
$$

with right adjoint

$$
\Pi_f^{\mho} : \quad \mathbf{Set(L)}/(A,\alpha) \quad \to \quad \mathbf{Set(L)}/(B,\beta)
$$
$$
((D,\delta) \xrightarrow{h} (A,\alpha)) \quad \mapsto \quad ((X,\xi) \xrightarrow{\hat{h}} (B,\beta))
$$

where the fiber of $X$ over a particular $b \in B$ is the product of the fibers over the $a$ in the preimage of $b$:

$$\hat{h}^{-1}(b) = \prod_{a \in f^{-1}(b)} h^{-1}(a).$$

A typical member of this fiber has the form $(d_a)_{a \in f^{-1}(b)}$; its membership is given by

$$\xi((d_a)_{a \in f^{-1}(b)}) = \beta(b) \wedge \bigwedge_{a \in f^{-1}(b)} (\alpha(a) \Downarrow \delta(d_a)).$$

Notice that when we restrict to unbalanced subobjects $f^{\mathbf{O}}$ is the same as $f^{\sqcap} : (B, \beta') \mapsto (A, \alpha(a) \sqcap \beta'(f(a)))$ and the right adjoint can be written as

$$\forall_f^{\sqcap} : (A, \alpha') \mapsto (B, b \mapsto \beta(b) \wedge \bigwedge_{\{a \mid f(a) = b\}} \alpha(a) \Downarrow \alpha'(a)).$$

There is no left adjoint to use as existential quantification.

The upshot of all of this is that $\mathbf{Set(L)}$ has a propositional logic which can be thought of as being based on quantum logic. When we move up to internal first order logic we get either a fragment because of the absence of existential quantifiers or a much more complicated structure because the existential quantifier is the left adjoint to a different substitution functor than the universal quantification is right adjoint to. The category does have the higher order internal logic given by weak representation of unbalanced subobjects, as do all categories of lattice valued fuzzy sets.

# 3   Quantale Valued Fuzzy Sets

Kimmo Rosenthal [7] defines a *quantale* as a complete lattice $\mathbf{Q}$ with an additional binary operation $\&$ which is associative and which distributes over arbitrary sups in either variable. As a consequence both of the functors $q\&-$ and $-\&q$ have right adjoints $(q \rightarrow_r (-)$ and $q \rightarrow_l (-)$, respectively). Quantales give a noncommutative analog to Heyting algebras.

Additional requirements can be (and often are) imposed:

- $\mathbf{Q}$ is *right-sided* if every element $q \in \mathbf{Q}$ has $q\&\top \leq q$;

- $\mathbf{Q}$ is *idempotent* if every $q \in \mathbf{Q}$ has $q\&q = q$;

- a *right unit* for $\mathbf{Q}$ is an element $1$ such that $q\&1 = q$ for all $q \in \mathbf{Q}$.

Similar definitions can be made using left instead of right.

**Proposition 3.1** *If $\mathbf{Q}$ is right-sided, left-sided, or idempotent then if $a \leq q$ and $b \leq q$ then $a\&b \leq q$.*

PROOF:

Because & preserves sups in both variables, it preserves order so

$$a\&b \le q\&q.$$

If **Q** is idempotent this is all we need to show that $a\&b \le q$. If **Q** is right-sided we use

$$q\&q \le q\&\top \le q;$$

if **Q** is left-sided, we put $\top$ in the other variable. ∎

We will need one of these conditions when we want structures on the slice categories or lattices of unbalanced subobjects, so let us give a name to quantale operations which restrict nicely:

**Definition 3.1** *The operation* & *is called inheritable if whenever* $a \le q$ *and* $b \le q$ *then* $a\&b \le q$.

It is worth noting that we have several times used a trick on the right adjoints to make them restrict to the lattices $\downarrow q$: if the right adjoint is $a \Rightarrow b$ then we restricted by taking $a \Rightarrow_q b = q \wedge (a \Rightarrow b)$. This worked on the right adjoints because the if we already know all values $x$ being considered have $x \le q$, $x \le a \Rightarrow_q b$ is the same as $x \le a \Rightarrow b$ and the properties of the implications (right adjoints to $a\&-$ or $a \sqcap -$) follow quickly. This trick will not work to produce the & on $\downarrow q$ for a non-inheritable & as the following example shows:
**Example:**

**Associativity does not truncate** This example resulted from discussion with J. Philip Kavanagh. Let **Q** be the closed extended real interval $[0, +\infty) \cup \{\infty\}$ equipped with the usual order on $[0, \infty)$ with $\infty$ larger than any finite element. We then define $a\&b = ab$ with $a\&\infty = \infty\&a = \infty$ if $a \ne 0$ and $0\&\infty = \infty\&0 = 0$. This is a commutative, associative operation with unit 1. It preserves arbitrary sups. Thus we have a quantale.

Now suppose we try to restrict to the interval $[0, 2]$. A direct approach will not give closure, so let us try truncating the operation

$$a\&_2 b = \min(2, ab).$$

This new operation will preserve sups in both variables, remains commutative, and still has 1 as unit. It is not, however, associative:

$$\begin{aligned}
(.5\&_2 1.8)\&_2 1.8 &= .9\&_2 1.8 \\
&= 1.62 \\
.5\&_2(1.8\&_2 1.8) &= .5\&_2 2 \\
&= 1.
\end{aligned}$$

Notice that in this example (and the less peculiar example of $[0, 1]$ with multiplication) the functors $-\&a$ and $a\&-$ do not preserve arbitrary infs, so there is no hope of a left adjoint.

The properties of the category $\mathbf{Set(Q)}$ mimic those of a quantale. They are given by the following theorem:

**Theorem 3.2** *If* $\mathbf{Q}$ *is a quantale then* $\mathbf{Set(Q)}$ *has an operation* $\textcircled{\&}$ *such that both of the functors* $-\textcircled{\&}(A, \alpha)$ *and* $(A, \alpha)\textcircled{\&}-$ *have right adjoints.*

PROOF:

The operation $\textcircled{\&}$ is defined by

$$(A, \alpha)\textcircled{\&}(B, \beta) = (A \times B, (a, b) \mapsto \alpha(a)\&\beta(b)).$$

Not surprisingly the adjoints are given by

$$(A, \alpha)\textcircled{\Diamond}_r(C, \gamma) = (\{f : A \to C\}, f \mapsto \bigwedge_{a \in A} \alpha(a) \to_r \gamma(f(a)))$$

and

$$(A, \alpha)\textcircled{\Diamond}_l(C, \gamma) = (\{f : A \to C\}, f \mapsto \bigwedge_{a \in A} \alpha(a) \to_l \gamma(f(a))).$$

∎

The $\textcircled{\&}$ operation is associative up to coherent isomorphism (i.e. the only difficulties with associativity are with the formation of the products in **Sets**, not with the membership function). It will have a unit given by $(\{*\}, 1)$ if $\mathbf{Q}$ has a unit 1. Either right-sidedness or left-sidedness in $\mathbf{Q}$ will be reflected in properties of $\textcircled{\&}$ using the object $(\{*\}, \top)$. Since the product in **Sets** is not idempotent, idempotency of $\&$ will not give idempotency of $\textcircled{\&}$.

To get decent properties at the level of slice categories and unbalanced sub-object lattices, we will need to make the assumption that $\&$ is inheritable. In particular, inheritability of $\&$ makes it so that the natural structure on the pull-back $B_f \times_g C$ induced by the inclusion in $(B, \beta)\textcircled{\&}(C, \gamma)$ has a natural $\mathbf{Sets(Q)}$ map into $(A, \alpha)$ if both $f : (B, \beta) \to (A, \alpha)$ and $g : (C, \gamma) \to (A, \alpha)$ are $\mathbf{Sets(Q)}$ maps. This gives an operation in the slice categories we can use to localize the structure of $\mathbf{Sets(Q)}$. Notice that, in general, the projection maps to $(B, \beta)$ and $(C, \gamma)$ need not be $\mathbf{Sets(Q)}$ maps.

**Theorem 3.3** *If* $\mathbf{Q}$ *is a quantale with an inheritable* $\&$, *then each of the functors*

$$(B, \beta) \xrightarrow{f} (A, \alpha) \underset{(A,\alpha)}{\&} (-)$$

*has a right adjoint*

$$(B, \beta) \xrightarrow{f} (A, \alpha) \underset{r}{\odot} {}^{(A,\alpha)}(-)$$

*and each of the functors*

$$(-) \underset{(A,\alpha)}{\&} (B, \beta) \xrightarrow{f} (A, \alpha)$$

*has a right adjoint*

$$(B, \beta) \xrightarrow{f} (A, \alpha) \underset{l}{\odot} {}^{(A,\alpha)}(-).$$

PROOF:

We we make the following definitions

$$\begin{pmatrix} (B, \beta) \\ f \downarrow \\ (A, \alpha) \end{pmatrix} \underset{r}{\odot} {}^{(A,\alpha)} \begin{pmatrix} (C, \gamma) \\ \downarrow g \\ (A, \alpha) \end{pmatrix} = \begin{pmatrix} (\{(a, h) | h : f^{-1}(a) \to g^{-1}(a)\}, \zeta) \\ \pi_1 \downarrow \\ (A, \alpha) \end{pmatrix}$$

where

$$\zeta(a, h) = \alpha(a) \wedge \bigwedge_{b \in f^{-1}(a)} (\beta(b) \to_r \gamma(h(b)))$$

and

$$\begin{pmatrix} (B, \beta) \\ f \downarrow \\ (A, \alpha) \end{pmatrix} \underset{l}{\odot} {}^{(A,\alpha)} \begin{pmatrix} (C, \gamma) \\ \downarrow g \\ (A, \alpha) \end{pmatrix} = \begin{pmatrix} (\{(a, h) | h : f^{-1}(a) \to g^{-1}(a)\}, \zeta) \\ \pi_1 \downarrow \\ (A, \alpha) \end{pmatrix}$$

where

$$\zeta(a, h) = \alpha(a) \wedge \bigwedge_{b \in f^{-1}(a)} (\beta(b) \to_l \gamma(h(b))).$$

As in the case of projectales showing that these give the right adjoints desired is easy. ∎

While we cannot, in general, use this adjointness to get a change of base functor with a right adjoint, we can get nice properties at the level of unbalanced subobject lattices:

**Theorem 3.4** *If & is inheritable, then each of the lattices $\mathcal{U}(A, \alpha)$ has a natural quantale structure.*

PROOF:

Since **Q** is a complete lattice, so will all of the $\mathcal{U}(A, \alpha)$'s be. What we need is the associative, sup preserving operation &. But for inheritable & this is easy:

$$(A, \alpha')\&(A, \alpha'') = (A, \alpha'\&\alpha'')$$

preservation of sups and associativity follow from the same properties in **Q**. All we need to worry about is closure, and that is precisely what inheritability guarantees.

The right adjoints are given by

$$(A, \alpha') \to_r (A, \alpha'') = (A, a \mapsto \alpha(a) \wedge (\alpha'(a) \to_r \alpha''(a))$$

and

$$(A, \alpha') \to_l (A, \alpha'') = (A, a \mapsto \alpha(a) \wedge (\alpha'(a) \to_l \alpha''(a)).$$

∎

# References

1 Joseph A. Goguen, Jr., *L-fuzzy sets*, Journal of Mathematical Analysis and its Appllications **18** (1967), 145–174.

2 Ulrich Höhle and Lawrence Neff Stout, *Foundations of fuzzy sets*, Fuzzy Sets and Systems **40** (1991), no. 2, 257–296.

3 G. Khatcherian, *Projectales*, Journal of Pure and Applied Algebra **74** (1991), 177–195.

4 Aleš Pultr, *Closed categories and L-fuzzy sets*, Vortrage zur Automaten und Algorithmentheorie, Technische Universität Dresden, 1975.

5 Aleš Pultr, *Fuzzy mappings and fuzzy sets*, Commentarii Mathematicae Universitae Carolinae **17** (1976), no. 3.

6 Aleš Pultr, *Remarks on dispersed (fuzzy ) morphisms*, Vorträge aus dem Problemseminar Algebraische Methoden und ihre Anwendungen in der Automatentheorie, Technische Universität Dresden, 1976.

7 Kimmo Rosenthal, *Quantales and their applications* , Pitman Research Notes in Mathematics, no. 234, Longman Scientific and Technical, Essex, 1990.

8 Lawrence N. Stout, *The logic of unbalanced subobjects in a category with two closed structures*, Applications of Category Theory to Fuzzy Subsets (S.E. Rodabaugh, E.P. Klement, U. Höhle ed.), Kluwer, 1992.

# IX

# Fuzzy logic and categories of fuzzy sets

## O. Wyler

## Introduction

This paper deals with three topics:

1. Fuzzy logic,

2. Categories of fuzzy sets,

3. Logic of fuzzy subsets.

While much of its contents can be found in my book [17] and in the existing literature, it also includes new results and a large amount of unpublished folklore. There is also a section discusssing a formal language for fuzzy logic, with interpretations of formulas. Thus I believe that this coherent and not too technical survey of fuzzy logic and categories of fuzzy sets is useful. I have tried to make the paper reasonably self-contained, except that I use the basic language of categories freely. Proofs in Sections 1 and 2 are mostly omitted; they are usually straightforward or can be found in the given references.

When L. ZADEH [18] introduced fuzzy sets, he regarded fuzzy sets essentially as "crisp" sets with a $[0,1]$-valued membership degree function. Membership degrees were soon perceived as truth-values, and this called for a fuzzy logic, with truth-values in the unit interval. ZADEH used Lukasiewicz logic for propositional connectives, without saying why, and most users of fuzzy logic have followed his example. There have been claims in the literature that Lukasiewicz logic *must* be used in certain applications, but these claims do not stand up to scrutiny.

As J. GOGUEN [6] soon pointed out, there is no mathematical need to use the real unit interval as set of truth-values, or to use Lukasiewicz logic. For technical reasons, truth-values must form a complete lattice, and preferably a complete Heyting algebra. This puts intuitionistic logic at our disposal, but it does not exclude non-intuitionistic propositional connectives and logics. We note that every order-complete chain, and in particular the real unit interval $[0,1]$, is a complete Heyting algebra. Open sets of a topological space form a complete

Heyting algebra, and every finite Heyting algebra is of this type, up to an isomorphism of Heyting algebras.

In some respects, logic and set theory are siamese twins. Set operations are based on logical connectives and quantifiers, and these connectives and quantifiers can be retrieved from the set operations. We cannot lay a foundation for fuzzy sets without fuzzy logic, and a language for fuzzy logic is based on a category of fuzzy sets.

Everybody working with fuzzy sets seems to agree that fuzzy sets are crisp sets with additional structure, and that this additional structure includes a degree of membership function with values in a complete lattice $H$. For the categorically minded, fuzzy sets are the objects of a category, and this category should be as set-like as possible. J. GOGUEN [6] was the first, but by no means the last author to present such a category, and it seems likely that there is no single category of fuzzy sets which satisfies all needs. There are basic questions which have not yet been answered to everybody's satisfaction. Two of these questions ask: how fuzzy should things be? Should equality be fuzzy or crisp? And should morphisms be crisp maps, extensional maps, or fuzzy functions? Another important question: what should the underlying fuzzy logic of our set theory be?

We do not try to answer the first two of these three questions; thus Section 2 presents six categories of fuzzy sets, each equipped with fuzzy logic, from which the reader can choose. On the other hand, we do not leave the choice of logics open; we base our categories on intuitionistic logic. This needs some discussion.

When we choose a lattice $H$ of truth-values for membership degrees, then meets and joins in $H$ define standard conjunction and disjunction. If $H$ is complete, then infima and suprema in $H$ define universal and existential quantifiers which generalize standard conjunction and disjunction. By basic principles, implication should be right adjoint to conjunction, and this right adjoint exists if $H$ is a Heyting algebra. The standard logic thus obtained is intuitionistic, it is always there, and it agrees with the basic categorical constructions. Other logics can also be obtained and used, depending on the choice of $H$, but basing a category of fuzzy sets on a non-intuitionistic logic seems to lead to complications. There may be gains justifying these complications, but I do not see them at this time. Another point is that categories with intuitionistic fuzzy logic may serve as models for categories with a non-standard fuzzy logic.

A sufficiently set-like category has an internal logic. This logic is intuitionistic when it exists, but it may not be what we want and need. For example, the internal logic of Goguen's category of $H$-valued fuzzy sets is always crisp, *i.e.* classical with just two truth-values. L.N. STOUT [16] has shown a way out of this seeming contradiction. A generalization of Stout's theory, based on a notion of fuzzy subset, was obtained in [17]. For the six categories constructed in Section 2, we present this theory in Section 3.

In Section 2, we justify or motivate axioms for fuzzy sets and maps by translating them informally into a first-order language. Section 4 tries to undergird this by introducing an appropriate first-order language, modelled on the MITCHELL–BÉNABOU language of topos theory, with interpretations in an underlying category of fuzzy sets. The description of such a language is of course just a starting point. Syntax and semantics of the language should be worked out, with particular attention to validity of statements, but this is beyond the scope of the present paper.

We use standard notations as much as possible, but a few remarks on notations may be in order.

For a product $A \times B$ in a category $\mathbf{C}$, and for morphisms $f : X \to A$ and $g : X \to B$, we denote by $\langle f, g \rangle$ the unique morphism of $\mathbf{C}$ characterized by

$$p \circ \langle f, g \rangle = f, \qquad q \circ \langle f, g \rangle = g,$$

for the projections $A \xleftarrow{p} A \times B \xrightarrow{q} B$ of the product.

We use the notation $S \subset T$, as introduced by G. PEANO [12], for subset inclusion, including the case $S \subset S$. There is usually no need for a "proper subset" notation.

Notations like $f(A)$ and $f^{-1}(B)$ for direct and inverse images are too ambiguous to be acceptable, and we replace them by the following. For a function $f : S \to T$ between sets, and for subsets $A$ of $S$ and $B$ of $T$, we denote by

$$f^{\to}(A) = \{ f(x) \mid x \in A \} \qquad \text{and} \qquad f^{\leftarrow}(B) = \{ x \in A \mid f(x) \in B \}$$

the direct image of $A$ and the inverse image of $B$ by $f$. This defines functions $f^{\to} : \mathsf{P}S \to \mathsf{P}T$ and $f^{\leftarrow} : \mathsf{P}T \to \mathsf{P}S$ between the powersets $\mathsf{P}S$ and $\mathsf{P}T$.

Parentheses in function-value notation $f(x)$ are often superfluous, and it can be convenient to omit them. We shall always omit unnecessary parentheses for arguments of functors, and we usually omit the parentheses for direct images $f^{\to}A$ and inverse images $f^{\leftarrow}B$.

# 1. Fuzzy Propositional Connectives and Quantifiers

**1.1. Generalities.** Fuzzy logic deals with statements which have truth-values in a complete lattice $H$, and preferably in a complete Heyting algebra. The standard choice for $H$ is the real unit interval $[0, 1]$. Nothing in the theory of fuzzy sets prevents the use of another lattice of truth-values; this may in fact be desirable for certain applications. Products $[0, 1]^n$ are an example, with lists $(t_1, \cdots, t_n)$ of real numbers $0 \leq t_i \leq 1$ as truth-values. Open sets of a topological space form a complete Heyting algebra, and Boolean algebras are special Heyting algebras. Decreasing subsets of a preordered set are open sets for a topology; this special case delivers all finite distributive lattices.

We do not deal with modal logic in this paper. Thus if $\diamond$ is a binary propositional connective, then the truth-value of a statement $\varphi \diamond \psi$ will depend only on the truth-values of $\varphi$ and $\psi$. This generalizes to other connectives, and it means that we can introduce propositional connectives as operations on the set $H$ of truth-values.

**1.2. Standard connectives.** We recall that a lattice is an ordered set, or "poset", in which every finite subset has an infimum, also called its *meet*, and a supremum or *join*. Thus every lattice $H$ has a greatest element (meet of the empty subset), which we denote by $\top$, and a least element $\bot$. These elements represent nullary propositional connectives *"true"* and *"false"*. Any two elements $p, q$ of $H$ have a meet $p \wedge q$ and a join $p \vee q$. Viewed as propositional connectives, $\vee$ is *standard disjunction*, and $\wedge$ *standard conjunction*.

A lattice $H$ is called *complete* if every subset of $H$ has a supremum in $H$, and a *Heyting algebra* if we can define *standard implication* $p \to q$ for all $p, q$ in $H$ by requiring

$$t \le p \to q \quad \text{iff} \quad t \wedge p \le q,$$

for all $t$ in $H$. This leads to *standard negation*, defined by

$$\neg p = p \to \bot$$

for $p \in H$. A complete lattice $H$ is a (complete) Heyting algebra if and only if $H$ satisfies the infinite distributive law

$$p \wedge \left( \bigvee_i q_i \right) = \bigvee_i (p \wedge q_i).$$

**1.3. Other connectives.** There is a general agreement that a fuzzy propositional connective, restricted to the crisp subalgebra $\{\bot, \top\}$ of $H$, should reduce to the corresponding crisp or "classical" connective. With this in mind; we define a *fuzzy conjunction* as a commutative and associative order preserving binary operation & on $H$ which satisfies $\top \,\&\, p = p = p \,\&\, \top$ for all $p \in H$.

There is also general agreement that *fuzzy implication* $\Rightarrow$ should be right adjoint to fuzzy conjunction &, i.e. that for $p, q$ in $H$, we must have

$$t \le p \Rightarrow q \quad \text{iff} \quad t \,\&\, p \le q,$$

for all $t$ in $H$. This determines $p \Rightarrow q$ uniquely. It is well known that a fuzzy conjunction & admits a right adjoint implication $p \Rightarrow q$, defined for all $p, q$ in $H$, if and only if & satisfies the infinite distributive law

$$(1) \qquad\qquad p \,\&\, \left( \bigvee_i q_i \right) = \bigvee_i (p \,\&\, q_i),$$

for all $p$ in $H$ and all families $(q_i)$ of elements in $H$. A fuzzy conjunction which satisfies this law is also called a *t-norm*.

*Fuzzy negation* $\sim p$ is always defined by

$$\sim p = p \Rightarrow \bot.$$

Thus $q \leq \sim p$ for $p, q$ in $H$ iff $p \& q = \bot$.

There is less agreement on how to connect a fuzzy disjunction with a fuzzy conjunction $\&$. We shall not need fuzzy disjunction; thus we do not discuss it here.

**1.4. Examples and comments.** For the unit interval $[0, 1]$, standard conjunction and implication are given by

(1) $\qquad p \wedge q = \min(p, q) \qquad$ and $\qquad p \rightarrow q = \begin{cases} 1 & \text{if } p \leq q, \\ q & \text{otherwise.} \end{cases}$

Lukasiewicz conjunction and implication are given by:

(2) $\quad p \& q = \max(p + q - 1, 0) \qquad$ and $\qquad p \Rightarrow q = \min(1, q - p + 1),$

and a third example is given by:

(3) $\qquad p \& q = pq \qquad$ and $\qquad p \Rightarrow q = \begin{cases} 1 & \text{if } p \leq q, \\ q/p & \text{otherwise.} \end{cases}$

In these three examples, conjunction is continuous, but implication is only continuous for Lukasiewicz logic. It is a reasonable conjecture that Lukasiewicz conjunction is the only $t$-norm in $[0, 1]$ for which both conjunction and implication are continuous.

This raises the question: how important is continuity of $[0, 1]$-valued propositional connectives? We observe that every $t$-norm is continuous from below by 1.3.(1). All implications satisfy the laws

$$\left( \bigvee_i p_i \right) \Rightarrow q = \bigwedge_i (p_i \Rightarrow q) \qquad \text{and} \qquad p \Rightarrow \left( \bigwedge_i q_i \right) = \bigwedge_i (p \Rightarrow q_i).$$

Thus $p \Rightarrow q$ is continuous from below in $p$, and continuous from above in $q$. We observe also that continuity only makes sense in the special case $H = [0, 1]$, where it is based on order. In the general theory, only order matters.

Standard conjunction is idempotent: $p \wedge p = p$ for $p \in H$. Non-standard conjunctions and $t$-norms usually are not idempotent, but this does not seem to be tremendously important. Non-commutative conjunctions and $t$-norms have also been proposed, but the ensuing complications in fuzzy logic have not been fully investigated. We would need two implications, one right adjoint to functors $p \& \, \text{---}$, and one right adjoint to functors $\text{---} \& p$, and the troubles just begin with this.

**1.5. Quantifiers.** F.W. LAWVERE [9] observed, and practitioners of fuzzy logic agree, that existential quantification is left adjoint, and universal quantification right adjoint, to substitution. This determines existential and universal quantifiers $\exists_f$ and $\forall_f$ for morphisms, not just for variables.

In the setting of fuzzy logic, mappings $S \to H$ for a set $S$ form a complete Heyting algebra $H^S$, with order and operations defined point-wise. For a mapping $f : S \to T$, substituting $f(x)$ with $x \in S$ for $y \in T$ means that we replace $h : B \to H$ by $hf : A \to H$. This defines an operator from $H^T$ to $H^S$, denoted by $f^{\leftarrow} : h \mapsto hf$, which preserves order, suprema and infima, and all propositional connectives defined pointwise.

We assign to $f : S \to T$ quantification maps

$$\exists_f : H^S \to H^T \qquad \text{and} \qquad \forall_f : H^S \to H^T,$$

left and right adjoint to $f^{\leftarrow}$, by requiring

$$\exists_f \alpha \leq \beta \iff \alpha \leq f^{\leftarrow}\beta \qquad \text{and} \qquad \beta \leq \forall_f \alpha \iff f^{\leftarrow}\beta \leq \alpha,$$

for $\alpha : S \to H$ and $\beta : T \to H$. As $\alpha \leq f^{\leftarrow}\beta$ iff always $\alpha(x) \leq \beta(y)$ for $y = f(x)$, we get

$$(1) \qquad \qquad (\exists_f \alpha)(y) = \bigvee_{f(x)=y} \alpha(x)$$

for $y \in T$. A similar argument shows that

$$(2) \qquad \qquad (\forall_f \alpha)(y) = \bigwedge_{f(x)=y} \alpha(x)$$

for $y \in T$.

We note that in particular

$$(\exists_q \alpha)(y) = \bigvee_{x \in S} \alpha(x,y) \qquad \text{and} \qquad (\forall_q \alpha)(y) = \bigwedge_{x \in S} \alpha(x,y),$$

for $y \in T$ and the projection $q : S \times T \to T$. Thus $\exists_q$ and $\forall_q$ are quantifieres $(\exists x)$ and $(\forall x)$.

**1.6. Formal laws.** With fuzzy implication $\Rightarrow$ right adjoint to fuzzy conjunction $\&$, every formal law for $\&$ produces a formal law for $\Rightarrow$. From $p \& \top = p = \top \& p$, we get

**1.6.1.** $\top \Rightarrow p = p$, and $p \Rightarrow q = \top$ iff $p \leq q$.

Since $p \& q$ is monotone increasing in $p$ amd $q$, we have

**1.6.2.** $p \& q \leq p \wedge q$, and $p \to q \leq p \Rightarrow q$.

The associative law for & is equivalent to

**1.6.3.** $(p \& q) \Rightarrow r = p \Rightarrow (q \Rightarrow r)$.

The idempotent law $p \& p = p$, which is usually not satisfied for $t$-norms, states that $p \leq p \Rightarrow q$ iff $p \leq q$, but there is no similar simple translation for the commutative law for &. This fact has led to the consideration of non-commutative $t$-norms. However, the commutative law for & is often used, as e.g. for the following formal laws.

**1.6.4.** $p \leq (p \Rightarrow q) \Rightarrow q$; in particular $p \leq {\sim}{\sim}p$.

PROOF. We claim that $p \& (p \Rightarrow q) \leq q$, and we observe that

$$(p \Rightarrow q) \& p \leq q \qquad \text{iff} \qquad p \Rightarrow q \leq p \Rightarrow q,$$

which is always true.

**1.6.5.** $p \& (p \Rightarrow q) \leq q$, and $(p \Rightarrow q) \& (q \Rightarrow r) \leq p \Rightarrow r$.

Fuzzy equivalence is defined by

$$p \Leftrightarrow q = (p \Rightarrow q) \& (q \Rightarrow p).$$

We note some basic properties of this connective.

**1.6.6.** (i) $p \Leftrightarrow q = \top$ iff $p = q$.
(ii) $p \Leftrightarrow q = q \Leftrightarrow p$.
(iii) $(p \Leftrightarrow q) \& (q \Leftrightarrow r) \leq p \Leftrightarrow r$.

We note that there are two kinds of logical equivalence for fuzzy logic. A statement "$\varphi$ iff $\psi$" means that $\varphi$ is valid iff $\psi$ is valid. Validity of a statement $\varphi \Leftrightarrow \psi$ is stronger; it means that $\varphi$ and $\psi$ always have the same truth-value. We also note that validity of statements $\varphi \Rightarrow \psi$ or $\varphi \Leftrightarrow \psi$ depends only on the truth-values of $\varphi$ and $\psi$; and not on the particular $t$-norm & used for $\Rightarrow$ and $\Leftrightarrow$.

We need standard connectives for two useful laws.

**1.6.7.** (i) ${\sim}(p \vee q) = {\sim}p \wedge {\sim}q$.
(ii) $p \wedge (p \rightarrow q) = p \wedge (p \leftrightarrow q) = p \wedge q$.

The infinite distributive laws for & and $\Rightarrow$ can be stated as equivalences:

**1.6.8.** If $x$ does not occur in a statement $\varphi$, then
(i) $\varphi \& (\exists x)\psi \Leftrightarrow (\exists x)(\varphi \& \psi)$,
(ii) $(\exists x)\psi \Rightarrow \varphi \Leftrightarrow (\forall x)(\psi \Rightarrow \varphi)$,
(iii) $\varphi \Rightarrow (\forall x)\psi \Leftrightarrow (\forall x)(\varphi \Rightarrow \psi)$,
for any statement $\psi$.

**1.7. Theorem.** *For every pullback square*

$$
\begin{array}{ccc}
P & \xrightarrow{\;v\;} & T \\
\downarrow{\scriptstyle u} & & \downarrow{\scriptstyle g} \\
S & \xrightarrow{\;f\;} & U
\end{array}
$$

*in* **Set**, *we have* $g^\leftarrow \exists_f = \exists_v u^\leftarrow$ *and* $g^\leftarrow \forall_f = \forall_v u^\leftarrow$.

PROOF. Let $\alpha : S \to H$. By the definitions,

$$
(g^\leftarrow \exists_f \alpha)(y) = \bigvee_{fx=gy} \alpha(x), \qquad \text{and} \qquad (\exists_v u^\leftarrow \alpha)(y) = \bigvee_{vz=y} \alpha(uz),
$$

for $y \in T$. Now $u(z) = x$ induces a bijection between $z \in P$ with $v(z) = y$, and $x \in S$ with $f(x) = g(y)$. Thus the two suprema range over the same values, proving the first assertion. Taking right adjoints, we get $f^\leftarrow \forall_g = \forall_u v^\leftarrow$. This is the second equation with $f$ and $g$, and $u$ and $v$, interchanged.

## 2. Categories of Fuzzy Sets

**2.1. Fuzzy and totally fuzzy sets.** Throughout this section, we use fuzzy logic informally, with truth-values in a complete Heyting algebra $H$. A *fuzzy set* $A$ will have fuzzy membership degrees $\varepsilon_A(x)$, and a *totally fuzzy set* $A$ will have fuzzy equality $\delta_A(x, y)$.

Thus we define an $H$-*valued fuzzy set* as a pair $A = (|A|, \varepsilon_A)$, consisting of a (crisp) set $|A|$ and a membership degree mapping $\varepsilon_A : |A| \to H$, with no further conditions. We regard $\varepsilon_A(x)$, for $x \in |A|$, as truth-value of a statement $x \in A$. An $H$-valued *totally fuzzy set* is defined as a pair $A = (|A|, \delta_A)$, consisting of a (crisp) set $|A|$ and a fuzzy equality mapping $\delta_A : |A| \times |A| \to H$, subject to the two conditions of symmetry and transitivity:

**2.1.1.** $\delta_A(x, y) = \delta_A(y, x)$,

**2.1.2.** $\delta_A(x, y) \wedge \delta_A(y, z) \leq \delta_A(x, z)$,

for all $x, y, z$ in $|A|$.

With $\delta_A(x, y)$ regarded as truth-value of a statement $x =_A y$, these axioms say that the statements

$$
x =_A y \iff y =_A x
$$

and
$$
x =_A y \wedge y =_A z \implies x =_A z
$$

are valid for variables $x, y, z$ of type $A$.

**2.2. $H$-sets and discrete $H$-sets.** A totally fuzzy set $A$ has a fuzzy set structure $\varepsilon_A$, given by putting

$$\varepsilon_A(x) = \delta_A(x, x)$$

for $x \in |A|$, or in other words by requiring that

$$x \in A \iff x =_A x,$$

is valid for a variable $x$ of type $A$. It follows easily that

$$\delta_A(x, y) \leq \varepsilon_A(x) \wedge \varepsilon_A(y),$$

for $x, y$ in $|A|$, or in other words that

$$x =_A y \implies x \in A \wedge y \in B,$$

is valid for variables $x, y$ of type $A$.

$H$-valued totally fuzzy sets will also be called *$H$-sets*.

On the other hand, we can regard fuzzy sets as totally fuzzy sets, with

$$\delta_A(x, y) = \begin{cases} \varepsilon_A(x) & \text{if } x = y, \\ \bot & \text{otherwise}. \end{cases}$$

We shall call $H$-sets with this property *discrete $H$-sets*.

**2.3. The category $\mathbf{Set}_{dc} H$.** This is the category introduced by J. GO-GUEN [6], with crisp equality and crisp morphisms. Its objects are discrete fuzzy sets, and a morphism $f : A \to B$ of $\mathbf{Set}_{dc} H$ is a mapping $f : |A| \to |B|$ of the underlying crisp sets with

$$\varepsilon_A(x) \leq \varepsilon_B(f(x)),$$

for all $x \in |A|$, or in other words such that

$$x \in A \implies f(x) \in B,$$

is valid for a variable $x$ of type $A$. Composition of morphisms is composition of the underlying mappings.

It is easily seen that $\mathbf{Set}_{dc} H$ is a topological category over sets in the sense of [1], with small fibres. Thus $\mathbf{Set}_{dc} H$ is complete and cocomplete, with limits and colimits lifted from sets. The forgetful functor from $\mathbf{Set}_{dc} H$ to sets has a concrete left adjoint $V$ and a concrete right adjoint $C$, assigning to every set $S$ the *void* fuzzy set $VS$ with $\varepsilon_{VS}(x) = \bot$, and the *crisp* set $CS$ with $\varepsilon_{CS}(x) = \top$, for $x \in S$.

It is well known, and proved e.g. in [17], that $\mathbf{Set}_{dc} H$ is a quasitopos. This fact does not help us much for fuzzy logic since $\mathbf{Set}_{dc} H$ shares with all topological categories over sets the property that its internal logic is crisp. Overcoming this handicap has been a strong motivation for STOUT's theory of fuzzy subsets.

**2.4. Set$_{tc}$ H: totally fuzzy sets and crisp maps.** We define a *map* or *crisp map* $f : A \to B$ of $H$-sets as a mapping $f : |A| \to |B|$ such that

$$\delta_A(x, x') \leq \delta_B(fx, fx')$$

for $x, x'$ in $|A|$, or in other words such that

$$x =_A x' \implies f(x) =_B f(x')$$

is valid for variables $x, x'$ of type $A$. It follows easily that

$$\varepsilon_A(x) \leq \varepsilon_B(fx), \quad \text{or} \quad x \in A \implies f(x) \in B,$$

for $x$ in $|A|$.

With composition of underlying mappings as composition, $H$-sets and their maps form a concrete category over sets which we denote by **Set**$_{tc}$ $H$. This is a topological category over sets, not previously discussed in the literature, with **Set**$_{dc}$ $H$ as a full concrete subcategory. The full embedding $I :$ **Set**$_{dc}$ $H \to$ **Set**$_{tc}$ $H$ has a concrete right adjoint $J$, with $JA$ for an $H$-set $A$ obtained by

$$\delta_{JA}(x, x') = \begin{cases} \varepsilon_A(x) & \text{if } x = x', \\ \bot & \text{otherwise.} \end{cases}$$

The full embedding $I$ also preserves all collectively injective initial sources, and hence all categorical limits. Thus $I$ also has a left adjoint (and left inverse) $K :$ **Set**$_{tc}$ $H \to$ **Set**$_{dc}$ $H$. If $\equiv$ is the equivalence relation on $|A|$ generated by the pairs $(x, x')$ with $\delta_A(x, x') \neq \bot$, then the unit $\eta_A : A \to IKA$ is the quotient mapping $|A| \to |A|/\equiv$, with

$$\varepsilon_{KA}(y) = \bigvee_{\eta_A(x)=y} \varepsilon_A(x)$$

for $y \in |KA|$.

**2.5. Finite products.** Every categogy of fuzzy sets has a terminal object, or empty product, which we denote by **1**, with $|\mathbf{1}|$ a singleton $\{\star\}$, with fuzzy membership $\varepsilon_A(\star) = \top$.

The product $A \times B$ of $H$-sets $A$ and $B$ is given by

$$|A \times B| = |A| \times |B|,$$

and
$$\delta_{A \times B}(\langle x, y \rangle, \langle x', y' \rangle) = \delta_A(x, x') \wedge \delta_B(y, y')$$

for $x, x'$ in $|A|$ and $y, y'$ in $|B|$. Projections of this product are the projections of $|A| \times |B|$.

It is easily seen that this defines products $A \times B$ in the categorical sense, for all six categories of fuzzy sets considered in this paper; see 2.16 below.

**2.6. Fuzzy relations.** We define a *fuzzy relation* on an $H$-set $A$ as a mapping $\alpha : |A| \to H$ which satisfies the following inequalities:

**2.6.1.** $\alpha(x) \leq \varepsilon_A(x)$ for $x \in |A|$, and

**2.6.2.** $\alpha(x) \wedge \delta_A(x, x') \leq \alpha(x')$ for $x, x'$ in $|A|$.

This means that we require the validity of

$$\alpha(x) \implies x \in A, \qquad \text{and} \qquad \alpha(x) \wedge (x =_A x') \implies \alpha(x'),$$

for variables $x, x'$ of type $A$ We note that 2.6.2 is equivalent to

$$\alpha(x) \wedge \delta_A(x, x') = \alpha(x') \wedge \delta_A(x, x'),$$

for $x, x'$ in $|A|$.

With pointwise order, $\alpha \leq \beta$ iff $\alpha(x) \leq \beta(x)$ for every $x \in |A|$, fuzzy relations on $A$ form a complete lattice which we denote by $H^A$. Arbitrary suprema $\bigvee_i \alpha_i$, and nonempty infima $\bigwedge_i \alpha_i$, in $H^A$ are obtained pointwise, and $\varepsilon_A$ is the top element of $H^A$. It follows easily that $H^A$ is a complete Heyting algebra, with implication given by

$$(\alpha \to \beta)(x) = \varepsilon_A(x) \wedge (\alpha(x) \to \beta(x)),$$

for every $x \in |A|$. We shall discuss this further in 3.8 and 3.9.

**2.7. Binary fuzy relations.** For $H$-sets $A$ and $B$, a *binary fuzzy relation* $\rho : A \to B$ is defined as a fuzzy relation $\rho$ on $A \times B$. We regard $\rho(x, y)$ as the truth-value of a statement $x\rho y$. The defining inequalities for relations then become statements

$$x\rho y \implies x \in A \wedge y \in B,$$

and
$$x\rho y \wedge x =_A x' \wedge y =_B y' \implies x'\rho y',$$

valid for variables $x, x'$ of type $A$ and $y, y'$ of type $B$.

Fuzzy relations $\rho : A \to B$ and $\sigma : B \to C$ can be composed, with

$$(\sigma \circ \rho)(x, z) = \bigvee_{y \in |B|} (\rho(x, y) \wedge \sigma(y, z))$$

for $x \in |A|$ and $z \in |C|$. This means that

$$x(\sigma \circ \rho)z \iff (\exists y)(x\rho y \wedge y\sigma z)$$

is valid, for variables $x, y, z$ of types $A, B, C$. Using the distributivity of $\wedge$ over $\bigvee$ in the complete Heyting algebra $H$, one sees easily that this defines a binary relation $\sigma \circ \rho : A \to C$. Composition is associative, and fuzzy equalities are identity relations $\delta_A : A \to A$. Thus we have a category of $H$-sets and fuzzy relations.

Every fuzzy relation $\rho : A \to B$ has a *dual fuzzy relation* $\rho^{\mathrm{op}} : B \to A$, given by

$$\rho^{\mathrm{op}}(y, x) = \rho(x, y),$$

for $x \in |A|$ and $y \in |B|$. Dual relations satisfy $(\rho^{\mathrm{op}})^{\mathrm{op}} = \rho$, and

$$(\sigma \circ \rho)^{\mathrm{op}} = \rho^{\mathrm{op}} \circ \sigma^{\mathrm{op}}$$

if either composition is defined.

Fuzzy relations on an $H$-set $B$ can also be regarded as binary fuzzy relations $B \to 1$. With this convention, we have a useful lemma.

**2.7.1. Lemma.** *If* $\rho, \sigma : A \to B$ *are binary fuzzy relations such that* $\beta \circ \rho = \beta \circ \sigma$ *for all* $\beta$ *in* $H^B$, *then* $\rho = \sigma$.

PROOF. For $\beta = \delta_B(y, -)$ with $y \in |B|$, the compositions are $\rho(-, y)$ and $\sigma(-, y)$.

**2.8. Special fuzzy relations.** For $H$-sets $A$ and $B$, binary fuzzy relations $\rho : A \to B$ form a complete Heyting algebra $H^{A \times B}$, with order and suprema defined point-wise. It is easily seen that composition of relations satisfies infinite distributive laws

$$\sigma \circ \left(\bigvee_i \rho_i\right) = \bigvee_i (\sigma \circ \rho_i) \quad \text{and} \quad \left(\bigvee_i \sigma_i\right) \circ \rho = \bigvee_i (\sigma_i \circ \rho),$$

valid whenever either side is defined.

We say that $\rho : A \to B$ is:

*single-valued* if always $\rho(x, y) \wedge \rho(x, y') \le \delta_B(y, y')$,
*injective* if always $\rho(x, y) \wedge \rho(x', y) \le \delta_A(x, x')$,
*total* if always $\bigvee_y \rho(x, y) = \varepsilon_A(x)$,
*surjective* if always $\bigvee_x \rho(x, y) = \varepsilon_B(y)$.

These properties are clearly dual in pairs. We note that a fuzzy relation $\rho : A \to B$ is

single-valued iff $\rho \circ \rho^{\mathrm{op}} \le \delta_B$,
injective iff $\rho^{\mathrm{op}} \circ \rho \le \delta_A$,
total iff $\rho^{\mathrm{op}} \circ \rho \ge \delta_A$,
surjective iff $\rho \circ \rho^{\mathrm{op}} \ge \delta_B$.

The first two inequalities are almost immediate; the other two inequalities take a bit longer to prove.

It follows easily that all four classes of fuzzy relations are closed under composition and include all equality relation. We note that $\rho \circ \rho^{\mathrm{op}} = \delta_B$ iff $\rho$ is single-valued and surjective, and that $\rho^{\mathrm{op}} \circ \rho = \delta_A$ iff $\rho$ is injective and total.

**2.9. Set$_{tf} H$: totally fuzzy sets and fuzzy functions.** For $H$-sets $A$ and $B$, we define a *fuzzy function* $\rho : A \to B$ as a single-valued and total fuzzy relation. Identity relations are fuzzy functions, and the relational composition of fuzzy functions $\rho : A \to B$ and $\sigma : B \to C$ is a fuzzy function $\sigma\rho : A \to C$. Thus $H$-sets and fuzzy functions define a category which we denote by $\mathbf{Set}_{tf} H$. This category was introduced by D. HIGGS in a widely circulated but never published preprint [7], and denoted by $\mathbf{Set}\, H$. It has also been studied in detail by M. FOURMAN and D. SCOTT in [4].

**2.10.** We note two useful special properties of fuzzy functions.

**Proposition.** (i) If $\rho : A \to B$ is a total and $\sigma : A \to B$ a single-valued fuzzy relation, with $\rho \leq \sigma$, then $\rho = \sigma$.

(ii) If $\sigma \circ \rho = \delta_A$ and $\rho \circ \sigma = \delta_B$, with $\rho : A \to B$ and $\sigma : B \to A$ both single-valued or both total fuzzy relations, then $\sigma = \rho^{\mathrm{op}}$, and $\rho$ and $\sigma$ are surjective and injective fuzzy functions.

PROOF. For (i), we have

$$\sigma \leq \sigma \circ \rho^{\mathrm{op}} \circ \rho \leq \sigma \circ \sigma^{\mathrm{op}} \circ \rho \leq \rho.$$

For (ii) with $\rho$ and $\sigma$ total, we have

$$\sigma \leq \rho^{\mathrm{op}} \circ \rho \circ \sigma = \rho^{\mathrm{op}}, \qquad \text{and} \qquad \rho^{\mathrm{op}} \leq \rho^{\mathrm{op}} \circ \sigma^{\mathrm{op}} \circ \sigma = \sigma.$$

The proof for $\rho$ and $\sigma$ single-valued is similar.

**2.11. $\mathbf{Set}_{te} H$: Totally fuzzy sets and extensional maps.** One trouble with maps of $H$-sets is that they are not always extensional, *i.e.* the statement

$$(\forall x)(x \in A \Longrightarrow f(x) =_B g(x))$$

may be valid for distinct maps $f, g : A \to B$. We say that crisp maps $f, g : A \to B$ are *extensionally equal* if the displayed statement is valid, *i.e.* if

$$\varepsilon_A(x) \leq \delta_B(f(x), g(x))$$

for all $x \in |A|$. Extensional equality clearly is an equivalence relation; we define an *extensional map* $[f] : A \to B$ as an equivalence class of a map $f : A \to B$ for this relation. We note that extensional maps were called crisp in [17]. With composition $[g][f] = [gf]$ if $gf$ is defined, totally fuzzy sets and extensional maps form a category which we denote by $\mathbf{Set}_{te} H$.

This category has been constructed in two different ways. From a map $f : A \to B$, we can construct the set $R_f$ of all pairs $(x, y)$ in $|A| \times |B|$ with $\varepsilon_A(x) \leq \delta_B(f(x), y)$. D. PONASSE [14, 15] characterized sets $R$ of this form by the following three properties:

(i) For $x$ in $|A|$, there is always $y$ in $|B|$ with $(x, y) \in R$,

(ii) If $(x, y) \in R$ and $(x', y') \in R$, then $\delta_A(x, x') \leq \delta_B(y, y')$,

(iii) If $(x, y) \in R$ and $\varepsilon_A(x) \leq \delta_B(y, y')$, then $(x, y') \in R$,

and used them as morphisms $R : A \to B$ of totally fuzzy sets. A set R satisfying these conditions always contains graphs of mappings $f : |A| \to |B|$. These mappings are crisp maps $f : A \to B$, with $R = R_f$, and $R_f = R_{f'}$ iff $f$ and $f'$ are extensionally equal. Composition is given by $R_g R_f = R_{gf}$, for maps $f : A \to B$ and $g : B \to C$.

PONASSE called $\mathrm{Set}_{te} J$, with $J$ denoting the complete Heyting algebra of truth-values, a category of totally fuzzy sets (ensembles totalement flous) and denoted it by $JTF$.

**2.12. Fuzzy functions induced by crisp maps.** The second construction is due to G.P. MONRO [11]. From a mapping $f : |A| \to |B|$, we construct $\langle f \rangle : |A| \times |B| \to H$ by

$$\langle f \rangle(x, y) = \varepsilon_A(x) \wedge \delta_B(f(x), y),$$

for $(x, y) \in |A| \times |B|$. We say that $\langle f \rangle$ is *induced* by $f$, and note the following result.

**Lemma.** $\langle f \rangle$ *is a fuzzy function from $A$ to $B$ if and only if $f$ is a map from $A$ to $B$, and $\langle f \rangle = \langle f' \rangle$, for maps $f, f' : A \to B$, iff $f$ and $f'$ are extensionally equal. Moreover, $\mathrm{id}_A$ induces $\delta_A$, and $\langle g \rangle \langle f \rangle = \langle gf \rangle$ if a composition $gf$ of maps is defined.*

MONRO defined the category $\mathbf{Mod}\, H$ of $H$-*valued models* as the subcategory of $\mathbf{Set}\, H = \mathbf{Set}_{tf} H$ with non-empty $H$-sets as objects, and with fuzzy functions $\langle f \rangle$ induced by crisp maps as morphisms. In view of the Lemma just stated, we replace extensional maps by induced fuzzy functions, replacing $[f] : A \to B$ by $\langle f \rangle : A \to B$ for a map $f : A \to B$. With these replacements, $\mathbf{Set}_{te} H$ without the empty set becomes MONRO's subcategory $\mathbf{Mod}\, H$ of $\mathbf{Set}\, H$. We note that

$$(\sigma \circ \langle f \rangle)(x, z) = \varepsilon_A(x) \wedge \sigma(f(x), z),$$

for $(x, z) \in |A| \times |C|$, if $f : A \to B$ is a crisp map and $\sigma : B \to C$ a fuzzy function.

**2.13. Singletons.** We say that a fuzzy relation $\beta$ on a totally fuzzy set $B$ is a *singleton* on $B$ if $\beta$, considered as a binary fuzzy relation $B \to 1$, is injective, *i.e.* if

$$\beta(y) \wedge \beta(y') \leq \delta_B(y, y')$$

for all $y, y'$ in $|B|$. With $\varepsilon_{SB}(\sigma) = \bigvee_{y \in |B|} \sigma(y)$, and

$$\delta_{SB}(\sigma, \tau) = \bigvee_{y \in |B|} (\sigma(y) \wedge \tau(y)) = \varepsilon_{SB}(\sigma) \wedge \bigwedge_{y \in |B|} (\sigma(y) \leftrightarrow \tau(y)),$$

for singletons $\sigma, \tau$ on $B$, singletons on $B$ form an $H$-set. We denote this $H$-set by $SB$.

Regarded as fuzzy relations $\beta : \mathbf{1} \to B$, singletons on $B$ are partial maps: single-valued but not necessarily total.

Every mapping $\delta_B(y, -) : |B| \to H$ is a singleton on $B$, and

$$s : y \mapsto \delta_B(y, -) : B \to SB$$

defines a map $s : B \to SB$, with

$$\delta_{SB}(s(y), \sigma) = \sigma(y) = \langle s \rangle(y, \sigma)$$

for $y \in |B|$, a singleton $\sigma$ on $B$, and the induced fuzzy function $\langle s \rangle$ of $s$. It follows that $\langle s \rangle$ is injective and surjective, and thus an isomorphism of $\mathbf{Set}_{tf} H$ with inverse $\langle s \rangle^{\mathrm{op}}$, but not necessarily an isomorphism in $\mathbf{Set}_{te} H$.

For a fuzzy function $\rho : A \to B$, the mappings $\rho(x, -) : |B| \to H$, for $x \in |A|$ are singletons on $B$. It is easily seen that this defines a map

$$r : x \mapsto \rho(x, -) : A \to SB,$$

with $\langle r \rangle(x, s(y)) = \rho(x, y)$ for $(x, y)$ in $|A| \times |B|$. Conversely, if $\varphi : A \to SB$ in $\mathbf{Set}_{tf} H$, then $\varphi = \langle s \rangle \circ \rho$ for a unique $\rho : A \to B$, and hence $\varphi = \langle r \rangle$ for the map $r : A \to SB$ obtained above from $\rho$. Thus every morphism $\varphi : A \to SB$ in $\mathbf{Set}_{te} H$ is induced by a map, and composition by $\langle s \rangle$ defines a bijection between morphisms $\rho : A \to B$ in $\mathbf{Set}_{tf} H$ and morphisms $\langle r \rangle : A \to SB$ in $\mathbf{Set}_{te} H$. This bijection is clearly natural in $A$; thus $H$-sets of singletons define a right adjoint of the embedding $\mathbf{Set}_{te} H \to \mathbf{Set}_{tf} H$.

**2.14. Six categories of fuzzy sets.** In addition to the four categories already described, we have a category $\mathbf{Set}_{de} H$ of discrete $H$-sets and extensional maps, and a category $\mathbf{Set}_{df} H$ of discrete $H$-sets and fuzzy functions. The latter category was introduced by M. EYTAN [3], who denoted it by $\mathbf{Fuz}\, H$.

The six categories can be arranged in a commutative diagram

$$(1) \qquad
\begin{array}{ccccc}
\mathbf{Set}_{dc} H & \longrightarrow & \mathbf{Set}_{de} H & \longrightarrow & \mathbf{Set}_{df} H \\
\downarrow & & \downarrow & & \downarrow \\
\mathbf{Set}_{tc} H & \longrightarrow & \mathbf{Set}_{te} H & \longrightarrow & \mathbf{Set}_{tf} H
\end{array}$$

of categories and functors. In this diagram, the vertical arrows are full embeddings, and the horizontal arrows are bijective on objects. The horizontal arrows at left are full, but not faithful, and the horizontal arrows at right are embeddings, but not full embeddings.

**2.15. Void fuzzy sets.** We say that an $H$-set $A$ is *void* if $\varepsilon_A(x) = \perp$ for all $|x| \in A$. A void $H$-set is necessarily discrete, and there is a void set $A$ with $|A| = S$ for every crisp set $S$. The empty $H$-set is void.

For a void $H$-set $A$ and an $H$-set $B$, there is exactly one fuzzy relation $\zeta$ : $A \to B$, with $\zeta(x, y) = \perp$ for all $(x, y)$ in $|A| \times |B|$, This relation is an injective fuzzy function, and surjective iff $B$ is void. Every mapping $f : |A| \to |B|$ is a map $f : A \to B$ with $\langle f \rangle = \zeta$. It follows that all void sets are isomorphic in $\mathbf{Set}_{tf} H$ and in $\mathbf{Set}_{df} H$, and initial objects of these categories.

If $\rho : A \to B$ is a fuzzy function with $B$ void, then $A$ is void, and $\rho$ an isomorphism in $\mathbf{Set}_{df} H$ and $\mathbf{Set}_{tf} H$.

The situation is more complicated for $\mathbf{Set}_{te} H$ and $\mathbf{Set}_{de} H$. In these categories, only the empty set is an initial object. All non-empty void sets are isomorphic, with exactly one morphism to any non-empty $H$-set, but there is no morphism from a non-empty object to the empty fuzzy set.

We conclude from this discussion that it is always safe, and in the extensional case definitely a simplification, to remove the empty fuzzy set from categories of fuzzy sets with extensional maps or fuzzy functions as morphisms. If we do so, we must modify the functors in 2.14 from categories with crisp maps to categories with extensional maps or fuzzy functions, by $\emptyset \mapsto \mathbf{0}$ for a specified non-empty void $H$-set $\mathbf{0}$.

**2.16. Adjunctions and finite limits.** We have already described in 2.4 the left and right adjoint functors of the full embedding $\mathbf{Set}_{dc} H \to \mathbf{Set}_{tc} H$, and in 2.13 the right adjoint $S$ of the embedding $\mathbf{Set}_{te} H \to \mathbf{Set}_{tf} H$.

Two crisp maps $f, g : A \to B$ of discrete $H$-sets are extensionally equal if and only if $f(x) = g(x)$ for all $x \in |A|$ with $\varepsilon_A(x) \neq \perp$. It follows easily that the functor $\mathbf{Set}_{dc} H \to \mathbf{Set}_{de} H$ has a left adjoint $R$, *if the empty set is removed from* $\mathbf{Set}_{de} H$, with $RA$ obtained for a non-empty discrete fuzzy set $A$ by removing all $x$ in $|A|$ with $\varepsilon_A(x) = \perp$, with $\varepsilon_{RA}(x) = \varepsilon_A(x)$ for the remaining elements of $|A|$. For an extensional map $\langle f \rangle : A \to B$, the map $R\langle f \rangle$ is the restriction of $f : |A| \to |B|$ to $|RA|$ and $|RB|$.

For special Heyting algebras $H$, there may be other adjunctions related to the diagram of 2.14, but we shall not discuss them.

Finite products are constructed in the same way in all six categories of 2.14, and hence preserved by all functors in the diagram of 2.14. For fuzzy functions $\rho : C \to A$ and $\sigma : C \to B$, we get $\langle \rho, \sigma \rangle : C \to A \times B$ by putting

$$\langle \rho, \sigma \rangle (z, x, y) = \rho(z, x) \wedge \sigma(z, y),$$

for $z, x, y$ in $|C| \times |A| \times |B|$. It follows immediately that $\langle \rho, \sigma \rangle$ is induced by the map $\langle f, g \rangle$ if $\rho$ and $\sigma$ are induced by maps $f : C \to A$ and $g : C \to B$. We shall see in 3.12 that equalizers, and hence also pullbacks, are constructed in the same way in the four categories with extensional maps or fuzzy functions as morphisms; thus the embedding functors between these categories preserve all finite limits. The embeddings $\mathbf{Set}_{dc} H \to \mathbf{Set}_{tc} H$ and $\mathbf{Set}_{dc} H \to \mathbf{Set}_{de} H$ have left adjoints and thus preserves all categorical limits. The following example shows that the functor $f \mapsto \langle f \rangle : \mathbf{Set}_{tc} H \to \mathbf{Set}_{te} H$ does not preserve equalizers or pullbacks.

**2.16.1. Example.** Let $B$ be a fuzzy set with two elements $0, 1$, with $\varepsilon_B(0) = \varepsilon_B(1) = \top$ and $\delta_B(0,1) \neq \bot$. There are two maps $f, g : \mathbf{1} \to B$, with an empty equalizer in $\mathbf{Set}_{tc} H$. By 3.12, the equalizer of $\langle f \rangle$ and $\langle g \rangle$ in $\mathbf{Set}_{te} H$ is a non-void singleton; thus the functor $f \mapsto \langle f \rangle$ does not preserve the equalizer of $f$ and $g$.

**2.17. Topoi and quasitopoi of fuzzy sets.** GOGUEN's category $\mathbf{Set}_{dc} H$ is a topological universe, a topological quasitopos over sets. Unfortunately, this makes the internal logic of $\mathbf{Set}_{dc} H$ crisp, hence useless for purposes of fuzzy logic. The category $\mathbf{Set}_{tc} H$ of $H$-sets and crisp maps is cartesian closed, but not a quasitopos.

As is well-known (see e.g. [7, 4, 17]), the category $\mathbf{Set}_{tf} H$ of $H$-sets and fuzzy functions is a topos, with $H$-valued internal logic and with very pleasing set-theoretic constructions. However, most practitioners of fuzzy logic prefer extensional maps to fuzzy functions, which brings us to $\mathbf{Set}_{te} H$. This is a quasitopos, with $H$-valued internal logic and equally pleasing set-theoretic constructions, and equivalent to the full subcategory of $\mathbf{Set}_{tf} H$ given by separated objects, in the topos-theoretic sense.

The other categories in 2.14.(1) are not topoi or quasitopoi, except for special Heyting algebras $H$ (see e.g. [13] or [2]).

**2.18. Two special cases.** If $H$ is a singleton, then $H$-sets and discrete $H$-sets just are sets, and maps just are mappings. Any two mappings $f : S \to T$ become extensionally equal; thus we have exactly one extensional map $f : S \to T$, except for $S$ non-empty and $T$ empty. For fuzzy functions, even this distinction disappears; we have exactly one $\rho : S \to T$ for any two sets $S$ and $T$.

The case $H = \{\bot, \top\}$ is a bit more interesting. In this case, an $H$-set $A$ is a triple $(|A|, S_A, \alpha_A)$, consisting of a set $|A|$, a subset $S_A$ of $|A|$, and an equivalence relation $\alpha_A$ on $S_A$, given by $x \in S_A$ iff $\varepsilon_A(x) = \top$, and $x \alpha_A y$ iff $\delta_A(x, y) = \top$. For discrete $H$-sets, $\alpha_A$ is the identity relation on $S_A$. Crisp maps $f : A \to B$ map $S_A$ into $S_B$ and preserve equivalence. A map $f : A \to B$ induces a mapping $[f] : S_A/\alpha_A \to S_B/\alpha_B$ of equivalence classes, and maps $f, g : A \to B$ are extensionally equal iff $[f] = [g]$. Fuzzy functions with non-empty codomain are induced by maps, and in addition there is a unique fuzzy function $Z \to \emptyset$ for every void $H$-set Z.

## 3. Fuzzy Subsets and their Logic

**3.1. Subsets and insertions.** Fuzzy relations on an $H$-set $A$ can also be viewed as fuzzy subset structures of $A$. We assign to every fuzzy relation $\alpha$ on $A$ a *fuzzy subset* $A\lceil\alpha$ of $A$, with

(i) $|A\lceil\alpha| = |A|$, and

(ii) $\delta_{A\lceil\alpha}(x, x') = \alpha(x) \wedge \delta_A(x, x') = \alpha(x') \wedge \delta_A(x, x')$ for $x, x'$ in $|A|$.

This equality relation is clearly symmetric and transitive, with

$$\varepsilon_{A\lceil\alpha}(x) = \alpha(x)$$

for $x \in |A|$, and $\mathrm{id}_{|A|}$ lifts to a map

$$j_\alpha : A\lceil\alpha \to A$$

which we call the *insertion* of $A\lceil\alpha$ into $A$.

This works equally well for all six categories $\mathbf{Set}_{xy} H$ introduced in Section 2, and we shall see that many properties of fuzzy subsets are shared by the six categories. We note that a fuzzy subset of a discrete $H$-set is again discrete.

**3.2. Categories $\mathcal{M}A$.** As already noted in 2.6, fuzzy subset structures of an $H$-set $A$ form a complete Heyting algebra $H^A$. The top element of $H^A$ is $\varepsilon_A$, with $A\lceil\varepsilon_A = A$ and insertion $\mathrm{id}_A$, and $A\lceil\perp$ is void for the bottom element. If $\alpha \leq \beta$ in $H^A$, then $A\lceil\alpha$ is a fuzzy subset of $A\lceil\beta$, with $j_\alpha = j_\beta j_{\alpha,\beta}$ for the insertion $j_{\alpha,\beta} : A\lceil\alpha \to A\lceil\beta$. Thus fuzzy subsets of an $H$-set $A$ and their subset insertions form a category, isomorphic to $H^A$ regarded as a category. We denote this category by $\mathcal{M}A$.

For the two categories with crisp maps, fuzzy subset insertions are monomorphic and epimorphic. Regarded as extensional maps or fuzzy functions, fuzzy subset insertions are injective (2.8), and hence monomorphic. Thus categories $\mathcal{M}A$ are full subcategories of the slice category $(\mathbf{Set}_{xy} H)/A$, in each of the six categories $\mathbf{Set}_{xy} H$.

In the three categories of discrete $H$ sets, every map $\mathrm{id}_{|A|} : C \to A$ is (or induces) a subset insertion. This is not the case for totally fuzzy sets, where $\langle\mathrm{id}_{|A|}\rangle$ need not be injective, and may be injective but not a subset insertion.

**3.3. Direct and inverse images.** For a map $f : A \to B$ and fuzzy subset structures $\alpha$ of $A$ and $\beta$ of $B$, we define the *image* $f^{\to}\alpha$ of $\alpha$ by $f$ by putting

$$(f^{\to}\alpha)(y) = \bigvee_{x\in|A|} (\alpha(x) \wedge \delta_B(f(x), y))$$

for $x \in |A|$, and the *inverse image* $f^{\leftarrow}\beta$ of $\beta$ by $f$ by

$$(f^{\leftarrow}\beta)(x) = \varepsilon_A(x) \wedge \beta(f(x)),$$

for $x \in |B|$. It is easily verified that $f^{\rightarrow}\alpha$ is a fuzzy subset structure of $B$, and $f^{\leftarrow}\beta$ a fuzzy subset structure of $A$. We note that

$$(f^{\rightarrow}\alpha)(y) = \bigvee_{f(x)=y} \alpha(x)$$

for a map $f : A \rightarrow B$ of discrete $H$-sets.

For a fuzzy function $\rho : A \rightarrow B$, we define the image $\rho^{\rightarrow}\alpha$ and the inverse image $\rho^{\leftarrow}\beta$ by putting

(1) $$(\rho^{\rightarrow}\alpha)(y) = \bigvee_{x \in |A|} (\alpha(x) \wedge \rho(x, y)),$$

(2) $$(\rho^{\leftarrow}\beta)(x) = \bigvee_{y \in |B|} (\rho(x, y) \wedge \beta(y)),$$

for $y \in |B|$ and $x \in |A|$ respectively. These again are fuzzy subset structures as desired.

For a map $f : A \rightarrow B$, the maps $f^{\rightarrow}$ and $f^{\leftarrow}$ clearly satisfy

(3) $$f^{\rightarrow} = \langle f \rangle^{\rightarrow} \quad \text{and} \quad f^{\leftarrow} = \langle f \rangle^{\leftarrow};$$

thus $f^{\rightarrow}$ and $f^{\leftarrow}$ are well defined for extensional maps. For a fuzzy function $\rho : A \rightarrow B$, the maps $\rho^{\rightarrow}\alpha$ and $\rho^{\leftarrow}\beta$ clearly preserve order and thus define functors

$$\rho^{\rightarrow} : \mathcal{M}A \rightarrow \mathcal{M}B \quad \text{and} \quad \rho^{\leftarrow} : \mathcal{M}B \rightarrow \mathcal{M}A.$$

If we regard $\alpha$ and $\beta$ as binary fuzzy relations $\alpha : A \rightarrow \mathbf{1}$ and $\beta : B \rightarrow \mathbf{1}$, then clearly

(4) $$\rho^{\rightarrow}\alpha = \alpha \circ \rho^{\mathrm{op}} \quad \text{and}' \quad \rho^{\leftarrow}\beta = \beta \circ \rho.$$

It follows that $\rho^{\rightarrow}$ and $\rho^{\leftarrow}$ are functorial, with

$$(\mathrm{id}_A)^{\rightarrow} = \mathrm{Id}\,\mathcal{M}A = (\mathrm{id}_A)^{\leftarrow}$$

for a fuzzy set $A$, and

$$(\sigma \circ \rho)^{\rightarrow} = \sigma^{\rightarrow} \circ \rho^{\rightarrow} \quad \text{and} \quad (\sigma \circ \rho)^{\leftarrow} = \rho^{\leftarrow} \circ \sigma^{\leftarrow}$$

if a composition $\sigma \circ \rho$ of fuzzy functions is defined.

By (3), the properties obtained in the preceding paragraph for fuzzy functions are also valid for crisp maps and extensional maps.

**3.4. Theorem.** *For a fuzzy function* $\rho : A \rightarrow B$ *and fuzzy subset structures* $\alpha$ *of* $A$ *and* $\beta$ *of* $B$*, the following are equivalent.*

(i) $\alpha(x) \wedge \rho(x, y) \leq \beta(y)$ *for all* $x \in |A|$ *and* $y \in |B|$.

(ii) $\alpha \le \rho^\leftarrow \beta$.

(iii) $\rho^\rightarrow \alpha \le \beta$.

(iv) $\rho \circ j_\alpha = j_\beta \circ \sigma$ for a fuzzy function $\sigma : A\lceil\alpha \to B\lceil\beta$.

If $\rho$ is induced by a map $f : A \to B$, then $\sigma$ in (iv) is induced by the map $f : A\lceil\alpha \to B\lceil\beta$.

By 3.3.(3), this result applies to all six categories of fuzzy sets.

PROOF. (i) $\Longleftrightarrow$ (iii) is immediate from the definition of $\rho^\rightarrow\alpha$.

If (ii) holds, then $\alpha\rho^{\mathrm{op}} \le \beta\rho\rho^{\mathrm{op}} \le \beta$ by 2.8, and (iii) holds. Conversely, if (iii) holds, then $\alpha \le \alpha\rho^{\mathrm{op}}\rho \le \beta\rho$, and (ii) is valid.

From the definitions, we have

$$(\rho \circ j_\alpha)(x, y) = \bigvee_{x' \in |A|} (\alpha(x) \wedge \delta_A(x, x') \wedge \rho(x', y))$$

$$= \alpha(x) \wedge \bigvee_{x' \in |A|} (\delta_A(x, x') \wedge \rho(x', y)) = \alpha(x) \wedge \rho(x, y),$$

and similarly $(j_\beta \circ \sigma)(x, y) = \sigma(x, y) \le \beta(y)$ if $\sigma$ in (iv) exists. Thus we have (i) $\Longleftrightarrow$ (iv).

If $\rho$ is induced by $f : A \to B$, then

$$\sigma(x, y) = \alpha(x) \wedge \beta(y) \wedge \delta_B(fx, y)$$

in (iv); thus $\sigma$ is induced by $f : A\lceil\alpha \to B\lceil\beta$ if (i)–(iv) are valid.

**3.5. Theorem.** For a map $f : A \to B$ or a fuzzy function $\rho : A \to B$, and for fuzzy subset structures $\beta$ of $B$ and $\alpha = \rho^\leftarrow\beta$ of $A$, the commutative square

(1)

$$
\begin{array}{ccc}
A\lceil\alpha & \xrightarrow{f} & B\lceil\beta \\
\downarrow{j_\alpha} & & \downarrow{j_\beta} \\
A & \xrightarrow{f} & B
\end{array}
\qquad \text{or} \qquad
\begin{array}{ccc}
A\lceil\alpha & \xrightarrow{\sigma} & B\lceil\beta \\
\downarrow{j_\alpha} & & \downarrow{j_\beta} \\
A & \xrightarrow{\rho} & B
\end{array}
$$

of 3.4.(iv) is a pullback square.

PROOF. For maps, with $\rho$ induced by $f : A \to B$, (1) lifts a pullback in **Set**; thus (1) is a pullback for $f : C \to B$ with $|C| = |A|$ and

$$\delta_C(x, x') = \delta_A(x, x') \wedge \beta(fx) \wedge \delta_B(fx, fx').$$

Since $\delta_A(x, x') \le \delta_B(fx, fx')$, this means that $C = A\lceil f^\leftarrow\beta$.

For fuzzy functions $\varphi : C \to A$ and $\psi : C \to B\lceil\beta$, with $\rho \circ \varphi = j_\beta \circ \psi$, we have

$$\varphi(z, x) \wedge \rho(x, y) \le \psi(z, x) \le \beta(y)$$

for $x \in |A|$, $y \in |B|$, $z \in |C|$, and hence

$$\varphi(z, x) \wedge \rho(x, y) \leq \rho(x, y) \wedge \beta(y).$$

Taking suprema $\bigvee_y$ and using totality of $\rho$, we get

$$\varphi(z, x) \wedge \varepsilon_A(x) = \varphi(z, x) \leq (\rho \leftharpoonup \beta)(x).$$

Thus $\varphi = j_\alpha \circ \chi$ for a fuzzy function $\chi : C \to A \lceil \alpha$, by 3.4 for $\mathrm{id}_C$ and $j_\alpha$. Then also $\psi = \sigma \circ \chi$ since $j_\beta$ is monomorphic; thus the Theorem is valid for fuzzy functions.

If $\varphi$ is induced by a map, then so is $\chi$ by 3.4, thus the Theorem is also valid for extensional maps.

**3.6. Covers.** Using a concept of [5], we say that a morphism $f : A \to B$ is a *cover* if $f^{\to} \varepsilon_A = \varepsilon_B$.

In $\mathbf{Set}_{tf} H$ and its subcategories $\mathbf{Set}_{te} H$, $\mathbf{Set}_{df} H$, $\mathbf{Set}_{de} H$, covers are the same as surjective morphisms. It follows that covers in these categories are epimorphic. Conversely, it can be proved that epimorphisms in $\mathbf{Set}_{tf} H$ and in $\mathbf{Set}_{te} H$ are covers. Epimorphisms in $\mathbf{Set}_{de} H$ and in $\mathbf{Set}_{df} H$ need not be covers.

Covers in $\mathbf{Set}_{dc} H$ and in $\mathbf{Set}_{tc} H$ can have any underlying mapping and thus need not be epimorphic. In $\mathbf{Set}_{dc} H$ covers are the same as final maps $f : A \to B$, i.e. maps with the following property: if a map $h : A \to C$ factors $h = gf$ at the set level, then $g : |B| \to |C|$ always lifts to a map $g : B \to C$. This is not the case for $\mathbf{Set}_{tc} H$. For example, if $A$ is an $H$-set, and if $C$ is the discrete $H$-set with $|C| = |A|$, and $\varepsilon_C(x) = \varepsilon_A(x)$ for $x \in |A|$, then the map $\mathrm{id}_{|A|} : C \to A$ is always a cover, but final only if $A$ is discrete, and $C = A$.

**3.7.** The following result shows that covers and fuzzy subset insertions define a factorization structure in the sense of [1], or a diagonal polarity in the sense of [17], for each of the six categories considered in Section 2.

**Theorem.** (i) *Every morphism* $\rho : A \to B$ *has a unique factorization* $\rho = j_\beta \circ e$ *into a cover* $e$ *followed by a fuzzy subset insertion. If* $\rho$ *is a fuzzy function induced by a map* $f : A \to B$, *then* $e$ *is induced by* $f : A \to B \lceil \beta$.

(ii) *Every commutative diagram*

$$
\begin{array}{ccc}
A & \xrightarrow{\ e\ } & C \\
\downarrow{\scriptstyle \rho} & & \downarrow{\scriptstyle \sigma} \\
B \lceil \beta & \xrightarrow{\ j_\beta\ } & B
\end{array}
,
$$

with $e$ a cover and $j_\beta$ a subset insertion, has a unique diagonal $\tau : C \to B\lceil\beta$ with $\tau \circ e = \rho$ and $j_\beta \circ \tau = \sigma$. If $\sigma$ is a fuzzy function induced by a map $g : C \to B$. then $\tau$ is induced by $g : C \to B\lceil\beta$.

PROOF. For (i), we must put $\beta = \rho^\to \varepsilon_A$, and then $\rho$ factors as claimed, by 3.4.

For (ii), we have $\varepsilon_C = e^\to \varepsilon_A$, and hence

$$\sigma^\to \varepsilon_C = \rho^\to \varepsilon_A \leq \beta.$$

But then $\sigma$ factors $\sigma = j_\beta \circ \tau$ by 3.4, with $\tau e = \rho$ since $j_\beta$ is monomorphic. If $\sigma$ is a fuzzy function induced by a map $g$, then $\tau$ is induced by $g$.

**3.7.1. Remark.** In the factorization $\rho j_\alpha = j_\beta \sigma$ of 3.4, the morphism $\sigma$ is a cover iff $\beta = \rho^\to \alpha$. Thus the left adjoint $f^\to$ or $\rho^\to$ of the pullback functor $f^\leftarrow$ or $\rho^\leftarrow$ is obtained by (cover, insertion) factorizations. This is a special case of a general result for factorization structures; see [1] or [17].

**3.8. Propositional connectives.** The top element of a Heyting algebra $H^A$ is $\varepsilon_A$, corresponding to $\mathrm{id}_A$ in $\mathcal{M}A$, and the bottom element has constant value $\perp$, corresponding in $\mathcal{M}A$ to the insertion of the void fuzzy subset of $A$ into $A$.

Unary and binary propositional connectives can be carried out pointwise in complete Heyting algebras $H^{|A|}$, and the result is natural in $A$ in the sense that

(1) $$f^\leftarrow (\beta \diamond \beta') = (f^\leftarrow \beta) \diamond (f^\leftarrow \beta'),$$

for $\beta, \beta'$ in $H^{|B|}$, a mapping $f : |A| \to |B|$ and a binary connective $\diamond$, with a similar formula for a unary connective. This follows immediately from the fact that $(f^\leftarrow \beta)(x) = \beta(fx)$ for $x \in |A|$.

Algebras $H^A$ need not be closed under pointwise connectives; thus a further step is needed. Since $H^A$ is closed under suprema in $H^{|A|}$, the embedding $I_A : H^A \to H^{|A|}$ has a right adjoint $M_A : H^{|A|} \to H^A$, called *modification*, with

$$M_A \alpha = \bigvee \{\gamma \in H^A \mid \gamma \leq \alpha\}$$

for $\alpha \in H^{|A|}$. Now

(2) $$\alpha \diamond \alpha' = M_A(I_A \alpha \diamond I_A \alpha'),$$

with the pointwise connective at right, is the best we can do. However, the naturality expressed by (1) is usually lost by this process.

For standard connectives, we can do better.

**3.9. Proposition.** *Standard conjunction and disjunction in $H^A$ are given by pointwise evaluation, and implication by*

(1) $$(\alpha \rightarrow \alpha')(x) = \varepsilon_A(x) \wedge (\alpha(x) \rightarrow \alpha'(x)),$$

*for $x \in |A|$. All standard connectives in algebras $H^A$ are preserved by inverse image functors.*

PROOF. The first part is clear for $\wedge$, and follows for $\vee$ immediately from the $(\vee, \wedge)$ distributivity of $H^A$. Now if

$$t \leq \delta_A(x, x') \wedge (\alpha(x) \rightarrow \alpha'(x)),$$

then we have

$$t \wedge \alpha(x') = t \wedge \alpha(x') \wedge \delta_A(x, x') = t \wedge \alpha(x) \wedge \delta_A(x, x')$$
$$\leq \alpha'(x) \wedge \delta_A(x, x') \leq \alpha'(x').$$

Thus $t \leq \alpha(x') \rightarrow \alpha'(x')$, so that $\alpha \rightarrow \alpha'$ defined by (1) is in $H^A$.

By 3.3.(3), it suffices for the second part to consider $\rho^\leftarrow : H^B \rightarrow H^A$ for a fuzzy function $\rho : A \rightarrow B$. Since $\rho^\leftarrow$ has a left adjoint and a right adjoint, by 3.11, it preserves top and bottom of $H^A$. For $\beta \wedge \beta'$ and $x \in |A|$, we have

$$(\rho^\leftarrow \beta)(x) \wedge (\rho^\leftarrow \beta')(x) = \bigvee_{y \in |B|} (\rho(x, y) \wedge \beta(y)) \wedge \bigvee_{y' \in |B|} (\rho(x, y') \wedge \beta(y'))$$

$$= \bigvee_{y, y'} (\rho(x, y) \wedge \rho(x, y') \wedge \beta(y) \wedge \beta'(y'))$$

$$= \bigvee_{y, y'} (\rho(x, y) \wedge \rho(x, y') \wedge \delta_B(y, y') \wedge \beta(y) \wedge \beta'(y'))$$

$$= \bigvee_{y, y'} (\rho(x, y) \wedge \beta(y) \wedge \beta'(y) \wedge \delta_B(y, y'))$$

$$= \bigvee_{y} (\rho(x, y) \wedge \beta(y) \wedge \beta'(y)) = (\rho^\leftarrow (\beta \wedge \beta'))(x).$$

using single-valuedness of $\rho$. The proof for $\beta \vee \beta'$ is similar, but simpler.

For implication, we have

$$\alpha \leq \rho^\leftarrow \beta \rightarrow \rho^\leftarrow \beta' \iff \alpha \wedge \rho^\leftarrow \beta \leq \rho^\leftarrow \beta' \iff \rho^\rightarrow (\alpha \wedge \rho^\leftarrow \beta) \leq \beta',$$

and

$$\alpha \leq \rho^\leftarrow (\beta \rightarrow \beta') \iff \rho^\rightarrow \alpha \leq \beta \rightarrow \beta' \iff \rho^\rightarrow \alpha \wedge \beta \leq \beta'.$$

Now the following Lemma, which is of independent interest, completes the proof.

**3.10. Lemma.** *For $\alpha$ in $H^A$, $\beta$ in $H^B$, and $\rho : A \rightarrow B$, we have:*

$$\rho^\rightarrow (\alpha \wedge \rho^\leftarrow \beta) = \rho^\rightarrow \alpha \wedge \beta.$$

PROOF. Starting with the value of the lefthand side at $y \in |B|$, we have

$$\bigvee_{x \in |A|} (\alpha(x) \wedge \rho(x,y) \wedge \bigvee_{y' \in |B|} (\rho(x,y') \wedge \beta(y')))$$

$$= \bigvee_{x,y'} (\alpha(x) \wedge \beta(y') \wedge \rho(x,y) \wedge \rho(x,y'))$$

$$= \bigvee_{x,y'} (\alpha(x) \wedge \beta(y') \wedge \rho(x,y) \wedge \rho(x,y') \wedge \delta_B(y,y'))$$

$$= \bigvee_{x} (\alpha(x) \wedge \rho(x,y) \wedge \beta(y)) = \beta(y) \wedge \bigvee_{x} (\alpha(x) \wedge \rho(x,y)),$$

which is $(\rho^{\rightarrow} \alpha \wedge \beta)(y)$.

**3.11. Quantifiers.** It is universally agreed, following [9], that universal quantifiers are right adjoint, and existential quantifiers left adjoint, to substitution, and it is also generally agreed that substitution is represented in categories by pullbacks of subobjects.

This means in our present context that substitution is given by pullback functors $f^{\leftarrow} : MB \to MA$, for morphisms $f : A \to B$, and that existential quantifiers are given by factorization functors

$$\exists_f = f^{\rightarrow} : MA \to MB.$$

For a fuzzy function $\rho : A \to B$, and for fuzzy subset structures $\alpha$ of $A$ and $\beta$ of $B$, we want

$$\beta \le \forall_\rho \alpha \iff \rho^{\leftarrow} \beta \le \alpha,$$

and this is the case iff

$$\rho(x,y) \wedge \beta(y) \le \alpha(x),$$

for all $x \in |A|$ and $y \in |B|$. Thus we must put

(1)
$$(\forall_\rho \alpha)(y) = \varepsilon_B(y) \wedge \bigwedge_{x \in |A|} (\rho(x,y) \to \alpha(x)),$$

for all $y \in |B|$. This defines the universal quantifier functor $\forall_\rho$.

Since $f^{\leftarrow} = \langle f \rangle^{\leftarrow}$ for a map $f$, we put

$$\forall_f = \forall_{\langle f \rangle}$$

for a map or extensional map $f$. We note that for a map $f : A \to B$ of discrete $H$-sets, this specializes to

$$(\forall_f \alpha)(y) = \varepsilon_B(y) \wedge \bigwedge_{fx=y} (\varepsilon_A(x) \to \alpha(x)).$$

By 3.3, the assignments $A \mapsto \mathcal{M}A$ and $f \mapsto f^{\leftarrow}$, or $\rho \mapsto \rho^{\leftarrow}$, define contravariant functors on the six categories of 2.14. It follows by adjunction that existential and universal quantifiers define covariant functors on these six categories.

**3.12. Equalizers.** For the topological categories $\mathbf{Set}_{dc}\, H$ and $\mathbf{Set}_{tc}\, H$, equalizers are embeddings and lifted from $\mathbf{Set}$. For the other four categories of 2.14, the following result describes equalizers.

**Proposition.** *For fuzzy functions* $\rho : A \to B$ *and* $\sigma : A \to B$, *the subset embedding* $j_\mu : A\lceil\mu \to A$ *with*

$$\mu(x) = \bigvee_{y \in |B|} (\rho(x,y) \wedge \sigma(x,y)),$$

*for* $x \in |A|$, *is an equalizer of* $\rho$ *and* $\sigma$.
*If* $\rho$ *and* $\sigma$ *are induced by maps* $f$ *and* $g$, *then*

$$\mu(x) = \varepsilon_A(x) \wedge \delta_B(fx, gx)$$

*for* $x \in |A|$, *and* $j_\mu$ *is an equalizer of* $\langle f \rangle$ *and* $\langle g \rangle$ *for extensional maps.*

PROOF. If $\rho\varphi = \sigma\varphi$ for $\varphi : C \to A$, then

$$\varphi(z,x) \wedge \rho(x,y) \leq \varphi(z,x) \wedge \bigvee_{x' \in |A|} (\varphi(z,x') \wedge \sigma(x',y))$$

$$= \bigvee_{x'} (\varphi(z,x) \wedge \varphi(z,x') \wedge \sigma(x',y))$$

$$\leq \bigvee_{x'} (\delta_A(x,x') \wedge \sigma(x',y)) = \sigma(x,y)$$

for $(x,y,z)$ in $|A| \times |B| \times |C|$, and

$$\varphi(z,x) \wedge \rho(x,y) = \varphi(z,x) \wedge \rho(x,y) \wedge \sigma(x,y)$$

follows. Taking suprema $\bigvee_y$ and using the totaltiy of $\rho$, we get $\varphi(z,x) \leq \mu(x)$. Thus $\varphi$ factors $\psi j_\mu$ by the proof of 3.4. Similar computations show that

$$\rho(x,y) \wedge j_\mu(x) = \sigma(x,y) \wedge j_\mu(x)$$

for $(x,y)$ in $|A| \times |B|$, so that $\rho j_\mu = \sigma j_\mu$.
For the second part, if $\rho = \langle f \rangle$ and $\psi = \langle g \rangle$, then

$$\mu(x) = \varepsilon_A(x) \wedge \bigvee_{y \in |B|} (\delta_B(fx,y) \wedge \delta_B(gx,y)) = \varepsilon_A(x) \wedge \delta_B(fx,gx)$$

for $x \in |A|$, and if $\varphi$ in the preceding paragraph is induced by $h : C \to A$, then $\psi$ is induced by $h : C \to A\lceil\mu$, by 3.4.

**3.13. Theorem.** *Let* **FS** *be one of the categories of 2.14. For a pullback square*

$$
(1) \qquad
\begin{array}{ccc}
P & \xrightarrow{\ v\ } & B \\
\downarrow{\scriptstyle u} & & \downarrow{\scriptstyle g} \\
A & \xrightarrow{\ f\ } & B
\end{array}
$$

*in* **FS**, *the following are equivalent.*

(i) $g^{\leftarrow} \circ \exists_f = \exists_v \circ u^{\leftarrow}$.  (ii) $g^{\leftarrow} \circ \forall_f = \forall_v \circ u^{\leftarrow}$.

(iii) $f^{\leftarrow} \circ \exists_g = \exists_u \circ v^{\leftarrow}$.  (iv) $f^{\leftarrow} \circ \forall_g = \forall_u \circ v^{\leftarrow}$.

(v) *The functor* **FS** $\to$ **Set**$_{tf}$ *H in 2.14 preserves the pullback* (1).

*Except for* **FS** $=$ **Set**$_{tc}$ *H , these conditions are satisfied for every pullback square in* **FS**.

The maps $f$ and $g$ of 2.16.1 provide an example of a pullback in **Set**$_{tc}$ *H* for which (i)–(v) are not satisfied.

PROOF. As in the proof of 1.7, (i) $\iff$ (iv) and (ii) $\iff$ (iii) by adjunction. It will be convenient to extend the notation $\langle f \rangle$ from maps to fuzzy function, putting $\langle f \rangle = f$ in this case. Then (iii) is valid, by 3.3.(4) and 2.7.1, iff

$$
\langle g \rangle^{\mathrm{op}} \circ \langle f \rangle = \langle v \rangle \circ \langle u \rangle^{\mathrm{op}}.
$$

(i) requires equality for the dual relations; thus (i) $\iff$ (iii).

Now the relations $\langle g \rangle^{\mathrm{op}} \circ \langle f \rangle$ and $\langle v \rangle \circ \langle u \rangle^{\mathrm{op}}$ can be regarded as fuzzy subset structures $\mu$ and $\nu$ of $A \times B$. For fuzzy functions, we get

$$
\mu(x, y) = \bigvee_{z \in |C|} (f(x, z) \wedge g(y, z))
$$

for $(x, y)$ in $|A| \times |B|$; this becomes

$$
\mu(x, y) = \varepsilon_A(x) \wedge \varepsilon_B(y) \wedge \delta_C(fx, gy)
$$

for crisp or extensional maps.

The map $\langle u, v \rangle : P \to A \times B$ in (1) is the equalizer of $fp$ and $gq$ for the projections $p$ and $q$ of $A \times B$. For **Set**$_{tf}$ *H* and its subcategories, this equalizer is given by the subset structure

$$
\bigvee_z ((fp)(x, y, z) \wedge (gq)(x, y, z)) = \bigvee_z (f(x, z) \wedge g(y, z));
$$

this is the structure $\mu$ of the preceding paragraph. We have $u = pj_\mu$ and $v = qj_\mu$, and thus

$$((\langle v \rangle \circ \langle u \rangle^{\mathrm{op}})(x, y) = \bigvee_{x', y'} (\delta_A(x, x') \wedge \mu(x', y') \wedge \delta_B(y, y')) = \mu(x, y).$$

Thus (i)–(iv) are valid in these categories.

For $\mathbf{Set}_{tc} H$, the pullback $P$ is lifted from sets, with

$$\delta_P(t, t') = \delta_A(ut, ut') \wedge \delta_B(vt, vt')$$

for $t \in P$. It follows for $\nu = \langle v \rangle \langle u \rangle^{\mathrm{op}}$ that

$$\nu(x, y) = \bigvee_{t \in |P|} (\delta_A(ut, x) \wedge \delta_B(vt, y)).$$

Now $h = \langle u, v \rangle$ defines an embedding $h : P \to A \times B$, with $\nu = h^{\to} \varepsilon_P$. The induced fuzzy function $\langle h \rangle$ is injective; thus $\langle h \rangle : P \to (A \times B)\lceil \nu$ is an isomorphism of $\mathbf{Set}_{tf} H$. This shows that $\nu = \mu$ iff the pullback (1) remains a pullback in $\mathbf{Set}_{tf} H$, proving (iii) $\Longleftrightarrow$ (v).

Except for $\mathbf{FS} = \mathbf{Set}_{tc} H$, the functor $f \mapsto \langle f \rangle : \mathbf{FS} \to \mathbf{Set}_{tf} H$ in 2.14 preserves all pullbacks, by 2.16.

**3.14. Powerset objects.** In category theory, powersets represent relations.

For a set $T$ with powerset $PT$, introduce a backward membership relation $\ni_T : PT \to T$ by putting $B \ni_T y$, for $y \in T$ and $B \subset T$, iff $y \in B$. If $\rho : S \to T$ is a relation, then $\rho = \ni_T \circ h$ for a mapping $h : S \to PT$ iff

$$y \in h(x) \iff x \rho y,$$

for $x \in S$ and $y \in T$. This mapping $h$ is called the *characteristic mapping* of $\rho$.

This can be generalized to categories with relations, which includes our six categories of fuzzy sets. We say that a relation $\ni_B : PB \to B$, in a category $\mathbf{C}$ with relations, *represents relations* with codomain $B$ if every relation $\rho : A \to B$ has a unique factorization $\rho = \ni_B \circ \langle h \rangle$ with $h : A \to PB$ in $\mathbf{C}$, and $\langle h \rangle$ the relation induced by $h$. In this situation, $PB$ is called a *powerset object* of $B$, with *membership relation* $\ni_B$. If the factorization always exists, but is not unique, then we speak of a *weak powerset object*.

For topological spaces with continuous maps, and with relations $\rho : X \to Y$ given by subspaces (and hence by subsets) of the product space $X \times Y$, a powerset object $PY$ is the powerset $P|Y|$ of the underlying set $|Y|$ of the space $Y$, with the indiscrete topology. This generalizes to all topological categories over sets.

Categories with relations were dicussed in a paper by A. KLEIN [8], for categories with (cover,insertion) factorization structures, with epimorphic covers and monomorphic insertions. This setting excludes fuzzy sets or $H$-sets with crisp maps. An exposition of relations in categories which includes these examples is planned, but cannot be accommodated in this paper.

**3.15. Fuzzy powersets.** Let **FS** be one of the six categories of 2.14, and let $\langle f \rangle$ be the fuzzy function induced by a morphism $f$ of **FS**, with $\langle f \rangle = f$ if the morphisms of **FS** are fuzzy functions. We define the *fuzzy powerset* $PB$ of an object $B$ of **FS** as the object with underlying set $H^B$ of fuzzy subset structures of $B$, with $\varepsilon_{PB}(\beta) = \top$ for $\beta \in H^B$. If the objects of **FS** are discrete $H$-sets, then $PB$ is a crisp discrete $H$-set. If the objects of **FS** are totally fuzzy sets, then we define $\delta_{PB}$ by

$$\delta_{PB}(\beta, \beta') = \bigwedge_{y \in |B|} (\beta(y) \leftrightarrow \beta'(y)),$$

for $\beta, \beta'$ in $H^B$. In other words, we require

$$\beta =_{PB} \beta' \; \leftrightarrow \; (\forall y)(\beta(y) \leftrightarrow \beta'(y)),$$

with $y$ a variable of type $B$.

For all six categories **FS**, we put

$$\ni_B (\beta, y) = \beta(y),$$

for $\beta \in H^B$ and $y \in |B|$.

We skip the proof that this always works, *i.e.* that $\delta_{PB}$ always is a fuzzy equality, and $\ni_B : PB \to B$ a fuzzy relation.

**Theorem.** *The fuzzy powersets just defined are powerset objects for* $\mathbf{Set}_{te} H$ *and* $\mathbf{Set}_{tf} H$, *and weak powerset objects for the other four categories of 2.14.*

PROOF. For a fuzzy relation $\rho : A \to B$ and $x \in |A|$, put

$$(\chi\rho)(x) = \rho(x, -) : |B| \to H.$$

This is in $H^B$, and it is easily verified that we have defined a map $\chi\rho : A \to PB$. Moreover by the definitions,

$$\ni_B ((\chi\rho)(x), y) = \rho(x, y)$$

for $(x, y) \in |A| \times |B|$. Thus $\ni_B \circ \langle \chi\rho \rangle = \rho$, and $PB$ is a weak powerset object in each of the six categories.

If $\rho = \exists_B \circ \varphi$ for a fuzzy function $\varphi : A \to PB$ in $\mathbf{Set}_{tf} H$, then

$$\rho(x, y) = \bigvee_\beta (\varphi(x, \beta) \wedge \beta(y))$$

for $(x, y)$ in $|A| \times |B|$. Thus $\varphi(x, \beta) \wedge \beta(y) \leq \rho(xy)$ for $\beta$ in $H^B$, and

$$\varphi(x, \beta) \wedge \rho(x, y) = \bigvee_{\beta'} (\varphi(x, \beta) \wedge \varphi(x, \beta') \wedge \beta'(y))$$

$$= \bigvee_{\beta'} (\varphi(x, \beta) \wedge \delta_{PB}(\beta, \beta') \wedge \beta'(y)) = \varphi(x, \beta) \wedge \beta(y).$$

It follows that

$$\varphi(x, \beta) \leq \varepsilon_A(x) \wedge \bigwedge_y (\rho(x, y) \wedge \beta(y)) = \langle \chi \rho \rangle(x, \beta),$$

for $(x, \beta)$ in $|A| \times H^B$. But then $\varphi = \langle \chi \rho \rangle$ by 2.10.(i), and $PB$ is a powerset object in $\mathbf{Set}_{tf} H$ and in $\mathbf{Set}_{te} H$.

## 4. A Language for Fuzzy Logic

**4.1. Outline.** We shall describe a language for fuzzy logic, consisting of typed terms and statements, and based on a category **FS** of $H$-valued fuzzy sets or totally fuzzy sets, where $H$ is a complete Heyting algebra of truth-values. We do not specify the requirements for **FS**, but all six categories introduced in Section 2 qualify. Types of terms will be objects of **FS**, and statements form a type $\Omega$ which need not be an object of **FS**. We shall also describe models of the language obtained from interpretations of its formulas, where formulas of the language are its terms and statements, but we shall not discuss validity of statements for these models.

The language to be described is a variant of the MITCHELL–BÉNABOU language of topos theory, as described e.g. in [17], and in a slightly different form in [10].

**4.2. Terms.** Terms are recursively defined as follow.

**4.2.1.** Every variable is a term, and there is a sufficiently large supply of variables of type $A$ for every object $A$ of **FS**.

**4.2.2.** There is a term $\star$ of type **1**, with **1** a fixed terminal object of **FS**.

**4.2.3.** For terms $s$ of type $S$ and $t$ of type $T$, there is a term $\langle s, t \rangle$ of type $S \times T$.

Generalizing 4.2.3 recursively, we get a term $\langle t_1, \cdots, t_n \rangle$ of type $\prod_{i=1}^n A_i$ for terms $t_i$ of types $A_i$, with $\langle \rangle = \star$ for $n = 0$.

**4.2.4.** For a term $t$ of type $A$ and a morphism $f : A \to B$ in **FS**, there is a term $f(t)$ of type $B$.

This last requirement may have to be modified if interpretations of statements with non-standard propositional connectives are desired.

**4.3. Statements.** Statements are recursively defined as follows.

**4.3.1.** For terms $s$ and $t$ of type $A$, there are atomic statements

$$(t) \in A \qquad \text{and} \qquad (s) =_A (t).$$

**4.3.2.** For a term $t$ of type $A$ and a fuzzy subset structure $\alpha$ of $A$, there is an atomic statement $\alpha(t)$.

**4.3.3.** The elements $\top$ and $\bot$ of $H$ are atomic statements.

If $\square$ is a unary propositional connective and $\varphi$ a statement, then $\square(\varphi)$ is a statement.

For statements $\varphi, \psi$ and a binary propositional connective $\diamond$, there is a statement $(\varphi) \diamond (\psi)$.

**4.3.4.** If $\varphi$ is a statement and $x$ a variable, then $(\forall x)(\varphi)$ and $(\exists x)(\varphi)$ are statements in which all occurences of the variable $x$ in $\varphi$ are replaced by links to the quantifier. Thus the variable $x$ does *not* occur in the quantified statements.

**4.4. Parentheses and Substitution.** In 4.2 and 4.3, we have put parentheses around all formulas which appear as building blocks of mor complex formulas. These parentheses can often be omitted, but we do not try to formulate rules for this. Of course, "Polish" notation would allow us to do away with parentheses altogether.

Formulas can be interpreted as rooted trees in the obvious way, with the operations introduced in 4.2 and 4.3 as nodes. Leaves of these trees are variables, links to quantifiers, the term $\star$, and nullary propositional connectives. Every node has a unique path to the root of the tree. If a leaf is a link to a quantifier $(\exists x)$ or $(\forall x)$, then the quantifier is in the path from the leaf to the root of the tree, and the first quantifier $(\exists x)$ or $(\forall x)$ in this path.

Our language allows unrestricted substitutions of terms of the same type for variables. Substituting a term $\Sigma$ of the same type for an occurrence of a variable $x$ in a formula $F$ means replacing the leaf $x$ in the tree for $F$ by the tree for $\Sigma$. We denote by $F[\Sigma \leftarrow x]$ (read "$\Sigma$ for $x$") the result of substituting the term $\Sigma$, of the same type as $x$, for all occurrences of the variable $x$ in the formula $F$. We note the following obvious fact.

**4.4.1.** If $u$ *is a variable of the same type as* $x$, *but not occurring in a formula* $F$, *then* $(F[u \leftarrow x])[\Sigma \leftarrow u]$ *is the same formula as* $F[\Sigma \leftarrow x]$.

**4.5. Products of types.** We denote by $T_x$ the type of a variable $x$. For a finite set $L$ of variables, we denote by $P_L$ "the" product $\prod_{x \in L} T_x$, with projections $\pi_x^L : P_L \to T_x$. We put in particular $P_{\{x\}} = T_x$ and $P_{\emptyset} = 1$. If $L \subset L'$, then $\pi_x^L \circ \pi_L^{L'} = \pi_x^{L'}$, for $x \in L$, defines a projection $\pi_L^{L'} : P_{L'} \to P_L$. In particular, $\pi_{\emptyset}^L$ is the unique morphism $P_L \to 1$.

It will also be convenient to put $Lx = L \cup \{x\}$ if $x \notin L$. Then $P_{Lx}$ is a product $P_L \times T_x$, with projections $\pi_L^{Lx}$ and $\pi_x^{Lx}$, and every element $v$ of $|P_{Lx}|$ can be regarded as a pair $(t, u)$ in $|P_L| \times |T_x|$, with $t = \pi_L^{Lx}(v)$ and $u = \pi_x^{Lx}(v)$.

If variables $x$ and $y$ have the same type and do not occur in $L$, then

$$\pi_L^{Lx} = \pi_L^{Ly} \circ \pi_{Ly}^{Lx}, \qquad \pi_x^{Lx} = \pi_y^{Ly} \circ \pi_{Ly}^{Lx},$$

for a morphism $\pi_{Ly}^{Lx} : P_{Lx} \to P_{Ly}$. This is obviously an isomorphism, with inverse $\pi_{Lx}^{Ly}$.

We note that always $\pi_L^{L''} = \pi_L^{L'} \circ \pi_{L'}^{L''}$ if the three projections are defined.

**4.6. Interpretations of terms.** Proceeding recursively in the obvious way, we assign to every formula $F$ a finite set of all variables occuring in $F$. This set, called the *support* of $F$, may be empty, as *e.g.* for the term $\star$ or a statement $(\exists x)(x \in A)$. If $F$ is a term of type $A$, and $L$ a finite set of variables containing the support of $F$, then we define the interpretation $|F|_L$ as a mapping $|P_L| \to |A|$, or as a crisp map $P_L \to A$ in **FS**, by the following rules.

**4.6.1.** $|x|_L = \pi_x^L$ for a variable $x$ in $L$.

**4.6.2.** $|\langle s, t \rangle|_L = \langle |s|_L, |t|_L \rangle : P_L \to A \times B$ if the interpretations $|s|_L : P_L \to A$ and $|t|_L : P_L \to B$ are defined.

**4.6.3.** $|\star|_L = \pi_{\emptyset}^L$, the unique mapping or morphism $P_L \to 1$.

**4.6.4.** $|f(t)|_L = f \cdot |t|_L$ if the righthand side is defined.

For a list $t_1, \cdots, t_n$ of terms of types $A_i$, it follows that

$$|\langle t_1, \cdots, t_n \rangle|_L = \langle |t_1|_L, \cdots |t_n|_L \rangle : P_L \to \prod_i A_i.$$

if the interpretations $|t_i|_L$ are defined.

**4.7. Heyting algebras** $[A, H]$. In order to interpret statements, we assign to every object $A$ of **FS** a complete Heyting algebra $[A, H]$ of mappings $\alpha : |A| \to H$, ordered pointwise and satisfying the following conditions.

**4.7.1.** $[A, H]$ is closed in $H^{|A|}$ under meets $\alpha \wedge \alpha'$, and under arbitrary suprema.

**4.7.2.** $[A, H]$ contains the algebra $H^A$ of fuzzy subset structures of $A$.

**4.7.3.** For every map $f : A \to B$ in **FS**, there is a composition map $f^{\leftarrow} : \beta \mapsto \beta \cdot f$ from $[B, H]$ to $[A, H]$ which preserves infima and suprema.

These maps are functorial, with

$$(\mathrm{id}_A)^{\leftarrow} = \mathrm{id}_{[A,H]}, \quad \text{and} \quad (gf)^{\leftarrow} = f^{\leftarrow} \cdot g^{\leftarrow}$$

if $gf$ is defined in **FS**.

As in 3.8, there are modification functors $M_A : H^{|A|} \to [A, H]$, left adjoint to the embeddings $I_A : [A, H] \to H^{|A|}$; thus we can use (1) and (2) in 3.8 to obtain propositional connectives in the algebras $[A, H]$.

Examples of algebras $[A, H]$ are the algebras $H^{|A|}$ and the algebras $H^A$ of 2.6. Other examples are the sets of mappings $\alpha : |A| \to H$ which satisfy just one of the conditions 2.6.1 and 2.6.2.

**4.8. Interpretations of statements.** An interpretation $|\varphi|_L$ of a statement $\varphi$ will be an element of the complete Heyting algebra $[P_L, H]$, as follows.

**4.8.1.** For terms $s, t$ of type $A$, with $|s|_L$ and $|t|_L$ defined, we put

$$|t \in A|_L = \varepsilon_A \cdot |t|_L, \quad \text{and} \quad |s =_A t|_L = \delta_A \cdot \langle |s|_L, |t|_L \rangle.$$

**4.8.2.** For a subset structure $\alpha$ of an object $A$ of **FS**, and for a term $t$ of type $A$ with $|t|_L$ defined, we put

$$|\alpha(t)|_L = \alpha \cdot |t|_L.$$

**4.8.3.** $|\top|_L$ and $|\bot|_L$ are the top and bottom elements of $[P_L, H]$. For statements $\varphi$ and $\psi$ with $|\varphi|_L$ and $|\psi|_L$ defined, we put $|\square \varphi|_L = \square |\varphi|_L$ for a unary propositional connective $\square$, and $|\varphi \diamond \psi|_L = |\varphi|_L \diamond |\psi|_L$ for a binary propositional connective $\diamond$.

**4.8.4.** If $x$ of type $A$ is not in $L$ and $|\varphi|_{Lx}$ is defined, then

$$|(\exists x)\varphi|_L = \exists_p |\varphi|_{Lx} \quad \text{and} \quad |(\forall x)\varphi|_L = \forall_p |\varphi|_{Lx},$$

for $p = \pi_L^{Lx}$, and we put

$$|(\exists x)\varphi|_{Lx} = |(\exists x)\varphi|_L \cdot \pi_L^{Lx} \quad \text{and} \quad |(\forall x)\varphi|_{Lx} = |(\forall x)\varphi|_L \cdot \pi_L^{Lx}.$$

**4.9. Discussion.** Our description of the formal language for fuzzy logic is not meant to be complete; other constructions can be added if they can be interpreted. Parentheses used in the description of 4.2 and 4.3 are often superfluous and should then be omitted.

For finite sets $L$ and $M$ of variables, with $|F|_L$ defined for a formula $F$ of the language and $L \subset M$, we need

$$(1) \qquad\qquad\qquad |F|_M = |F|_L \cdot \pi_L^M.$$

We also want

(2)
$$|F[\Sigma \hookleftarrow y]|_L = |F|_{Ly} \cdot \langle \mathrm{id}_{P_L}, |\Sigma|_L \rangle ,$$

if the substitution at left and the interpretations in the righthand side are defined, with $y$ not in $L$, and not occurring in the term $\Sigma$.

These laws are proved recursively, by induction over the length of the tree corresponding to $F$. We omit the details, noting only three points.

The first point is that (2) is automatically valid if $y$ does not occur in $F$, so that $F[\Sigma \hookleftarrow y]$ is $F$ and $|F|_{Ly} = |F|_L \cdot \pi_L^{Ly}$.

For propositional connectives, we need naturality, such as

$$(\alpha \cdot \pi_L^M) \diamond (\beta \cdot \pi_L^M) = (\alpha \diamond \beta) \cdot \pi_L^M$$

for a binary connective $\diamond$, and $\alpha, \beta$ in $[P_L, H]$.

For quantifiers $(\exists x)$ and $(\forall x)$, we need 3.13.(i)–(iv) for a pullback square 3.13.(1) with $f = \pi_L^{Lx}$ or $g = \pi_L^{Lx}$. We note that such pullbacks are preserved by every functor which preserves finite products.

**4.10. More Heyting algebras.** Every linearly ordered set with a least element and a greatest element is a Heyting algebra, with standard implication given by 1.4.(1).

Open sets in a topological space, ordered by set inclusion, form a complete Heyting algebra, with set intersections as meets and set uions as suprema. Up to isomorphism, every finite Heyting algebra is of this kind. In this example, the negation of an open set is the interior of its complements.

Every Boolean algebra is a Heyting algebra.

# BIBLIOGRAPHY

[1] J. ADÁMEK, H. HERRLICH, G.E. STRECKER, *Abstract and Concrete Categories.* Wyley, New York etc. (1990).

[2] J.C. CARREGA, The categories Set$H$ and Fuz$H$ . *Fuzzy Sets and Systems* **9** (1983), 227–332.

[3] M. EYTAN, Fuzzy sets, a topos-logical point of view. *Fuzzy Sets and Systems* **5** (1981), 47–67.

[4] M.P. FOURMAN and D.S. SCOTT, Sheaves and logic. *Applications of Sheaves. Proceedings of the Durham Conference.* Lecture Notes in Math. **753** (1979), pp. 302–401.

[5] PETER J. FREYD, ANDRÉ ŠČEDROV, *Categories, Allegories.* North-Holland Elsevier, Amsterdam (1990).

[6] J.A. GOGUEN, *L*-Fuzzy Sets. *Jour. Math. Anal. Appl.* **18** (1967), 145–174.

[7] DENIS HIGGS, A category approach to Boolean–valued set theory. *Preprint*, 1973.

[8] A. KLEIN, Relations in categories. *Illinois Jour. of Math.* **14** (1970), 536–550.

[9] F.W. LAWVERE, Adjointness in foundations. *Dialectica* **23** (1969), 281–296.

[10] S. MACLANE and I. MOERDIJK, *Sheaves in Set Theory and Logic. A first introduction to topos theory.* Springer-Verlag, New York etc. (1992).

[11] G.P. MONRO, Quasitopoi, logic and Heyting-valued models. *Jour. Pure Applied Algebra,* **42** (1986), 141–164.

[12] GIUSEPPE PEANO, *Formulaire de Mathématiques.* Torino, vol. 1 1895, vol. 2 1898, vol. 3 1901, vol. 4 1902, vol. 5 1908.

[13] ANDREW M. PITTS, Fuzzy sets do not form a topos. *Fuzzy Sets and Systems* **8** (1982), 101–104.

[14] D. PONASSE, Une nouvelle conception des ensembles flous. *BUSEFAL* **17** (1984), 4–9.

[15] D. PONASSE, Categorical studies of fuzzy sets, *Fuzzy Sets and Systems* **28** (1988), 235–244.

[16] L.N. STOUT, The logic of unbalanced subobjects in a category with two closed structures. *Applications of Category Theory to Fuzzy Subsets,* Kluwer, Dortrecht (1992), pp. 73–105.

[17] OSWALD WYLER, *Lecture Notes on Topoi and Quasitopoi.* World Scientific Publishing Co., Singapore, 1991.

[18] L.A. ZADEH, Fuzzy sets. *Information Control* **8** (1965), 338–353.

# Part C

# General Aspects Of Non–CLassical Logics

# X

# Prolog extensions to many–valued logics

## F. Klawonn

The aim of this paper is to show that a restriction of a logical language to clauses like Horn clauses, as they are used in Prolog, applied to [0,1]–valued logics leads to calculi with a sound and complete proof theory. In opposition to other models where generally the set of axioms as well as the deduction schemata are enriched we restrict ourselves to a simple modification of the deduction rules of classical logic without adding new axioms.

In our model the truth values from the unit interval can be interpreted in a probabilistic sense, so that a value between 0 and 1 is not just intuitively interpreted as a 'degree of truth'.

# 1 Introduction

N. Rescher [23] pointed out that there are at least three different approaches to the field of many–valued logic, namely

- the metalogical viewpoint, which is mainly concerned with proof theoretic and algebraic aspects of logical systems as for example described in [22],

- the semantical standpoint, from which N. Rescher's book is written, where the set of truth values is enriched with values like *undetermined* or more abstract values like '0.5',

- and the practical view, which concentrates on applications of many–valued systems for example in physics as indicated in [14].

In this paper we emphasize the semantical viewpoint and focus our attention to applications in the domain of approximate reasoning. We deliberately restrict our investigations to [0,1]–valued logics, so that we are enabled to provide probabilistic interpretations for our concepts.

In the field of approximate reasoning it is very common to attach a (truth) value to a (logical) formula, expressing for instance a degree of truth, possibility, necessity, plausibility, or belief. This weighting or valuation of formulae enforces an extension of the logical language in order to be able to express the truth value attached to a formula. Although the definition of the notion of a model or an interpretation (i.e. the semantical part) of such a language is straight forward, the rules for logical deduction have to be modified and new rules have to be added for the sake of completeness. Examples for such extensions can be found in [21, 16, 17, 18, 19]. The completeness results of these papers are obtained for the price of a complex deduction mechanism, that guarantees completeness, but does not provide efficient methods for finding proofs. Therefore, these approaches are very valuable from a theoretical point of view, but are subject to limitations for practical applications.

Another problem for some applications is a missing interpretation for the truth values between 0 and 1. It is often not enough to understand truth values in an intuitive sense as degrees of truth, possibility, necessity, plausibility, or belief. These notions without making more precise, what the meaning of a certain degree of 0.8 is, and when we should attach this degree to a formula instead of the degree 0.7, can cause undesired results in applications or may even lead to the rejection of such approaches according to the inherent arbitrariness in the choice of the numbers (degrees).

A simple way to overcome these problems is to interpret the unit interval as an ordinal scale as proposed by Dubois and Prade in [5] for their possibilistic logic. Disadvantages of such an approach are that values specified by different persons cannot be compared and that the richer structure of the unit interval is reduced to a linear ordering, although generally more than a simple ranking is associated with numbers between zero and one.

In this paper we present the following approach. As in many other models for approximate reasoning on a logical basis, we consider the unit interval as the set of 'truth values'. In Section 2 a purely formal approach is described without discussing an interpretation of the 'truth values'. In order to avoid a complicated proof theory we restrict our considerations to a subset of a first order logical language. We only admit formula that are similar to Horn clauses, so that we obtain a language suitable for 'fuzzy' Prolog including some completeness results.

Section 3 is devoted to possible interpretations of the truth values. We provide probabilistic models that can be used for an underlying semantics of the truth values. It turns out that the probabilistic interpretation can also be applied to possibilistic logic, so that we obtain an equivalence between fuzzy Prolog based on the Gödel implication, possibilistic Prolog, and a probabilistic model.

# 2   Extending Prolog to [0,1]–valued Logics

We consider a first order logical language $L$ containing the logical connectives $\rightarrow$, a set $\bigoplus = \{\oplus_1, \ldots, \oplus_n\}$ of binary connectives, and the universal quantifier $\forall$. The set of truth values is the unit interval [0,1]. The valuation function associated to the logical connective $\rightarrow$ is either

$$/\varphi \rightarrow \psi/ = \min\{1 - /\varphi/ + /\psi/, 1\}  \qquad \text{(Lukasiewicz implication)}$$

or

$$/\varphi \rightarrow \psi/ = \left\{ \begin{array}{ll} 1 & \text{if } /\varphi/ \leq /\psi/ \\ /\psi/ & \text{otherwise.} \end{array} \right. \qquad \text{(Gödel implication)}$$

If $\rightarrow$ is intended to be the Lukasiewicz implication we write $L_L$ for $L$, in case of the Gödel implication we write $L_G$. For the connectives in $\bigoplus$ we only assume that the corresponding valuation functions are continuous and non–decreasing in both arguments. Examples for such operators are continuous $t$–norms and $t$–conorms. For the universal quantifier we define $/(\forall x)(\varphi(x))/ = \inf_x\{/\varphi(x)/\}$.

The following definition describes a restricted subset of the logical language $L$, which generalizes the notion of Horn–clauses for our purposes.

**Definition 2.1** *An implication clause is a closed well formed formula of the form*

$$(\forall x_1) \ldots (\forall x_k)(\varphi \rightarrow \psi) \tag{1}$$

*or*

$$(\forall x_1) \ldots (\forall x_k)(\psi), \tag{2}$$

*where $\psi$ is an atomic formula with no other free variables than $x_1, \ldots, x_k$. $\varphi$ is a formula containing only connectives belonging to $\bigoplus$ and no quantifiers.*

A rule base in Prolog consists of rules and facts. Such a rule base is interpreted as a set of axioms known to be true. Instead of a crisp set of axioms (facts and rules) $A \subseteq L$ as in classical logic or Prolog, we consider a mapping $a : L \rightarrow [0,1]$ assigning to each logical formula $\varphi \in L$ a lower bound $a(\varphi)$ for the truth value of $\varphi$. For classical logic $a$ would correspond to the characteristic function of the given axioms. In practical applications $a(\varphi)$ will in general only be specified for some $\varphi \in L$, whereas for all other $\psi \in L$ the default lower bound zero is assumed, i.e. $a(\psi) = 0$. Since we want to restrict our considerations to implication clauses, we allow a lower bound greater than zero only for implication clauses.

**Definition 2.2** *A mapping $a : L \rightarrow [0,1]$ is called regular, if only implication clauses belong to the support of $a$, i.e. $a(\varphi) > 0$ implies that $\varphi$ is an implication clause.*

Our intention is to keep the lower bounds specified by $a$ out of the logical language. For this reason, we also use only the classical deduction schemata, i.e. modus ponens and substitution. Of course, we have to compute corresponding lower bounds for formulae involved in the deduction procedure. This means that, instead of adding a new valid formula in each deduction step as in classical logic, we improve the lower bound for a formula in a deduction step. Formally, a deduction step derives from $a : L \to [0,1]$ a mapping $b : L \to [0,1]$, where $b \geq a$. This motivates the following definition for the inference procedure.

**Definition 2.3** *Let* $a, b : L_L \to [0,1]$ *be regular.*

(i) *$b$ is directly derivable from $a$ if*

   (a) *there exists an implication clause* $(\forall x_1)\ldots(\forall x_k)(\varphi \to \psi)$ *in* $L_L$ *such that*

   (a1) *If $\psi_0$ is an implication clause with free variables $x_1,\ldots,x_r$ and $\psi_0 \neq \psi$, then* $a((\forall x_1)\ldots(\forall x_r)(\psi_0)) = b((\forall x_1)\ldots(\forall x_r :)(\psi_0))$.

   (a2)

$$b((\forall x_1)\ldots(\forall x_k)(\psi)) = \max\{/(\forall x_1)\ldots(\forall x_k)(\varphi)/_a \quad (3)$$
$$+a((\forall x_1)\ldots(\forall x_k)(\varphi \to \psi))$$
$$-1, a((\forall x_1)\ldots(\forall x_k)(\psi))\},$$

   *where the value* $/(\forall x_1)\ldots(\forall x_k)(\varphi)/_a$ *is obtained by considering the Herbrand universe of $L_L$ and valuating atomic formulae according to $a$,*

   *holds or*

   (b) *there exists an implication clause* $(\forall x_1)\ldots(\forall x_k)(\chi)$ *and terms* $t_{i_1},\ldots,t_{i_r}, (i_1,\ldots,i_r \in \{1,\ldots,k\})$ *without free variables such that for the formula $\chi'$, which is obtained by substituting $x_{i_j}$ $(j = 1,\ldots,r)$ by $t_{i_j}$ in $\chi$ and quantifying over the remaining free variables,*

$$b(\chi') = \max\{a((\forall x_1)\ldots(\forall x_k)(\chi)), a(\chi')\}. \quad (4)$$

   *is satisfied.*

(ii) *$b$ is derivable from $a$ $(a \lhd b)$ if there is a regular sequence $a_0,\ldots,a_n : L_L \to [0,1]$ where $a_{k+1}$ is directly derivable from $a_k$ for each $k \in \{0,\ldots,n-1\}$ and $b \leq a_n$ holds.*

(iii) *The mapping* $\mathrm{th}^{(a)} : L_L \to [0,1]$ *is given by*

$$\mathrm{th}^{(a)}(\varphi) = \begin{cases} \sup\{b(\varphi) \mid a \lhd b\} & \text{if } \varphi \text{ is an implication clause} \\ & \text{of the form (2)} \\ 0 & \text{otherwise.} \end{cases}$$

We use the same terminology as in definition 2.3 for $L_G$. But we have to replace (3) by

$$b((\forall x_1)\ldots(\forall x_k)(\psi)) = \max\Big\{ \min\{/(\forall x_1)\ldots(\forall x_k)(\varphi)/_a, \qquad (5)$$
$$a((\forall x_1)\ldots(\forall x_k)(\varphi \to \psi))\},$$
$$a((\forall x_1)\ldots(\forall x_k)(\psi)) \Big\}.$$

The application of modus ponens and the substitution of free variables by other terms is formalized in (i)(a) and (i)(b), respectively, in the above definition. (ii) describes the application of a finite number of deduction steps of the form introduced in (i). Finally, (iii) specifies what can be obtained from $a$ if we allow an arbitrary number of deduction steps.

For the semantical or model–theoretic part of our $[0, 1]$–valued logic we define accordingly to the interpretation of $a$ as a specification of lower bounds for the truth values the mapping $\mathrm{Th}^{(a)}$ that describes the consequences for the lower bounds for all formulae induced by $a$.

**Definition 2.4** *Let $a : L \to [0, 1]$. Let $/\varphi/_I$ denote the truth value the formula $\varphi$ obtains under the $[0,1]$–valued interpretation $I$. $I$ is called compatible with $a$ if $/\varphi/_I \geq a(\varphi)$ holds for all $\varphi \in L$.*
*$\mathrm{Th}^{(a)} : L \to [0, 1]$ denotes the infimum over all $[0,1]$–valued interpretations of $L$, that are compatible with $a$.*

We write $/\varphi/$ instead of $/\varphi/_I$ if it is clear to which interpretation $I$ we refer.

**Theorem 2.5 (Soundness of $L_L$)** *Let $a : L_L \to [0, 1]$ be regular. For all closed formulae $\varphi \in L_L$*
$$\mathrm{th}^{(a)}(\varphi) \leq \mathrm{Th}^{(a)}(\varphi)$$

*holds.*

**Proof.** We have to prove that direct derivability preserves compatibility. Let $I$ be an interpretation compatible with $a$ and let $b$ be directly derivable from $a$. Case 1. $b$ is obtained from $a$ by applying (3).
From the compatibility of the interpretation $I$ with $a$ and the monotonicity of the valuation functions in $\oplus$, we derive

$$/(\forall x_1)\ldots(\forall x_k)(\psi)/ \geq \max\Big\{ \begin{array}{l} /(\forall x_1)\ldots(\forall x_k)(\varphi)/ \\ +/(\forall x_1)\ldots(\forall x_k)(\varphi \to \psi)/_a - 1, \\ /(\forall x_1)\ldots(\forall x_k)(\psi)/ \end{array}\Big\}$$

$$\geq \max\Big\{ \begin{array}{l} /(\forall x_1)\ldots(\forall x_k)(\varphi)/_a \\ +/(\forall x_1)\ldots(\forall x_k)(\varphi \to \psi)/_a - 1, \\ /(\forall x_1)\ldots(\forall x_k)(\psi)/_a \end{array}\Big\}$$

$$= b((\forall x_1)\ldots(\forall x_k)(\psi)).$$

Case 2. $b$ is obtained from $a$ by applying (4).
For the same reasons as in case 1, we obtain

$$
\begin{aligned}
/\chi'/ &= \max\{/(\forall x_1)\ldots(\forall x_k)(\chi)/, /\chi'/\} \\
&\geq \max\{a((\forall x_1)\ldots(\forall x_k)(\chi)), a(\chi')\}.
\end{aligned}
$$

$\square$

**Theorem 2.6 (Soundness of $L_G$)** *Let $a : L_G \to [0,1]$ be regular. For all closed formulae $\varphi \in L_L$*

$$
\text{th}^{(a)}(\varphi) \leq \text{Th}^{(a)}(\varphi)
$$

*holds.*

**Proof.** The proof is analogous to the proof of Theorem 2.5, except that we have to consider equation (5) instead of (3) in case 1. $\square$

**Theorem 2.7 (Completeness of $L_L$)** *Let $a : L_L \to [0,1]$ be regular and let $\psi$ be an implication clause of the form (2). Then*

$$
\text{th}^{(a)}(\psi) \geq \text{Th}^{(a)}(\psi).
$$

*holds.*

**Proof.** Let $U$ be the Herbrand universe of $L_L$. We show that the Herbrand interpretation $I$ induced by $\text{th}^{(a)}$ is compatible with $a$. For implication clauses $\chi$ of the form (2) the definition of $\text{th}^{(a)}$ yields

$$
/\chi/ \geq a(\chi).
$$

Thus, we only have to consider implication clauses like $(\forall x_1)\ldots(\forall x_k)(\varphi \to \psi)$ of the form (1) where

$$
/(\forall x_1)\ldots(\forall x_k)(\varphi \to \psi)/ < a((\forall x_1)\ldots(\forall x_k)(\varphi \to \psi)).
$$

There exists a tuple $u = (u_1, \ldots, u_k) \in U^k$ such that

$$
/\varphi(u) \to \psi(u)/ < a((\forall x_1)\ldots(\forall x_k)(\varphi \to \psi)) \tag{6}
$$

holds where $\varphi(u)$ and $\psi(u)$ are obtained by substitution of $x_1, \ldots, x_k$ by $u_1, \ldots, u_k$ in $\varphi$ and $\psi$, respectively. By applying part (i)(b) of Definition 2.3 to the formula $(\forall x_1)\ldots(\forall x_k)(\varphi \to \psi)$ by substituting $x_1, \ldots, x_k$ by $u_1, \ldots, u_k$ we derive $b : L_L \to [0,1]$ directly from $a$. Then we have

$$
\begin{aligned}
1 &\geq b(\varphi(u) \to \psi(u)) \\
&\geq a((\forall x_1)\ldots(\forall x_k)(\varphi \to \psi)) \\
&> /\varphi(u) \to \psi(u)/ \\
&= \min\{1 - /\varphi(u)/ + /\psi(u)/, 1\} \\
&= 1 - /\varphi(u)/ + /\psi(u)/. \tag{7}
\end{aligned}
$$

(7) implies

$$/\varphi(u)/ + b(\varphi(u) \rightarrow \psi(u)) - 1 > /\psi(u)/.$$

According to the continuity of the valuation functions for the connectives in $\bigoplus$ and the definition of $\text{th}^{(a)}$ and $/./$ there exists $b' : L_{\text{L}} \rightarrow [0, 1]$ directly derivable from $a$ such that

$$/\varphi(u)/_{b'} + b(\varphi(u) \rightarrow \psi(u)) - 1 > /\psi(u)/ \qquad (8)$$

holds. Let $\tilde{b} = \max\{b, b'\}$. Obviously, $\tilde{b}$ is also derivable from $a$. Taking (8) into account we get

$$/\varphi(u)/_{\tilde{b}} + \tilde{b}(\varphi(u) \rightarrow \psi(u)) - 1 > /\psi(u)/.$$

Therefore, there exists $\hat{b} : L_{\text{L}} \rightarrow [0, 1]$ directly derivable from $\tilde{b}$ by Definition 2.3(i)(a) such that

$$\begin{aligned}\hat{b}(\psi(u)) &= /\varphi(u)/_{\tilde{b}} + \tilde{b}(\varphi(u) \rightarrow \psi(u)) - 1 \\ &> /\psi(u)/.\end{aligned}$$

This leads to the contradiction

$$/\psi(u)/ = \text{th}^{(a)}(\psi(u)) \geq \hat{b}(\psi(u)) > /\psi(u)/.$$

$\square$

**Theorem 2.8 (Completeness of $L_G$)** *Let $a : L \rightarrow [0, 1]$ be regular and let $\psi$ be an implication clause of the form (2). Then*

$$\text{th}^{(a)}(\psi) \geq \text{Th}^{(a)}(\psi).$$

*holds.*

**Proof.** The proof is analogous to that of Theorem 2.7 except for modifications induced by the differing valuation functions for the Lukasiewicz– and the Gödel implication. Note, that in the proof of Theorem 2.7 we only needed the continuity of the valuation functions for the connectives in $\bigoplus$, but not for the Lukasiewicz implication. Therefore, the discontinuity of the Gödel implication does not lead to any problems. $\square$

Theorems 2.5 – 2.8 show that what we can derive by the deduction steps defined in Definition 2.3 coincides with what is deducible from $a$ in the model–theoretic sense of Definition 2.4. This result holds for $L_{\text{L}}$ as well as for $L_G$. Note, that we allow an infinite number of deduction steps according to the supremum in Definition 2.3. Therefore, it is possible that the value $\text{Th}^{(a)}(\varphi)$ can only be approximated (with arbitrary exactness) when we only allow a finite number of deduction steps.

The possibility of an infinite number of deduction steps is also considered in [16, 21]. But a number of additional axiom schemata and inference rules is needed for the completeness results in these papers.

# 3   A Probabilistic Interpretation for Prolog Extensions

The previous section was devoted to a purely formal approach to $[0, 1]$–valued Prolog without giving an interpretation of the truth values. In the following we provide a formal framework in which the truth values originate from probabilities.

The probabilistic setting for our investigations is related to probabilistic logic [15], but does generalize the assumption of probabilistic logic that the probability for a formula plus the probability for its negation sum up to one to the weaker requirement that the sum is at most one. This corresponds to the idea that there are some 'worlds' in which the formula $\varphi$ is known to be true, some different 'worlds' in which the negation of $\varphi$ holds, and other 'worlds' in which nothing is known about $\varphi$.

Instead of the term '(possible) worlds' we will use the notion of a (consideration) context in the following, since our probabilistic model is motivated by the context model [9, 6], which was introduced as an integrating model for vagueness and uncertainty and later on also adopted for logical approaches [8].

**Definition 3.1** *Let $L$ be the set of (closed) well formed formulae (wff's) of a first order predicate language $L$ and let $\mathcal{C} = (C, \mathcal{A}, P)$ be a probability space with $\sigma$-algebra $\mathcal{A}$ together with a mapping $\mu : C \to 2^L$ s.t.*

*(i) for all $c \in C$ : $\mathrm{TH}(\mu(c)) = \{\varphi \in L \mid \mu(c) \vdash_L \varphi\} = \mu(c)$*

*(ii) for all $c \in C$ : $\bot \notin \mu(c)$, (where $\bot \leftrightarrow \varphi \wedge \neg\varphi$)*

*(iii) for all $\varphi \in L$ : $\{c \in C \mid \varphi \in \mu(c)\} \in \mathcal{A}$.*

*Then $(\mathcal{C}, \mu)$ is called a context evaluation of $L$.*

$\mathrm{TH}(\mu(c))$ denotes the set of (classical) logical consequences of the set $\mu(c)$.

$C$ can be understood as a set of contexts or possible worlds. $\mu(c)$ represents the set of formulae that are known in context $c \in C$. It is assumed that all possible deductions are carried out in $c$ (condition (i)), that $\mu(c)$ is consistent (condition (ii)), and that we can assign a number $P_\mu(\varphi)$ (i.e. a probability) to each formula $\varphi \in L$ due to the measurability condition (iii) via

$$P_\mu(\varphi) = P(\{c \in C \mid \varphi \in \mu(c)\}).$$

$P_\mu(\varphi)$ is the probability for those contexts in which $\varphi$ is known to be true.

In the same way as we have defined compatibility of (logical) interpretations with a mapping $a : L \to [0, 1]$, that specifies lower bounds for the truth values (compare Definition 2.4), we introduce the notion of compatibility for context evaluations.

**Definition 3.2** *Let $a : L \to [0, 1]$.*

(i) *A context evaluation $(\mathcal{C}, \mu)$ of $L$, where $\mathcal{C} = (C, \mathcal{A}, P)$ is a probability space, is compatible with $a$ if*

$$\text{for all } \varphi \in L : a(\varphi) \le P_\mu(\varphi)$$

(ii) *The mapping $\text{Th}_a : L \to [0, 1]$ is given by*

$$\text{Th}_a(\varphi) = \inf\Big\{ P_\mu(\varphi) \mid ((C, \mathcal{A}, P), \mu) \text{ is a context evaluation}$$
$$\text{compatible with } a \Big\},$$

*where $\inf \emptyset = 1$.*

Note, that $\text{Th}_a$ is not truth-functional.

## 3.1 A Probabilistic Interpretation for Prolog Based on the Lukasiewicz Implication

The following theorem shows that the logic $L_L$ can be understood as a cautious interpretation compared to the concept of context evaluations. In other words, if we use the inference procedure for $L_L$, which was introduced in Definition 2.3, we obtain a sound but not complete proof theory for our probabilistic interpretation.

**Theorem 3.3** *Let $a : L_L \to [0, 1]$ be regular. Let $\bigoplus = \{\wedge, \vee\}$ where we associate the valuation functions*

$$\begin{aligned}
/\varphi \wedge \psi/ &= \max\{/\varphi/ + /\psi/ - 1, 0\}, \quad \text{and} \\
/\varphi \vee \psi/ &= \max\{/\varphi/, /\psi/\}
\end{aligned}$$

*with $\wedge$ and $\vee$, respectively. If $\varphi$ is an atomic formula with free variables $x_1, \ldots, x_n$, then*

$$\text{th}^{(a)}((\forall x_1) \ldots (\forall x_n)(\varphi)) \le \text{Th}_a((\forall x_1) \ldots (\forall x_n)(\varphi))$$

*holds.*

**Proof.** By induction we prove that for any $b$ such that $a \lhd b$, and for all atomic formulae with free variables $y_1, \ldots, y_k$

$$b((\forall y_1) \ldots (\forall y_k)(\psi)) \leq \mathrm{Th}_a((\forall y_1) \ldots (\forall y_k)(\psi)) \tag{9}$$

holds.

By definition we have $a \leq \mathrm{Th}_a$, which gives us the basis for the induction. Now we have to consider $a'$ derivable from $a$ in $n$ steps and $b$ directly derivable from $a'$ and a context evaluation $\left((C, \mathcal{A}, P), \mu\right)$ compatible with $a$. By the hypothesis of the induction we obtain that $\left((C, \mathcal{A}, P), \mu\right)$ is also compatible with $a'$. For the induction step we have to prove that $\left((C, \mathcal{A}, P), \mu\right)$ is compatible with $b$. There are two possibilities of deriving $b$ directly from $a'$.

Case 1: $b$ is obtained from $a'$ by substitution of quantified variables by terms containing no free variables, for instance by replacing $y_i$ by the term $t$.

$$b((\forall y_1) \ldots (\forall y_{i-1})(\forall y_{i+1}) \ldots (\forall y_k)(\psi(y_1, \ldots, y_{i-1}, t, y_{i+1}, \ldots, y_k)))$$
$$= a'((\forall y_1) \ldots (\forall y_k)(\psi(y_1, \ldots, y_k)))$$

For all other formulae $b$ coincides with $a'$. Obviously,

$$a'((\forall y_1) \ldots (\forall y_k)(\psi(y_1, \ldots, y_k)))$$
$$\leq P_\mu \Big( \{c \in C \mid ((\forall y_1) \ldots (\forall y_k)(\psi(y_1, \ldots, y_k))) \in \mu(c)\} \Big)$$
$$\leq P_\mu \Big( \{ \quad c \in C \mid$$
$$((\forall y_1) \ldots (\forall y_{i-1})(\forall y_{i+1}) \ldots (\forall y_k)$$
$$(\psi(y_1, \ldots, y_{i-1}, t, y_{i+1}, \ldots, y_n))) \in \mu(c)\} \Big)$$

holds.

Case 2: $b$ is obtained from $a'$ by the application of a deduction rule of the form

$$(\forall y_1) \ldots (\forall y_k)(\chi \to \psi),$$

where $\chi$ and $\psi$ contain at most $y_1, \ldots, y_n$ as free variables. Furthermore, $\psi$ is an atomic formula, whereas $\chi$ can be composed of atomic formulae and the connectives $\wedge$ and $\vee$. In the following, we use $(\forall y)$ as an abbreviation for $(\forall y_1) \ldots (\forall y_k)$.

We carry out an induction on the number of connectives in $\chi$. For the basis of the induction $\chi$ is an atomic formula. Since $(\forall y)(\chi)$ together with $(\forall y)(\chi \to \psi)$ implies $(\forall y)(\psi)$, we derive

$$P_\mu((\forall y)(\psi)) \geq P_\mu \Big( \{c \in C \mid ((\forall y)(\chi)) \in \mu(c) \text{ and } ((\forall y)(\chi \to \psi)) \in \mu(c)\} \Big)$$
$$\geq P_\mu((\forall y)(\chi)) + P_\mu((\forall y)(\chi \to \psi)) - 1$$
$$\geq a'((\forall y)(\chi)) + a'((\forall y)(\chi \to \psi)) - 1 \tag{10}$$
$$= b((\forall y)(\psi)).$$

Now let $\chi = \chi_1 \vee \chi_2$ or $\chi = \chi_1 \wedge \chi_2$. This implies

$$P_\mu(\chi) \geq \max\{P_\mu(\chi_1), P_\mu(\chi_2)\} \quad \text{or}$$

$$P_\mu(\chi) \geq P_\mu(\chi_1) + P_\mu(\chi_2) - 1, \tag{11}$$

respectively. In the same way as in (10) we obtain

$$P_\mu((\forall y)(\psi)) \geq b((\forall y)(\psi)).$$

$\square$

The following example shows that equality in Theorem 3.3 is not satisfied in general, i.e. that the proof theory is indeed incomplete.

**Example 3.4** Let $L_L$ be the propositional calculus induced by the propositional constants $\varphi_0, \chi_0$, and $\psi_0$. $a : L_L \to [0,1]$ is given by

$$a(\varphi) = \begin{cases} 0.5 & \text{if } \varphi = \varphi_0 \\ 1 & \text{if } \varphi = (\varphi_0 \to \chi_0) \text{ or } \varphi = (\varphi_0 \wedge \chi_0 \to \psi_0) \\ 0 & \text{otherwise.} \end{cases}$$

This implies $\text{th}^{(a)}(\psi_0) = 0$, since the interpretation with $/\varphi_0/ = /\chi_0/ = 0.5$ and $/\psi_0/ = 0$ is compatible with $a$. For a context evaluation $\big((C, \mathcal{A}, P), \mu\big)$ compatible with $a$ we have

$$\begin{aligned} P_\mu(\psi_0) &\geq P\Big(\{c \in C \mid \varphi_0, (\varphi_0 \to \chi_0), (\varphi_0 \wedge \chi_0 \to \psi_0) \in \mu(c)\}\Big) \\ &\geq P\Big(\{c \in C \mid \varphi_0 \in \mu(c)\}\Big) \\ &\geq a(\varphi_0) \\ &= 0.5, \end{aligned}$$

since

$$P\Big(\{c \in C \mid (\varphi_0 \to \chi_0) \in \mu(c)\}\Big) = P\Big(\{c \in C \mid (\varphi_0 \wedge \chi_0 \to \psi_0) \in \mu(c)\}\Big) = 1.$$

Therefore, $\text{Th}_a(\psi_0) \geq 0.5$ holds.

## 3.2 A Probabilistic Interpretation for Possibilistic Logic and Gödel Prolog

The incompleteness result of the previous subsection can be amended if we restrict the set of contexts. We now allow only context evaluations that are nested. This means that the contexts can be understood as a linearly ordered set of more and more speculative contexts, where the set of true formulae in a context becomes larger with the speculative level of the context.

**Definition 3.5** *A context evaluation $\left((C, \mathcal{A}, P), \mu\right)$ of a first order language $L$ is nested if there exists a subset $C_0 \subseteq C$ s.t.*

*(i)* $P(C_0) = 1$

*(ii)* *for all $c, d \in C_0$:* $\left(\mu(c) \subseteq \mu(d) \text{ or } \mu(d) \subseteq \mu(c)\right).$

If we consider nested context evaluations, we have to modify the notion of $\text{Th}_a$, i.e. which minimal restrictions are induced by a specification $a : L \to [0, 1]$ of lower bounds for the probabilities for formulae.

**Definition 3.6** *Let $a : L \to [0, 1]$. The mapping $\text{Th}_a^{(\text{poss})} : L \to [0, 1]$ is given by*

$$\text{Th}_a^{(\text{poss})}(\varphi) = \inf\left\{P_\mu(\varphi) \mid \quad ((C, \mathcal{A}, P), \mu) \text{ is a nested}\right.$$
$$\left. context\ evaluation\ compatible\ with\ a\right\}.$$

Before we prove that the proof theory for $L_G$ is sound and complete with respect to nested context evaluations, we make a short excursus to possibilistic logic. The aim of this excursus is to show that nested context evaluations provide an appropriate interpretation for possibilistic logic so that our final result is the equivalence between possibilistic, nested context, and the $[0, 1]$–valued Gödel Prolog. In order to simplify and clarify the necessary terminology for possibilistic logic, we give here slightly modified definitions compared to the originals from Dubois, Lang, and Prade [3, 1, 2, 5].

**Definition 3.7** *Let $L$ be the language of a first order predicate logic. A possibility measure $\Pi$ on $L$ is a mapping $\Pi : L \to [0, 1]$ with the following properties.*

*(i)* $\Pi(\bot) = 0$

*(ii)* $\Pi(\top) = 1$

*(iii)* *For all $\varphi, \psi \in L$: $\Pi(\varphi \lor \psi) = \max\{\Pi(\varphi), \Pi(\psi)\}$*

*(iv)* $(\exists x : \varphi(x)) \in L \quad \Rightarrow \quad \Pi(\exists x : \varphi(x)) = \sup\{\Pi(\varphi(d)) \mid d \in D\}$ *where $D$ is the Herbrand universe of $L$.*

*(v)* *For all $\varphi, \psi \in L$:*
$$\left(\text{If } (\varphi \leftrightarrow \psi) \text{ is a tautology then, } \Pi(\varphi) = \Pi(\psi) \text{ holds.}\right)$$

$\Pi(\varphi)$ is interpreted as the degree to which $\varphi$ is considered to be possible. The corresponding necessity measure $N$ is given by

$$N : L \to [0, 1], \quad \varphi \mapsto 1 - \Pi(\neg\varphi)$$

where $N(\varphi)$ is understood as the degree to which $\varphi$ is necessarily true.

$N$ also satisfies conditions (i), (ii), and (v) of the above definition. The axioms (iii) and (iv) have to be replaced by the dual axioms, i.e. $\vee$, $\exists$, max, and sup should be substituted by $\wedge$, $\forall$, min, and inf, respectively. If $N$ is a necessity measure, the corresponding possibility measure is obtained by $\Pi(\varphi) = 1 - N(\neg\varphi)$.

**Definition 3.8** *Let* $a : L \to [0, 1]$ *be regular. The mapping*

$$\mathrm{Th}_a^{(\mathrm{poss})} : L \to [0, 1]$$

*is defined by*

$$\mathrm{Th}_a^{(\mathrm{poss})}(\varphi) = \inf\Big\{ P_\mu(\varphi) \mid \quad C = (C, \mathcal{A}, P) \text{ and}$$
$$(C, \mu) \text{ is a context evaluation}$$
$$\text{compatible with } a \Big\}.$$

The following three theorems elucidate the connection between possibilistic logic and nested context evaluations.

**Theorem 3.9** *Let* $N$ *be a necessity measure on* $L$. *Then there exists a nested context evaluation* $\Big((C, \mathcal{A}, P), \mu\Big)$ *such that*

$$N = P_\mu$$

*holds.*

**Proof.** Define

$$\mu :]0, 1] \to 2^L, \quad \alpha \mapsto \{\varphi \in L \mid N(\varphi) \geq \alpha\}.$$

If $N(\varphi) \geq \alpha$ and $N(\varphi \to \psi) = N(\neg\varphi \vee \psi) \geq \alpha$ hold, then

$$\begin{aligned}
\alpha &\leq \min\{N(\varphi), N(\neg\varphi \vee \psi)\} = N(\varphi \wedge (\neg\varphi \vee \psi)) \\
&= N(\varphi \wedge \psi) = \min\{N(\varphi), N(\psi)\} \leq N(\psi).
\end{aligned} \quad (12)$$

According to the conditions (i), (ii), (iv), and (v) for necessity measures we derive by exploiting (12) that $\mathrm{TH}(\mu(\alpha)) = \mu(\alpha)$ holds and that $\mu(\alpha)$ is consistent. Let $C =]0, 1]$ and let $P$ the probability measure which corresponds to the uniform distribution on $]0, 1]$. Let $C = \Big(]0, 1], \mathcal{B}(]0, 1]), P\Big)$ where $\mathcal{B}(]0, 1])$ is the Borel $\sigma$–algebra on $]0, 1]$. For $\varphi \in L$ we have

$$\{c \in ]0, 1] \mid \varphi \in \mu(c)\} = ]0, N(\varphi)].$$

Therefore, $(C, \mu)$ is a context evaluation of $L$. Since $P\big(]0, 1]\big) = 1$ and by the definition of $\mu$, we obtain that $(C, \mu)$ is nested.

Let $\varphi \in L$.

$$P_\mu(\varphi) = P\Big(\{c \in ]0, 1] \mid \varphi \in \mu(c)\}\Big) = P\Big(]0, N(\varphi)]\Big) = N(\varphi)$$

$\square$

**Theorem 3.10** Let $\big((C, \mathcal{A}, P), \mu\big)$ be a nested context evaluation of $L$. Then $P_\mu$ is a necessity measure.

**Proof.** Let $C_0 \subseteq C$ such that $P(C_0) = 1$ is satisfied and for all $c, d \in C_0$: $\big(\mu(c) \subseteq \mu(d)$ or $\mu(d) \subseteq \mu(c)\big)$ holds. Since $\mu(c)$ is consistent for all contexts $c \in C$ and because $\mu(c)$ contains at least all tautologies of $L$, we obtain $P_\mu(\bot) = 0$ and $P_\mu(\top) = 1$. Condition (v) for necessity measures is fulfilled according to $\mu(c) = \mathrm{TH}(\mu(c))$ (for all contexts $c \in C$).

For $\chi \in L$ we define $C_\chi = \{c \in C_0 \mid \chi \in \mu(c)\}$. Thus $P_\mu(\chi) = P(C_\chi)$ holds because of $P(C_0) = 1$.

Let $\varphi, \psi \in L$. In case of $C_\varphi \not\subseteq C_\psi$, there exists a context $c \in C_0$ with $\varphi \in \mu(c)$ and $\psi \notin \mu(c)$. For any context $d \in C_\psi$ we have $\mu(d) \not\subseteq \mu(c)$. Since $\big((C, \mathcal{A}, P), \mu\big)$ is nested, $\mu(c) \subseteq \mu(d)$ follows and therefore also $\varphi \in \mu(d)$. Thus, for the case $C_\varphi \not\subseteq C_\psi$ the inclusion $C_\psi \subseteq C_\varphi$ holds. Analogously, we obtain $C_\varphi \subseteq C_\psi$ for the case $C_\psi \not\subseteq C_\varphi$. In any case we have $C_\varphi \subseteq C_\psi$ or $C_\psi \subseteq C_\varphi$.

Without loss of generality let $C_\varphi \subseteq C_\psi$.

$$
\begin{aligned}
P_\mu(\varphi \wedge \psi) &= P\Big(\{c \in C_0 \mid (\varphi \wedge \psi) \in \mu(c)\}\Big) \\
&= P\Big(\{c \in C_0 \mid \varphi \in \mu(c) \text{ and } \psi \in \mu(c)\}\Big) \\
&= P(C_\varphi \cap C_\psi) \\
&= P(C_\varphi) \\
&= \min\{P(C_\varphi), P(C_\psi)\}
\end{aligned}
$$

Now we consider $(\forall x)(\varphi(x)) \in L$. Let $U$ be the Herbrand universe of $L$. $U$ is a countable set. Let $U = \{u_n \mid n \in \mathbb{N}\}$. We define

$$C^{(n)} = \bigcap_{i=0}^{n} C_{\varphi(u_i)}.$$

$C^{(n)}$ is a non–increasing sequence of sets and

$$\bigcap_{i=0}^{\infty} C^{(i)} = C_{(\forall x)(\varphi(x))}$$

holds. This implies

$$P\Big(C_{(\forall x)(\varphi(x))}\Big) = P\Big(\lim_{n\to\infty}\bigcap_{i=0}^{n} C^{(i)}\Big) = \lim_{n\to\infty} P(C^{(n)})$$

$$= \lim_{n\to\infty} P\Big(C_{\varphi(u_0)\wedge\ldots\wedge\varphi(u_n)}\Big)$$

$$= \lim_{n\to\infty} \min\{P_\mu(\varphi(u_0)),\ldots,P_\mu(\varphi(u_n))\}$$

$$= \inf\{P_\mu(\varphi(u_n)) \mid n \in \mathbb{N}\}.$$

This proves that $P_\mu$ is a necessity measure. □

**Theorem 3.11** *Let* $a : L \to [0,1]$ *be regular. If there is at least one nested context evaluation compatible with* $a$*, then* $\mathrm{Th}_a^{(\mathrm{poss})}$ *is a necessity measure on* $L$*.*

**Proof.** According to Theorem 3.10 $\mathrm{Th}_a^{(\mathrm{poss})}$ is the infimum of a set of necessity measures, which is obviously also a necessity measure. □

In [4, 5] an interpretation of possibility theory and possibilistic logic on the basis of consonant or nested worlds was already suggested and the parallels to Spohn's generalized possible world model [25] were indicated by a purely qualitative interpretation, i.e. the unit interval was only considered as a linear ordering. Refraining from the rich structure of the unit interval by concentrating on the linear ordering imposes two drawbacks for possibility theory. On the one hand, the connection between possibility and necessity measures is based on subtraction which is not a property inherent in linear orderings. On the other hand, generally we associate a quantitative and not only a qualitative ranking with numbers from the unit interval. The interpretation of possibilistic logic in the light of nested context evaluations provides a meaningful, quantitative interpretation of the numbers.

Now, after we have clarified the connection between possibilistic logic and nested context evaluations, we can proof that nested context evaluations also provide a model or interpretation for 'Gödel Prolog', i.e. for $L_G$.

**Theorem 3.12** *Let* $a : L_G \to [0,1]$ *be regular. Let* $\bigoplus = \{\wedge, \vee\}$ *where we associate the valuation functions* min *and* max *to* $\wedge$ *and* $\vee$*, respectively. If* $\varphi$ *is an atomic formula with free variables* $x_1,\ldots,x_n$*, then*

$$\mathrm{th}^{(a)}((\forall x_1)\ldots(\forall x_n)(\varphi)) = \mathrm{Th}_a^{(\mathrm{poss})}((\forall x_1)\ldots(\forall x_n)(\varphi))$$

*holds.*

**Proof.** We again abbreviate $(\forall y_1)\ldots\forall(y_k)$ by $(\forall y)$.
As the first step of the proof we show by induction over the number of (direct)

derivation steps that

$$b((\forall y)(\psi(y))) \leq \text{Th}_a^{(\text{poss})}((\forall y)(\psi(y))) \tag{13}$$

holds for all atomic formulae $\psi$ with free variables $y_1, \ldots, y_k$ and for all $b$ with $a \vartriangleleft b$.

The basis of the induction is given by $a((\forall y)(\psi(y))) \leq \text{Th}_a^{(\text{poss})}((\forall y)(\psi(y)))$. Now let $a'$ be derivable from $a$ in $n$ steps. By the induction hypothesis $a'$ satisfies (13). We have to consider a nested context evaluation $\big((C, \mathcal{A}, P), \mu\big)$ compatible with $a$ (and according to $a \vartriangleleft a'$ also compatible with $a'$) and $b : L \to [0, 1]$ directly derivable from $a'$.

There are two possibilities to obtain $b$ from $a'$.

Case 1 can be treated analogously to case 1 in the proof of Theorem 3.3.

For the second case we have to replace the inequality (10) by

$$
\begin{aligned}
P_\mu((\forall y)(\psi)) &\geq P_\mu\big((\forall y)(\chi)) \wedge ((\forall y)(\chi \to \psi))\big) \\
&= \min\Big\{ P_\mu((\forall y)(\chi)), P_\mu((\forall y)(\chi \to \psi)) \Big\} \\
&\geq \min\Big\{ a'((\forall y)(\chi)), a'((\forall y)(\chi \to \psi)) \Big\}
\end{aligned}
\tag{14}
$$

in case 2 of the proof of Theorem 3.3. (14) is satisfied, since $\big((C, \mathcal{A}, P), \mu\big)$ is a nested context evaluation and therefore, by Theorem 3.10 a necessity measure.

We also have to replace the inequality (11) by

$$P_\mu(\chi) \geq \min\{P_\mu(\chi_1), P_\mu(\chi_2)\}. \tag{15}$$

Since $\big((C, \mathcal{A}, P), \mu\big)$ is nested, (15) is satisfied. All together we obtain

$$\text{th}^{(a)}((\forall y)(\psi)) \leq \text{Th}_a^{(\text{poss})}((\forall y)(\psi)). \tag{16}$$

The opposite inequality to (16) is also fulfilled, since the interpretation (on the Herbrand universe) induced by $\text{th}^{(a)}$ is by definition a necessity measure compatible with $a$, to which we obtain a corresponding nested context evaluation by Theorem 3.9, so that this context evaluation contributes to the infimum for $\text{Th}_a^{(\text{poss})}$.                                                                     □

# 4   Conclusions

We have introduced extensions of Prolog to $[0, 1]$-valued logics. We carry out these extensions on the basis of simple mappings $a : L \to [0, 1]$ that can be understood as 'fuzzy' Prolog programs. In opposition to other approaches, we keep the truth values out of the logical language, so that we are able to define simple inference mechanisms, which lead to the soundness and completeness results presented in Section 2.

From the formal point of view these results are satisfactory, but they do not provide any hints for the interpretation of the truth values from the unit interval. To fill this gap of missing semantics for the truth values, we introduced the notion of context evaluations based on the idea of a set of consideration contexts or possible worlds weighted by a probability measure.

It turned out that the general idea of context evaluations provides an interpretation of 'fuzzy' Prolog based on the Lukasiewicz implication with a sound but incomplete proof theory. The restriction to nested context evaluations yields an interpretation for a 'fuzzy' Prolog based on the Gödel implication with a sound and complete proof theory. Nested context evaluations are also a possible interpretation for possibilistic logic, so that we obtain the equivalence of 'fuzzy' Prolog based on the Gödel implication, possibilistic Prolog, and the interpretation in the light of nested context evaluations.

We did not use the resolution principle [24] as the inference mechanism in our extensions of Prolog. This would only make sense if the connectives in $\bigoplus$ can be interpreted as conjunctions (i.e. if they are $t$-norms). Indeed, in this case we would obtain the same results if we would use the resolution principle modified with respect to the given lower bounds of the truth value. But even in this case, the resolution principle would not yield a severe improvement, since it is an efficient method for finding one proof, where we have to consider all proofs due to the supremum in Definition 2.3 (iii).

The advantage of our approach to $[0, 1]$-valued extensions of Prolog compared to other fuzzy Prolog systems [10, 13, 7, 12, 26] is the clearly defined semantics of the truth values.

# References

[1] D. Dubois, J. Lang, H. Prade, Automated Reasoning Using Possibilistic Logic: Semantics, Belief Revision and Variable Certainty Weights. Proc. 5th Workshop on Uncertainty in Artificial Intelligence, Windsor, Ontario (1989), 81–87.

[2] D. Dubois, J. Lang, H. Prade: Fuzzy Sets in Approximate Reasoning, Part 2: Logical Approaches. Fuzzy Sets and Systems 40 (1991), 203–244.

[3] D. Dubois, H. Prade, An Introduction to Possibilistic and Fuzzy Logics. In: P. Smets, E.H. Mamdani, D. Dubois, H. Prade (eds.), Non-Standard Logics for Automated Reasoning. Academic Press, London (1988), 287-326.

[4] D. Dubois, H. Prade, Fuzzy Sets, Probability and Measurement. European Journal of Operational Research 40 (1989), 135–154.

[5] D. Dubois, H. Prade, Epistemic Entrenchment and Possibilistic Logic. Artificial Intelligence 50 (1991), 223–239.

[6] J. Gebhardt, R. Kruse, A Possibilistic Interpretation of Fuzzy Sets by the Context Model. Proc. International Conference on Fuzzy Systems, IEEE, San Diego (1992), 1089–1096.

[7] M. Ishizuka, N. Kasai, Prolog–ELF Incorporating Fuzzy Logic. Proc. 9th IJCAI Conference, Los Angeles (1985), 701–703.

[8] F. Klawonn, J. Gebhardt, R. Kruse: Logical Approaches to Uncertainty and Vagueness in the View of the Context Model. Proc. International Conference on Fuzzy Systems, IEEE, San Diego (1992), 1375–1382.

[9] R. Kruse, J. Gebhardt, F. Klawonn, Reasoning with Mass Distributions and the Context Model. In: R. Kruse, P. Siegel (eds.), Symbolic and Quantitative Approaches to Uncertainty. Springer–Verlag, Berlin (1991), 81–85.

[10] R.C.T. Lee, Fuzzy Logic and the Resolution Principle. J. Association of Computation and Machines 19 (1972), 109–119.

[11] J.W. Lloyd: Foundations of Logic Programming (2nd ed.). Springer, Berlin (1987).

[12] T.P. Martin, J.F. Baldwin, B.W. Pilsworth, The Implementation of FProlog – A Fuzzy Prolog Interpreter. Fuzzy Sets and Systems 23 (1987), 119–129.

[13] M. Mukaidono, Z.L. Shen, L. Ding, Fundamentals of Fuzzy Prolog. Int. J. Approximate Reasoning 3 (1989), 179–193.

[14] D. Mundici, Ulam Games, Lukasiewicz Logic, and AF $C^\star$–Algebras. Fundamenta Informaticae 18 (1993), 151–161.

[15] N.J. Nilsson, Probabilistic Logic. Artificial Intelligence 28 (1986), 71–87.

[16] V. Novák, On the Syntactico–Semantical Completeness of First–Order Fuzzy Logic, Part I: Syntax and Semantics. Kybernetika 26 (1990), 47–66.

[17] V. Novák, On the Syntactico–Semantical Completeness of First–Order Fuzzy Logic, Part II: Main Results. Kybernetika 26 (1990), 134–154.

[18] V. Novák, The Unifying Role of Fuzzy Logic in Fuzzy Set Theory. Annales Univ. Sci. Budapest, Sect. Comp. 12 (1991), 187–194.

[19] V. Novák, On the Logical Basis of Approximate Reasoning. In: V. Novák, J. Ramík, M. Mareš, M. Černy, J. Nekola (eds), Fuzzy Approach to Reasoning and Decision Making. Academia, Prague (1992), 17–27.

[20] I.P. Orci, Programming in Possibilistic Logic. Int. J. Expert System 2 (1989), 79–96.

[21] J. Pavelka, On Fuzzy Logic I, II, III. Z. Math. Logik Grundlag. Math. 25 (1979), 45–52, 119–134, 447–464.

[22] H. Rasiowa, R. Sikorski, The Mathematics of Metamathematics (3rd ed.). Polish Scientific Publishers, Warszawa (1970).

[23] N. Rescher, Many–Valued Logic. McGraw–Hill, New York (1969).

[24] J.A. Robinson: A Machine–Oriented Logic Based on the Resolution Principle. Journ. of the Association for Computing Machinery 12 (1965), 23–41.

[25] W. Spohn, Ordinal Conditional Functions: A Dynamic Theory of Epistemic States. In: W.L. Harper, B. Skyrms (eds.), Causation in Decision, Belief, Change, and Statistics, Vol.2. Reidel, Dordrecht (1987), 105–134.

[26] M. Umano, Fuzzy Set Prolog. Proc. 2nd IFSA Congress, Tokyo (1987), 750–753.

# XI

# Epistemological aspects of many–valued logics and fuzzy structures

## L.J. Kohout

## 1    Introduction

The aim of this paper is to analyze the epistemological and ontological status of some basic notions of logic that correspond to the essential intuitions of logicians that may contain important necessary concepts, without which the field would lose its character and justification for its autonomy. In this analysis we pay particular attention to the epistemology and ontology of fuzzy systems and their link with many-valued logics.

## 2    What is Logic?

### 2.1    Algebraic vs Conceptual Issues in Many-Valued Logics

Why does Encyclopedia Britanica classify non-classical logic as applied? Presumably, this is based on the assumption that there is one logic – but various definitions of logical systems. Primarily, we have truth and falsity, and any other values do not have logical, but some epistemological, or possibly applied interpretation. Does this caricature that I have just painted, have any scientific or scholarly substance? From antiquity it has been accepted that logic is concerned with form. Is "form" synonymous with "formal"? Is logic just concerned with the notions of falsity and truth? What else is involved? These questions one has to ask and ought to make an attempt to answer if one deals with many-valued logics. Otherwise one would just deal with some kind of more or less abstract algebra.

One can characterize as "abstract" that approach to many–valued logic which proceeds on the basis that we are dealing with assignment of values of some unspecified kind, that are not *truth*-values at all. As Rescher puts it ([57], p. 103), this misses an important issue:

we overlook wholly the relevance of 'values' at issue to semantical considerations regarding truth and falsity: The assignment of values to propositions is simply viewed as a formal exercise, a (possibly very useful) abstract
sort of symbolic game". Indeed, it is very useful in demonstration on non-
derivability. But is this all there is to logic? Only the proof-theoretic
issues?

Arto Salomaa has written [58]:

The problem of finding an interpretation for the truth-values is still far
from a satisfactory solution, at least in the general $n$-valued case. This
need not deter us. ... in the formal development of many–valued logics
the semantical meaning of truth-values is quite unessential.

But Rescher objects:

Without an interpretation that gives some semantic rationale for its truth-
values, somehow connecting them with the conception of *truth status* of
propositions, a many-valued system may be an interesting abstract mechanism, but deficient in its claim to the rubric of a 'logic'.

This discussion of the 1950s and early 1960s concerning the question of what
are proper concerns in many-valued logics domain has also been paralleled more
recently in the field of fuzzy logics and mathematics. The stakes are, however,
higher in the fuzzy domain. There, it is not just the question of concern with
the algebra, possibly with some added ontological issues involving the status of
the truth. The problem of fuzziness also involves the question of epistemology.

The epistemological question, which is at the roots of what I have called *the
schism of the third decade* [39] of the evolution of fuzzy sets was put forward
succintly by Ellen Hisdal [26]:

Is fuzzy set theory a purely formal, syntactic, mathematical theory? Or
does it include a second branch which is concerned with the connection
between the theory, the logic of language, and applications?

The discussion about what the acceptable properties of the fuzzy set are, ought
to be determined by the properties of the studied phenomenon itself, by the way
it manifests itself in various sciences and the domains of human activity. So some
extramathematical, perhaps systemic assumptions must enter here. One cannot
do philosophy by rejecting the philosophical concepts, or logic while refusing to
accept at least some implicitly present ontological principles of logic, or to have a
meaningful theory of probability without at least implicitly acknowledging some
extra-mathematical concept of uncertainty. This applies even more strongly in
the case when one is studying the phenomenon of fuzziness.

## 2.2  Interplay of Formal Techniques and Conceptual Issues

Sometimes, mathematicians make over-simplifying ontological and epistemological assumptions concerning logics. For example, look at the foundational use of many-valued logic systems supporting the construction of fuzzy mathematical and inferential systems. Often, various systems of fuzzy mathematics are developed that create the impression that we deal just with another branch of conventional mathematics, with its meta-framework solely based on certain rather restrictive assumptions of the Fregean logics [1].

On the other hand, some critics (e.g. Susan Haack [25]) tacitly assume that Tarski's theory of truth is adequate even when dealing with problems of *identification* and *observation* of structures, whether this may be the case or not, disregarding those epistemological features of fuzzy systems that have a systemic origin.

Another, rather extreme attitude is to reject any mathematical results that do not stem from some extra-mathematical considerations. I quote [68]:

> In order to take shortcuts and speed up the development of applications, fuzzy set researchers have resorted to *fuzzification*. ... Yet the insights achieved by employing such strategy are truly insignificant. Fuzziness is injected at a superficial level of formalizm, resulting models do not emerge from any deep understanding of human observation and communication but from simple manipulation of symbols. .... It is of course harder to build new, behaviourally grounded and humanistically oriented models. Mathematicians and engineers are not particularly equipped for such a task. (p. 2720)

Yet, fuzzification is an important logical-epistemological tool, when applied appropriately [3], [4].

Contemporary mathematical logic is conveniently classified into the parts listed below, which extend into the many-valued domain of fuzzy structures by means of judicious fuzzification. It can be seen that Zadeh and his disciples attempted to fuzzify with success some of these, now classical parts of mathematical logic. Some selected references of such attempts are listed together with the classification below.

- Recursive functions.

- Set theory.

- Arithmetics.

- Quantification & identity.

---

[1] The term 'non-Fregean logic' is used in two distinct, but related senses. The 1st sense: Frege's principle of compositionality does not hold. The 2nd sense is that of Suszko; concern with the logics where the axiom $(A \equiv B) \vee (A \equiv \neg B)$ that he called 'Fregean' does not hold.

- Propositional logic.

Although the above hierarchy covers what is known as mathematical logic – the logic intimately linked with the foundations of mathematics and computation, other approaches to logic stem from the linguistic philosophy [60] and the linguistics proper [51]. So, in any foundational studies, attention has to be paid also to these.

Höhle and Stout ask a pertinent question in the context of foundational studies, and offer an answer [28]:

> What should the study of foundations of fuzzy sets offer? Certainly it should place fuzzy sets in a longer and broader tradition of many-valued mathematics ... but it must also speak to the needs of the practitioner of applied fuzzy set theory. A foundation for fuzzy set theory should provide a rigorous base for the actual practice of those applying the theory. ... people working with fuzzy sets want to use them for practical purposes ... These practitioners need a fuzzy set theory which is robust ... not particularly sensitive to the details of the model and connectives used but flexible enough so that the model can be tuned to provide high levels of performance. Thus a foundation for fuzzy sets needs to provide for a variety of connectives while clarifying the bounds on choices available.
>
> The second property that foundation should have is elegance. ... We can also ask if a foundation can take into account the 'linguistic variables' and experimental, computational approach [2] suggested by Zadeh in Seattle in 1989.

The suggestions are a good start, but in my opinion, one has to go even further. One has to build on the algebraic strength of many-valued logic, but also learn from its failure to tap the conceptual and formal resources of contemporary philosophical logic.

The foundations of fuzzy sets, logics and systems contain some general systemic concepts that run across the boundary between theory and methodology. Although the initial motivation came from Systems Science through the important work of Zadeh that predated his first paper on fuzzy sets in 1965, the field has become rather fragmented in the last decade, losing to a great extent its initial cross disciplinary character. There is also a wide gap between mathematical and philosophical formal logic. Mathematical theory of General Systems has some features that may help to bridge this gap by mediating communication between the two disparate logic disciplines. Also the notions of *dynamics, stability, approximation, optimization* etc. may provide a fertile ground for formalization employing the notion of many-valuedness; in particular in the form of many-valued logic based algebraic theories of relations.

So, in a foundational analysis we have to distinguish sharply not only

---

[2]It should be pointed out that we have proposed [33], [5] and practised such an experimental approach supported by a sound mathematical theory for a number of years [43], [11]

1. mathematical questions,

2. logical questions,

3. ontological, epistemological and metaphysical questions,

but also look at their interrelationship, with particular emphasis on many-valued systems. For example, there are some interesting links between the mathematical and logical features of fuzzy structures of any kind and the ontological and epistemological questions of the foundational concepts [36]. In order to bring these out explicitly, we need to employ an adequate method of conceptual analysis. In (1) we deal with the structure, in (2) we add to the structure the logical form. Dealing with logical form should involve not only the usual syntactic formal aspects of inference, but also metalogical question of distinct *logical roles*, the same syntactic or algebraic structure can play in different contexts of various systems. In (3) we deal with the problem of ontology, epistemology of the primitive concepts and perhaps, with some *minimal* metaphysics of the systems involved. One should also deal with the questions of selection and justification of the appropriate meaning of the concepts employed.

We have also to add the problematics of methods of enquiry and problem solving. This provides us with a conceptual framework, on the backloth of which we shall judge the issues dealt with in the remaining portions of this chapter.

# 3   From Two-Valued to Many-Valued Structures

In the realm of many-valued logics, we have hinted at two interdependent but not entirely overlapping themes, one adumbrated by the comments of Salomaa, the other by the attitude of Rescher, that will undoubtedly carry over into the domain of fuzzy logic and set structures. Namely: (1) *In the formal development of many-valued logics, there are many algebraic issues where the truth values and their semantic meaning are quite unessential.* (2) On the other hand, *once the truth values are introduced, without looking at alternative interpretations that give some semantic rationale for connecting the valuations with the conception of truth status of the propositions, the many-valued abstract structure can hardly be considered to be logic.* In this section we shall firstly, look at some less conventional but very interesting normal forms of functionally complete many-valued logic algebras. Secondly, we shall introduce the issue of valuation, which will be looked at again in the subsequent sections, this with some semantic and epistemological rationale providing the evaluative criteria for the abstract mathematical structures.

## 3.1  A repertory of complete many-valued logic normal forms

The most important point that emerges when one contrasts Salomaa's and Rescher's views on logic (cf. Sec. 2 above) is that one has to find a dividing line. The essential point that Salomaa makes is that even the algebraic part has something to offer. The question however remains, how much it can offer and what is missing if one restricts himself to just that part.

Important structural relationships that provide the algebraic backbone of various logics are contained in their normal forms. It is possible to generalize from two-valued normal forms to many-valued normal forms in various ways. We shall discuss here one such generalization, namely partial functionally complete Pinkava algebras (Pi-algebras) which offer some interesting insights. Two highlights will emerge of our discussion. (1) Even in the simple two-valued case, the normal forms of generalized Pi-algebras subsume not only the conjunctive and disjunctive normal form but also the implication, equivalence, exclusive-or and other normal forms in one unifying pattern. (2) Two distinct general n-valued connectives may "collapse" into a single connective when one sets n=2. Viewed the other way, a two-valued connective may "bifurcate" into two distinct types of connectives when more than two values are used. This bifurcation of structures and concepts is an interesting phenomenon that accompanies fuzzification of two-valued structures. We shall see something similar later, in the case of disjointness of two power sets, where several distinct definitions of disjointness become equivalent in the two-valued case.

It is fashionable to think about fuzzy connectives in terms of associative structures with additional properties, the so called t-norms and conorms. However, by restricting the choice of logic connectives just to associative pairs of norms and conorms one loses a great deal of algebraic information about the classes of other interesting systems that is contained in the normal forms such as (non-associative) groupoids.

After presenting a brief outline of the Pinkava algebras in their original form [55] thus demonstrating a variety of possible extension that two valued logic can take, we shall discuss our generalization of the Pinkava algebras presented in abstract algebraic form. We shall see that intricate syntactic and semantic links can be offered by such algebraic forms, subsuming in their richness even the currently fashionable associative connectives (e.g. t-norms and conorms). This will complement more semantically oriented picture given later, of the groups of transformations of the connectives participating in the Checklist paradigm (cf. section 6).

### 3.1.1  Many-Valued Families of the Pinkava Logic Algebras

Definition 1: *Pinkava algebras* [53],[54].
The Pinkava n-valued family logical calculi Pi= $\{ A, \Box, \diamond, \odot, \hookrightarrow \}$ consists of the

partially defined connectives operating on the value-set $\{\ 0, 1, 2, 3, \ldots k\text{-}1\ \}$.

$$v_i \diamond v_j = \begin{cases} 0 & \text{iff} & v_i = 0 \\ 1 & \text{if} & v_i \neq 0 \,\& v_j = 1 \\ & & \text{otherwise undefined} \end{cases}$$

$$v_i \square v_j = \begin{cases} 0 & \text{if} & v_i = 0 \\ v_i & \text{if} & v_i \neq 0 \,\& v_j = 1 \\ & & \text{otherwise undefined} \end{cases}$$

$$v_i \odot v_j = \begin{cases} v_i & \text{if} & v_i = 0 \\ & & \text{otherwise undefined} \end{cases}$$

$$v^{\hookrightarrow} = v + 1 \ (mod \ k)$$

$$\Psi_\kappa = \begin{cases} 1 & \text{iff} & v = \kappa \\ 0 & \text{iff} & v \neq \kappa \end{cases}$$

Note that the carrier and the characteristic functions also can be generated in the Pinkava logic algebras by the connectives. For example, the characteristic function $\Psi$ can be generated by $\diamond$ [38].

Theorem 2: *Functionally complete normal forms*
Any k-valued logic function that is obtained by a completion of the partially defined Pinkava connectives of the type $\{\odot, \square, \diamond, \hookrightarrow\}$ defined above is functionally complete and can be expressed in the following canonical normal form.

$$f(v_1, v_2, v_3, \ldots v_n) = \bigodot_{f(v_1, v_2, \ldots v_n) \neq 0} [c_\gamma \square (\diamond_{s=1}^n \Psi_{\alpha_s}(v_s))]$$

It is illuminating to look at two-valued well known special instances of logic connectives and classify them in terms of Pi-algebra connective types. For instance, the $\odot$, which is partial, offers two distinct completions: either Boolean (inclusive) OR $\vee$ or EXCLUSIVE-OR (nonequivalence) $\oplus$. Although the connectives $\square$ and $\diamond$ are identical in the two-valued case, both forming Boolean $\wedge$, they extend each to a *distinct* partial connective for $n > 2$. This is because each of these connectives plays a *different role* in the normal form, serving a different purpose.

This demonstrates an interesting phenomenon: the two-valued $\wedge$ bifurcates into two distinct many-valued connectives $\square, \diamond$ for higher values. Although the two distinct types, $\square, \diamond$ collapse algebraically into the $\wedge$ in the binary case, $\wedge$

plays a different role in each distinct place of the binary normal form; so that by this role the types can be distinguished even when the connectives become indistinguishable by their binary values. Each distinct role the binary connective plays in the normal form is comparable to the role of the general type that it replaces in that context, and of which it is a special collapsed instance. This additional **syntactic** information that the normal form carries can be further explored. We shall return to this point again in section 3.1.3, where we provide a fuller classification of two-valued normal forms, this time in the framework of Pi-algebras.

To summarize: increasing the number of values above two brings in a number of different extensions, leading to several distinct partial structures which determine interesting types of many-valued operations. The *syntactic* information contained in the *normal forms* denotes *semantic objects* formed by the families of algebraic components. These structures that can be dealt with in a more transparent way if one turns to abstract algebraic generalizations of the Pinkava logic forms by means of groupoids, as we shall do now.

### 3.1.2   Families of Pi-Algebras and their Functionally Complete Normal Forms

This section presents algebraic generalizations of the Pinkava algebras and their abstract normal forms [40],[38]. These generalizations will be called Pi-algebras in the sequel. The connectives involved are partial non-associative non-commutative general (algebraic) groupoids in their most general form. Associative connectives, such as the t-norms and conorms are their special instances.

Definition 3: *Functionally complete families of Pi-algebras* (Kohout [33],[40],[38]) Let *PI* be an algebra with carrier $P$ such that $PI = < P, \diamond, \square, \odot, \Phi >$ where:

1. $< P, \diamond >$ is an arbitrary groupoid with zero $z_\diamond$, without divisors of zero, and with the almost absorbing element $a_\diamond$ such that $a_\diamond \diamond p = p \diamond a_\diamond = a_\diamond$ for every $p \in P, p \neq z_\diamond$.

2. $< P, \odot >$ is an arbitrary groupoid with the unit $e_\odot$

3. $< P, \square >$ is an arbitrary groupoid with a right zero $z_{r\square}$ and a right unit $e_{r\square}$.

4. $\Phi$ is a discrete cyclic shift function $\Phi : P \longrightarrow P$ satisfying the following conditions: Given a discrete cyclic order[3] of P, then for every $p \in P$ it holds that $p \prec \Phi(p)$ and $\Phi^0(p) = p$, $\Phi^{k+1}(p) = \Phi(\Phi^k(p))$.

---

[3] For the axioms of a cyclic ordering, conditions of uniqueness, definition of the predecessor and other details not required in the present discussion see [17],[38].

In the above definition $a \prec b$ means that $a$ is a *direct predecessor* of $b$. The situation is straightforward for any arbitrary finite carrier $P$. For countable infinite carrier $P$, the definition of the direct predecessor using the cyclic order can be replaced by a group of permutations, where the distinguished permutations[4] satisfy the transformation rules (interrelationships) between $\delta_1, \delta_2, \delta_2^*$ as defined below.

**Definition 4**: *advance*
Let $p_1, p_2 \in P$, and $\Phi$ be a cyclic shift function. Then the *advance* $\delta$ from $p_1$ to $p_2$ with respect to $\Phi$ is the least ordinal such that $\Phi^\delta(p_1) = p_2$). We write $\delta_\Phi(p_1) = p_2$). The advance $\delta^*$ denotes the inverse of $\delta$.

**Theorem 5**: *Canonical normal forms* (Kohout [40],[38])
Any function f on the carrier $P$ in a Pi-algebra can be expressed in its canonical normal form:

$$f(v_1, v_2, v_3, \ldots) = \underset{\forall f(v_1, v_2, \ldots) | f \neq (e_\odot)}{\bigodot} \Phi^{\delta_2}\{\Phi^{\delta_2^*}(c_\gamma) \square (\lozenge_{s=1}^{card\ P} \Phi^{\delta_1} [\Psi_{\alpha_s}(v_s)])\}$$

where $\delta_1 \overset{\text{def}}{=} (a_\diamond, e_{r\square})$, $\delta_2 \overset{\text{def}}{=} (z_\square, e_\odot')$.

**Theorem 6**: *Functional completeness* (Kohout [40],[38])
Any Pi-algebra is *functionally complete* if given the advance $\delta_1 = \delta(a_\diamond, e_{r\square})$, it also holds that $\delta_1 = \delta(z_\diamond, z_\square)$.

Richness of structures subsumed in the Pi-algebras normal forms stands out better if we rewrite the normal forms in the Polish notations. This makes clearer the purely syntactic aspects of the normal forms and reveals some other, less conventional but interesting specific normal forms that are usually overlooked.

Let us introduce the following abbreviations: $\zeta := \Phi^{\delta_1}$, $\eta := \Phi^{\delta_2}$, $\bar{\eta} := \Phi^{\delta_2^*}$; $\spadesuit v := [\lozenge_{s=1}^{card\ P} \psi_{a_s}(v_s)]$. Further, let us abbreviate the $\Theta, \Xi$ and $\Upsilon$ type of operations as follows.

$\Xi$ -type: $\sharp := \square xy$, $\sharp_\zeta := \square x\zeta y$, $\sharp^\eta := \eta \square xy$, $\sharp_\zeta^\eta := \eta \square x\zeta y$;
$\Theta$-type: $\odot := \odot xy$, $\odot_\eta := \odot \eta x \eta y$;
$\Upsilon$-type: $\spadesuit := \spadesuit v$, $\spadesuit^\eta := \eta \spadesuit v$

**Definition 7**: *Instantation of general types:*
The sequence of the general type[5] $\lambda u \lambda v.[\ldots \Theta \Xi u \Upsilon v \ldots]$ represents one of the

---

[4]This, however, eliminates the generative capability of the cyclic order, and hence is not constructive in the general case. Naturally, one would like to retain the constructive advantage of the cyclic order. Singling out the distinguished permutations also involves the axiom of choice in the general case. It is not clear to me whether within a non-conventional set axiomatization (such as Quine's New Foundation or Vopěnka's AST) or for higher cardinalities there exists an infinite discrete generalized topology which could contain a cyclic order.

[5]The symbols $\lambda u \lambda v.[\ldots]$ represent the usual abstraction of lambda calculus, thus making

following instantiations:

$$\lambda u \lambda v.[\ldots \odot \sharp_\zeta^\eta u \spadesuit v \ldots] \quad \lambda u \lambda v.[\ldots \odot_\eta \sharp_\zeta u \spadesuit v \ldots]$$

$$\lambda u \lambda v.[\ldots \odot \sharp^\eta u \spadesuit^\zeta v \ldots] \quad \lambda u \lambda v[\ldots \odot_\eta \sharp_\zeta u \spadesuit v \ldots]$$

**Theorem 8:** *Functional completeness*
General syntax of the PI-algebra normal form is given by the following formula written in the Polish notation:

$$\lambda w_1 \lambda w_2 \ldots \lambda w_\omega.[\Theta \Xi w_\omega \Upsilon_\omega \ldots$$

$$\ldots \Theta \Xi w_i \Upsilon_i \Theta \Xi w_{i-1} \Upsilon_{i-1} \ldots \Theta \Xi w_3 \Upsilon_3 \Theta \Xi w_2 \Upsilon_2 \Theta \Xi w_1 \Upsilon_1]$$

When all the sequences in this formula are replaced by any of the instances from the above Def. 7 at the places of their occurrence, the family of Pi-logic algebras *remains functionally complete.* ¶
This feature is used for generating the specific instances of two-valued complete systems of connectives listed in Table 1 below.

### 3.1.3   Diversity of structures

The main question to ask is, what structures and restrictions of two-valued logics should be preserved and carried over into the extensions and generalizations of many-valued logics. The Pi-logic algebras are partial systems that put under one roof a wide variety of families of functionally complete many-valued logical systems. Thus they offer a useful framework in which various generalizations and extensions can be carried out.

   In the table below, selected cases of functionally complete two-valued systems of connectives are listed. This incomplete listing provides a partial taxonomy of the two-valued logic by the types $\odot, \Box, \diamond$ with respect to the canonical normal forms of Theorem 8 above. It can be seen again that two different connective types often collapse into the same algebraic form in the two-valued case. This provides us with important information, if we require that fuzzy algebraic and logic structures should default to the classical propositional connectives for the set $\{0, 1\}$. It helps also to separate conceptually the many-valued families of connectives that have the same default values for the two-valued case. These would be difficult to distinguish conceptually otherwise, when *only the default* values for these families are available.

Table 1: Functionally complete bases of connectives for 2-valued normal forms, classified in terms of the types $\{\odot, \Box, \diamond\}$

---

unambiguous the more common informal $[\ldots \Theta \Xi \_ \Upsilon \_ \ldots]$. Note that the use of the Polish notation is essential as it removes arbitrary restrictions on the generality of applicative operations and the notational dependency that a more usual mathematical notation such as $\lambda u \lambda v.[\{\Theta[(u, \Upsilon), v]\}]$ would introduce when dealing with the replacements of Polish strings of a general type by their specific instantiations.

| $\odot$ | V | $\oplus$ | $\equiv$ | V | $\wedge$ | V | $\downarrow$ | $\mid$ | $\equiv$ | $\downarrow$ |
|---|---|---|---|---|---|---|---|---|---|---|
| $\square$ | $\wedge$ | $\wedge$ | V | V | $\wedge$ | $\leftarrow$ | $\rightarrow$ | $\rightarrow$ | $\rightarrow$ | $\nrightarrow$ |
| $\diamond$ | $\wedge$ | $\wedge$ | V | $\mid$ | $\downarrow$ | $\wedge$ | V | $\mid$ | $\mid$ | $\downarrow$ |
| $\delta_2$ | 0 | 0 | 0 | 0 | 0 | 1 | 1 | 1 | 1 | 1 |
| $\delta_1$ | 0 | 0 | 0 | 1 | 1 | 0 | 0 | 1 | 1 | 1 |

We have seen that the canonical normal forms capture important algebraic features of logics and *indicate by their syntax* the **role** each component of the normal form plays within the overall structure of the form. The partial taxonomy of Table 1 shows that the Pi-connectives, despite their strong partiality, provide rich information about interrelationship of distinct systems that can be successfully extracted. This is unfortunately lost by the specialization of connectives to associative pairs of t-norms and conorms. To appreciate and explore more fully the richness of Pi-algebras, one has to look at their taxonomy in general many-valued case. For a more detailed taxonomy see [38] Fig. 8.1.2, or [33].

### 3.1.4  Algebraic structures vs. logical concepts

In the previous section, we were concerned with the generalizations and extensions of logic algebras, with the connectives being just partial groupoids, and the set of values of the algebra (the carrier) that was not ordered but was completely unstructured. We have seen that several conceptually distinct structures collapse into one algebraic form in the two-valued case. Yet these are distinguishable by *the context in which they stand*, and by the different *role* they play within the normal form. This feature becomes useful in fuzzification of algebraic structures of mathematical logic because different contexts support different logical concepts.

The algebraic form, of course, neither defines an inferential system, nor deals with the notion of truth in logic. In order to deal with these matters, one has to introduce further conceptual refinements which, however, can be supported algebraically. For example, any arbitrary subset of the carrier of the Pinkava algebras can be used as the set of designated elements (e.g. truth-values) under very mild restrictions; the algorithms exist for finding tautologies with respect to the arbitrary designated set, given some logical matrix [54]. The main point to be made here is that only when the set of designated values is *decided* upon, an *algebra turns into logic*. There are other conceptual issues in the foundations of logic which may also be important in the context of fuzzy logics. As these are often neglected by mathematical logicians, we have to turn to a brief examination of the issues of philosophical logics to find more about these issues.

# 4   On the relationship of Mathematical and Philosophical Logic

## 4.1   Some similarities and differences

It will be helpful to look first at the similarities and differences of 'mathematical' and 'philosophical' formal logic. The techniques may be similar, but the issues of concern are different. Mathematical logic is structured around systems of increasing intricacy: propositional logic, identity and quantifiers, formalization of arithmetic, set theory, recursive functions, and around techniques involved: model theory, formal theory of proof, category theory, combinatorial logic etc. Philosophical logic on the other hand pays attention to such issues as what is proposition, analycity, necessity, existence, truth, meaning, reference, paradox, and more recently, the problem of vagueness has become an issue as well. Model theory is employed in both branches, but the goals are different.

One might object that such concepts of philosophical logic as those listed above are irrelevant to the mathematics of the fuzzy domain. Let us look just at one example to demonstrate that this is not so. Take the distinction between à priori and à posteriori truth which is an important distinction, indispensable when one is concerned with empirical investigations and their logical structure. Thus, *à priori truth* is defined as the truth known independently of any experience of how things are in the world. On the other hand, *à posteriori truth* is the truth known only on the basis of empirical investigation; one which is known, and can only be known, as a result of experience [24]. Such a distinction has the direct relevance for the choice of the appropriate logic when used as a knowledge representation tool. For example, it has been demonstrated *by computational experiments*, that abstract space generated by the deducibility or refutability relation has different characteristic structure for different problem domains [46], such as mathematical group theory, electronic switching circuit design, of Artificial intelligence "block problem", despite of the fact that these were axiomatized within the same classical two-valued logic. This has direct impact on the efficiency of e.g. resolution theorem provers, as different consequential structures require different theorem proving strategies for the proofs to be successful. What is interesting in the context of fuzzy sets is the fact that the approximate structures of such strategies can be extracted by fuzzy relational algorithms based on appropriate fuzzy closures and interiors [15],[43].

It is, however, not just the problem of efficiency, as by restricting the problem appropriately, an undecidable problem may become decidable, or exponentially complete complexity might be be reduced to NP complete complexity etc. This may also have a direct impact on the choice of a many-valued base logic for the fuzzy system, the purpose of which is to formalize approximations of precise reasoning caused by the loss of information variety (cf. Sec. 6 below on the Checklist paradigm). Furthermore, different many-valued base logics 'give birth' to different fuzzy set theories (cf. Sec. 5 below), where the structure of the power set (partially determined by its generating implication-ply operator) has crucial importance. Again, it can be shown that the choice of logic structures has to match the structure of the problem. An inappropriate choice may lead to the distortion of the knowledge that is to be represented as was demonstrated computationally, by detecting the actual hidden mathematical relational properties of empirical data [16] thus extracting the actual degree of distortion. One can safely say that structures with different algebraic properties are needed, not only to reflect the concrete differences in particular samples of data within a specific knowledge domain, but also to capture the intrinsic structural and formal differences between the distinct knowledge domains.

Although only very recent computational techniques have allowed us to commence empirical investigations into the adequacy of different types of logic to different domains of *wissenschaftliche*[6] knowledge [7],[15], [43], careful epistemologico-logical analysis of the work of Thomas Masaryk shows that he already anticipated this in 1885 [49],[50][7]. The lesson that one can learn from Masaryk is how to link the theory and practice of logic. Thus the advancement of many-valued logics is not just the issue of manufacturing the sufficient supply of many-valued logic based mathematical structures to be handed to the 'practitioners' but it also involves deep epistemological issues that might deeply influence the development of the adequate formalisms in the first instance.

## 4.2 On the theories of truth

### 4.2.1 Conceptual aspects of truth

Classification of truth is one of the major concerns of philosophy. This, of course, reflects in the concerns of philosophical logic. Various notions connected with the notion of truth are carefully examined by the logicians working in the domain of philosophical logic. Such studies are difficult and demanding conceptually, because the notions examined belong to different but interlinked conceptual categories. One has to distinguish semantics from metaphysics and epistemology. Such distinctions are usually neglected by mathematical logicians. Yet probability and fuzzy set theory and logic crucially depends on such distinctions. Such a classification is, however, complicated by the fact that it involves logical notions of diverse character that cannot be fully elucidated just in a structural,

---

[6] In this context the English 'equivalent' scientific I find to be utterly inadequate, as it excludes such pursuits as scholarly analysis of literary texts, etc. that are relevant to the broader definition of knowledge discussed here.

[7] This was connected with Masaryk arguing the need of introducing different degrees of empirical support [48],[47] for different parts of knowledge extracted from different scientific domains. When all these parts are utilized together in some explanatory scientific model [34],[37], the need for different logics in different 'wissenschaftliche' domains becames apparent. An adequate formalization of Masaryk's approach would undoubtedly bring out some non-Fregean ingredients of his approach.

mathematical way. Let us take a simple example:

- *analycity* is a semantic notion.

- *necessity* is a metaphysical notion.

- *a priority* is an epistemological notion.

One may ask *what is for a proposition* (statement, sentence, belief) *to be true*? The *pragmatic, coherence, correspondence, redundancy* and *semantic* theories of truth are the major theories that deal with this question ([24], ch. 5 & 6).

**Pragmatic theory**:  true beliefs are those which are fruitful in terms of experience.

**Coherence theory**:  truth consists in a relation of coherence between propositions (beliefs, etc.) in a set, such that a proposition is false when it fails to fulfil the conditions of coherence that the other mutually coherent members fulfil.

**Correspondence theory**:  a proposition is true when it corresponds to the facts (facts: the way the things are in the world).

**Redundancy theory**:  because '*p* is true' and '*p*' means the same, '*p* is true' is redundant.

**Semantic theory**:  truth is considered to be a property of sentences expressed in terms of a recursive notion of satisfaction as a relation between sentences of a given language and a domain characterised by a description in a metalanguage.

Philosophical content of the theories of truth is dismissed by the majority of mathematicians developing mathematical theories of various logics. This is done with good reason in the realm of Fregean logic, where the necessary groundwork was done by Frege himself and brought to great formal perfection by Tarski. When introducing new paradigms, such as many valued and fuzzy logics, meticulous care has, however, to be exercised in dealing with these questions. For this reason we shall turn now to discussing the crucial distinctions on which the Fregean logics are based.

The notion of truth in Tarski's consequence system was purely formal, not paying any attention to the distinction between a *sentence* and *proposition*. Most philosophical logicians make a distinction between a sentence and a proposition, although a number of mathematical logicians find this distinction incomprehensible and unnecessary. We shall see that it is connected with the notion of the truth, and that without making this distinction some of the listed theories of truth would run even into much more serious philosophical difficulties than they are already in. Because fuzzy sets and logic systems purport to be not only about formal algebraic structures but also claim their competence in handling uncertainty, vagueness and inference with approximate truth (whatever that may be), one has to pay careful attention to these issues, otherwise one would deal not with fuzzy logics but with algebras devoid of any epistemological or ontological meaning. We shall now discuss the reasons for introducing the sentence – proposition distinction, also noting the difficulties that introduction of this distinction brings about.

**Sentence – proposition distinction:**    The most commonly held view is that *declarative sentences* express *propositions* to state that some predicate holds of a subject or that certain items are related in a certain way. This is for the following reasons:

1. Sentences may be *meaningless* or *nonsensical*, expressing nothing.

2. The same sentence may be used by different people, or by the same person on different occasions, to state what is *true on these occasions* but false on others; e.g. 'I have a hedache'.

3. Propositions are something common to sentences 'It's raining', 'Es regnet', 'Prši', 'Il plent'.

For all these reasons, *propositions*, **not sentences** are true or false.

The crucial question as to what a sentence expresses, looms large in the foundations of mathematical logic, although it is often dismissed as unnecessary 'philosophical baggage' by a contemporary mathematical logician. The reason for this is that the ingenious foundational work of Frege has been absorbed by the mathematical logician's often subcoscious conceptual equipment, which when surfaces is considered just 'folklore'. On the other hand, in the foundational studies of philosophical logic and in the philosophy of mathematics and philosophy of language, Frege's contribution is regarded as one of the most fundamental contributions of the end of the second millennium. It, however, took over a hundred years for it to gain this just recognition.

As fuzzy logics are supposed to handle not only the conventional but also the non-Fregean issues, it is essential for a 'fuzzy logician' to have a grasp of at least some Fregean issues and keen awareness of what are the essential non-Fregean issues. These are closely linked with the question what sentences express, what is their meaning, sense, reference, intension, extension ... and such like. I shall briefly survey here only a small part of the bare essentials of this topic that is important for further discussion.

There are two important families of notions:  (1) the **meaning** of a term, and (2) the range of items into which it applies. The most important, although not entirely controversial are the three related pairs tabulated below.

| **Meaning of a term** | **The range to which a term applies** |
| --- | --- |
| sense | *reference* |
| connotation | *denotation* |
| intension | *extension* |

Frege, in order to start his syntactic formalization of logic, had to deal with the problems of *sense – reference*. In his famous example, he noted that the planet called by the ancients *Phosphorus* we would call the "morning star" (m.s.), and the planet called *Hesperus* and we call the "evening star" (e.s.) **denote** the same physical entity, namely the planet Venus. Frege said that m.s. and e.s. have the same *reference* but that they differ in *sense*.

So, Frege's choice was to identify the **sense** of a sentence with the *proposition* it expresses, and the reference of the same sentence to identify with 'The True' or with 'The False'. This leads to further notions: Expressions that have the same meaning (synonyms) are called *intensionally* equivalent (**IEq**); expressions with the same reference (or extension) are *extensionally* equivalent (**EEQ**). Thus we have: IEq → EEq, but it does **not** hold that EEq → IEq.

**Non-Extensional Context.** The sentences according to Frege are complex species of names. The sentences used non-extensionally, Frege identified with the sense of the words *'the proposition that ....'*. Frege's motivation was to preserve Leibniz's notion of *intersubstitutivity* of co-referential terms *salva veritate* ('saving the truth'). This is important, because if two terms $A$, $B$ have the same *reference*, then the term $A$ can be *substituted* for term $B$ in any sentence in which $B$ occurs *without changing* the truth-value[8] of the sentence (cf. [24]).

Problems with the Fregean framework, however, appear when intersubstitutivity *does not* occur *salva veritate*. This was pointed out by Quine [56]: taking two sentences, the equality *Cicero = Tully* and the sentence *'Cicero' contains 6 letters* as true, the substitution of *Tully* for *Cicero* renders the second sentence false.

The fuzzy logics, as adumbrated by Bellmann and Zadeh are local and context dependent. This combined with the fact that the epistemology and ontology of fuzzy logics starts from the semantic side, by exhibiting the objects that the syntax should refer to, should be sufficient reason for displaying great caution in syntactic formalization of such structures. Two things can go wrong with this formulation: either the formalization is inconsistent, or the consistent formalization does not reflect faithfully the local "intended model" that is embedded in a context-dependent way in some hyper-structure. The former is generally acknowledged, the latter is often dismissed using the argument that it is the mathematical form of fuzzy logics, not the content which the form expresses, that matters.

### 4.2.2 Formal aspects of truth

Frege's framework for logic dealt with the logic as a language that has sense and denotation relating to some extralinguistic entities, that is to the propositions and to the Truth. Tarski carried this further by providing the formalization of truth for the formal languages in a paper first published in 1920s [62], ch. 8. He thus formulated a rigorous metamathematical version of the semantic theory

---

[8]Hence the Fregean axiom $(A \equiv B) \vee (A \equiv \neg B)$, when Frege's convention of reference is added.

of truth in which he captured the notion of truth in a formal meta-language; hence also formalizing the notion of denotation and paving the way to his later model-theoretic framework.

Before discussing Tarski's theory of truth, we shall look at a way of turning logic algebra into logic proper, by providing the syntactic system with its denotation – the semantics of logic matrices (as formulated by Lukasiewicz and Tarski). Let us take the following collection of axioms and meta-axioms for two-valued propositional logic [62], ch. XI.

A1. $0 < card\ S \leq \aleph_0$
A2. $x, y \in S \Rightarrow \neg x \in S, x \to y \in S$
A3. $L \subseteq S$
A4. $x, y, z \in S \Rightarrow ((\neg x \to x) \to x) \in L, \quad (x \to (\neg x \to y)) \in L,$
     $(x \to y) \to ((y \to z) \to (x \to z)) \in L$
A5. $x, x \to y \in L \Rightarrow y \in L$

In the above, A1, A2 A3 and A5 belong to the meta-system, while A4 is the list of the Lukasiewicz axioms for the propositional logic. A5 captures the inference rule *Modus ponens*. It is clear that the theorems of the propositional calculus (object language expressions) are derived from the three expressions of A4 by the Modus ponens A5 and (unformalized) rule of substitution. These coincide with all tautologies, i.e. the class of expressions that attain the constant value 1 (truth) under all substitutions.

Thus the link with the Fregean object 'The Truth', hence also with the Fregean semantics is provided conveniently by means of a logical matrix (cf. Tarski and Lukasiewicz (1930) [62], ch.IV): *A logical matrix is a quadruple* $\mathcal{M} =< A, B, f, g >$ *consisting of two disjoint sets A and B, a function f of two variables and g of one variable, both into $A \cup B$ and with the domains covering all the elements of $A \cup B$.* The set $B$ is, of course, the set of designated elements, the 'truth set'.

Now, given the set of formulas in $S$ e.g. A4 above, the valuation $h$ of these into the logical matrix $\mathcal{M}$ has to be provided: If $x \in S$, $y \in S$, then $h(x \to y) = f(h(x), h(y))$   if $x \in S$ then $h(\neg x) = g(h(x))$. The sentence $x$ is *satisfied* by the matrix $\mathcal{M}$, in symbols $x \in \mathcal{S}(\mathcal{M})$, if the formula $h(x) \in B$ holds for every $h$ of $\mathcal{M}$. For example, the system of Lukasiewicz $L$ given by the axioms A4 of the object language listed above, is satisfied by the matrix $\mathcal{M}$ such that $A = \{0\}$, $B = \{1\}$, where $f$ is the usual two-valued implication operator and $g$ is the usual negation.

The notions of logical matrix $\mathcal{M}$, valuation $h$, of the choice of the set of designated elements $B$ as well as the definition of satisfaction are, of course, well known. The point worth making though is that the set of sentences $L$ belongs to the *object language* OL, the matrix $\mathcal{M}$ to a *meta-language* ML and the satisfaction set $\mathcal{S}(\mathcal{M})$, strictly speaking into a *meta-metalanguage* MML. The definition of truth in the semantic theory of truth that Tarski provides involves the formalization of this relationship, including the adequate description of the syntax of OL in a formalized ML. So we can turn now to a brief discussion of Tarski's theory of truth.

An algebraic valuation is not adequate exemplification of what the truth is, although it can serve as a very useful formal metamathematical technique for providing algebraic semantics. Tarski's objective was to find a satisfactory definition of truth [61],[62],[24],[25] that would be (a) materially adequate, (b) formally correct, (c) which would do the justice to the classical Aristotelian notion of truth.

A definition of the statement 'true sentence' for a given language L given in the meta-language M of L should entail all sentences of *M* of the form:

*S is a true sentence of L if and only if p,*     (T)

where 'S' is the name or a structural description of a sentence in L, and 'p' is the translation of that sentence into M. This so called **convention** T is the required condition of material adequacy: any acceptable definition of truth should have as a consequence all the instances of the schema (T). A definition of truth must, however, provide more – a proof of formal correctness in respect of both: (i) the structure of the language, (ii) the concepts involved in the definition, given in the metalanguage. Hence, (iii) both the object language L and the meta-language should be formally specifiable. To achieve this, one needs (a) expected stock of expressions, (b) names of these expressions, (c) sentence terms like 'true', referring to its sentences, (d) tacit assumption that determine the use of 'true' can be asserted in that language itself.

The more "philosophical" part of Tarski's theories are closely linked with an abstract formalization of the notion of *consequence* and are closely linked with his "calculus of systems", that formalizes the concept of equivalence of distinct logics [62]. In this calculus, Tarski formalized the intuitive idea that *distinct logics* are *equivalent as deductive systems* if their sets of *consequences* are identical.

The consequence operator of Tarski's calculus **Cn** forms a *generalized* topological space $< P, \mathbf{u} >$, where:

- The set of all well-formed formulas of a logic is taken as the *carrier P*. The set of all *consequences* generated by the inference rules of this logic belongs to the power set of the carrier.

In order to deal with Tarski's approach in a systematic way, generalized topologies (in the sense of Čech-Koutský) are briefly introduced.

A *generalised topological system* $\mathcal{C}_X$ is a family of subsets of $X$ that is generated by a *generalised closure operator* **u**. Let $\mathcal{P}(X)$ designate the power set of $X$, that is the set of all subsets of $X$. Then the generalised closure operator itself is defined as a meta-relation:

**u**: $\mathcal{P}(X) \longrightarrow \mathcal{P}(X)$.

Axioms are then introduced, to restrict **u** to specialize to various systems of generalized topologies. Some of useful axioms are listed below.

**I**: if $A \subseteq X$, then $A \subseteq \mathbf{u}(A)$;
**M**: if $A \subseteq B \subseteq X$ then $\mathbf{u}(A) \subseteq \mathbf{u}(B)$;
**O**: $\mathbf{u}(0)=0$, where 0 is the empty set.
**U**: For any $A \subseteq X$, $\mathbf{u}(A) = \mathbf{u}(\mathbf{u}(A))$.

Further technical details concerning generalised closure systems are contained in Kohout ([35], [38]).

The system, satisfying **I, M, U** axioms, captures adequately all the aspects of consequences of non-modal reasoning. This comes from the fact that in this system **U** is equivalent to **U'**: $M \subseteq \mathbf{u}(N)$ and $N \subseteq \mathbf{u}(P)$ implies $M \subseteq \mathbf{u}(P)$. Let us look at Tarski's system of 1930 in order to get a clearer idea how a concrete meta-aximatisation looks like[9]

A1.   card $S \leq \aleph_0$
A6*.  If $x, t \in S$ then $\neg x \in S, x \to y \in S$
A2.   If $X \subseteq S$, then $X \subseteq \mathbf{Cn}(X) \subseteq S$.
A3.   If $X \subseteq S$, then $\mathbf{Cn}(\mathbf{Cn}(X)) = \mathbf{Cn}(X)$.
A4.   If $X \subseteq S$, then $\mathbf{Cn}(\mathbf{Cn}(X)) = \bigcup_{Y \subseteq X} \mathbf{Cn}(Y)$, Y finite.
Th1.  If $A \subseteq B \subseteq S$, then $\mathbf{Cn}(A) \subseteq \mathbf{Cn}(B)$.
A7*  If $X \subseteq S, y \in S, z \in S$ and $y \to z \in \mathbf{Cn}(X)$, then $z \in \mathbf{Cn}(X \cup \{y\})$.
A8*  If $X \subseteq S, y \in S, z \in S$ and $z \in \mathbf{Cn}(X \cup \{y\})$ then $y \to z \in \mathbf{Cn}(X)$.
A9* If $x \in S$, then $\mathbf{Cn}(\{x, \neg x\}) = S$.
A10* If $x \in S$, then $\mathbf{Cn}(\{x\}) \cap \mathbf{Cn}(\{\neg x\}) = \mathbf{Cn}(\emptyset)$.

Above, A2, A3, Th1 are the **I, U** and **M** axioms of generalized topology. A1 and A6* form an algebra of formulas $< S; \neg, \to >$. Meta-axioms A1 – A8 capture the implicational fragment of logic of Hilbert. In the system with A1 – A10, the set **Cn**( $\emptyset$) is equivalent to the two-valued logic axioms of Tarski–Lukasiewicz paper listed above.

When we deal with stability and instability induced by some time-dependent changes, it might be necessary to relax the **U** axiom in some instances (Kohout [35]). To conclude this excursion into the use of generalized topologies in logic, let us look briefly at topological properties of some modal logics.

| Axiom of Generalized Topology | Consequence Space of Logics: | General Topologies |
|---|---|---|

---

[9] The numbering of axioms retained as in Tarski [62]. **Cn** denotes the consequence operator.

| MIO | Logic satisfying Tarski's | General Čech |
|-----|--------------------------|-------------|
|     | semantic theory of truth | Closure |
| A   | Modal algebra |  |
| AI  | (Modal) Epistemic algebra |  |
| AIO | Extension algebra | Čech Closure |
|     | Modal system T |  |
| MIU (O) | Calculus of systems (Tarski) |  |
| AIOU | Closure algebra, Modal S4 | Kuratowski axioms |
| AIOUK |  | $T_0$ - Kolmogorov space |
| ABIOUH |  | $T_2$ - Hausdorf space |
| ABIOUR |  | $T_3$ - Regular space |

It can be seen that the axioms[10] of generalized "topologies" as conceived by Čech for Boolean lattices and extended by Koutský to general lattices characterise in an interesting way various logic systems in terms of spaces of their consequence relations. The subsets in the powerset mapping $\mathbf{u}$: $\mathcal{P}(X) \longrightarrow \mathcal{P}(X)$ are however crisp. So the validity of the theorems of various many-valued logic systems is crisp too. Indeed, this corresponds to the general practice to take as the designated element of the logic matrix just the top of the valuation lattice. For example, in the case of axiomatization of Lukasiewicz logic by Wajsberg this choice represented the value 1 of [0,1]. To talk about the *fuzzy degree of validity* of a theorem or about the degree of truth to which $B$ is a consequence of $A$, one has to deal with *fuzzy* elements of the power sets of the consequence spaces, that is to fuzzify the Tarski's calculus of systems.

Because the consequence space of logic in Tarski's framework of calculus of systems is generated by the interaction of two different ingredients, namely of the sentences of the object language and the inference rules expressed in a meta-language, one has to ask the question as what does it mean to fuzzify either of these. This leads us to the matters to be examined in the sections that follow. Section 5 is concerned with fuzzification of the logical formulas of the *object language* which in turn reflects in the properties fuzzy elements of power sets that form the fuzzified consequences in a meta-language. This leads to closer examination of various power set systems, and with the study of the impact of different connectives on the characteristics of the "fuzziness" of power sets. Hence, understanding the nature of various characteristics of fuzziness of power sets is essential for the apprisal of the effect of fuzzification of the consequence space of Tarski's calculus of logics. This space, however, reflects not only the fuzziness induced by the elements of the object language, but also that generated by the elements of a metasystem, namely by the inference rules that act upon the formulas of the object language.

---

[10]For the definitions of **B**, **H**, **K** and **R** see e.g. [38] or [35]

# 5  On Fuzzy Object- and Meta-Systems of Logics

## 5.1  From Crisp to General Fuzzy Consequence Spaces

General form of the calculus of systems representing consequence spaces of logics in an abstract way was adumbrated in Section 4.2.2. Examples of concrete logical systems shown there demonstrate how Tarski's calculus of systems depicts a concrete logic theory. Link of an object system with its meta-systemic counterpart – the abstract consequence space – is provided by generalized closure operations generated by inference rules. In this setting, sentences of object language are represented by the points of the carrier $X$ of the closure (consequence) space $< X, \mathbf{u} >$. The points that conform with **MIOU** closure axioms are potential candidates for being included as *simple parts* of logical theories. Further restrictions on **MIOU** spaces determine a specific logic theory.

Tarski's calculus of systems in its original form provides the framework that has more potential for extension into fuzzy realm than its specialised forms which are extensively used in conventional treatements of model theory. Such conventional restrictions must be lifted in order to provide generalizations and extensions that have the capability to capture extended fuzzy systems containing possibilities and other modalities. Indeed, as we have seen above, even the most general form of Tarski's framework is not sufficient for generalised closure spaces of epistemic and modal algebras. One has to move outside it, to the Čech closures and such like, which drop the **U** axiom.

Many-valued extensions perform fuzzification adequately only if these include crisp logical structures as a special case. We shall look now at some basic features of the notion of *theory* as it is used in a restricted form by standard model theoretic approaches in order to see which properties the extensions should retain and which should be dropped.

General consequence space can capture a greater variety of logical features than appear in the conventional definitions of the notion of *theory*. One needs all this generality, and more, when dealing with richer systems in which the positive and negative inference are separated from each other and also a provision for accomodating the paraconsistency is required.

We shall provide a reference point to our previous discussions of philosophical logic by referring to standard practices of the contemporary mathematical logic before embarking on a more detailed discussion of fuzzy systems. The standard presentations (e.g. Chang and Kreisler [18] p. 123) define *theory* and its link with closure and axiomatics as follows.

(i) A set $\Gamma$ of sentences is called theory.
(ii) A theory is closed  *iff* every consequence of $\Gamma$ belongs to $\Gamma$.
(iii) $\Delta$ is a set of axioms  *iff* $\Gamma$, $\Delta$ have the same consequences.

From the point of view of philosophical logic, the above formulation is problematic, as it works directly with the notion of *sentences*. Comparing different systems from the semantic point of view of fuzziness may be unduly preempted by premature ontological commitment on the propositional level, identifying sentences with propositions or statements. Using the notion of statement (as Haskel Curry did) may be more appropriate in our context. In Curry's view, a theory is *"a certain non-void definite class of elementary statements. A statement is a class of sentences equivalent in some sense, just as expressions are classes of equiform expressions. This gives us free parameters, without comitting ourselves to particular philosophy [19]."* Of course, one does not yet have a definite criterion for specifying the equivalence at this point. However, free parameters allow us at this point to provide for a variety of different characterisations of fuzzy systems when fuzzifying theories. These characterisations can be achived by imposing a varying range of different conditions of formal, epistemological or ontological character on fuzzy statements. This is also important if one is translating from logical systems to their equivalents in category theory or combinatory logic/$\lambda$-calculus formulation.

We have seen that the consequence space is generated from the object language sentences by those elements of the meta-system that function as inference rules. The points belonging to a specific theory enter as premisses and conclusions in the inference rules. What conventional properties have to be dropped to allow for adequate fuzzification of consequence spaces? The standard definitions of a closure with respect to a set of rules are as follows ([1] pp. 741-2).

(i) A rule is a pair $(X, x)$ where $X$ is a set of premisses and $x$ is a conclusion, usually written as $X \vdash x$.
(ii) If $\Phi$ is a set of rules, then a set $A$ is $\Phi$-closed if $\Phi : X \vdash x \& X \subseteq A$ implies $x \in A$.
(iii) If $\Phi$ is a set of rules, then $I(\Phi)$, the set inductively defined by $\Phi$, is given by $I(\Phi) = \bigcap\{A | A \text{ is } \Phi \text{ closed}\}$.

In the above, $A$ belongs to the elements of the closure (consequence) space. It is clear, that (iii) leads to the preservation of Frege's principle of compositionality, in most cases, assuming that Tarski's theory of truth holds. To provide an appropriate fuzzification, the elements of the crisp power set of the Tarski closure space have to be replaced by the elements of an appropriate fuzzy power set. Furthermore, the rules of inference have to be appropriately fuzzified. In fuzzified rules the *degree of assertion* of the set of premisses $A$ has to be distinguished from the degree to which the rule, i.e, the pair $(X, x)$ *holds*. So at this level of generality, one has to seek a meta-formula determining the degree of equality of the fuzzily asserted premiss, say $X'$ with the fuzzy relation $(X, x)$ specifying the rule. This leads to the employment of an *afterset* formula based on the *square* relational composition $X' \Box (X, x)$, which itself does not need to be associative [14]. This is the point where theory of categories breaks down, and one has to work directly with square and triangle relational products [43], together with an appropriate theory of generalized morphisms [10]. Alternatively, an appropriate generalization, say L-categories as developed by Bandler [2] have to be employed[11].

## 5.2 The Need for Meta-Meta Level in Fuzzification Process

Distinction between meta-language and object language systems is important, but often it is not clearly stated. A pertinent comment comes from [3], [4]:

> Logic has long been beset with the often-muddied distinction between (i) inferences made in a *meta-language* from statements in an *object language* on the one hand, and on the other. (ii) the formation in the object language *itself of* an implicative combination of *its own statements*. Both the need for this distinction and the difficulty of keeping to it become more acute in the fuzzy environment.

In the context of fuzzy logics, an additional meta-meta level is required because the concepts of validity and satisfaction semantically belong to the meta-meta level at which one deals with their degrees. Hence a triad, the *object* language level (OL), the *metalanguage* level (ML) and the *meta-metalevel system* (MML) is needed.

Object Language contains formulas (sentences) of a chosen system of logic. Meta-Language consists of a syntactic part containing the inference rules operating on the elements of the object language, as well as the semantic translation of the object language expressions themselves into a meta-language. This two-level structure is insufficient for *expressing the degree* to which inference rules hold – therefore an additional meta-meta system is introduced. On this meta-meta level, the interplay of the syntactic part of the metalanguage with the semantic notions characterising an object language is captured in a suitable Meta-Meta-System in which meta-semantics is interpreted. *Intention* structures [38] have to be added to the meta-metalevel system when dealing with the epistemological concept of agency, as one should if involved with fuzzy logics.

---

[11]MacLane, in a personal communication to Bandler suggested that (generally non-associative) L-category is superfluous, as it is possible to find a structure that is equivalent to it up to isomorphism. This equivalent structure consists of a classical category together with functors over it. Although this observation of MacLane is mathematically correct, it misses the essential epistemological point: the reason for using L-categories is that one wants to capture not only the construction of a category, but also describe possible failures of such a construction. This can be hypostatised in L-categories by disproving the possibility of matching the objects with morphisms. The primitives of which specific L-categories are built when employed for modelling of systems possessing explicit epistemology, may be given by epistemological reasons a priori. In this case, the failure of hypostatization may be more interesting than its success. Of course, when one knows the structure of the functors then L-categories are not needed and classical categories would suffice. One may, hovever, need to approximate and identify adaptively the new functors emerging in a changing structure. In this case L-categories are essential

## 5.3   Comparative Evaluation of Alternative Many-Valued Systems of Power Sets

A variety of choices of *multi-valued logic implication operator* (Bandler and Kohout [13]) makes for a variety in semantics of the predicate __⊆ __ as this is determined amongst other parameters by the implication operator that appears in the semantic formula defining a subset. This makes it possible to obtain a whole *repertory of alternative power sets* as extensions of the classical crisp power set. All many-valued power set theories considered in the sequel will default into the crisp set theory for the boundary values of subset membership (that is, the values 0 and 1).

This plurality of fuzzy set systems and relations that is thus made available is important for at least two reasons:

1. This provides a viable choice of alternatives for dealing with uncertainty and imprecision, supplying a whole family of set systems, each of which has some *well-defined specific* and *desirable properties* that may be needed in a particular use.

2. The *unifying logical framework* in which they are defined offers a semantic link between various alternative systems, such as probability theory, fuzzy set theory of Zadeh [65], and other different many-valued logic based set theories (Klaua [30],[31], Zadeh [67]) and semantically defined power set systems (Bandler and Kohout [4],[3]).

Differences and similarities of semantic properties of power sets have to be evaluated from the point of view of contributing to the epistemology of fuzziness by some well defined evaluarive criteria. Opening a new field of comparative theoretical **and empirical** studies dealing with the evaluation of a whole spectrum of alternative fuzzy set theories and computational algorithms poses, of course, new questions. One should ask:

1. What are *the essential differences* between the newly available alternatives, in both their theoretical aspects, and in their applications?

2. How do we determine which system is *the best* in a particular theoretical use or an empirical application; and what are the optimality criteria for defining appropriately what is "the best"?

Systems of fuzzy power sets play the crucial role in fuzzifying the consequence spaces of logic theories. For this reason a comparative semantic evaluation of fuzzy properties of various power set extensions of fuzzy subsets is essential for good understanding the semantics of approximate reasoning. This evaluation cannot be haphazard. Some *normative concepts* that form the backbone of the epistemology which can yield adequate evaluation criteria, is needed for the goal in hand. We shall briefly discuss an early attempt which is not necessarily bound to a specific fuzzy set theory as it also provides useful semantic information needed for building a fuzzy theory of types.

In 1978, Bandler and Kohout [3], [4] examined the *semantic* effect of many-valued logic connectives on the semantic properties of fuzzy power set constructs. Starting from the definition of a fuzzy power set in a meta-meta system they employed several semantic characteristics for semantic evaluation of object language logical formulas in which various second order expressions involving power set constructs were employed. The notion of *crispness* and its dual *fuzziness* that Bandler and Kohout introduced in their 1978 paper [3] were important, new, evaluative notions. These notions played the crucial role in the comparative evaluation of the characteristics of fuzzy power sets of various kinds that Bandler and Kohout attempted.

The main goal of the inquiry in [3], [4] was to define some meaningful criteria by which to evaluate the fuzzification effect – to determine the effects of the choice among the (object language → ply) operators upon the *theory* of fuzzy sets. This aim was dictated by the "... *present (i.e. in 1978) need for a 'suitable' generalization of the internal implication operator in the object language ... Where A,B are fuzzy subsets of a crisp universe U, it has become almost traditional to assert $A \subseteq B$ iff $\forall x \in U, \mu_a x \leq \mu_b x$.*"

In order to deal with fuzzification of power sets not only on the formal but also on the epistemological level, Bandler and Kohout introduced

1. Semantic assumptions specifying the fuzzification process.

2. Criteria for evaluation of essential features of fuzzy sets that yield the parameters by which fuzzy characteristics of alternative power sets are compared.

3. Questioning strategy determining to which object language sentences the evaluative criteria will be applied to.

**Semantic assumptions** for fuzzification of a power set were as follows:

1. There are at least 2 levels, namely the level of:

    (a) fuzzy subsets, with individuals a,b,c ... (variables x, y, z, ... ) and fuzzy subsets denoted by A, B, C, ... ;

    (b) of power set 'objects' $\mathcal{P}(X)$.

2. The levels are bound by the following restrictions:

    (a)  $\pi(\mathbf{A} \subseteq \mathbf{B}) = \mu_{\mathcal{P}(B)} A$     $\mu_{\mathcal{P}(B)} A = \bigwedge_{x \in U} (\mu_{\mathbf{A}} x \to \mu_{\mathbf{B}} x)$

The degree to which A is a subset of B is given by (2).

The link of object- and meta-levels is depicted in Table 2. Ten different ply operators → that are listed in [3], [4] were used in this study.

Table 2:

| Level: | Propositional Part | | | | | Fuzzy Subset Part | | |
|---|---|---|---|---|---|---|---|---|
| Object | $\neg$ | $\vee$ | $\wedge$ | $\{\rightarrow\}_1^n$ | | $X^C$ | $\cap$ | $\emptyset$ |
| Meta | $1-a$ | max | min | $\{f_i(x,y)\}_1^n$ | $\in$ | | | |

**Characteristics parameters of fuzzy power set extensions.** Semantic evaluation of interesting OL sentences makes sense only if some well defined evaluative criteria (normative evaluative concepts) of essential features of fuzzy sets are applied to investigated power set extensions. Here, crispness, fuzziness, hight and plinth of a fuzzy set were used for this purpose. We use these evaluative characteristics on the meta-meta level (MML), thus parametrizing the semantic formulas[12] on the meta-level (ML).

1. The crispness of $a \in V$ is $\kappa a = a \vee (1-a)$.

2. The crispness of a fuzzy subset B is $\kappa B = \bigvee \kappa \mu_B x$ (by harsh criterion).

3. The dual concept of fuzziness: $\phi a = 1 - \kappa a, \ldots$

4. Crispness by Mean criterion:
$$\kappa_m B = \frac{\sum_U \kappa \mu_B x}{cardsupB}$$

5. The height of a fuzzy set $hB = \bigvee_U \mu_B x$.

6. The plinth $pB = \bigwedge_U \mu_B x$.

7. $pB = 1 - hB^c$, etc.        $\kappa B = p(B \cup B^c) = 1 - h(B \cup B^c)$

**Interesting sentences of OL**    Meaningful questions have to be formulated, determining what constructs of object language should be evaluated. The following queries (i.e. questions) were evaluated semantically with respect to the criteria given above:

| OL-formula | Question |
|---|---|
| '$\delta(a \rightarrow a)$' | strong, moderate tautology? |
| '$\delta(a \leftrightarrow \neg a)$' | strong, moderate contradiction |
| $\delta(A \subseteq B)$ | degree of the set inclusion |
| $\delta(A \equiv B)$ | degree of the set equality |
| $\delta(\_disj_1\_)$ | degree of disjointness of the 1st kind |
| $\delta(\_disj_2\_)$ | degree of disjointness of the 2nd kind |
| $\delta(A \subseteq B^c)$ | degree to which a set is contained in its complement |

---

[12] Using these characteristics of fuzzy sets at the meta-meta level as done by Bandler and Kohout is fundamentally different from a later use of one of these parameters (namely fuzziness) on the meta-level to characterise an arbitrary fuzzy subset, as done later independently by others, e.g. by Yager and also by Higashi and Klir.

Looking at the typical result of [3],[4], we can see that the resulting expression giving the degree of the OL–formulas often depends on *crispness* or *height* as one of the parameters. For example, the fuzzy value of the OL–formula $a \leftrightarrow \neg a$ is $2(1 - \kappa a)$ for the Lukasiewicz implication operator and $1 - \kappa a$ for the Kleene–Dienes operator.

A small sample of other typical Bandler and Kohout [3], [4] for selected implication operators is depicted in Tables 3, 4 and 5. L, KD, G43 denote Lukasiewicz, Kleene–Dienes and Goguen–Gaines operators, respectively. S* is given by the formula $a \rightarrow b = 1$ if $a \leq b$, b otherwise. The degree of disjointness of the 1st and the 2nd kind is computed from the formulas
$\delta(A disj_1 B^c) = \delta(A \subseteq B^c) \wedge \delta(B \subseteq A^c)$ and $\delta(A disj_2 B^c) = \delta((A \wedge B) \equiv \emptyset)$.

It can be seen again that the height $h(\_)$ and the crispness $\kappa(\_)$ of some subexpressions of a given OL–formula play an important role as parameters characterising the "commitment to fuzziness" of many–valued logic connectives that are used in the systems of base logics underlying various fuzzy set theories.

Table 3: (Degree of disjointness in the first sense)

Here $T = \{x | \mu_A x < 1 - \mu_B x\}$ , and the abbreviations $\bigwedge_U$ for $\bigwedge_{x \in U}$ and $\bigwedge_T$ for $\bigwedge_{x \in T}$ are used.

1. $S^\#$   $\pi_1(A \ disj_1 \ B) = \begin{cases} 1 & \text{always except} \\ 0 & \text{iff } \exists x, \ \mu_A x = 1 \text{ but } \mu_B x \neq 0 \text{ or vice versa.} \end{cases}$

2. $S$   $\pi_2(A \ disj_1 \ B) = \begin{cases} 1 & \text{iff } \forall x, \ \mu_A x \leq 1 - \mu_B x \\ 0 & \text{otherwise.} \end{cases}$

3. $S^*$   $\pi_3(A \ disj_1 \ B) = \begin{cases} 1 & \text{iff } \forall x, \ \mu_A x \leq 1 - \mu_B x \\ \bigwedge_T ((1 - \mu_A x) \wedge (1 - \mu_B x)) & \text{otherwise} \\ 0 & \text{iff } \exists x, \mu_A x = 1 \text{ but } \mu_B x \neq 0 \text{ or vice versa.} \end{cases}$

4. $G43$   $\pi_4(A \ disj_1 \ B) = \begin{cases} 1 & \text{iff } \forall x, \ \mu_A x \leq 1 - \mu_B x \\ \bigwedge_U \left(1 \wedge \frac{1 - \mu_B x}{\mu_A x} \wedge \frac{1 - \mu_A x}{\mu_B}\right) & \text{in all cases} \\ 0 & \text{iff } \exists x, \ \mu_A x = 1 \text{ but } \mu_B x \neq 0 \text{ or vice versa.} \end{cases}$

5. $L$   $\pi_5(A \ disj_1 \ B) = \begin{cases} 1 & \text{iff } \forall x, \ \mu_A x + \mu_B x \leq 1 \\ 1 \wedge (2 - \bigvee_U (\mu_A x + \mu_B x)) & \text{in all cases} \\ 0 & \text{iff } \exists x, \ \mu_A x = 1 \text{ and } \mu_B x = 1. \end{cases}$

6. $KD$   $\pi_6(A \ disj_1 \ B) = \begin{cases} 1 & \text{iff } \forall x, \ \mu_A x \wedge \mu_B x = 0 \\ 1 - \bigvee_U (\mu_A x \wedge \mu_B x) = \\ = 1 - h(A \cap B) & \text{in all cases} \\ 0 & \text{iff } \exists x, \ \mu_A x = 1 \text{ and } \mu_B x = 1. \end{cases}$

Table 4: (Degree of disjointness in the second sense)

The expression $\pi(A \ disj_2 \ B)$ takes the following values in the given systems:

1. $S^\#$ $\begin{cases} 1 & \text{always except} \\ 0 & \text{iff } \exists x, \ \mu_A x = 1 \text{ and } \mu_B x = 1. \end{cases}$

2. $S$ $\begin{cases} 1 & \text{iff } \forall x, \ \mu_A x = 0 \text{ or } \mu_B x = 0 \\ 0 & \text{otherwise.} \end{cases}$

3. $S^*$ $\begin{cases} 1 & \text{iff } \forall x, \ \mu_A x = 0 \text{ or } \mu_B x = 0 \\ 0 & \text{otherwise.} \end{cases}$

4. G43 $\begin{cases} 1 & \text{iff } \forall x, \ \mu_A x = 0 \text{ or } \mu_B x = 0 \\ 0 & \text{otherwise.} \end{cases}$

5. L $\begin{cases} 1 & \text{iff } \forall x, \ \mu_A x = 0 \text{ or } \mu_B x = 0 \\ 1 - h(A \cap B) & \text{in all cases} \\ 0 & \text{iff } \exists x, \ \mu_A x = 1 \text{ and } \mu_B x = 1. \end{cases}$

6. $KD$ $\begin{cases} 1 & \text{iff } \forall x, \ \mu_A x = 0 \text{ or } \mu_B x = 0 \\ 1 - h(A \cap B) & \text{in all cases} \\ 0 & \text{iff } \exists x, \ \mu_A x = 1 \text{ and } \mu_B x = 1. \end{cases}$

Table 5: (Degree to which a set is disjoint from its complement, in the two senses)

| | $\pi(A \ disj_1 \ A^c)$ | $\pi(A \ disj_2 \ A^c)$ |
|---|---|---|
| 1. $S^\#$ | 1 always. | 1 always. |
| 2. $S$ | 1 always. | $\begin{cases} 1 & \text{iff } \grave{A} \text{ is crisp} \\ 0 & \text{otherwise.} \end{cases}$ |
| 3. $S^*$ | 1 always. | $\begin{cases} 1 & \text{iff } A \text{ is crips} \\ 0 & \text{otherwise.} \end{cases}$ |
| 4. G43 | 1 always. | $\begin{cases} 1 & \text{iff } A \text{ is crips} \\ 0 & \text{otherwise.} \end{cases}$ |
| 5. L | 1 always. | $\kappa A.$ |
| 6. $KD$ | $\kappa A.$ | $\kappa A.$ |

We shall see later that it is important not to conflate two distinct ontological concepts appearing within a particular mathematical or logic system, when it happens that these are logically identical despite their **structural difference**. Indeed, the identity (that is collapse) of two distinct concepts may be a definitional feature of a particular system. This is intimately connected with the principle of bifurcation, that was demonstrated in the previous sections by examples drawn from the domain of the Pinkava functionally complete logic algebras. See later e.g. the equality [degree of possibilty = height of a fuzzy set; a numerical equality characterising a particular fuzzy power set]

This has provided a useful taxonomy of various power set systems. It can be seen that the evaluative principles used in this kind of system classification are not as arbitrary as some seem to suggest. One can say that there are three essential steps in this evaluation: (1) selection of evaluative concepts; (2) application of the selected evaluative concepts to individual systems; (3) comparison of results and apprisal of bifurcations and 'collapses' of the evaluative concepts that have been detected.

### 5.3.1 Construction of general fuzzy possibilistic families of power sets  ·

In their 1978 paper [3] Bandler and Kohout distinguished the meaning of OL formulas involved in power set construction from their denotation – the power set objects, giving the following three meaning postulates:

1. Connection between $\subseteq$ *and* $\rightarrow$:
   $A \subseteq B$ means $(b \in A \rightarrow b \in B)$.

2. Subset relation as a construct build of $\in$ and $\mathcal{P}(B)$:
   $A \subseteq B$ means $A \in \mathcal{P}(B)$.

3. Binding two meanings of $A \subseteq S$ yields the third one:
   $A \in \mathcal{P}(B)$ means $(b \in A \rightarrow b \in B)$.

These three meaning postulates were powerful enough to generate a family of power set objects parametrized by $\rightarrow$. The employment of these meaning postulates has, however, another facet that leads directly to the meta-concept of "possibilistic construct". If the above "meaning postulates" are used as the restricting conditions in Zadeh's calculus of fuzzy restriction [66], they yield a *generic possibilistic*[13] *specification* of a large family of power set systems [3], [4] (of which, of course, the crisp power set is a special instance). A brief discussion of this point may have some merit in providing better insight into the matter of how the "meaning postulates" are involved.

A family of fuzzy power set systems as defined in [3], [4] can be viewed as a linguistic proposition of the form p:= $< Q \subseteq S$ is $Q \in \mathcal{P}(S) >$. The meaning assigned to it is determined by the *possibilistic* postulate (1) and the *restriction* postulate (2) of the following form:

1. $\pi(Q \subseteq S) = \mu_{\mathcal{P}(S)}Q$

---

[13] As explicitely defined by Zadeh [67], and also used implicitly by Gaines and Kohout [23],[22] is a meta-notion which adds an ontological dimension to the meaning of a fuzzy power set. Thus in this usage the term possibilistic has a more universal meaning than the standard notion of possibility employed in the so called "theory of possibility" [32], [21].

2. $R_{\subseteq}(A(\mu_Q x, \mu_S x)) = \mu_{\mathcal{P}(S)} Q = \bigwedge_{x \in U}(\mu_Q x \to \mu_S x)$.

The possibilistic postulate (1) above is a particular instance of the linguistic sentence *sent* := $X$ *is* $F$, making a statement about the relationship of the whole component *sent* to the assumed parts of it, the individual names $X$ and $F$. Clearly, $X$ is the name of a construct (object) and $F$ is the label of a fuzzy subset of the universe of discourse $U$. The restriction postulate (2) above is a particular instance of Zadeh's relational assignment equation. In its general form, the assigned meaning is defined by the relational assignement equation $\phi(sent) : R(A(X)) = F$; where $A$ is an *implied attribute*[14] of $X$, i.e. an attribute which is implied by $X$ and $F$. $R$ denotes a fuzzy restriction on $A(X)$ *to which* the value of $F$ is assigned by $\phi(sent)$. In the possibilistic construction of power sets by Bandler and Kohout, $F$ is not just a fuzzy subset ( a first order construct), as in the definition by Zadeh [67] of the possibility distribution and the associated possibility measure. Instead, it is a functor over a 2-ary fuzzy relation $\bigwedge_{x \in U}(\mu_S x \to \mu_Q x)$, that is a *second order* logic construct. In its full generality, the relations in Zadeh's calculus of restriction must be given by their intension and their reducibility to the *satisfaction graphs* (cf. Bandler and Kohout [14]) must be proved. For further discussion of related issues see [39].

# 6 The Checklist Paradigm Semantic Model

## 6.1 Logic of Approximation and Loss of Information Variety

It was shown elsewhere [8], how intervals have come naturally when the rules of inference were extended to many-valued logics. In the present section we shall show, how an appropriate epistemological meaning can be given to both, the fuzzy membership function and the interval-valued inference, and some specific metamathematical results explicitly derived. This exposition abstracts the structure of the epistemological framework of the approach yielding the *checklist paradigm* more explicitly than previous papers, which, however, contained all the mathematical details and proofs. In the early 1980s accounts by Bandler and Kohout of the checklist paradigm, the conceptual side was also implicitly present, but only the mathematical parts were presented in a formal, technical way. Current trend on the emphasis *only* on the formalism to the detriment of clear statement of ontological meaning and epistemological content (often dismissed by mathematicians an unnecessary "philosophy") lead to obfuscation of the essential conceptual points of Bandler's and Kohout's work by the authors

---

[14] The implied attribute $A$ is a relation linking the two fuzzy sets $Q$ and $S$, restricting the full carrier of the fuzzy universe $U$ by imposing additional constraints, special relational properties that form a universal closure. These, of course, will be 'induced' also on the power set level. In the case described above, $A$ is the identity relation, but a relation with other properties can be employed instead.

of some review papers in which this work was discussed and the ontology of it was misrepresented. For this reason I adopt relational machinery to emulate within a relational structure some more subtle epistemological points that were previously expressed in clear but perhaps too concise way in plain English. This presentation, although less elegant to a philosophical logician has, however, the additional advantage for the international scientific community. The reader of other native tongue than English, who has not been extensively trained in the more subtle but perhaps too esoteric ways of philosophical English can appreciate the essence of the argument, as it is captured in the relational constructs.

Let us consider first a quaternary relation in its predicate form, relating the entities of four kinds: (1) Agent/observer $\Theta$, (2) observed object $x$, (3) an abstract construct $\mathcal{D}$ and (4) construct characteristics $p^{\mathcal{D}}$:

Object _____ can be identified with a construct/descriptor named _____ by force of the description characteristics _____ assessed by the observer _____.

Disregarding the question of observers in this paper, let us call an *abstract checklist* [6] the ternary subrelation of the above quaternary one, which does not contain the fourth "observer" slot. Then the valuation function or the 'abstract score' of the checklist is given by the $\lambda$–abstraction

$$\lambda x.[\mu(x \, \mathcal{E}\mathcal{Q} \, D) = \Omega_{\forall i}\nu(x R p_i^{\mathcal{D}})]$$

It is clear that $\mu(x \, \mathcal{E}\mathcal{Q} \, D)$ gives the degree of equivalence [36] of the object $x$ to the abstract construct $D$, relative to the chosen characteristic features $p^D$ of $D$.

Thus, this *abstract checklist* [6],[12] consists of an abstract construct $\mathcal{D}$ together with a finite or infinite list of propositional characteristics $p^D = (p_1^D, p_2^D, \ldots p_k^D, \ldots)$ of some kind, characterising the construct. As it is clear from the (semiotic) meaning of the predicate, this checklist is made in order to determine whether or not a selected object that is to be assessed, can be identified with the abstract construct. The fuzzy identity is brought about by the force of agreement of some of the features of the selected object with the features listed in the checklist for the abstract construct itself. The items on the list of characteristics that apply to the object are then asserted. This assessment is then *summarised* by 'measuring' the size of the portion of the list that contains the items that have been aserted, relative to the total size of the list of characteristics. Let us assume that this is done by some measure, say $\Omega$, the exact form of which is to be determined later and call this summarisation 'a score'.

Now suppose the same kind of checklist for construct $D$ is used twice, filled out for two objects say $a$ and $b$. The least one can do is to record the observer's two total scores for these objects, $r_1$ and $c_1$ respectively. In formal notation, we have: $r_1 = \Omega_{\forall i}\nu(a R p_i^D)$, $c_1 = \Omega_{\forall i}\nu(b R p_i^D)$. We also assume that the assessemnt for the negative items are available, namely $r_0 = \Omega_{\forall i}\nu((\neg a) R p_i^D)$ and $c_0 = \Omega_{\forall i}\nu((\neg b) R p_i^D)$. Let us call this the *coarse structure*.

At the other extreme of detail, one could compare, item by item, the terms on the checklist to which the observer assented for the two objects $a$ and $b$. The fine characteristics of the assessment of both object $a$ and $b$ can be compared, computing the degree to which $\nu(aRp_i^D)$ and $\nu(bRp_i^D)$ are both asserted; $\nu((\neg a)Rp_i^D)$ and also the degree to which $\nu(bRp_i^D)$ are both asserted; similarly for $a$ and $\neg b$, and also for $\neg a$ and $\neg b$ and only after this is done, to compute the scores by $\Omega$ as follows:

$$\alpha_{00} = \Omega_{\forall i}\nu(a\,AND\,b)Rp_i^D)\ \alpha_{01} = \Omega_{\forall i}\nu(\neg a\,AND\,b)Rp_i^D$$
$$\alpha_{10} = \Omega_{\forall i}\nu(a\,AND\,\neg b)Rp_i^D\ \alpha_{11} = \Omega_{\forall i}\nu(a\,AND\,b)Rp_i^D$$

If $\Omega$ is an additive measure, the solution of the constraint inequalities can be arranged with advantage into a constraint table [6] (see Fig. 1), with $A$ standing for the first assessment and $B$ for the second. The resultant table summarizes the item-by-item information in a way which tells us much more than merely the two scores $r_1$ and $r_2$ which can be read from the margins. The information given inside adds significantly to that given by the margins alone. There are four pieces of information inside the table, namely $\alpha_{00}, \alpha_{01}, \alpha_{10}, \alpha_{11}$, and three in the margins, $r_1, c_1$ and the additive grand total $\Omega(r,c)$. The question **To what extent can the extra information inside the table be reconstructed from a knowledge of the margins?** is the central question of the paradigm [6], [8].

## 6.2   The Checklist Paradigm Inequalities

Now, consider a single checklist used to give a degree of assent to two different objects $a$ and $b$, where these abstract objects $a$, $b$ are both some propositions (cf. Bandler and Kohout [6]). For the general case, formally, let $F$ be any logical two-valued propositional function connecting the propositional forms of relations $[\nu(aRp_i^D)]$ **F** $[\nu(bRp_i^D)]$ component wise. This composed fine structure will induce a many-valued counterpart, a connective **CON**. This connective couples the two propositions $a$ and $b$, the exact value of which can be computed from the fine structure:

$$\mu[(a\ \mathcal{E}Q\ D)CON(b\ \mathcal{E}Q\ D)] = \Omega_{\forall i}\nu[(aRp_i^D)\mathbf{F}(bRp_i^D)]$$

where $\mu(a\ \mathcal{E}Q\ D|p^D)$ is the score of $a\ \mathcal{E}Q\ (D)$ with respect to the family of characteristics $p^D$.

Given three distinct $a_1$, $a_2$ and $b$, for the values of the corresponding fine structures may or may not equal:
$\mu[(a_1\ \mathcal{E}Q\ D)CON(b\ \mathcal{E}Q\ D)] \neq \mu[(a_2\ \mathcal{E}Q\ D)CON(b\ \mathcal{E}Q\ D)]$, while simultaneously $\Omega_{\forall i}\nu(a_1Rp_i^D) = \Omega_{\forall i}\nu(a_2Rp_i^D)$ may hold. Hence any specific single many-valued CON cannot distinguish two fine structures, as these both collapse into one coarse structure. When the fine structure is substituted by the

coarse structure of two interacting propositions, this represents *loss of the information variety* in Ashby's sense:

Fine structure $\longrightarrow$ *Loss of information variety* $\longrightarrow$ Coarse structure

The previous question about the relationship of the fine and coarse structures can now be reformulated.

Problem 1: Approximation of the fine structure of the checklist interaction by a "best" many-valued connective acting over the coarse structure.

*Let us assume that interaction of two propositions a and b is determined by the logical type of a two-valued connective F operating over the characteristics of a given abstract checklist, thus yielding the fine structure of this interaction. Let us further assume that this fine structure is not available, and that the the value of the interaction of a and b has to be* **approximated** *by some, (yet unknown) many-valued connective instead. Then one may ask:*

> *(a) What many-valued connective using only the information of the coarse structure will yield the value that is the closest to the exact value of the score contained in the now inaccessible fine structure?*

> *(b) If (a) is not possible, can bounds be imposed on the values of the score computed from the fine structure of interaction? Can these be computed from the coarse structure by a pair of à priori predetermined many-valued connectives?*

Epistemologically, (a) is asking for the best MVL estimator of the fine structure value, and (b) for the logic of approximation of the fine structure that is interval-based. We can now turn to surveying the solutions of this problem, to be worked out under some more specific meta-logic conditions, yielding thus concrete mathematical results.

Bandler and Kohout have shown elsewhere [6],[8],[9],[12]) what are the bounds the *coarse* structure imposes upon the fine structure. These bounds depend on the specific forms of additional constraints imposed on the fine and coarse structure and the link between these. The constraints formally specify the laws under which the loss of information variety occurs, while restricting the information variety by replacing the fine structure by the coarse. By putting the bounds on the variation of the exact values $EXVAL(a,b) = \mu[(a \; \mathcal{EQ} \; D)\mathbf{CON}(b \; \mathcal{EQ} \; D)]$ we obtain the inequality:

$$\mu(a\mathrm{CONTOP}b) \geq \mathrm{EXVAL} \; (a.b) \geq \mu(a\mathrm{CONBOT}b)$$

Thus the exact value EXVAL(a, b) is replaced by an appropriate summarization function $G$ that yields the TOP and BOT bounds. This aggregation function depends on two components, namely the selection function $F$ and aggregation measure $m_i$: $G(F(k,l, TCON), m_i)$. $F$ selects the appropriate arguments for $G$ and $G$ yields the bounds CONTOP and CONBOT on $EXVAL$ $(a.b)$ when combined with the aggregation measure $m_i$.

When a list with a finite number of items is employed for the checklist characteristic of the abstract construct, score of the object can be obtained by dividing the "yes" answers by the total number of the descriptor terms on the checklist, thus taking for $\Omega_{\forall i}(\ldots) := 1/n \sum_{\forall i}(\ldots)$. As this computation corresponds to the sigma count used in fuzzy logic, it is natural to interpret the proportion thus scored as a *fuzzy degree* of the agreement of the features of the object with the abstract construct $D$ specified by the checklist. In this case, the aggregation measure yields the bounds on all possible values of the constraint table (cf. Figure 1 of [12]), spliting the table into two pairs the so called *Maxdiag* and *Mindiag* pair, each pair giving the extreme values (the bounds). This makes it possible to distinguish two types of interval based connectives that interact, Max- and Mindiag, thus providing another taxonomy for the connectives (see [8], [9], [12]).

Let us look now at some typical results [6]. Selecting implication operator as the logical type of the checklist connective and choosing as the aggregation function $m_1(F) = 1 - (\alpha_{10}/n)$ we obtain $min(1, 1 - a + b) \geq m_1(PLY) \geq max(1 - a, b)$. The *plytop* is the Lukasiewicz implication operator, while the *plybot* is the Kleene-Dienes operator.

Choosing the connective type **AND** one obtains [8]: $min(a, b) \geq m_1(AND) \geq max(0, a + b - 1)$. The **OR** yields: $min(a + b, 1) \geq m_1(OR) \geq max(a, b)$. The last two inequalities are formally identical with those of Schweizer and Sklar [59] giving the bounds on *copulas* which play an important role in their theory of *norms* and *conorms*. Surprisingly, these checklist paradigm bounds also coincide with Novák's recent derivation of bounds on fuzzy sets approximating classes of Vopěnka's Alternative Set Theory [52]. Hisdal derives the same inequalities as the bounds on some connectives of her TEE model and comments on a possible link (c.f. Appendix A2 "The TEE model and Bandler and Kohout's checklist paradigm in [27]). Yet all these models are neither formally nor epistemologically identical. This indicates the need for a more precise meta- and metametalogical formulation of many-valued based mathematical systems, that would include in their full definition a part formulating their "mathematical epistemology".

Taking a specific constraint measure, Bandler and Kohout obtained specific results also for other connectives. For example, $m_1$ gave pairs *conbot* $\leq$ *contop* for all 16 logical types of connective, some of which are listed below. For the exhaustive listing of all 16 connectives see e.g. [8],[12].

Logical Type:          Valuation:

$\neg(a \rightarrow b)$             $max(b - a) \leq min(1 - a, b)$

$\neg(a \leftarrow b)$             $max(a - b) \leq min(a, 1 - b)$

$a \equiv b$               $max(a + b - 1, 1 - (a + b)) \leq min(1 - a + b, 1 - b + a)$

$a \text{ EOR } b$           $max(a - b, b - a) \leq min(a + b, 2 - (a + b))$

$\neg(a \text{ OR } b)$         $max(0, 1 - a - b) \leq min(1 - a, 1 - b)$

$\neg(a \text{ AND} b)$         $max(1 - a, 1 - b) \leq min(1, 2 - a - b)$

Again, like with the powers sets (c.f. Sec. 5) the concept of fuzziness is an important evaluation criterion of the results obtained. In this case it specifies the *bands of imprecision* of an interval logic expression. If we define the *unnormalized fuzziness of x* (cf. Bandler and Kohout 1978 [3]) as $\phi x = min(x, 1 - x)$ then for $x$ in the range $[0, 1]$, $\phi x$ is in the range $[0, .5]$, with value 0 iff and only iff x is *crisp*, and value .5 iff $x$ is .5. We have:

Gap Theorem – Bandler and Kohout (1986), [12]:

$a$ ANDTOP $b$ - $a$ ANDBOT $b =$
$a$ ORTOP $b - aORBOTb =$
$a$ PLYTOP $b - a$ PLYBOT $b = min(\phi a, \phi b)$.

$a$ IFFTOP $b - a$ IFFBOT $b =$
$a$ EORTOP $b - a$ EORBOT $b = 2min(\phi a, \phi b)$.

So the interval between the TOP connective and the bottom connective is directly linked to the concept of fuzziness $\phi$. Hence the *margins of imprecision* can be directly measured by the degree of $\phi$.

## 6.3   Other Checklist Paradigm Models

Let us define $u_{lk} = \alpha_{lk}/n$.

Measures other than $m_1$ yield other interesting results as demonstrated by Bandler and Kohout in their 1980 paper [6]. If for **F** an implication operator type of connective is chosen again, but this time evaluation "by performance" using $m_2 = u_{11}/(u_{10} + u_{11})$ is performed, this yields the inequality

$$min(1, b/a) \geq m_2(F) \geq max(0, (a + b - 1)/a)$$

where *plytop* is in this instance the well-known G43 implication of Goguen-Gaines (cf. eg. [6]).

Still another contracting measure $m_3 = u_{11} \bigvee (u_{00} + u_{01})$ yields [6]

$$max[min(a, b), 1 - a] \geq m_3(F) \geq max(a + b - 1, 1 - a).$$

The lower contrapositivization of $m_3$ yields the measure $m_4 = (u_{11} \bigvee (u_{00} + u_{01}) \bigvee (u_{00} \bigvee (u_{01} + u_{11})$ gives the following bounds [6]:

$$min[max(a + b - 1, 1 - a), max(b, 1 - a - b)] \leq m_4 \leq min[max(1 - a, b), \kappa a, \kappa b]$$

The measure $m_5 = m_2 \bigvee u_{00} + u_{11}$ yields:

$$max[min(1, b/a), 1 - a] \geq m_5 \geq max[(a + b - 1)/a, 1 - a]$$

For the proofs of the results presented in this section and further explanation see ([6], sections 5 and 6).

## 6.4   Collaps of Intervals into Points when Additional Probabilistic Assumptions Are Introduced

So far, **no** probabilistic assumption were used. Bandler and Kohout, however, also asked *probabilistic questions* about the ways the fine structure can be characterised [6], Sec. 7 . When only the row and column totals $r_i, c_j$ of the fine structure are known (see Fig. 1 of [6]), one can ask what are the *expected values* for the $\alpha_{ij}$. This involves the notion of *probability*, in addition to the other assumptions employed. We may ask then, how does this additional statistical assumption manifest itself within the fine structure, and also what are its probabilistic characteristics. This leads the way to introducing the following additional assumption (cf. [6], Sec. 7) : *Let the ways in which numbers can be distributed within the cells of the fine structure (so as to give the fixed coarse totals) constitute a hypergeometric distribution.* Then the means of the distribution, for each cell, give the *expected configuration* of the fine structure. Surprisingly, involving the expected value causes the interval to collapse into a single point – the *expected value* giving the value of the **mid** connective.

The inequalities determining the interval of approximation which is bounded by the pair $BOTCON \leq TOPCON$ of the same logical type now turn into equalities:

$$(\alpha_{ij}/\alpha_{ik}) = (c_j/c_k); \qquad (\alpha_{ij}/\alpha_{hk}) = (r_i/r_h)$$

This is one of the specific solutions of of Problem 1 (a) of Section 6.2.

What are the expected values for various measures $m_i$ and various logical types of connectives? For measure $m_1$ we have the following.

| Logical Type: | Expected Value: | | |
|---|---|---|---|
| a OR b | $ab$ | a AND b | $a + b - ab$ |
| $a \rightarrow b$ | $1 - a + ab$ | $a \leftarrow b$ | $1 - b + ab$ |
| $\neg(a \rightarrow b)$ | $a(1 - b)$ | $\neg(a \leftarrow b)$ | $(1 - a)b$ |
| $a \equiv b$ | $(1 - a)(1 - b) + ab$ | a EOR b | $(1 - a)b + a(1 - b)$ |
| $\neg(a$ OR $b)$ | $(1 - a)(1 - b)$ | $\neg(a$ AND $b)$ | $1 - ab$ |

For the exhaustive listing of all 16 **mid** (expected value) connectives see e.g. [8],[12].

For the implication operators defined by other measures, the expected values are given by the theorem below.

Theorem 9: (Bandler and Kohout, 1980 – [6] Th. 7.1)

The values of various measures for the expected configuration of constraint tables, supplying the valuation of midply implication operators are:

(1) $midply(m_1; a, b) = 1 - a + ab$,

(2) $midply(m_2; a, b) = b$,

(3) $midply(m_3; a, b) = ab \bigvee (1 - a)$,

(4) $midply(m_4; a, b) = (ab \bigvee (1 - a)) \bigwedge ((1 - a)(1 - b) \bigvee b)$

(5) $midply(m_5; a, b) = max(1 - a, b)$

The first result gives us an implication operator KDL that is worthy of attention[15]. Formula (5) is of high interest on two counts: it shows the Kleene-Dienes (KD) operator, not as a lower bound of $m_1$ where no probabilistic assumptions were employed (cf. Sec. 6.2 above), but as the value in the *expected case* of a *different* measure $m_5$ (cf. Theorem 6.3, part (5) of [6]. The Lukasiewicz operator is involved in a similar way in the following theorem.

Theorem 10: *Values of contrapositivized measures*

Let the expected values be given by the following formulas: $expected(m_2) = em_2 = a.b/a$; $expected(\neg m_2) = \neg em_2 = (1 - a).(1 - b)/(1 - b)$ to which upper contrapozitivation and lower contrapozitivization is applied.

(a) The upper contrapozitivization of $em_2$ yields:

$em_5 = ORBOT(em_2, \neg em_2) = max(em_2, \neg em_2) = a \rightarrow_6 b$;

$em_6 = ORTOP(em_2, \neg em_2) = min(1, em_2 + \neg em_2) = a \rightarrow_5 b$.

(b) The lower contrapozitivization yields:

$em_8 = ANDTOP(em_2, \neg em_2) = min(em_2, \neg em_2) = min(b, 1 - a)$;

$em_9 = ANDBOT(em_2, \neg em_2) = max(0, em_2 + \neg em_2 - 1 = max(0, b - a)$.

The second way of obtaining Klene-Dienes ply is via the upper contrapozitivization of EZ (involving $m_3$), using $aORBOTb = max(a, b)$ as the contrapozitivizing connective (see [6] Sec.5).

The third way was that given in Section 6.2 – derived form $m_1$, but *without* any explicit probabilistic assumptions added.

These examples show clearly that the *roles* which may be played by one specific connective in the three distinct logical systems are indeed formally and epistemologically *very different*. For this reason it is absolutely essential to judge the properties of the whole system of connectives *jointly together* with the theory within which these connectives are employed. This has also significance

---

[15] When we wrote the paper [6] we named it Kleene-Dienes-Łukasiewicz (KDL) as it ensues by the "probabilistic collapse" of the interval [Klene-Dienes ply, Łukasiewicz ply] into the KDL ply, i.e. point. Later, we have learned that this implication operator was introduced earlier by Hans Reichenbach via different epistemological considerations concerned with the logics of quantum physics

for the theory and practice of approximate reasoning as such demarcation helps in the apprisal of similarities and differences of various logical systems when used for a particular purpose.

As shown above, the bounds of interval-valued inference system produced by $m_1$ are given by the pair of implication operators $(\to_6, \to_5)$ (cf. Section 6.2 above and [6]). With additional statistical assumptions imposed, this interval induced by $m_1$ "collapses' into a single point, determined by the implication operator $\to_{5.5}$ determining the expected values. The question arises whether using this connective combined with other $m_1$ generated checklist paradigm connectives would satisfy the axioms of conditional probabilistic logic. The answer to this question is, however, negative [42]. This follows from the results of investigation of validity of individual axioms for conditional probability. Starting from Popper's axioms for conditional probability, it was demonstrated that for the $\to_{5.5}$ implication operator combined with some other checklist paradigm connectives, not all axioms for conditional probability hold, although the system still remains "probability like". In order to employ the checklist paradigm connectives in Popper's system of axioms, however, an appropriate extension of these has to be made. This extension has to distinguish between the object and meta role of the valuation formulas for probabilities. Let us list Popper's axiom first and then explain the essential distinctions that we have introduced in greater detail.

Definition *Popper's axioms for conditional probability*

A1. Existence:           $(\exists C)(\exists D)\mu(A/B) \neq \mu(C/D)$.
A2. Substitutivity:      If $\forall C, \mu(A/C) = \mu(B/C)$ then $\forall D, \mu(D/A) = \mu(D/B)$.
A3. Reflexivity:         $\mu(A/A) = \mu(B/B)$.
B1. Monotonicity:        $\mu(A\&B/C) \leq \mu(A/C)$.
B2. Multiplicativity:    $\mu(A\&B/C) = \mu(A/B\&C) \wedge \mu(B/C)$.
C. Complementation:      If $\mu(A/A) \neq \mu(B/A)$ then $\mu(A/A) = \mu(C/A) \vee \mu(\overline{C}/A)$.

The above axioms were refined in the following way. Distinguishing between the object and meta level and also carefully noticing at which of these levels in the above axioms each valuation took effect, led to the conceptual separation of the different formulas that had the identical valuation (cf. Kohout and Bandler [42] originally. This justified imposing of *two different valuations* on that formula in Popper's system of axioms which appeared to play two different roles.

On the object level the valuation of $\mu(A\&B) = \mu(a).\mu(b)$ remains the same as that of Popper; on the meta-level, a new symbol for meta-connective *wedge* is distinguished: $\mu(A) \wedge \mu(B) = \mu(a).\mu(B)$;

It can be seen that not all axioms of conditional probabilities are satisfied when selecting a system involving $\{\&, \wedge, \vee\}$, where $a\&b = a.b, a\wedge b = max(0, a + b - 1), a \vee b = a + b$, which postulates a different valuation for $\&$ and $\wedge$. The *complementation* axiom does not hold for this extended system. It can, however be replaced by a pair of weaker axioms, namely [42]:

$$C1: \mu(\overline{A}/1) = \mu(C/A) \wedge \mu(\overline{C}/A), \quad C2: \mu(A/A) \geq \mu(C/A) \wedge \mu(\overline{C}/A)$$

It is interesting to note the logical role of C2. It generates the valuation of the singleton of the power set, as in $A/A$ the sign $/$ denotes the ply operator. This is another justification of our view that conflating the connectives of the level of fuzzy subsets (called level 1 in Section 5) with level 2 of power set connectives for mathematical convenience (e.g. to get residuation) may be epistemologically justified only in special cases.

# 7 The Checklist Paradigm Semantics – Conclusion

## 7.1 Link of the Checklist Paradigm to Fuzzy Sets

The checklist paradigm is compatible with a number of ontologies, here we look at the set theoretical one. Given some objects, potential elements of a class, the characteristic function $h_A a \in \{0,1\}$ classifies whether or not an element $a$ belongs to a named class $A$. So more formally, what is involved in set theory is a repertory of object language components: elements $a$, $b$, $c$, $d$, ... and classes $A$, $B$, $C$, $D$, ..., a predicate $\_ \in \_$ symbolizing the statement $\_belongs$ $to$ $a$ $named$ $class\_$; the metalanguage construct, the valuation (characteristic, choice) $h_A a$ from $h(\_ \in \_)$ to $\{0,1\}$. In 1965, Zadeh [65] extended $h(\_ \in \_)$ to $[0,1]$.

It is clear that the checklist can accomodate the conept fuzzy 'subset' (i.e. class) as a special case of wider checklist paradigm fuzzy semantics, as the following correspondence table shows.

| General Checklist | Fuzzy Class ('subset') |
| --- | --- |
| Object x | Element of fuzzy subset x |
| Construct $\mathcal{C}$ | Class A |
| $\lambda x.[\mu(x \, \mathcal{EQ} \, D)]$ | $\lambda x.\mu_A x := \lambda x.[\pi(x \in A)]$ |

where the fuzzy membership function is given by the formula $\lambda x.[\mu(x \, \mathcal{EQ} \, D) = \Omega_{\forall i} \nu(x R p_i^D)]$

It can be seen from the above table that the 'subset' $A$ which enters the predicate $\pi(x \in A)$ is a special instance of the abstract checklist construct $D$. In order to complete the picture, one has to supply the semantic meaning of the $\_ \in A$ predicate. This comes from the checklist general expression $\lambda x.[\mu(x \, \mathcal{EQ} \, D) := \Omega_{\forall i} \nu(x R p_i^D)]$ determining the fuzzy membership function.

The list of class properties is something hidden in the inner structure of the propositional form of the predicate/relation $\_\_ \in A$. The result of the action of filling the blank checklist, that is used to assess to what degree $(x \, \mathcal{E}Q \, D)$ holds is made explicit by the *second order* relational composition, ( the "mean square product" [6]) employed in the following formula:

$$\mu(x \, \mathcal{E}Q \, D) := \Omega_{\forall i} \nu(x R p_i^D) = x(R_\nu' \Box R_\nu^T A =$$

$$= \bigwedge_{\forall i} (x R_\nu' p_i^A \equiv p_i^A R_\nu^T A) = \| \sum_{\forall i} x R_\nu' p_i^A \equiv p_i^A R_\nu^T A \|$$

where $x R_\nu' p_i^A := \nu[x R' p_i^A]$, etc. the symbol $\| \ldots \|$ represents suitable normalization, for the finite number $n$ of properties it is $1/n \sum \ldots$.

## 7.2   General Epistemological Remarks

The checklist paradigm is embodiment of the the following general meta-principle: *a system of logic connectives is formed by a specific family of connectives together with some common process/structure/principles that involve the said family of connectives in some unifying way, causing these to interact.* To make this workable, one has to state more precisely what these common unifying principles are[16]. In the checklist paradigm semantic model we use 2 basic unifying principles: (i) approximation (contraction) measures, (ii) transformations of logical types of connectives leading to a global characterisation of logics by their groups of transformations [44].

# 8   Semantics of Meta-Rules for Generating Contrapositive Ply Operators

Checklist paradigm is applicable not only to the components of the object language such as logical operators and connectives, but also at the meta-level, thus providing an interval logic based semantics for various rules of inference. We get the following theorems [12].
Four Checklist Theorems for Inference Rules.

For each rule of inference, given the *premisses* that satisfy the stated *consistency condition*, the values of *conclusion* are *subject to constraints* as listed below:

---

[16]Because a mathematician uses a special language where the name and the entity it denotes obey the law that a computer scientist would call referential transparency (cf. functional languages such as ML), it usually does not do much harm to mix 'use' and 'mention' in an informal argument, however deplorable this may be to the philosophers - logicians of Quine's persuasion. This is however one occasion where one has to be careful. The conceptual definition is required, of an open concept that might close when subject to further precisation.

| RULE: | Given | Consistency | Enforced values |
|---|---|---|---|
| NAME: | premisses | condition | of conclusion |
| **Modus** | $r = m(A \rightarrow B)$ | $r \geq 1 - a$ | $r - (1 - a) \leq b \leq r$ |
| **Ponens** | $a = m(A)$ | | |
| | | | |
| **Modus** | $r = m(A \rightarrow B)$ | $b \leq r$ | $1 - r \leq a \leq 1 - (r - b)$ |
| **Confirmans** | $b = m(B)$ | | |
| | | | |
| **Modus** | $r = m(A \rightarrow B)$ | $r \geq b$ | $r - b \leq \neg a \leq r$ |
| **Tollens** | $\neg b = m(\text{not-B})$ | | |
| | | | |
| **Denial** | $r = m(A \rightarrow B)$ | $1 - a \leq r$ | $1 - r \leq \neg b \leq 2 - (r + a)$ |
| | $\neg a = m((not - A))$ | | |

For the proofs and further results see [45] and [12].

Assertion and rejection of a proposition belong both to the domain of meta-operations. In those kinds of logic, for which the mutual duality of assertion and rejection are their meta-properties the object language defined ply operators will be *contrapositive*. The duality of inference rules (such as modus ponens and tollens) is assured by the contrapozitive property of the ply operator [3],[6],[4] that enters the rules. A ply operator is contrapositive, if its valuation satisfies the semantic equality $a \rightarrow b = \neg b \rightarrow \neg a$.

# 9 The Global Transformations of Some Checklist Paradigm Generated Systems of Many-Valued Logics

Properties of abstract transformations of Pi-normal forms are reflected in transformations between connectives defined of specific MVL systems. How these can be captured by groups of transformations will be briefly outlined.

We have shown [8], [12] that for a given contraction/approximation measure, there are 16 inequalities linking the TOP and BOT types of connectives. Important transformations linking TOP and BOT logical types of connectives can be found and their global structure further investigated.

It is known that the global structure of systems of various logic connectives can be investigated by looking at abstract group properties of the group of their transformations. Let us recall that a realization of an abstract group is any group of concretely performable operations which has the same algebraic structure as the given abstract group. It is well known that any abstract group can be concretely realized by a family of permutations. The global structure imposed on many-valued connectives by certain type of checklist paradigm contracting measures is captured by the $S_{2 \times 2 \times 2}$ group [44] characterising algebraically their transformations.

## 9.1   The Piaget Group of Transformations

### 9.1.1   Klein Abstract Group and Its Realization by the Piaget Group of Logic Transformations

Let 4 transformations on basic propositional functions $f(x, y)$ of 2 arguments be given as follows:
$$I(f) = f(x, y), \quad D(f) = \neg f(\neg x, \neg y), \quad C(f) = f(\neg x, \neg y), \quad N(f) = \neg f(x, y).$$
In the set of the above transformations $T_p = \{I, D, C, N\}$ the individual transformations are called *identity, dual, contradual, negation* transformation, respectively. It is well known that for the crisp (2-valued) logic these transformations determine the Piaget group [55]. This group of transformations is a realization of abstract Klein 4-element group. It is also known that the Piaget group of transformation is satisfied by some many-valued logics (cf. Turksen [64],[63], Dubois and Prade [20], Kandel [29]). All the logics described in these just quoted references, link only separately within each group of transformations, the connectives of (what we call) the *conbot* or *contop* family. The structures described by these authors do not provide a crosslink, a **mutual coupling** of the *conbot* and *contop* connectives. The checklist paradigm, however, generates also these richer cases. Let us look at some useful examples of such crosslinks now.

### 9.1.2   A Mutual Link of TOP and BOT Connectives by the Piaget Group of Logic Transformations Generated by the Checklist Paradigm

Given a logical connective (or a set of connectives) **CON**, the set *closed* with respect to some family of logic transformations $T$ will be denoted $C(T; \textbf{CON})$. This will also be written $C_T(\textbf{CON})$ where appropriate.

Example:    $C(T_P; \{\equiv_{TOP}, \bigoplus_{BOT}\}) = \{\equiv_{TOP}, \bigoplus_{BOT}, \equiv_{BOT}, \bigoplus_{TOP}\}$.
Here (cf. Bandler and Kohout [8], [12]) $a \equiv_{TOP} b = min(1 - a + b, 1 - b + a)$, $a \bigoplus_{BOT} b = max(a - b, b - a)$. $a \equiv_{BOT} b = min(1 - a - b, a + b - 1)$, $a \bigoplus_{TOP} b = max(a + b, 2 - a - b)$
(for $T_p$ as defined in [44]).

Theorem 11:

The system of connectives $\{\equiv_{TOP}, \bigoplus_{BOT}, \equiv_{BOT}, \bigoplus_{TOP}\}$ obeys the Piaget group of transformations. Hence it possesses the abstract structure of the Klein 4-element group.

## 9.2   An 8-Element Group of Logic Transformations

Adding new non-symmetrical transformations to those defined by Piaget enriches the algebraic structure of logical transformations. [41],[44].

Adding $LC(f) = f(\neg x, y), RC(f) = f(x, \neg y), LD(f) = \neg f(\neg x, y), RD(f) = \neg f(x, \neg y)$ to the above defined four symmetrical transformations we obtain a new 8-element group of transformations $T = \{I, D, C, N, LC, RD, LC, RD\}$. The corresponding abstract group can be characterised as follows.

**Theorem 12:**   The 8-element group $\{T, *\}$ that captures the structure of of the above defined logic transformations is commutative and its elements satisfy the following equations: $LC=LD^*N=RC^*C=RD^*N$,
$RC=LD^*D=LC^*C=RD^*N$, $LC^*RC=C$, $LC^*LD=N$, $LC^*RD=D$,
$LD=LC^*N=LD^*C=RC^*D$, $RD=RC^*N=LC^*D=LD^*C$;
$N^2 = C^2 = D^2 = LC^2 = LD^2 = RC^2 = RD^2 = I.$
The above 8-element group is called $S_{2\times2\times2}$ group in the standard terminology of group theory.

## 9.3   On Some Checklist Paradigm Generated Realizations of the $S_{2\times2\times2}$ group

The $S_{2\times2\times2}$ group also brings some order to the connectives generated by the contraction measure $m_2$ (cf. Section 6.3 above) as Theorem 13 below clearly indicates. For the proof of this theorem see Kohout and Bandler [44].

**Theorem 13:**   The closed set of connectives generated from the $a \rightarrow_4 b = min(1, b/a)$ by the transformation $T$ is listed below. This set of connectives together with $T$ is a realization of the abstract group $S_{2\times2\times2}$.

$g_1 = I(\rightarrow_4) = min(1, b/a)$
$g_2 = C(\rightarrow_4) = min(1, 1 - b/1 - a)$
$g_3 = D(\rightarrow_4) = max(0, b - a/1 - a)$
$g_4 = N(\rightarrow_4) = max(0, a - b/a)$
$g_5 = LC(\rightarrow_4) = min(1, b/1 - a)$
$g_6 = LD(\rightarrow_4) = max(0, 1 - a - b/1 - a)$
$g_7 = RC(\rightarrow_4) = min(1, 1 - b/a)$
$g_8 = RD(\rightarrow_4) = max(0, a + b - 1/a).$

For $m_1$ generated connectives we have the following two theorems.

**Theorem 14:**   The closed set $C\{T; \rightarrow_6\}$ generated by the connective $a \rightarrow_6 b = max(1 - a, b)$ represents a concrete realization of the abstract group $S_{2\times2\times2}$ under the family of transformations $T$.

**Theorem 15:**   The closed set $C\{T; \rightarrow_5\}$ generated by the connective $a \rightarrow_5 b = min(1, 1 - a + b)$ represents a concrete realization of the abstract group $S_{2\times2\times2}$ under the family of transformations $T$.

Also these theorems demonstrate that $\equiv$ and $\subseteq$ are assumimg independent epistemological features and there is scope for $\equiv$ independent from that customarily defined by $\wedge \rightarrow$ in a power set construction.

Note that the pair of implication operators $(\rightarrow_6, \rightarrow_5)$ represents the bounds of interval-valued inference system produced by $m_1$ (cf. Sec 6.2 above and [6]). The pair of implication operators $(g_1, g_8)$ of Theorem 13, on the other hand, represents the bounds of interval-valued inference system produced by $m_2$.

# References

[1] P. Aczel. An introduction to inductive definitions. In J. Barwise, editor, *Handbook of Mathematical Logic*, chapter C.7, pages 739–782. North-Holland, Amsterdam, 1977.

[2] W. Bandler . Some esomathematical uses of category theory. In : Klir, G. (ed.), *Applied General Systems Research: Recent Developments and Trends*, Plenum Press, New York 1978, pages 243-255.

[3] W. Bandler and L.J. Kohout. Fuzzy relational products and fuzzy implication operators. In *International Workshop on Fuzzy Reasoning Theory and Applications*, London, September 1978. Queen Mary College, University of London.

[4] W. Bandler and L.J. Kohout. Fuzzy power sets and fuzzy implication operators. *Fuzzy Sets and Systems*, 4:13–30, 1980. Reprinted in: *Readings in Fuzzy Sets for Intelligent Systems*, pages 88–96, D. Dubois, H. Prade and R. Yager (eds.), Morgam Kaufmann Publishers, San Mateo, Calif.

[5] W. Bandler and L.J. Kohout. Fuzzy relational products as a tool for analysis and synthesis of the behaviour of complex natural and artificial systems. In P.P. Wang and S.K. Chang, editors, *Fuzzy Sets: Theory and Applications to Policy Analysis and Information Systems*, pages 341–367. Plenum press, New York and London, 1980.

[6] W. Bandler and L.J. Kohout. Semantics of implication operators and fuzzy relational products. *Internat. Journal of Man-Machine Studies*, 12:89–116, 1980. Reprinted in Mamdani, E.H. and Gaines, B.R. eds. *Fuzzy Reasoning and its Applications*. Academic Press, London, 1981, pages 219-246.

[7] W. Bandler and L.J. Kohout. Fast fuzzy relational algorithms. In A. Ballester, D. Cardús, and E. Trillas, editors, *Proc. of the Second Internat. Conference on Mathematics at the Service of Man*, pages 123–131, Las Palmas, 1982. (Las Palmas, Canary Islands, Spain, 28 June - 3 July), Universidad Politechnica de las Palmas.

[8] W. Bandler and L.J. Kohout. Unified theory of multiple-valued logical operators in the light of the checklist paradigm. In *Proc. of the 1984 IEEE Conference on Systems, Man and Cybernetics*, pages 356–364, New York, 1984. IEEE.

[9] W. Bandler and L.J. Kohout. The interrelations of the principal fuzzy logical operators. In M.M. Gupta, A. Kandel, W. Bandler, and J.B. Kiszka, editors, *Approximate Reasoning in Expert Systems*, pages 767–780. North-Holland, Amsterdam, 1985.

[10] W. Bandler and L.J. Kohout. On the general theory of relational morphisms. *International Journal of General Systems*, 13:47–66, 1986.

[11] W. Bandler and L.J. Kohout. A survey of fuzzy relational products in their applicability to medicine and clinical psychology. In L.J. Kohout and W. Bandler, editors, *Knowledge Representation in Medicine and Clinical Behavioural Science*, pages 107–118. an Abacus Book, Gordon and Breach Publ., London and New York, 1986.

[12] W. Bandler and L.J. Kohout. The use of checklist paradigm in inference systems. In C.V. Negoita and H. Prade, editors, *Fuzzy Logic in Knowledge Engineering*, chapter 7, pages 95–111. Verlag TÜV Rheinland, Köln, 1986.

[13] W. Bandler and L.J. Kohout. Fuzzy implication operators. In M.G. Singh, editor, *Systems and Control Encyclopedia*, pages 1806–1810. Pergamon Press, Oxford, 1987.

[14] W. Bandler and L.J. Kohout. Relations, mathematical. In M.G. Singh, editor, *Systems and Control Encyclopedia*, pages 4000 – 4008. Pergamon Press, Oxford, 1987.

[15] W. Bandler and L.J. Kohout. Special properties, closures and interiors of crisp and fuzzy relations. *Fuzzy Sets and Systems*, 26(3):317–332, June 1988.

[16] B. Ben-Ahmeida, L.J. Kohout, and W. Bandler. The use of fuzzy relational products in comparison and verification of correctness of knowledge structures. In L.J. Kohout, J. Anderson, and W. Bandler, editors, *Knowledge-Based Systems for Multiple Environments*, chapter 16. Ashgate Publ. (Gower), Aldershot, U.K., 1992.

[17] E. Čech. *Point Sets*. Academia, Prague, 1969.

[18] C.C. Chang and H.J. Keisler. *Model Theory*. North-Holland, Amsterdam, 1973.

[19] H.B. Curry. *Foundations of Mathematical Logic*. Dover, Nwe York, 1977.

[20] D Dubois and H. Prade. *Fuzzy Sets and Systems: Theory and Applications.* Academic Press, New York, 1980.

[21] D Dubois and H. Prade. *Possibility Theory.* Plenum Press, New York, 1988.

[22] B.R. Gaines and L.J. Kohout. The logic of automata. *Internat. Journal of General Systems*, 2:191–208, 1975.

[23] B.R. Gaines and L.J. Kohout. Possible automata. In *Proc. of 1975 Internat. Symposium on Multiple-Valued Logic*, pages 183–196, New York, May 1975. IEEE 75CH0959.

[24] A.C. Grayling. *An Introduction to Philosophical Logic.* The Harwester Press, Sussex, 1982.

[25] S. Haack. *Philosophy of Logic.* Cambridge University Press, Cambridge, 1978.

[26] E. Hisdal. The philosophical issues raised by fuzzy set theory. *Fuzzy Sets and Systems*, 25(3):349–356, 1988.

[27] E. Hisdal. Infinite-valued logic based of two-valued logic and probability. pt. 1.4 the TEE model fro grades of membership. ISBN 82-7368-054-1 140, University of Oslo, Inst. of Informatics, Box 1080, Blindern Oslo 3, Norway, October 1990.

[28] U. Höhle and L.N. Stout . Foundations of fuzzy sets, *Fuzzy Sets and Sytsems* **40** (1991), 257–296.

[29] A. Kandel. *Fuzzy mathematical techniques with applications.* Addison Wesley, Reading, Mass., 1986.

[30] D. Klaua. Über einen Ansatz zur mehrwertigen Mengenlehre. *Monatsb. Deutsch. Akad. Wiss. (Berlin)*, 7.859–867, 1965.

[31] D. Klaua. Einbettung der klassischen Mengenlehre in die mehrwertige. *Monatsb. Deutsch. Akad. Wiss. (Berlin)*, 9:258–272, 1967.

[32] G.J. Klir and T.A. Folger. *Fuzzy Sets, Uncertainty, and Information.* Prentice Hall, Englewood Cliffs, N.J., 1988.

[33] L. Kohout. The Pinkava many-valued complete logic systems and their application to the design of many-valued switching circuits. In D.C. Rine, editor, *Proceedings of 1974 Internat. Symposium on Multiple-Valued Logic*, pages 261–284, New York, 1974. (West Virginia University, May, 1974), IEEE 74CH08945.

[34] L.J. Kohout. Towards systems that preserve quality of human life: On systemic approach of Thomas G. Masaryk. Presented at ICSRIC'88 (4th International Conference on Systems Research, Informatics and Cybernetics, Baden Baden, Germany, 15-21 August 1988).

[35] L.J. Kohout. Generalised topologies and their relevance to general systems. *Internat. Journal of General Systems*, 2:25–34, 1975.

[36] L.J. Kohout. Theories of possibility: Meta-axiomatics and semantics. *Fuzzy Sets and Systems*, 25:357–367, 1988.

[37] L.J. Kohout. Systems, Technologies and Quality of Human Life. In G.E. Lasker, editor, *Second Internatinal Congress on Systems Research, Informatics and Cybernetics*, Germany, August 1989. A Keynote Invited Openning Address.

[38] L.J. Kohout. *A Perspective on Intelligent Systems: A Framework for Analysis and Design*. Chapman and Hall & Van Nostrand, London & New York, 1990.

[39] L.J. Kohout. Quo vadis fuzzy systems: A critical evaluation of recent methodological trends. *Internat. J. of General Systems*, 19(4):395–424, 1991.

[40] L.J. Kohout. *Methodological Foundations of the Study of Action*. Ph.D. Thesis, University of Essex, U.K., January 1978.

[41] L.J. Kohout and W. Bandler. Checklist paradigm and group transformations. *Technical Note EES-MMS-ckl91.2*, 1979. Dept. of Electrical Engineering, University of Essex. U.K.

[42] L.J. Kohout and W. Bandler. Axioms for conditional inference: probabilistic and possibilistic. In A. Ballester, D. Cardús, and E. Trillas, editors, *Proc. of the Second Internat. Conference on Mathematics at the Service of Man*, pages 413–414, Las Palmas, 1982. (Las Palmas, Canary Islands, Spain, 28 June - 3 July), Universidad Politechnica de las Palmas.

[43] L.J. Kohout and W. Bandler. Fuzzy relational products in knowledge engineering. In V. Novák et al., editors, *Fuzzy Approach to Reasoning and Decision Making*, pages 51–66. Academia and Kluwer, Prague and Dordrecht, 1992.

[44] L.J. Kohout and W. Bandler. How the checklist paradigm elucidates the semantics of fuzzy inference. In *Proc. of the IEEE Internat. Conference on Fuzzy Systems 1992*, pages 571–578. IEEE, New York, 1992.

[45] L.J. Kohout and W. Bandler. Modes of plausible reasoning viewed via the checklist paradigm. In *Proc. of the IPMU'92 Conference.* Palma de Mallorca, Spain, 1992.

[46] L.J. Kohout and Y.G. Kim . Generating control strategies for resolution-based theorem provers by means of triangle products and relational closures. In : Lowen,R. and Roubens,M. (eds.), *Fuzzy Logic: State of the Art*, Kluwer, Dordrecht, 1993, pages 181–192.

[47] T.G. Masaryk. *Dav. Hume's Skepsis und die Wahrscheininlichkeitrechnung.* Carl Konegen, Wien, 1883.

[48] T.G. Masaryk. *Počet pravděpodobnosti a Humeova skepse.* J. Otto, Prague, 1883.

[49] T.G. Masaryk. *Základové Konkretné Logiky.* Bursík and Kohout, Prague, 1885.

[50] T.G. Masaryk. *Versuch einer concreten Logik.* Konegen, Vienna, 1887. (Translation of Masaryk (1885); enlarged and revised. Reprinted 1969).

[51] R. Montague. *Formal Philosophy.* Yale University Press, New Haven, 1974. Selected Papers of Richard Montague. Edited and with an introduction by R. H. Thomason.

[52] V. Novák. On the position of fuzzy sets in modelling of vague phenomena. In R. Lowen and M. Roubens, editors, *IFSA '91 Brussels, vol. Artificial Intelligence*, pages 165–167. Internationsl Fuzzy Systems Association, 1991.

[53] V. Pinkava. On a class of functionally complete multi-valued logical calculi. *Studia Logica*, 2:201–212, 1978.

[54] V. Pinkava. On potential tautologies in k-valued calculi. In P.P. Wang, editor, *Fuzzy Sets: Theory and Applications to Policy Analysis and Information systems*, pages 77–86. Plenum Press, New York, 1980.

[55] V. Pinkava. *Introduction to Logic for System Modelling.* Gordon and Breach, London and New York, 1988.

[56] W.V. Quine. *From a Logical Point of View.* Harper & Row, New York, 1961. 2nd edition. Ch. VIII: Reference and Modality.

[57] N. Rescher. *Many-Valued Logic.* McGraw-Hill, New York, 1969.

[58] A. Salomaa. On many-valued systems of logic. *Ajatus*, 22:115–159, 1959.

[59] B. Schweizer and A. Sklar. *Probabilistic metric spaces.* North Holland, New York, 1983.

[60] P.F. Strawson. *Introduction to Logical Theory.* University Paperbacks, Methuen, London, 1963.

[61] A. Tarski. The semantic conception of truth and the foundations of semantics. *Philosophy and Phenomenological Research*, 4:341–376, 1944.

[62] A. Tarski. *Logic, Semantics and Meta-Mathematics.* Hacket Publishing Company, Indianapolis, 1983. Papers from 1923 to 1938. Translated by J.H. Woodger. 2nd edition edited and introduced by J. Corcoran.

[63] I.B. Turksen. Containment and Klein groups of fuzzy propositions. *Working Paper 79-010, Dept. of Industrial Eng., University of Toronto, Canada.*

[64] I.B. Turksen. Klein groups in fuzzy inference. In *Proc. of the American Control Conference*, pages 556–560. American Automatic Control Council, 1984.

[65] L.A. Zadeh. Fuzzy sets. *Information and Control*, 8:338–353, 1965.

[66] L.A. Zadeh. Calculus of fuzzy restrictions. In L.A. Zadeh, K.S. Fu, K. Tanaka, and M. Shimura, editors, *Fuzzy Sets and Their Applications to Cognitive and Decision Porcesses*, chapter 1, pages 1–39. Academic Press, New York, 1975.

[67] L.A. Zadeh. Fuzzy sets as a basis for a theory of possibility. *Fuzzy Sets and Systems*, 1:3–28, 1978.

[68] M. Zelený. Fuzzy sets: precision and relevancy. In G.E. Lasker, editor, *Applied Systems and Cybernetics. Vol. VI, Fuzzy Sets and Systems, Possibility Theory and Special Topics in Systems Research*, pages 2718–2721. Pergamon Press, New York and Oxford, 1980.

# XII

# Ultraproduct theorem and recursive properties of fuzzy logic

## V. Novák

First–order fuzzy logic should be a formal basis of many considerations in fuzzy set theory and approximate reasoning. For example, the inference in the latter can be understood to be a sequence of inferences in a certain fuzzy theory given by a fuzzy set of special axioms. For various purposes, it may be useful to have an analogue of the famous Los' ultraproduct theorem also in fuzzy logic. We introduce elements of model theory for fuzzy logic and prove the ultraproduct theorem for it.

Another problem concerns recursive properties of fuzzy logic. We analyze the Scarpellini's result which states that fuzzy theory is not recursively enumerable. We prove that a set of syntactic conclusions of the set of formulas without symbols for irrational numbers from $[0, 1]$ is weakly recursively enumerable.

## 1 INTRODUCTION

In this paper, we will deal with some deeper mathematical properties of fuzzy logic. First of all, let us elucidate the therm "fuzzy logic". In its original meaning, fuzzy logic is a logic with more than two truth values. Later on, after famous L.A. Zadeh's paper [15], fuzzy logic became a synonym for "approximate reasoning". Let us also remark that now, by fuzzy logic people understand in a very broad sense a theory of fuzzy sets, approximate reasoning as well as many-valued logic.

In this paper, we will deal with *many-valued* logic and call it *fuzzy logic* since, as demonstrated, e.g., in [8, 9] a logical part of approximate reasoning can be formulated as a special (fuzzy) theory of many-valued (fuzzy) logic and thus, the term "fuzzy logic" should include many-valued logic as well. However, it is not the aim of this paper to deal with approximate reasoning.

Fuzzy (= many-valued, in the sequel) logic has its roots in early 30's, especially in works of famous Polish logician J. Lukasiewicz. He himself took his

theory as an interesting possibility and did not claim more sophisticated applications. The significance of many-valued logic became apparent only after the
discovery of fuzzy sets in connection with the possibility to model the vagueness
phenomenon.

There are many works devoted to fuzzy logic. Two of them, however, are
direct predecessors of the theory presented in this paper, namely the papers
by J. A. Goguen [5] and J. Pavelka [12]. Especially the latter one presents a
self-contained theory of propositional many-valued calculus with many general
notions (e.g., many-valued rules of inference), proved the completeness property
of its semantics and also its uniqueness on a linearly ordered set of truth values
(up to isomorphism). His work has been continued in [7] where all the notions
have been extended to first-order (fuzzy) logic and a generalization of Gödel's
completeness theorem was proved.

In this paper, we will overview the mentioned results in fuzzy logic, discuss
some questions concerning its recursive properties and present theorems, that,
in a certain sense, generalize the well known classical Los' ultraproduct theorem
to fuzzy logic.

## 2   A SHORT OVERVIEW OF FUZZY LOGIC

In this section, we briefly present the main notions and theorems of fuzzy logic.
If a proof of a lemma or theorem is omitted then it can be found in [7].

There are good reasons (cf. [5, 6, 7, 12] and others) to assume that the set
of truth values forms a complete, infinitely distributive, *residuated lattice*

$$\mathcal{L} = \langle L, \vee, \wedge, \otimes, \rightarrow, \mathbf{1}, \mathbf{0} \rangle \tag{1}$$

where $\mathbf{0}$, $\mathbf{1}$ are the smallest and the greatest elements respectively. The operations $\otimes$, $\rightarrow$ are binary operations of *Lukasiewicz multiplication* (sometimes
called also bold multiplication) and *residuation* (Lukasiewicz implication), respectively, with the following properties:

(a) $\langle L, \otimes, \mathbf{1} \rangle$ is a commutative monoid.

(b) The operation $\otimes$ is isotone in both variables and $\rightarrow$ is antitone in the first
     variable and isotone in the second one.

(c) The *adjunction property*

$$a \otimes b \leq c \quad \text{iff} \quad a \leq b \rightarrow c$$

holds for every $a, b, c \in L$.

Furthermore, we assume that $\mathcal{L}$ is either the interval $[0, 1]$ or a finite chain
$\mathcal{L} = \{\mathbf{0} = a_0 \leq \ldots \leq a_m = \mathbf{1}\}$ and put

$$a \otimes b = 0 \vee (a + b - 1) \tag{2}$$
$$a \rightarrow b = 1 \wedge (1 - a + b) \tag{3}$$

if $L = [0, 1]$, and

$$a_k \otimes a_p = a_{\max(0, k+p-m)} \tag{4}$$
$$a_k \rightarrow a_p = a_{\min(m, m-k+p)} \tag{5}$$

if $L$ is a finite chain where $0 \leq k, p \leq m$.

We will also use the operations

$$a \leftrightarrow b := (a \rightarrow b) \wedge (b \rightarrow a)$$
$$\neg a := a \rightarrow 0$$

and call *biresiduation* and *negation*, respectively.

Many properties of this structure can be found in the cited literature. Let us stress that the above defined structure preserves all the properties that have been intuitively required in the whole literature on fuzzy sets.

One of the strongest arguments in favour of the choice of residuation (implication) operation (3) follows from the following fact proved in [12].

**Theorem 1** *Let $\mathcal{L}$ be a complete residuated chain. Let a topology $\tau$ be given by the open basis*

$$B = \{\{x \in L; \ a < x < b\}; \ a, b \in L\}.$$

*If the operation $\rightarrow$ is not continuous with respect to $\tau$ then fuzzy logic cannot be syntactico-semantically complete.*

Due to this theorem, if $L = [0, 1]$ then our only choice of implication operation is the operation (3) since otherwise the completeness of our logical system fails.

Note, that residuated lattice is a special case of MV-algebra. Thus, it is possible to consider $\mathcal{L}$ to form an MV-algebra instead of the residuated lattice. Some of the definitions presented further would have to be modified but the final result concerning the case when $\mathcal{L}$ is a chain would remain unchanged. An interesting case when $\mathcal{L}$ is not a chain and not a Boolean lattice, however, is still not resolved.

Besides basic operations defined above, it is possible to enrich the lattice $\mathcal{L}$ by additional $n$-ary operations $c : L \longrightarrow L$. In [12, 11] it is demonstrated that every reasonable operation $c$ on $L$ should fulfil the following *fitting condition*. There are non-zero natural numbers $k_1, \ldots, k_n$ such that

$$(a_1 \leftrightarrow b_1)^{k_1} \otimes \cdots \otimes (a_n \leftrightarrow b_n)^{k_n} \leq c(a_1, \ldots, a_n) \leftrightarrow c(b_1, \ldots, b_n) \tag{6}$$

holds for every $a_1, \ldots, a_n, b_1, \ldots, b_n \in L$, where the power is taken with respect to the operation $\otimes$. We say that the operation $c$ is *logically fitting* if it fulfils (6).

The fitting condition ensures us that fuzzy logic preserves all the properties presented in the sequel. Therefore, we must require any additional operation

$c$ to be logically fitting as well. Note that each basic operation $\vee, \wedge, \otimes, \rightarrow$ is logically fitting.

The possibility to enrich $\mathcal{L}$ by additional operations is very important feature of fuzzy logic since in applications, we often need more operations than only the basic ones. Therefore, it is very favourable that we may consistently introduce additional operations without harming important properties of fuzzy logic. Among them, we stress that any formula with additional connectives keeps the properties of equivalence and also, that the completeness property is not harmed. Since fitting condition is not too much restrictive, first-order fuzzy logic is general enough to cover most of the systems of fuzzy logic presented in the literature.

Due to the previous discussion, the structure of truth values is assumed to be an *enriched residuated lattice*

$$\mathcal{L} = \langle L, \vee, \wedge, \otimes, \rightarrow, \{c_i; \ i \in Jop\}, \mathbf{0}, \mathbf{1} \rangle \tag{7}$$

which is, moreover a finite or uncountable chain (in the latter case, $L = [0, 1]$). The $Jop$ is an index set for the additional $n_i$-ary operations $c_i$ on $L$ which are logically fitting (cf. [6, 12]).

## 2.1 Language

The *language* of first-order fuzzy logic consists of:

(i) Variables $x, y, \ldots$.

(ii) Constants $\mathbf{c}, \mathbf{d}, \mathbf{r}, \ldots$.

(iii) $n$-ary functional symbols $f, g, \ldots$.

(iv) Symbols for truth values $\{a; \ a \in L\}$.

(v) $n$-ary predicate symbols $p, q, \ldots$.

(vi) A binary connective $\Rightarrow$ and a set $\{c_j; \ j \in Jop\}$ of additional $n_j$-ary connectives.

(vii) A symbol for a general quantifier $\forall$.

(viii) Auxiliary symbols.

Terms are defined as usual.

### Formulas

(a) A symbol $a$ for a truth value $a \in L$ is a (atomic) formula.

(b) If $t_1, \ldots, t_n$ are terms and $p$ an $n$-ary predicate symbol then $p(t_1, \ldots, t_n)$ is a (atomic) formula.

(c) If $A, B, A_1, \ldots, A_n$ are formulas then $A \Rightarrow B$, $c_j(A_1, \ldots, A_n)$, $j \in J$op and $(\forall x)A$ are formulas.

We introduce the following abbreviations of formulas:

$$
\begin{array}{rcll}
\neg A & := & A \Rightarrow 0 & \text{(negation)} \\
A \vee B & := & (A \Rightarrow B) \Rightarrow B & \text{(disjunction)} \\
A \wedge B & := & \neg((A \Rightarrow B) \Rightarrow \neg A) & \text{(conjunction)} \\
A \,\&\, B & := & \neg(A \Rightarrow \neg B) & \text{(Lukasiewicz conjunction)} \\
A \Leftrightarrow B & := & (A \Rightarrow B) \wedge (B \Rightarrow A) & \text{(equivalence)} \\
(\exists x)A & := & \neg(\forall x)\neg A & \text{(existential quantifier)} \\
A^k & := & \underbrace{A \,\&\, A \,\&\, \cdots \,\&\, A}_{k-\text{times}} & \text{(power)}
\end{array}
$$

A set of all the terms of a language $J$ is denoted by $M_J$ and a set of all the formulas by $F_J$ .

Analogously to classical logic we introduce the notions of *free* and *bound* variables and a substitutible term. If $t$ is a term and $A$ a formula then $A_x[t]$ is a formula resulting from $A$ when substituting the term $t$ instead of each free occurrence of $x$ in $A$. By $t(x_1, \ldots, x_n)$ or $A(x_1, \ldots, x_n)$ we denote a term or formula whose all variables occur among $x_1, \ldots, x_n$, respectively.

## 2.2  Semantics

A *structure* for the language $J$ of first-order fuzzy logic is a tuple

$$ \mathcal{D} = \langle D, p_D, \ldots, f_D, \ldots, u, v, \ldots \rangle $$

where $D$ is a set, $p_D \underset{\sim}{\subseteq} D$, ... are $n$-ary relations assigned to each $n$-ary predicate symbol $p, \ldots$, and $f_D$ are ordinary $n$-ary functions on $D$ assigned to each $n$-ary functional symbol $f$. Finally, the $u, v, \ldots \in D$ are elements which are assigned to each constant $\mathbf{u}$, $\mathbf{v}$ of the language $J$.

*Truth valuation of formulas*

Let $\mathcal{D}$ be a structure for the language $J$. A *truth valuation* of formulas in $D$ is a function

$$ \mathcal{D} : F \longrightarrow L \tag{8} $$

which assigns a truth value to every formula $A \in F_J$ as follows.

Let $d_1, \ldots, d_n \in D$ be a sequence of elements. Then we define a value of a term $t(x_1, \ldots, x_q)$ on this sequence,

$$ t[d_1, \ldots, d_q] \in D, $$

in the same way as in classical logic (cf. [4]). We only underline that $t = \boldsymbol{r}$ ($\boldsymbol{r}$ is a constant) implies $t[d_1, \ldots, d_q] = d_r$ where $d_r$ is an interpretation of the constant $\boldsymbol{r}$ in $\mathcal{D}$.

Furthermore, we define a truth value of a formula on a sequence $d_1, \ldots, d_q \in D$ by:

(i) $\mathcal{D}(a)[d_1, \ldots, d_q] = a, \quad a \in L.$

(ii) $\mathcal{D}(p(t_1, \ldots, t_n))[d_1, \ldots, d_q] = p_D(t_1[d_1, \ldots, d_q], \ldots, t_n[d_1, \ldots, d_q]).$

(iii) $\mathcal{D}(A \Rightarrow B)[d_1, \ldots, d_q] = \mathcal{D}(A)[d_1, \ldots, d_q] \rightarrow \mathcal{D}(B)[d_1, \ldots, d_q].$

(iv) $\mathcal{D}(c_j(A_1, \ldots, A_n)[d_1, \ldots, d_q] =$
$$= c_j(\mathcal{D}(A_1)[d_1, \ldots, d_q], \ldots, \mathcal{D}(A_n)[d_1, \ldots, d_q]), \quad j \in J \text{op}.$$

(v) $\mathcal{D}((\forall x_i)A)[d_1, \ldots, d_q] = \bigwedge_{d \in D} \mathcal{D}(A)[d_1, \ldots, d_{i-1}, d, d_{i+1}, \ldots, d_q].$

**Lemma 1**

(a) *Given a term* $t(x_1, \ldots, x_q)$ *and* $d_1, \ldots, d_r, e_1, \ldots, e_s \in D$, $q \leq r, s$ *and* $d_i = e_i$ *if* $x_i$ *is a free variable in* $t$. *Then*
$$t[d_1, \ldots, d_q] = t[e_1, \ldots, e_s].$$

(b) *Let* $A(x_1, \ldots, x_q)$ *be a formula and* $d_i = e_i$ *if* $x_i$ *is a free variable in* $A$. *Then*
$$\mathcal{D}(A)[d_1, \ldots, d_q] = \mathcal{D}(A)[e_1, \ldots, e_s].$$

PROOF: The proof of (a) is identical with that of analogous statement in classical logic.

(b) For $A := a$ is the statement obvious. By (a), we also have
$$
\begin{aligned}
\mathcal{D}(p)[d_1, \ldots, d_r] &= p_D(t_1[d_1, \ldots, d_r], \ldots, t_n[d_1, \ldots, d_r]) \\
&= p_D(t_1[e_1, \ldots, e_s], \ldots, [e_1, \ldots, e_s]).
\end{aligned}
$$

If $A := B \Rightarrow C$ or $A := c(A_1, \ldots, A_n)$ then we obtain the statement using the inductive assumption. Finally,
$$
\begin{aligned}
\mathcal{D}((\forall x_i)A)[d_1, \ldots, d_r] &= \bigwedge_{d \in D} \mathcal{D}(A)[d_1, \ldots, d_{i-1}, d, d_{i+1}, \ldots, d_r] \\
&= \bigwedge_{e \in D} \mathcal{D}(A)[e_1, \ldots, e_{i-1}, e, e_{i+1}, \ldots, e_s] \\
&= \mathcal{D}((\forall x_i)A)[e_1, \ldots, e_s].
\end{aligned}
$$
$\square$

On the basis of this lemma, we may define the truth valuation (8) by
$$\mathcal{D}(A(x_1, \ldots, x_n)) = \bigwedge_{d_1, \ldots, d_q \in D} \mathcal{D}(A(x_1, \ldots, x_n))[d_1, \ldots, d_q] \qquad (9)$$

for arbitrary $q \geq n$.

Due to the lack of place we omit the proof of the following lemma.

**Lemma 2** *Let $A(x_1, \ldots, x_n)$ be a formula and $t_1, \ldots, t_n$ terms such that no variable occurring in them is bound in $A$. Then*

$$\mathcal{D}(A)[d_1, \ldots, d_q] = \mathcal{D}(A_{x_1, \ldots, x_n}(t_1[d_1, \ldots, d_q], \ldots, t_n[d_1, \ldots, d_q]))$$

*holds for every sequence $d_1, \ldots, d_q \in D$.*

From the definition of the truth valuation we immediately obtain

$$
\begin{aligned}
\mathcal{D}(A \wedge B)[d_1, \ldots, d_q] &= \mathcal{D}(A)[d_1, \ldots, d_q] \wedge \mathcal{D}(B)[d_1, \ldots, d_q] \\
\mathcal{D}(A \vee B)[d_1, \ldots, d_q] &= \mathcal{D}(A)[d_1, \ldots, d_q] \vee \mathcal{D}(B)[d_1, \ldots, d_q] \\
\mathcal{D}(A \,\&\, B)[d_1, \ldots, d_q] &= \mathcal{D}(A)[d_1, \ldots, d_q] \otimes \mathcal{D}(B)[d_1, \ldots, d_q] \\
\mathcal{D}(A^k)[d_1, \ldots, d_q] &= (\mathcal{D}(A)[d_1, \ldots, d_q])^k \\
\mathcal{D}(\neg A)[d_1, \ldots, d_q] &= \neg \mathcal{D}(A)[d_1, \ldots, d_q] = \mathcal{D}(A)[d_1, \ldots, d_q] \rightarrow 0 \\
\mathcal{D}(A \Leftrightarrow B)[d_1, \ldots, d_q] &= \mathcal{D}(A)[d_1, \ldots, d_q] \leftrightarrow \mathcal{D}(B)[d_1, \ldots, d_q] \\
\mathcal{D}((\exists x_i)A)[d_1, \ldots, d_q] &= \bigvee_{d \in D} \mathcal{D}(A)[d_1, \ldots, d_{i-1}, d, d_{i+1}, \ldots, d_q]
\end{aligned}
$$

for all $d_1, \ldots, d_q \in D$.

Let $X \subseteq F$ be a fuzzy set of formulas. Then the fuzzy set of *semantic consequences* of the fuzzy set $X$ is

$$(C^{sem}X)A = \bigwedge\{\mathcal{D}(A); \mathcal{D} \text{ is a structure for } J \text{ and } (\forall B \in F_J)(X(B) \le \mathcal{D}(B))\}$$

where $X(B) \in L$ is the grade of membership of $B$ in $X$. It can be proved that $C^{sem}$ is a closure operation on $L^{F_J}$.

A formula $A$ is and $a$-*tautology* if $a = (C^{sem}\emptyset)A$ and we write $\models_a A$. If $a = 1$ then we write simply $\models A$.

**Lemma 3**

(a) $\models A \Rightarrow B$  *iff*  $\mathcal{D}(A) \le \mathcal{D}(B)$,

(b) $\models A \Leftrightarrow B$  *iff*  $\mathcal{D}(A) = \mathcal{D}(B)$

*holds in every structure $\mathcal{D}$.*

The following are the most important schemata of tautologies which take the role of logical axioms.

(T1) $\models (a \Rightarrow b) \Leftrightarrow \overline{(a \to b)}$

  where $\overline{a \to b}$ denotes the symbol (atomic formula) for the truth value $a \to b$ when $a$ and $b$ are given.

(T2) $\models A \Rightarrow A$

(T3) $\models A \Rightarrow 1$

(T4) $\models ((A \& B) \Rightarrow C) \Leftrightarrow (A \Rightarrow (B \Rightarrow C))$

(T5) $\models (A \& B) \Leftrightarrow (B \& A)$

(T6) $\models (A \Rightarrow B) \Rightarrow ((B \Rightarrow C) \Rightarrow (A \Rightarrow C))$

(T7) $\models (A \Rightarrow B) \vee (B \Rightarrow A)$

(T8) $\models (A \vee B)^n \Rightarrow (A^n \vee B^n), \quad n \in \mathbb{N}^+$

(T9) $\models (\forall x) A \Rightarrow A_x[t]$

  for any term $t$.

(T10) $\models (\forall x)(A \Rightarrow B) \Leftrightarrow (A \Rightarrow (\forall x) B)$

  provided that $x$ is not free in $A$.

(T11) $\models (\exists y)(A_x[y] \Rightarrow (\forall x) A)^n$

(T12) $\models (A \Rightarrow a_k) \vee (a_{k+1} \Rightarrow A)$ for $k < m$

  if $L$ is a finite chain.

(T13) $\models ((a \Rightarrow B)^n \Rightarrow b) \Rightarrow ((a' \Rightarrow B) \Rightarrow b')$

  for $b < b' < 1$, $0 < a' < a$, $na + b < na' + b'$ if $L = [0, 1]$.

(T14) $\models ((A \Rightarrow a)^n \Rightarrow b) \Rightarrow ((A \Rightarrow a') \Rightarrow b')$

  for $b < b' < 1$, $a < a'$, $na' - b' \leq na - b$ if $L = [0, 1]$.

A specific feature of fuzzy logic is evaluation of its syntax. This means that at each step of the derivation, the resulting formula is evaluated by some truth value (we will call it *syntactic truth value*). This is necessary because in general, we start from a fuzzy set of axioms, i.e., there may be axioms taken as not fully convincing. Note, that the necessity to evaluate syntax is not as strange in comparison with classical logic as it might seem at first glance. Each derivation step in the latter must be valid, i.e., we must not derive false formulas from true ones if we use sound rules of inference. Hence, the syntax in classical logic is evaluated as well but the evaluation is always 1 (true) and this fact is not explicitly stressed. The syntactic truth values accompany all the formulas but they do not belong to the language. We may think about modification of the definitions of the language and formulas accordingly to include syntactic truth values.

Necessity to evaluate syntax leads to the concept of many-valued rules of inference. An $n$-ary *rule of inference* $r$ is a couple

$$r = \langle r^{syn}, r^{sem} \rangle$$

where $r^{syn}$ is its *syntactic part* which is a partial $n$-ary operation on $F_J$ and $r^{sem}$ is a *semantic part* which is an $n$-ary operation on $L$ preserving arbitrary non-empty joins in each argument (semicontinuity).

A fuzzy set $X \subseteq F_J$ is *closed* with respect to $r$ if

$$X(r^{syn}(A_1, \ldots, A_n)) \geq r^{sem}(X(A_1), \ldots, X(A_n))$$

holds for all $A_1, \ldots, A_n \in F_J$ for which $r^{syn}$ is defined. A rule of inference is *sound* if

$$\mathcal{D}(r^{syn}(A_1, \ldots, A_n)) \geq r^{sem}(\mathcal{D}(A_1), \ldots, \mathcal{D}(A_n))$$

holds for all the truth valuations $\mathcal{D}$.

Note that soundness is an indispensable property of rules of inference. If we did not require it then we might derive formulas with nonzero syntactic truth evaluation which are not true in any structure, i.e., a rubbish.

The rules of inference are usually written in the form

$$r : \frac{A_1, \ldots, A_n}{r^{syn}(A_1, \ldots, A_n)} \left( \frac{a_1, \ldots, a_n}{r^{sem}(a_1, \ldots, a_n)} \right)$$

where $a_i \in L$ are truth valuations of the respective formulas $A_i$, $i = 1, \ldots, n$.

We introduce the following rules of inference:

(a) *Modus ponens*

$$r_{MP} : \frac{A, A \Rightarrow B}{B} \left( \frac{a, b}{a \otimes b} \right).$$

(b) *$a$-lifting rule*

$$r_{Ra} : \frac{B}{a \Rightarrow B} \left( \frac{b}{a \to b} \right).$$

(c) *Generalization*

$$r_G : \frac{A}{(\forall x)A} \left( \frac{a}{a} \right).$$

It can be demonstrated that the operation $\otimes$ in the rule of Modus ponens cannot be replaced by $\wedge$.

Let $X \subseteq F_J$ be a fuzzy set of formulas. Then

$$(C^{syn}X)A = \bigwedge \{U(A); \ U \subseteq F_J, \ U \text{ is closed with respect to all } r \in R$$
$$\text{and } A_L, X \subseteq U\}$$

defines a fuzzy set of *syntactic consequences* of the fuzzy set $X$.

A *proof* of a formula $A$ from the fuzzy set $X$ is a sequence

$$w := A_0\,[a_0; P_0], A_1\,[a_1; P_1], \ldots, A_n\,[a_n; P_n]$$

such that $A_n$ is $A$ and $P_i$, $i \leq n$ is LA or SA if $A$ is a logical or a special axiom respectively, or P is $r$ if $A$ is a formula

$$r^{syn}(A_1, \ldots, A_n), \quad i_1, \ldots, i_n < i$$

and $r_i$ is an $n$-ary sound rule of inference. Furthermore,

$$a_i = \mathrm{Val}_X(w_{(i)})$$

is the *value* of the proof

$$w_{(i)} := A_0\,[a_0; P_0], A_1\,[a_1; P_1], \ldots, A_i\,[a_i; P_i]$$

defined as follows:

$$\mathrm{Val}_X(w_{(i)}) = \begin{cases} A_L(A_i) \text{ if } P_i = \text{LA i.e., } A_i \text{ is a logical} \\ \qquad\qquad\qquad\qquad \text{axiom} \\[6pt] X(A_i) \text{ if } P_i = \text{SA i.e., } A_i \text{ is a special} \\ \qquad\qquad\qquad\qquad \text{axiom} \\[6pt] r_i^{sem}(\mathrm{Val}_X(w_{(i_1)}), \ldots, \mathrm{Val}_X(w_{(i_n)})) \\ \qquad \text{if } A_i = r_i^{syn}(A_{i_1}, \ldots, A_{i_n}) \end{cases}$$

Note that the definition of a proof is an exact expression of the concept of evaluated syntax.

**Theorem 2**

$$(C^{syn}X)A = \bigvee\{\mathrm{Val}_X(w);\ w \text{ is a proof of } A \text{ from } X \subseteq F_J\}.$$

It follows from this theorem that if we find a proof, say $w$, of a formula $A$ then we only know that the degree in which it is a theorem is greater than or equal to $\mathrm{Val}_X(w)$. If $\mathrm{Val}_X(w) \neq 1$ then it is difficult to assure ourselves that we cannot find a proof with a greater value.

The *syntax* $\langle A_L, R \rangle$ of first-order fuzzy logic consists of:

(a) The *fuzzy set* $A_L$ of *logical axioms* defined as follows:

$$A_L a = a, \quad a \in L,$$
$$A_L(a \Rightarrow b) = a \to b, \quad a, b \in L,$$

$$A_L(A) = \begin{cases} 1 & \text{if } A \text{ is any of the formulas of the form} \\ & \text{(T1) — (T11) and either (T13), (T14) if} \\ & L = [0,1] \text{ or (T12) if } L \text{ is a finite chain,} \\ 0 & \text{otherwise.} \end{cases}$$

In the case when $A := B \Leftrightarrow C$ we understand that both

$$A_L(B \Rightarrow C) = 1$$

as well as

$$A_L(C \Rightarrow B) = 1.$$

(b) The set $R$ of rules of inference contains at least the rules from the set

$$\bar{R} = \{r_{MP}, r_G, \{r_{Ra}; a \in L\}\}.$$

A *theory* $T$ in the language $J$ of first-order fuzzy logic (a *fuzzy theory*) is a triple

$$T = \langle A_L, A_S, R \rangle$$

where $\langle A_L, R \rangle$ is the above defined syntax of fuzzy logic and $A_S \subseteq F_J$ is a fuzzy set of special axioms. By $J(T)$ we denote the language of fuzzy theory. A *fuzzy predicate calculus* is the fuzzy theory with $A_S = 0$.

Note that the definition of a fuzzy theory does not exclude the possibility to add further (sound) rules of inference, if it is necessary.

Let $\mathcal{D}$ be a structure for $J(T)$. Then $\mathcal{D}$ is a *model* of the theory $T$, $\mathcal{D} \models T$, if

$$A_S(A) \leq \mathcal{D}(A)$$

holds for every $A \in F_{J(T)}$. It follows from the definition of logical axioms that

$$A_L(A) \leq \mathcal{D}(A)$$

holds in any structure $\mathcal{D}$ for every formula $A \in F_{J(T)}$. Then

$$(C^{sem} A_S)A = \bigwedge \{\mathcal{D}(A); \mathcal{D} \models T\}.$$

If $(C^{sem} A_S)A = a$ then the formula $A$ is *true* in the degree $a$ in the theory $T$ and we write

$$T \models_a A.$$

If $(C^{syn} A_S)A = a$ then $A$ is a *theorem* in the degree $a$ of the theory $T$ and we write

$$T \vdash_a A.$$

We write $T \vdash A$, $T \models A$ instead of $T \vdash_1 A$, $T \models_1 A$, respectively and say that $A$ is a theorem (true) of the theory $T$.

It follows from the definition that in fuzzy predicate calculus

$$\vdash_a a \quad \text{and} \quad \models_a a$$

holds for every $a \in L$. If $w$ is a proof in theory $T$ then we write $\text{Val}_T(w)$ for its value. If $T$ is predicate calculus then we omit the subscript $T$.

**Theorem 3 (validity theorem)** *If $T \vdash_a A$ and $T \models_b A$ then $a \leq b$ holds for every formula $A$.*

Due to this theorem, we may derive formulas whose syntactic evaluation is at most as big as their semantic interpretation.

A theory $T$ is *contradictory* if there is a formula $A$ and proofs $w_1$ and $w_2$ of $A$ and $\neg A$, respectively, such that

$$\mathrm{Val}_T(w_1) \otimes \mathrm{Val}_T(w_2) > 0.$$

It is *consistent* in the opposite case. Obviously, if $T$ is contradictory then $T \vdash_a A$, $T \vdash_b \neg A$ and $a \otimes b > 0$.

**Theorem 4** *A theory $T$ is contradictory iff $T \vdash A$ holds for every formula $A \in F_{J(T)}$.*

It follows from this theorem that a contradictory fuzzy theory collapses into a contradictory theory of classical logic.

**Theorem 5 (closure theorem)** *Let $A \in F_{J(T)}$ and $A'$ be its closure. Then*

$$T \vdash_a A \quad \text{iff} \quad T \vdash_a A'.$$

A language $J'$ is an *extension* of $J$ if $J \subseteq J'$. Obviously, $F_J \subseteq F_{J'}$. Let $T = \langle A_L, A_S, R \rangle$, $T' = \langle A'_L, A'_S, R \rangle$ be theories in the respective languages. Put $\bar{A}_S(A) = A_S(A)$ if $A \in F_J$ and $\bar{A}_S(A) = 0$ otherwise. If

$$\bar{A}_S \subseteq A'_S$$

then $T'$ is an *extension* of $T$. To simplify the notation, we will write $A_S$ instead of $\bar{A}_S$ and understand that $A_S(A) = 0$ for all $A \in F_{J'} - F_J$.

The extension $T'$ is a *conservative extension* of $T$ if $T' \vdash_b A$ and $T \vdash_a A$ implies $a = b$ for every formula $A \in F_{J(T)}$. The extension $T'$ is a *simple extension* of $T$ if $J(T') = J(T)$.

The following is a generalization of an important deduction theorem of classical logic.

**Theorem 6 (deduction theorem)** *Let $A$ be a closed formula and $T' = T \cup \{1/A\}$.*

(a) *If $T \vdash_a A^n \Rightarrow B$ and $T' \vdash_b B$ for some $n$ then $a \leq b$.*

(b) *To every proof $w'$ of $B$ in $T'$ there are an $n$ and a proof $w$ of $A^n \Rightarrow B$ in $T$ such that*
$$\mathrm{Val}_T(w') = \mathrm{Val}_T(w).$$

**Corollary 1** *Let $\mathcal{L}$ be a finite chain. Then there is an $n$ such that*

$$T \vdash_a A^n \Rightarrow B \quad \text{iff} \quad T' \vdash_a B$$

**Theorem 7** *Let $T$ be a consistent theory, $T \vdash (\exists x)(A(x))^n$ for every $n$, and $t \notin J(T)$ be a new constant. Then the theory*

$$T' = T \cup \{1/A_x[t]\}$$

*in the language $J(T) \cup \{t\}$ is a conservative extension of the theory $T$.*

The following theorems are generalizations of the classical Gödel's completeness theorems.

Let $A_0$ be a chosen closed formula and let $T \vdash_a A_0$. Put

$$b = \begin{cases} a & \text{if } \mathcal{L} \text{ is a finite chain} \\ c, & c > a \text{ if } L = [0,1] \text{ and } a < 1 \\ 1 & \text{if } a = 1 \end{cases} \quad . \tag{10}$$

**Theorem 8 (completeness theorem II)** *A fuzzy theory $T$ is consistent iff it has a model. If $T$ is consistent then to every $A \in F_{J(T)}$ and $b$ defined in (10) there is a model $\mathcal{D}$ such that $\mathcal{D}(A) \leq b$.*

**Theorem 9 (completeness theorem I)** *Let $T$ be a consistent theory. Then*

$$T \vdash_a A \quad \text{iff} \quad T \models_a A$$

*holds true for every formula $A \in F_{J(T)}$.*

The proofs of these theorems are complicated and they are based on deep algebraic properties of the structure of closed formulas. The closure theorem then enables us to state the completeness theorems for all the formulas.

It can be demonstrated (see [12]) that fuzzy logic cannot be complete, if the set of truth values is only an infinitely countable chain. If it is a finite or uncountable chain then the structure (1) with the operations (2), (3) or (4), (5) is the only plausible one provided that we demand fuzzy logic to be syntactico–semantically complete (i.e., any other structure preserving these properties is isomorphic with this one). If $\mathcal{L}$ is not a chain and not a Boolean lattice then the answer is not yet known.

Let us stress that first-order fuzzy logic is equivalent with classical logic if $L = \{0, 1\}$.

Let us now define a set of special constants for all the formulas of the form $(\forall x)A$ in a well known classical way. A fuzzy theory $T$ is of *Henkin type* (shortly, *Henkin fuzzy theory*) if the formulas

$$A_x[r] \Rightarrow (\forall x)A(x)$$

are taken as special axioms with membership degree equal to **1** for all formulas $A(x)$, where $r$ is a special constant for $(\forall x)A(x)$. Let $\mathcal{D}$ be a model of a Henkin theory $T$ and $r$ a special constant for a formula $A$. Then there is an element $d_0 \in D$ such that $\mathcal{D}(r) = d_0$ and

$$\mathcal{D}(A_x[r])[d_0] = \bigwedge_{d \in D} \mathcal{D}(A)[d, d_1, \ldots, d_q]$$

holds for every sequence $d_1, \ldots, d_q$.

We have proved in [7] that to each consistent fuzzy theory $T$, there is a conservative extension $T'$ which is a Henkin fuzzy theory.

**Lemma 4** *Let $T'$ be a conservative Henkin extension of a fuzzy theory $T$. Then to every model $\mathcal{D} \models T$ there is an extension $\mathcal{D}' \models T'$ such that $\mathcal{D}(A) = \mathcal{D}'(A)$ holds for every formula $A \in F_{J(T)}$.*

PROOF: We put $D' = D \cup \{r; \ r \in K\}$ where $K$ is a set of special constants. Functions will be defined by extending $f_D$ into $f_{D'}$ in such a way that $\mathrm{rng}(f_{D'}) = \mathrm{rng}(f_D)$.

Let $p(x_i)$ be an atomic formula. Then we put

$$p_{D'}[d'_1, \ldots, d'_q] = p_D[d_1, \ldots, d_q] \qquad (11)$$

where $d'_j = d_j$ if $d'_j \notin K$, $j = 1, \ldots, q$. Furthermore, $d'_j = d$ for some $d \in D$ if $j \neq i$ or $d'_j$ is not a special constant $r$ for $(\forall x_i)p$, otherwise $d'_j = r$.

Let $r$ be a special constant for $(\forall x_i)B$ of the level $n > 0$ and assume that the value of $\mathcal{D}'$ has already been defined for all the formulas with special constants of the level smaller than $n$. Then we define

$$\mathcal{D}'(B_{x_i}[r])[d'_1, \ldots, d'_{i-1}, r, d'_{i+1}, \ldots, d'_q] = \mathcal{D}((\forall x_i)B)[d_1, \ldots, d_q].$$

Let $A \in F_{J(T)}$ be atomic. Then $\mathcal{D}'(A)[d'_1, \ldots, d'_q] = \mathcal{D}(A)[d_1, \ldots, d_q]$ is immediate. For $A := B \Rightarrow C$, $A := c(A_1, \ldots, A_n)$ follows $\mathcal{D}'(A) = \mathcal{D}(A)$ from inductive assumption.

Let $A := (\forall x_i)B$ where $B \in F_{J(T)}$ and $r$ be a special constant for $A$. Then

$$
\begin{aligned}
\mathcal{D}'((\forall x_i)B)[d'_1, \ldots, d'_q] &= \bigwedge_{d' \in D'} \mathcal{D}'(B)[d'_1, \ldots, d'_{i-1}, d', d'_{i+1}, \ldots, d'_q] \\
&= \bigwedge_{d \in D} \mathcal{D}(B)[d_1, \ldots, d_{i-1}, d, d_{i+1}, \ldots, d_q] \wedge \\
&\quad \wedge \bigwedge_{r' \in K} \mathcal{D}'(B)[d'_1, \ldots, d'_{i-1}, r', d'_{i+1}, \ldots, d'_q] \\
&= \mathcal{D}'(B_x[r])[d'_1, \ldots, d'_{i-1}, r, d'_{i+1}, \ldots, d'_q] \\
&= \mathcal{D}((\forall x_i)B)[d_1, \ldots, d_q]
\end{aligned}
$$

by the definition and inductive assumption. We have proved that

$$\mathcal{D}'(A)[d'_1, \ldots, d'_q] = \mathcal{D}(A)[d_1, \ldots, d_q]$$

for every $A \in F_{J(T)}$ and a sequence $d'_1, \ldots, d'_q \in d'$ which gives $\mathcal{D}'(A) = \mathcal{D}(A)$. $\qquad \square$

# 3  AXIOMATIZABILITY OF FUZZY THEORIES

Recursive properties of mathematical theories is an important problem which gives us information about their tractability. In 1962, B. Scarpellini published a paper [13] in which he proved that infinite–valued predicate Łukasiewicz logic is not axiomatizable[2] First–order fuzzy logic presented in the previous section has much common with this logic. Therefore, we will deal with the problem of its axiomatizability in this section. We will especially consider the case when $L = [0, 1]$ since for $L$ finite, the recursive properties of fuzzy logic coincide with those of classical logic.

First, we will reformulate the Scarpellini's result in our terms.

**Theorem 10 (Scarpellini)** *A set of formulas of first-order fuzzy logic such that $\vdash_0 B$ is not recursively enumerable.*

PROOF: We will outline the main ideas from the paper [13]. Note that the language of first–order fuzzy logic differs from that of classical logic only by atomic formulas $a, a \in L$ and by additional connectives which play no important role in this place. Therefore, it is possible to consider certain formulas of classical logic and translate them into special formulas of the fuzzy one. Let $A^{(2)}$ be a formula of classical logic. If it is of the form $B^{(2)} \Rightarrow C^{(2)}$ then it is understood as $\neg B^{(2)} \vee C^{(2)}$. The formula $A^{(2)}$ is assigned a formula

$$Q := (A \,\&\, \neg a) \wedge K \wedge H$$

where $a$ is a truth value symbol, $K$ and $H$ are specific formulas and $A$ is exactly $A^{(2)}$ provided that all the occurrences of "$\Rightarrow$" are replaced by "$\neg \cdot \vee \cdot$". Then, a $\mathcal{D}'$ is constructed to every $\mathcal{D}$ such that $\mathcal{D}(Q) > 0$ holds iff $\mathcal{D}'$ has finite support and $\mathcal{D}' \models A^{(2)}$. Thus, the following lemma is proved.

**Lemma 5** *A formula $A^{(2)}$ is finitely satisfiable in $\mathcal{D}'$, $\mathcal{D}' \models A^{(2)}$ iff $\mathcal{D}(Q) > 0$.*

Following the fact (see [14]) that a set of all the formulas not finitely satisfiable is not recursively enumerable we can proceed as follows.

Let $g$ be a Gödel number of $A^{(2)}$ and $m(g)$ a Gödel number of the corresponding formula $Q$ ($m$ is a recursive function). Let a set of all the formulas $B$ such that $\mathcal{D}(B) = 0$ for some $\mathcal{D}$ be recursively enumerable. Then there is a recursive predicate $S(u, v)$ such that

$$(\exists v) S(u, v) \quad \text{iff} \quad \mathcal{D}(B) = 0.$$

---

[2]Note that the term "axiomatizable theory" is used in two meanings. First, it means that theorems of the given theory are effectively provable (they form a recursively enumerable set). Sometimes, however, it is used in the sense of having the completeness property, i.e., to keep the balance between syntax and semantics. We do not use this term in the second meaning in this paper.

Due to Lemma 5, every formula $A^{(2)}$ of classical logic is not finitely satisfiable if $(\exists v)S(m(g), u)$ holds where $g$ is its Gödel number. But this means that the set of all such $A^{(2)}$ is recursively enumerable which contradicts to [14]. Hence, such $S(u, v)$ does not exist. If there is a model $\mathcal{D}$ such that $\mathcal{D}(B) = 0$ then, by definition, $\models_0 B$ and by Validity theorem, $\vdash_0 B$ which concludes the proof of the theorem.                                                                                                    □

By this theorem, we cannot effectively decide that, given a formula $B$, it is not a theorem (i.e., a theorem in the degree 0) of the predicate calculus. Let us remark that this is not very surprizing conclusion since fuzzy logic is more complicated and recursivity requirements are too strong for it. The other reason is that there is an uncountable number of formulas in fuzzy logic due to the requirement that the fuzzy set of atomic formulas $\{a/a;\ a \in [0,1]\}$ is a (fuzzy) subset of a (fuzzy) set of logical axioms.

To cope with this situation, we have to reduce the fuzzy set of logical axioms and to weaken the notion of recursive enumerability which is done in the works of L. Biancino and G. Gerla [1, 2].

Given a fuzzy theory $T = \langle A_L, A_S, R \rangle$. Put

$$I = \{a/a;\ a \in [0,1] - \mathbb{Q}\}$$

where $\mathbb{Q}$ is a set of rational numbers. We define a fuzzy theory

$$T' = \langle A'_L = A_L - I, A_S, R \rangle \tag{12}$$

which is obtained from $T$ by removing of all the logical axioms being names of truth values which are not rational numbers. Recall that the set $[0,1] \cap \mathbb{Q}$ is dense in $[0,1]$.

**Lemma 6** *To every proof $w_A$ of a formula $A$ in $T$ there is a set of proofs*

$$W = \{w'_A;\ w'_A \text{ is a proof of } A \text{ in } T'\}$$

*such that*

$$\mathrm{Val}_T(w_a) = \bigvee_{w'_A \in W} \mathrm{Val}_{T'}(w'_A).$$

PROOF: Let the length of the proof $w_A$ be 1. If $A$ is not a symbol for a truth value then we put $W = \{w_A\}$.

Let $A := a$, $a \in [0,1]$. If $a \in \mathrm{Supp}(A_L - I)$ then, again, we put $W = \{w_A\}$. Let $a \in \mathrm{Supp}(I)$, i.e.,

$$w_A := a\,[a \in [0,1] - \mathbb{Q}; \mathrm{LA}].$$

Consider a proof

$$w_{c,A} := c\,[c, \mathrm{LA}],\ c \Rightarrow a\,[c \to a; r_{Rc}],\ a\,[c \otimes (c \to a); r_{MP}]$$

where $c \in [0,1] \cap \mathbb{Q}$. Obviously, $w_{c,A}$ is a proof of $\boldsymbol{a}$ ( $:=A$) in $T'$. Put $W = \{w_{c,A}; \ c < a\}$. Then

$$\bigvee_{w \in W} \mathrm{Val}_{T'}(w) = \bigvee \{c \otimes (c \to a); \ c < a\} = \bigvee \{c; \ c < a\} = a$$

since $c \to a = 1$.

Assume that the proposition holds for every proof $w_A$ of the length smaller than $n$. We denote $B := r^{syn}(A_1, \ldots, A_m)$ and let the proof $w_B$ have length $n$,

$$w_B := \ldots, w_{A_1}\,[a_1], \ldots, w_{A_m}\,[a_m], \ldots B\,[r^{sem}(a_1, \ldots, a_m); r]$$

where $w_{A_i}$ are proofs of $A_i$ of the length smaller than $n$ and $a_i = \mathrm{Val}_T(w_{A_i})$, $i = 1, \ldots, m$. Due to the inductive assumption

$$a_i = \bigvee_{w_i \in W_i} \mathrm{Val}_{T'}(w)$$

for some sets $W_i$ of proofs of the formulas $A_i$ in $T'$, $i = 1, \ldots, m$. Then we may put

$$w := w_1, \ldots, w_m, B\,[r^{sem}(\mathrm{Val}_{T'}(w_1), \ldots, \mathrm{Val}_{T'}(w_m)); r]$$

where $w_i \in W_i$, $i = 1, \ldots, m$ and

$$W = \{w; \ w_1 \in W_1, \ldots, w_m \in W_m\}.$$

Then

$$\begin{aligned}
\bigvee_{w \in W} \mathrm{Val}_{T'}(w) &= \\
&= \bigvee \{r^{sem}(\mathrm{Val}_{T'}(w_1), \ldots, \mathrm{Val}_{T'}(w_m)); \ w_1 \in W_1, \ldots, w_m \in W_m\} \\
&= r^{sem}(\bigvee_{w_1 \in W_1} \mathrm{Val}_{T'}(w_1), \ldots, \bigvee_{w_m \in W_m} \mathrm{Val}_{T'}(w_m)) \\
&= r^{sem}(a_1, \ldots, a_m) \\
&= \mathrm{Val}_T(w_B)
\end{aligned}$$

since $r^{sem}$ is an upper semicontinuous operation. □

**Theorem 11** *For every formula $A \in J(T)$ and a theory $T'$ given in (12),*

$$T \vdash_a A \qquad \textit{iff} \qquad T' \vdash_a A$$

*holds.*

PROOF: Let

$$a = \bigvee \{\mathrm{Val}_T(w_A); \ w_A \text{ is a proof of } A \text{ in } T\}.$$

By Lemma 6, to every proof $w_A$ there is a set $W_{w_A}$ such that

$$\mathrm{Val}_T(w_A) = \bigvee_{w \in W_{w_A}} \mathrm{Val}_{T'}(w).$$

Hence, the set of all the proofs of $A$ in $T'$ is equal to

$$W = \bigcup \{W_{w_A}; \ w_A \text{ is a proof of } A \text{ in } T\}$$

and we obtain

$$a = \bigvee_{w \in W} \mathrm{Val}_{T'}(w)$$

which gives the first implication. The converse implication immediately follows if we realize that every proof $w_A$ of $A$ in $T'$ is also a proof of $A$ in $T$.     □

It follows from this theorem that we may restrict ourselves only to fuzzy theories with a countable support of the fuzzy set of logical axioms. Therefore, in the sequel, we will assume that both *logical* $A_L$ as well as *special axioms* $A_S$ always have a countable support. We will denote by $\bar{F}_J$, $\bar{F}_J \subseteq F_J$, a set of all the formulas in which no $a \in \mathrm{Supp}(I)$ occurs. Hence, $\bar{F}_J$ is countable.

Now, let us turn our attention to recursive properties of fuzzy logic. Let $X$ be coded by numbers of $\mathbf{N}$. Then a fuzzy set $B \subseteq X$ is *weakly recursively enumerable* if there is a recursive function $b : X \times \mathbf{N} \longrightarrow [0,1] \cap \mathbb{Q}$ such that

$$Bx = \bigvee \{b(x,n); \ n \in \mathbb{N}\}, \qquad x \in X.$$

By $\bar{L}$ we denote a dense countable subset of $L$.

The following theorem due to L. Biacino and G. Gerla [1] characterizes weakly recursive enumerable fuzzy sets.

**Theorem 12 (Biacino, Gerla)** *Let*

$$\beta = \bigvee \{\alpha \in \bar{L}; \ \alpha < \beta\}$$

*or let $L = \bar{L}$ be a finite chain. Then the following is equivalent:*

*(a) $B \subseteq X$ is weakly recursively enumerable.*

*(b) A set*

$$K(B) = \{(x, \alpha) \in X \times \bar{L}; \ B(x) > \alpha\}$$

*is recursively enumerable.*

*(c) There is a recursive function $h : \bar{L} \longrightarrow \mathbb{N}$ such that*

$$B_\alpha = W_{h(\alpha)}$$

*where $W_{h(\alpha)}$ is a domain of some $h(\alpha)$-th partially recursive function from $X$ to $\bar{L}$.*

A fuzzy set $B : X \longrightarrow \bar{L}$ is recursively enumerable if its membership function is a partial recursive function. A recursively enumerable fuzzy set is weakly recursive enumerable.

Note that all the operations $\vee, \wedge, \otimes, \rightarrow$ are computable in $\bar{L}$. Similarly, $r^{sem}_{MP}$, $r^{sem}_G$, $\{r^{sem}_{Ra}; \ a \in \bar{L}\}$ are computable. The same will be assumed for any rule of inference.

**Lemma 7** *The fuzzy set of logical axioms $A_L$ is recursively enumerable.*

PROOF: By the assumption, $A_L : \bar{F}_J \longrightarrow \bar{L}$ where both $\bar{F}_J$ as well as $\bar{L}$ can be coded by natural numbers. Assume that this is already done using the Gödel's construction. If $g$ is a Gödel's number of a logical axiom not being a symbol for truth value then $A_L(g) = \mathbf{1}$. If $g_a$ is a Gödel's number of $a \in \bar{L}$ then $A_L(g_a) = g_a$, otherwise $A_L(g) = \mathbf{0}$. Hence, $A_L$ is a recursive function. $\square$

It trivially follows from this lemma that $A_L$ is a weakly recursively enumerable fuzzy set.

The following theorem is based on the results of L. Biacino and G. Gerla.

**Theorem 13** *Let $T$ be a fuzzy theory. If $A_S$ is weakly recursively enumerable then the fuzzy set of syntactic conclusions $C^{syn} A_S \subseteq F_{J(T)}$ is weakly recursively enumerable.*

PROOF: We will prove that there is a recursive function $h : \bar{F}_{J(T)} \times \mathbb{N} \longrightarrow \bar{L}$ such that

$$(C^{syn} A_S)B = \bigvee_{n \in \mathbb{N}} h(B, n)$$

holds for every $B \in \bar{F}_{J(T)}$.

Let a proof $w$ in $T$ have a length 1. Then its last formula is a logical or a special axiom and

$$\mathrm{Val}(w) = \bigvee_{n \in \mathbb{N}} g(w)$$

for some recursive function $g$, by assumption.

Let the inductive assumption hold, the length of $w$ be $m$ and its value be obtained from the values of proofs $w_1, \ldots, w_k$ of the length smaller than the length of $w$ using a rule $r$. Then

$$
\begin{aligned}
\mathrm{Val}(w) &= r^{sem}(\mathrm{Val}(w_1), \ldots, \mathrm{Val}(w_k)) \\
&= r^{sem}( \bigvee_{n_1 \in \mathbb{N}} g_1(w_1, n_1) \ldots, \bigvee_{n_k \in \mathbb{N}} g_k(w_k, n_k)) \\
&= \bigvee_{n_1, \ldots, n_k \in \mathbb{N}} r^{sem}(g_1(w_1, n_1), \ldots, g_k(w_k, n_k)) \\
&= \bigvee_{n \in \mathbb{N}} g(w, n)
\end{aligned}
$$

since $r^{sem}$ is, by assumption, computable and preserves unions.

Let all the proofs $w_B$ of $B$ be numbered by natural numbers $p \in \mathbb{N}$. We denote them by $w_{B,p}$. By the previous part of the proof,

$$
\mathrm{Val}(w_{B,p}) = \bigvee_{q \in \mathbb{N}} g(w_{B,p}, q).
$$

Then

$$
\begin{aligned}
(C^{syn})B &= \bigvee_{p \in \mathbb{N}} \mathrm{Val}(w_{B,p}) \\
&= \bigvee_{p \in \mathbb{N}} \bigvee_{q \in \mathbb{N}} g(w_{B,p}, q) \\
&= \bigvee_{n \in \mathbb{N}} h(B, n)
\end{aligned}
$$

for some recursive function $h$.                                                        □

By this theorem, we demonstrated that first–order fuzzy logic has analogous recursive properties as classical logic, but in a weaker sense. However, this concerns only formulas from $\bar{F}_{J(T)}$. We may state only some approximation propositions about formulas from $F_{J(T)} - \bar{F}_{J(T)}$ since there is uncountably many of them.

# 4   ELEMENTS OF MODEL THEORY

## 4.1   Few fundamental notions

In this section, we present some direct generalizations of classical notions of model theory.

Two models $\mathcal{D}$ and $\mathcal{D}'$ are *isomorphic*, $\mathcal{D} \cong \mathcal{D}'$, if there is a one–to–one mapping

$$
g : D \longrightarrow D'
$$

such that the following hods for all $x_1, \ldots, x_n \in D$:

(i) For each couple of functions $f_D$ in $D$ and $f_{D'}$ in $D'$ assigned to a functional symbol $f \in J$

$$g(f_D(x_1, \ldots, x_n)) = f_{D'}(g(x_1), \ldots, g(x_n)).$$

(ii) For each couple of fuzzy relations $p_D$ in $D$ and $p_{D'}$ in $D'$ assigned to a predicate symbol $p \in J$

$$p_D(x_1, \ldots, x_n) = a \quad \text{iff} \quad p_{D'}(g(x_1), \ldots, g(x_n)) = a$$

where $a \in L$.

(iii) For each couple of constants $u$ in $D$ and $u'$ in $D'$ assigned to a constant symbol $u \in J$

$$g(u) = u'.$$

Two models $D$ and $D'$ are *elementary equivalent*, $D \equiv D'$, if

$$D(A) = D'(A) \tag{13}$$

holds for every formula $A \in J(T)$.

$D$ is a *submodel* of $D'$, $D \subset D'$ if $D \subseteq D'$, $f_D = f_{D'}|D^n$, and $p_D = p_{D'} \cap D^n$. The $D$ is an *elementary submodel* of $D'$, $D \prec D'$, if, moreover, (13) holds for every formula $A$. The model $D'$ is an *extension* or *elementary extension* of $D$, respectively.

The following is immediate.

**Lemma 8**

*(a) If $D \cong D'$ then $D \equiv D'$.*

*(b) If $D \subset D'$ then $D(A) = D'(A)$ holds for every formula without variables.*

*(c) If $D \prec D'$ then $D \equiv D'$.*

Let $D_1 \subset D_2 \subset \ldots \subset D_\alpha \subset \ldots$ be a chain of models, $\alpha < \xi$ for some ordinal number $\xi$. Then

$$D = \bigcup_{\alpha < \xi} D_\alpha$$

is a model in which $D = \bigcup_{\alpha < \xi} D_\alpha$ and each $f_D$ is a union of all $f_{D_\alpha}$, $\alpha < \xi$. Furthermore, we extend $p_{D_\alpha}$ into $D$ by putting

$$p_D(d_1, \ldots, d_n) = \bigvee_{\alpha < \xi} p_{D_\alpha}(d_1, \ldots, d_n)$$

for each predicate symbol and a each sequence $d_1, \ldots, d_n \in D$, where we put $p_{D_\alpha}(d_1, \ldots, d_n) = 0$ if it is undefined.

**Theorem 14** *Let*

$$\mathcal{D}_1 \prec \mathcal{D}_2 \prec \ldots \prec \mathcal{D}_\alpha \ldots$$

*be an elementary chain of models, $\alpha < \xi$ for some ordinal number $\xi$. Then*

$$\mathcal{D} = \bigcup_{\alpha < \xi} \mathcal{D}_\alpha$$

*is an elementary extension of each $\mathcal{D}_\alpha$, i.e., $\mathcal{D}_\alpha \prec \mathcal{D}$ for every $\alpha$.*

PROOF: If $A := a$ then $\mathcal{D}(A) = \mathcal{D}_\alpha(A) = a$ for every $\alpha < \xi$.
Let $A := p(x_1, \ldots, x_n)$, $t_1, \ldots, t_n$ be terms and $\alpha < \xi$. Then

$$
\mathcal{D}(p(t_1, \ldots, t_n)) = \bigwedge_{d_1, \ldots, d_q \in \bigcup_{\alpha < \xi} D_\alpha} \left( \bigvee_{\alpha < \xi} \mathcal{D}_\alpha(p(t_1, \ldots, t_n))[d_1, \ldots, d_q] \right)
$$

$$
= \bigwedge_{d_1, \ldots, d_q \in D_\alpha} \mathcal{D}_\alpha(p(t_1, \ldots, t_n))[d_1, \ldots, d_q]
$$

since $p_{\mathcal{D}_\alpha}(t_1, \ldots, t_n)[d_1, \ldots, d_q] = p_{\mathcal{D}_{\alpha+1}}(t_1, \ldots, t_n)[d_1, \ldots, d_q]$ for every $\alpha < \xi$ and $d_1, \ldots, d_q \in D_\alpha$.

For $A := B \Rightarrow C$ and $A := \dot{c}(A_1, \ldots, A_n)$ we obtain the statement by inductive assumption.

Finally, let $A := (\forall x_i)B$ and $\alpha < \xi$. Then

$$
\mathcal{D}(A) = \bigwedge_{d_1, \ldots, d_q \in \bigcup_{\alpha < \xi} D_\alpha} \bigwedge_{d \in \bigcup_{\alpha < \xi} D_\alpha} \mathcal{D}(B)[d_1, \ldots, d_{i-1}, d, d_{i+1}, \ldots, d_q]
$$

$$
= \bigwedge_{\alpha < \xi} \left( \bigwedge_{d_1, \ldots, d_q \in D_\alpha} \bigwedge_{d \in D_\alpha} \mathcal{D}_\alpha(B)[d_1, \ldots, d_{i-1}, d, d_{i+1}, \ldots, d_q] \right)
$$

$$
= \bigwedge_{d_1, \ldots, d_q \in D_\alpha} \bigwedge_{d \in D_\alpha} (\mathcal{D}_\alpha(B)[d_1, \ldots, d_{i-1}, d, d_{i+1}, \ldots, d_q])
$$

by the inductive assumption.                                                              □


**Theorem 15 (Compactness)** *A fuzzy theory $T$ has model iff each its finite subtheory $T' \subseteq T$ has a model.*

PROOF: The implication left to right is immediate. Conversely, let each $T'$ be consistent and have a model $\mathcal{D}' \models T'$. We prove that $T$ is also consistent. Let $T$ be contradictory. Then there is and $a \in L - \{1\}$ and a proof $w$ of $a$ such that

$$\text{Val}(w) > a.$$

Let $B_S = \{B_1, \ldots, B_n\}$ be all the special axioms in $w$. Put $A'_S = A_S | B_S$. Obviously, $\mathrm{Supp}(A'_S) = B_S$ is finite, i.e., the corresponding fuzzy theory $T'$ is contradictory and has no model — a contradiction. Hence, we conclude that $T$ has a model by completeness theorem. □

## 4.2  Ultraproduct theorem

In this section, we will deal with sets of models and prove an analogy of the famous Los' ultraproduct theorem for fuzzy logic.

Let $\mathcal{L}$ be a residuated lattice and $I$ an index set. Consider a power

$$\mathcal{L}^I = \langle L^I, \vee, \wedge, \otimes, \rightarrow, \mathbf{1}^I, \mathbf{0}^I \rangle \tag{14}$$

where $L^I$ is a power of $L$ (a set of functions $I \longrightarrow L$) with the operations $\vee, \wedge, \otimes, \rightarrow$ defined by components, i.e., for example,

$$(a \vee b)(i) = a(i) \vee b(i) \qquad i \in I,$$

and similarly for the other operations. By $a^I \in L^I$ we denote an element such that $a(i) = a$ for all $i \in I$. It is easy to verify that (14) is a residuated lattice. Moreover, since $\mathcal{L}$ is complete, $\mathcal{L}^I$ is complete, as well.

A set $G \subseteq \mathcal{L}^I$ is a *filter* if the following holds:

(i) $\mathbf{0}^I \notin G$.

(ii) $x \in G$ and $x \leq y$ implies $y \in G$.

(iii) If $x, y \in G$ then $x \otimes y \in G$.

A maximal filter will be called an *ultrafilter*.
Sometimes, we will require also the following condition

(iv) If $P \subseteq G$ then $\bigwedge P \in G$.

Given a filter $P$, we define the following property:

$$\mathbf{1}^I \in G \text{ and } a^I \notin G \text{ for any } a < 1. \tag{15}$$

**Lemma 9** *Let $G$ be a filter with the property (15). Then it is possible to extend it into an ultrafilter with the same property.*

PROOF: By Zorn's lemma. □

**Lemma 10** *Let $G \subseteq \mathcal{L}^I$ be an ultrafilter with the property (15). Then*

$$x \notin G \text{ iff there are } y \in G, c < 1 \text{ and } n > 0 \text{ such that } x^n \otimes y \leq c^I.$$

PROOF:   Let $x \in G$ and $y, c, n$ fulfil the right hand condition. Then $x^n \otimes y \leq c^I$ and since $x^n, y \in G$, we have $c^I \in G$ due to the assumption that $G$ is a filter — a contradiction.

Conversely, let there be no $c < 1$ such that $x^n \otimes y \leq c^I$. Then

$$G' = \{z;\ (\exists n)(\exists y \in G)(x^n \otimes y \leq z)\}$$

is a filter with the property (15) and $G \cup \{x\} \subseteq G'$. Since $G$ is maximal, we have $G = G'$, i.e., $x \in G$.                                                                                    □

**Lemma 11** *Let $G \subseteq \mathcal{L}^I$ be an ultrafilter with the property (15). Then to every $x \in G$ there is $i \in I$ such that $x(i) = 1$.*

PROOF:   Let there be $x \in G$ such that $x(i) < 1$ for every $i \in I$. Since $\otimes$ is nilpotent, it is possible to find $n > 0$, $y \in G$ and $c < 1$ such that $x^n \otimes y \leq c^I$ which gives $x \notin G$ due to Lemma 10 — a contradiction.                                           □

Let $G$ be an ultrafilter with the property (15). Put

$$x \sim y \qquad \text{iff} \qquad x \leftrightarrow y \in G. \tag{16}$$

**Lemma 12** *The relation $\sim$ given in (16) is a congruence on $\mathcal{L}^I$. If $G$ has, moreover, a property (iv) then $\sim$ is also a congruence with respect to $\bigwedge$.*

PROOF:   The first part of this lemma follows from the properties of $G$ and biresiduation $\leftrightarrow$ (cf. [6, 7, 12]).

Let $x_j \leftrightarrow y_j \in G$, $j \in J$ hold for some index set $J$. Then $\bigwedge_{j \in J}(x_j \leftrightarrow y_j) \in G$ and since $\bigwedge_{j \in J}(x_j \leftrightarrow y_j) \leq \bigwedge_{j \in J} x_j \leftrightarrow \bigwedge_{j \in J} y_j$ we obtain that $\bigwedge_{j \in J} x_j \sim \bigwedge_{j \in J} y_j$.                                                                                    □

We denote by $L^I|G$ a factor set on $L^I$ with respect to $\sim$. The equivalence class of $x \in L^I$ is denoted by $[x]$.

Note that $G$ must be an ultrafilter with the property (15) since otherwise Lemma 10 would not hold and moreover, since $b^I \leftrightarrow a^I = c^I$ for some $c \leq 1$, $[a^I] = [b^I]$ might hold for some $a \neq b$.

The operations $\vee, \wedge, \otimes, \rightarrow, \bigwedge$ can be defined on $L^I|G$ as usual. We obtain a residuated lattice which will be denoted by $\mathcal{L}^I|G$.

**Theorem 16** *Let $G$ be an ultrafilter on $L^I$ with the properties (iv) and (15). Then $\mathcal{L}^I|G \cong \mathcal{L}$.*

PROOF:

First, we prove that to every $[x] \in L^I|G$ there is $a \in L$ such that $[x] = [a^I]$.

Let $x \leftrightarrow a^I \notin G$ for every $a \in L$. Then there are $y_a \in G$, $n_a > 0$ and $c_a < 1$ such that

$$(x \leftrightarrow a^I)^{n_a} \otimes y_a \leq c_a^I.$$

Due to the property (iv),

$$y = \bigwedge \{y_a;\ a \in L\} \in G$$

and

$$(x \leftrightarrow a^I)^{n_a} \otimes y \le (x \leftrightarrow a^I)^{n_a} \otimes y_a \le c_a^I.$$

for every $a$. By Lemma 11, there is $i \in I$ such that $y(i) = 1$. Choose $a \in L$ such that $x(i) = a$. Then $(x(i) \leftrightarrow a)^{n_a} \otimes y(i) = 1 > c_a$ — a contradiction. Hence, there is $a$ such that

$$x \leftrightarrow a^I \in G,$$

i.e., $x \in [a^I]$.

Now, let us define two functions $f : L \longrightarrow L^I|G$ and $g : L^I|G \longrightarrow L$ by

$$f(a) = [a^I], \qquad a \in L$$

and

$$g([a^I]) = a.$$

Then

$$g(f(a)) = g([a^I]) = a$$

and

$$f(g([a^I])) = f(a) = [a^I].$$

We have obtained $L^I|G \cong L$. Furthermore,

$$g([a^I] \vee [b^I]) = g([a^I \vee b^I]) = g([(a \vee b)^I]) = a \vee b$$

and analogously for the operations $\wedge, \otimes$ and $\rightarrow$ because $\sim$ is a congruence with respect to these operations.

Let $P \subseteq L^I|G$ and denote $P' = \{a;\ [a^I] \in P\}$. Then

$$g\left(\bigwedge P\right) = g\left(\left[\bigwedge \{a^I;\ [a^I] \in P\}\right]\right) = g\left(\left[\left(\bigwedge \{a;\ a \in P'\}\right)^I\right]\right) = \bigwedge \{a;\ a \in P'\}$$

because $\sim$ is a congruence also with respect to $\bigwedge$. □

Let us now demonstrate the connection between the filters in $\mathcal{L}^I$ and in $I$. Let $x \in \mathcal{L}^I$ and

$$B_x = \{i \in I;\ x(i) = 1\}$$

Obviously, $B_0 = \emptyset$. If $G \subseteq L^I$ then we denote

$$H_G = \{B_x;\ x \in G\}. \tag{17}$$

**Lemma 13** *If $G$ is a filter then $H_G$ is a filter. If $G$ is a maximal filter then $H_G$ is also a maximal filter.*

PROOF: First, let us realize that given $B \subseteq I$, there is $x \in L^I$ such that $B_x = B$. Furthermore, if $B \subseteq B' \subseteq I$ then there are $x \leq y$ such that $B_x = B$ and $B_y = B'$. Indeed, if $j \notin B'$ then $x(j) < 1$ and thus, it will do to put $y(j) = a \in L$ such that $x(j) \leq a < 1$.

Now, $\emptyset \notin H_G$ since $0^I \notin G$. Let $x \in G$, $B_x \in H_G$ and $B_x \subseteq B$. Then there is $y \geq x$ such that $B_y = B$. Since $y \in G$, we obtain $B \in H_G$.

Let $B_x, B_y \in H_G$. Then

$$B_x \cap B_y = \{i; \ x(i) = 1 \text{ and } y(i) = 1\} = \{i; \ (x \wedge y)(i) = 1\} = B_{x \cap y}.$$

Since $x \wedge y \in G$, we have $B_x \cap B_y \in H_G$.

Let $G$ be maximal and $H$ be a filter on $I$ such that $H_G \subseteq H$. Choose $B \in H$. Then $B \subseteq I$ and there is $y$ such that $B = B_y$. By Lemma 10, $y \notin G$ if there are $z \in G$, $n > 0$ and $a < 1$ such that

$$y^n \otimes z \leq a^I. \tag{18}$$

But $B_z \in H$ and since $H$ is a filter, $B_z \cap B_y \neq \emptyset$, i.e., there is $i \in I$ such that $z(i) = 1$ and $y(i) = 1$ which contradicts to (18).  □

The rest of this section will be devoted to the formulation and proof of the ultraproduct theorem. By $T$, we denote a fixed consistent first–order fuzzy theory.

Let us be given a set of models

$$\{\mathcal{D}_i; \ \mathcal{D}_i \models T, i \in I\} \tag{19}$$

where $I$ is some index set. Analogously as in classical logic, we construct a new structure $\mathcal{D}_G$.

Let

$$D = \prod_{i \in I} D_i$$

where $D_i$, $i \in I$ are supports of the corresponding models $\mathcal{D}_i$. Given a term $t$, we denote by $\mathcal{D}(t)$ an $I$-tuple

$$\mathcal{D}(t) = \langle \mathcal{D}_i(t); \ i \in I \rangle$$

and

$$\mathcal{D}(p(t_1, \ldots, t_n)) = \langle p_{\mathcal{D}_i}(\mathcal{D}_i(t_1), \ldots, \mathcal{D}_i(t_n)); \ i \in I \rangle \in L^I$$

where $p_{\mathcal{D}_i}$ are fuzzy relations assigned to predicate symbol $p$ in the respective models $\mathcal{D}_i$. Similarly, $\mathcal{D}(A)$ denotes an element $\langle \mathcal{D}_i(A); \ i \in I \rangle \in L^I$.

Let $G \subseteq L^I$ be an ultrafilter and $H_G$ the adjoint ultrafilter from (17). We define

$$d \approx d' \quad \text{iff} \quad \{i; \ d(i) = d'(i)\} \in H_G$$

for every $d, d' \in D$. Obviously, $\approx$ is an equivalence on $D$. We put

$$D_G = D| \approx .$$

This factor set will serve us as a support of the structure $\mathcal{D}_G$. Its elements, equivalence classes, will be denoted by $|d|, d \in D$.

Given a function symbol $f$. We put

$$f_{\mathcal{D}_G}(|d_1|, \ldots, |d_n|) = |\langle f_{\mathcal{D}_i}(d_1(i), \ldots, d_n(i)); i \in I \rangle| = |d| \quad \text{iff}$$
$$\{i; f_{\mathcal{D}_i}(d_1(i), \ldots, d_n(i)) = d(i)\} \in H_G. \tag{20}$$

Let $p$ be a predicate symbol. Then we interpret $p$ in $\mathcal{D}_G$ by a fuzzy relation

$$p_{\mathcal{D}_G}(|d_1|, \ldots, |d_n|) = a \quad \text{iff} \quad p_{\mathcal{D}}(d_1, \ldots, d_n) \leftrightarrow a^I \in G \tag{21}$$

where $p_{\mathcal{D}_G}, p_{\mathcal{D}}$ are interpretations of the predicate symbol $p$ (i.e., fuzzy relations) in $D_G$ and $D$, respectively and $|d_j| = \mathcal{D}_G(t_j) = |\mathcal{D}(t_j)|, j = 1, \ldots, n$.

**Lemma 14** *The definitions (20) and (21) do not depend on the choice of the representatives from $|d|$.*

PROOF: The proof of (20) is identical with the proof of the analogous statement in classical logic.

Let $d_j, d_j', j = 1, \ldots, n$ be representatives of the same equivalence class and $\mathcal{D}_G(t_j) = |d_j| = |d_j'|$. Then

$$\mathcal{D}_G(p(t_1, \ldots, t_n)) = a = p_{\mathcal{D}_G}(|d_1|, \ldots, |d_n|) = p_{\mathcal{D}_G}(|d_1'|, \ldots, |d_n'|).$$

Due to the definition, this holds iff $p_{\mathcal{D}}(d_1, \ldots, d_n) \leftrightarrow a^I \in G$. Since

$$p_{\mathcal{D}}(d_1, \ldots, d_n) \leftrightarrow p_{\mathcal{D}}(d_1', \ldots, d_n') \in G,$$

we obtain $p_{\mathcal{D}}(d_1', \ldots, d_n') \leftrightarrow a^I \in G$ using the properties of $\leftrightarrow$ and the fact that $G$ is a filter. □

The structure $\mathcal{D}_G$ for an ultrafilter $G$ is now defined as

$$\mathcal{D}_G = \langle D_G, p_{\mathcal{D}_G}, \ldots, f_{\mathcal{D}_G}, \ldots, u_G, \ldots \rangle \tag{22}$$

where $p_{\mathcal{D}_G}$ are fuzzy relations defined in (21), $f_{\mathcal{D}_G}$ functions defined by (20), and $u_G = |\mathcal{D}(u)|$ for a constant $u$. We will call $\mathcal{D}_G$ an *ultraproduct* of $\{\mathcal{D}_i; i \in I\}$.

The following is a fuzzy analogy of the Los' ultraproduct theorem.

**Theorem 17** *Let $G \subseteq L^I$ be an ultrafilter on $L^I$ with the properties (iv) and (15) and $\mathcal{D}_G$ the structure (22). Then*

*(a)*

$$\mathcal{D}_G(t) = |d| \quad \text{iff} \quad \{i; \mathcal{D}_i(t) = d(i)\} \in H_G$$

*holds for every term $t$.*

(b)

$$\mathcal{D}_G(A) = a \quad iff \quad \mathcal{D}(A) \leftrightarrow a^I \in G$$

holds for every formula A.

PROOF: Due to Lemma 13, $H_G$ is a filter and so the proof of (a) is identical with the proof of analogous statement in classical logic.

(b) Given a formula A, we have $\mathcal{D}(A) = \langle \mathcal{D}_i(A); \ i \in I \rangle \in L^I$. The right hand side of the statement is then equivalent with $\mathcal{D}(A) \in [a^I]$. Taking a canonical epimorphism $a^I \mapsto [a^I]$ and using Theorem 16 we conclude that this is equivalent with $\mathcal{D}_G(A) = a$.                                                             □

Theorem 17 holds provided that the ultrafilter G has a property (iv). This assumption may be avoided if we deal with Henkin extension of the fuzzy theory.

On the basis of Lemma 4, given a set (19) of models of the fuzzy theory T, we extend it conservatively to a Henkin theory $T_H$ and each model $\mathcal{D}_i \models T$ to a model $\mathcal{D}_i' \models T_H$ such that

$$\mathcal{D}_i'(A) = \mathcal{D}_i(A)$$

for every formula $A \in F_{J(T)}$. This leads to the following theorem.

**Theorem 18** *Let $G \subseteq L^I$ be an ultrafilter on $L^I$ with the property (15) and $\{\mathcal{D}_i'; \ i \in I\}$ be a set of Henkin extensions of the respective models $\mathcal{D}_i$ from (19). Let $\mathcal{D}_G$ be an ultraproduct of $\{\mathcal{D}_i'; \ i \in I\}$. Then*

(a)

$$\mathcal{D}_G(t) = |d| \quad iff \quad \{i; \ \mathcal{D}_i(t) = d(i)\} \in H_G$$

holds for every term t without variables.

(b)

$$\mathcal{D}_G(A) = a \quad iff \quad \mathcal{D}(A) \leftrightarrow a^I \in G$$

holds for every formula $A \in F_{J(T)}$.

PROOF: (a) If t is a constant then $\mathcal{D}_G(t) = |d|$ iff $\mathcal{D}'(t) \in |d|$ iff $\{i; \ \mathcal{D}_i'(t) = \mathcal{D}_i(t) = d(i)\} \in H_G$ where $d(i)$ is an interpretation of the constant t in the model $\mathcal{D}_i$. We may further proceed by induction on the length of t.

(b) For $A := \boldsymbol{a}$, $a \in L$ is the statement obvious. Let $A := p(t_1, \ldots, t_n) \in F_{J(T)}$. By Lemma 4, we have

$$\mathcal{D}_G(A) = a \quad iff \quad \mathcal{D}'(A) \leftrightarrow a \in G \quad iff \quad \mathcal{D}(A) \leftrightarrow a \in G.$$

For $A := B \Rightarrow C$ and $A := c(B_1, \ldots, B_n)$ the statement holds by the inductive assumption, the properties of $\leftrightarrow$ (fitting condition), and the fact that G is a filter.

Let $A := (\forall x_i)B(x)$ and $r$ be a special constant for $A$. Then we have

$$
\begin{aligned}
\mathcal{D}_G(A) = a \quad &\text{iff} \quad \mathcal{D}'((\forall x_i)B) \leftrightarrow a^I \in G \\
&\text{iff} \quad \bigwedge_{\substack{d_1,\ldots,\ldots,d_q \in D \\ d_i = r}} \mathcal{D}'(B)[d_1,\ldots,d_{i-1},r,d_{i+1},\ldots,d_q] \leftrightarrow a^I \in G \\
&\text{iff} \quad \mathcal{D}'(B_x[r]) \leftrightarrow a^I \in G \\
&\text{iff} \quad \mathcal{D}((\forall x_i)B) \leftrightarrow a^I \in G.
\end{aligned}
$$

$\square$

In this theorem, we have used special properties of Henkin extension of a fuzzy theory. However, the ultraproduct is constructed from Henkin extensions of the models from the given set of models and not from the given models themselves.

# 5 CONCLUSION

In this paper, we have briefly presented a first–order fuzzy logic and discussed its recursive properties. The recursive properties of fuzzy logic are weaker than those of classical logic and we have to confine ourselves only to the formulas containing symbols $a$ for truth values $a \in [0,1]$ being rational numbers.

In Section 4, we have presented an outline of the model theory in fuzzy logic and proved a fuzzy analogy of the Los' ultraproduct theorem in two versions. The first version assumed that the given ultrafilter is closed with respect to meets and the second one uses properties of Henkin extension of a fuzzy theory.

The theory presented in this paper may have important impact on the theory of approximate reasoning. Using the ideas formulated in [8, 9], it is possible to study properties of truth values during the inference in approximate reasoning.

## References

[1] Biacino, L. and G. Gerla: *Weak Decidability and Weak Recursive Enumerability for L-subsets*, preprint No. 45, Università Degli Studi di Napoli, Napoli 1986.

[2] Biacino, L. and G. Gerla: *Recursively Enumerable L-sets*, Zeit. Math. Logic. Grundl. Math. 1988.

[3] Chang, C. C. and H. J. Keisler: *Model Theory*, North–Holland, Amsterdam 1973.

[4] Chang, C. C. and H. J. Keisler: *Continuous Model Theory*, Princeton University Press, Princeton 1966.

[5] Goguen, J. A.:*The logic of inexact concepts*, Synthese **19**(1968-69), 325–373.

[6] Novák, V.: *Fuzzy Sets and Their Applications*, Adam–Hilger, Bristol, 1989.

[7] Novák, V.: *On the Syntactico-Semantical Completeness of First–Order Fuzzy Logic. Part I — Syntactical Aspects; Part II — Main Results*. Kybernetika **26**(1990), 47–66; 134–154.

[8] Novák, V.: *Fuzzy logic as a basis of approximate reasoning*, in L. A. Zadeh and J. Kacprzyk, Eds.: Fuzzy Logic for the Management of Uncertainty. J. Wiley and Sons, New York 1992.

[9] Novák, V.: *On the logical basis of approximate reasoning*, in V. Novák, J. Ramík, M. Mareš, M. Černý and J. Nekola, Eds.: Fuzzy Approach to Reasoning and Decision Making. Kluwer, Dordrecht 1992.

[10] Novák, V.: *Fuzzy Sets, Fuzzy Logic and Natural Language*. Int. J. of General Systems **20**(1991), 83–97.

[11] Novák, V. and W. Pedrycz: *Fuzzy Sets and t-norms in the Light of Fuzzy Logic*, Int. J. Man.-Mach. Stud. **29**(1988), 113–127.

[12] Pavelka, J.: *On fuzzy logic I, II, III*, Zeit. Math. Logic. Grundl. Math. **25**(1979), 45–52, 119–134, 447–464.

[13] Scarpellini, B.: *Die Nichaxiomatisierbarkeit des unendlichwertigen Prädikatenkalküls von Łukasiewicz*, J. of Symbolic Logic **27**(1962), 159–170.

[14] Trachtenbrot, B. A.: *Impossibility of the algorithm of the decidability problem on finite classes*, Doklady Akademii Nauk SSSR **70**(1950), 569–572 (in Russian).

[15] Zadeh, L. A.: *The concept of a linguistic variable and its application to approximate reasoning I, II, III*, Inf. Sci.,**8**(1975), 199–257, 301–357;**9**(1975), 43–80.

# Bibliography

[1990] J. Adámek, H. Herrlich and G.E.Strecker .
*Abstract and Concrete Categories* (John Wiley & Sons , New York 1990).

[1977] P. Aczel .
*An introduction to inductive definitions* ,"Handbook of Mathematical Logic" (ed. J. Barwise) Chapter C.7, pages 739–782. (North-Holland, Amsterdam, 1977).

[1988] M. Anderson and T. Feil .
*Lattice Ordered Groups: an Introduction* (D. Reidel Publishing Company, Dordrecht, 1988).

[1975] M.A. Arbib and E.G. Manes .
*Arrows, Structures and Functors. The Categorical Imperative* (Academic Press , New York 1975).

[1970] R. Balbs and A. Horn .
*Injective and projective Heyting algebra*, Trans. Amer. Soc. **148** (1970), 549–559.

[1980] W. Bandler and L.J. Kohout .
*Fuzzy power sets and fuzzy implication operators*, Fuzzy Sets and Systems 4 (1980),13–30.
(Reprinted in: *Readings in Fuzzy Sets for Intelligent Systems*, D. Dubois, H. Prade and R. Yager (eds.), Morgam Kaufmann Publishers, San Mateo, Calif., in press)

[1980] ———.
*Fuzzy relational products as a tool for analysis and synthesis of the behaviour of complex natural and artificial systems*, in "Fuzzy Sets: Theory and Applications to Policy Analysis and Information Systems"

(ed. P.P Wang and S.K. Chang), pages 341–367. (Plenum press, New York and London, 1980).

[1980] ――――.
*Semantics of implication operators and fuzzy relational products*, Internat. Journal of Man-Machine Studies **12** (1980), 89–116.
(Reprinted in Mamdani, E.H. and Gaines, B.R. eds. *Fuzzy Reasoning and its Applications*, Academic Press, London, 1981, pages 219-246)

[1986] ――――.
*On the general theory of relational morphisms*, International Journal of General Systems **13** (1986), 47–66.

[1986] ――――.
*The use of checklist paradigm in inference systems*, in "Fuzzy Logic in Knowledge Engineering" (eds. C.V. Negoita and H. Prade) chapter 7, pages 95–111 (Verlag TÜV Rheinland, Köln, 1986).

[1988] ――――.
*Special properties, closures and interiors of crisp and fuzzy relations*, Fuzzy Sets and Systems **26** (1988), 317–332.

[1986] M. Barr .
*Fuzzy set theory and topos theory*, Canad. Math. Bull **29** (1986), 501–508.

[1985] M. Barr and C. Wells .
*Toposes, Triples and Theories*, Grundlehren der mathematischen Wissenschaften **278** (Springer–Verlag 1985).

[1964] L.P. Belluce .
*Further results on infinite valued predicate logic*, Journal of Symbolic Logic **29** (1964), 69–78.

[1986] ――――.
*Semi–simple algebras of infinite valued logic and bold fuzzy set theory*, Canad. J. Math. **38** (1986), 1356–1379.

[1992] ――――.
*Semi–simple and complete MV–algebras*, Algebra Universalis **29** (1992), 1–9 .

[1963] L.P. Belluce and C.C. Chang .
*A weak completeness theorem for infinite valued first–order logic*, Journal of Symbolic Logic **28** (1963), 43–50.

[1988] L. Biacino and G. Gerla .
*Recursively enumerable L-sets*, Zeit. Math. Logik u. Grundl. Math. (1988).

[1977] A. Bigard, K. Keimel, and S. Wolfenstein .
*Groupes et Anneaux Reticules*, Lecture Notes in Mathematics **608** (Springer–Verlag, Berlin, New York 1977).

[1973] G. Birkhoff .
*Lattice Theory* , Amer. Math. Soc. Colloquium Publications, third edition (Amer. Math. Soc., RI, 1973).

[1936] G. Birkhoff and J.v. Neumann .
*The logic of quantum mechanics*, Ann. of Math.,Ser. 2, **37**(1936), 823–843.

[1966] R. Bull .
*MPS as the formalization of an intuitionistic concept of modality*, Journal of Symbolic Logic **31** (1966), 609–616.

[1958] C.C. Chang .
*Algebraic analysis of many valued logics*, Trans. Amer. Math. Soc. **88** (1958), 467 – 490 .

[1959] ———.
*A new proof of the completeness of the Lukasiewicz axioms*, Trans. Amer. Math. Soc. **93** (1959), 74 – 80.

[1966] C.C. Chang and H.J. Keisler .
*Continuous Model Theory*, Annals of Math. Studies **58** (Princeton University Press 1966).

[1973] ———.
*Model Theory*, (North-Holland, Amsterdam, 1973).

[1986] U. Cerruti and U. Höhle .
*An approach to uncertainty using algebras over a monoidal closed category*, Supplemento ai Rendiconti del Circolo Matematico di Palermo (Serie II) **12** (1986), 47–63.

[1991] R. Cignoli .
*Complete and atomic algebras of the infinite-valued Lukasiewicz logic*, Studia Logica **50** (1991), 375–384.

[1992] ———.
Free lattice-ordered abelian groups and varieties of MV algebras, in :
"Proc. Latin American Symp. Logic, Bahia Blanca 1992", Notas de
Logica Matematica, Univ. Nacional del Sur, Bahia Blanca, Argentina,
1994.

[1991] R. Cignoli, A. Di Nola, and A. Lettieri .
Priestly duality and quotient lattices of many-valued algebras, Rend.
Circ. Mat. Palermo (2) **40** (1991), 371–384.

[1990] P. Conrad and J. Martinez .
Complemented lattice-ordered groups, Indag. Math. **3** (1990), 281–298.

[1992] J.L. Coulon, J. Coulon and U. Höhle .
Injectivité dans la catégorie des ensembles fortement M–valués com-
plets, C.R. Acad. Sci. Paris **314** (1992), Série I, 591–594.

[1977] H.B. Curry .
Foundations of Mathematical Logic, (Dover, New York, 1977).

[1988] D Dubois and H. Prade .
Possibility Theory, (Plenum Press, New York, 1988).

[1974] L. Esakia .
Topological Kripke model, Dokl. Sov. Akad. **214**, 2, 298–301.

[1989] ———.
The provability status of intuitionistic logic with the maximality principle (Preprint, 1989).

[1988] ———.
The provability logic with quantifier modalities, IV Soviet-Finnish Symposium on Intensional Logic and Logical Structure of Theory, "Metsniereba", Tbilisi, (1988), 4–10.

[1973] L. Esakia and R. Grigolia .
The criterion of Brouwerian and closure algebras to be finitely generated, Bull. Sect. Logic 6, 2, (1973).

[1979] M. Fourman and D.S. Scott .
Sheaves and logic, in : "Applications of Sheaves" , Lecture Notes in
Mathematics **753**, 302 – 401 (Springer–Verlag 1979).

[1990] Peter J. Freyd and André Ščedrov .
Categories, Allegories (North-Holland Elsevier, Amsterdam 1990).

[1938] O. Frink .
*New algebras of logic*, Amer. Math. Monthly **45** (1938), 210 – 219 .

[1975] B.R. Gaines and L.J. Kohout .
*The logic of automata*, Internat. Journal of General Systems **2** (1975), 191–208.

[1987] J.Y. Girard .
*Linear logic*, Theor. Comp. Sci. **50** (1987), 1–102.

[1992] D. Gluschankof.
*Prime deductive systems and injective objects in the algebras of Lukasiewicz infinite valued calculi*, Algebra Universalis **29** (1992), 354–377.

[1993] ———— .
*Cyclic ordered groups and MV-algebras*, Czechoslovak Math. J. **43** (1993), 249–263.

[1967] J.A. Goguen .
*L-Fuzzy sets* , J. Math. Anal. Appl. **18** (1967), 145 – 174 .

[1969] ———— .
*The logic of inexact concepts*, Synthese **19** (1968-69), 325–373.

[1979] R. Goldblatt .
*Topoi : The Categorial Analysis of Logic* (North–Holland , Amsterdam 1979).

[1989] S. Gottwald .
*Mehrwertige Logik* (Akademie–Verlag, Berlin 1989).

[1982] A.C. Grayling .
*An Introduction to Philosophical Logic* (The Harwester Press, Sussex, 1982).

[1987] R. Grigolia .
*Free Algebras of Non–Classical Logics*, Monograph, "Metsniereba", Tbilisi, 1987

[1978] S. Haack .
*Philosophy of Logic* (Cambridge University Press, Cambridge, 1978).

[1] P. Halmos .
*Algebraic Logic* (Chesea Publ. Comp., N. Y., 1962).

[1960] R. Harrop .
*Concerning formulas of types $A \to B \vee C$, $A \to (\exists x)B(x)$ in intuition-istic formal system*, Journal of Symbolic Logic **25** (1960), 27–32.

[1963] L.S. Hay .
*Axiomatization of infinite-valued predicate calculus*, Journal of Symbolic Logic **28** (1963), 77–86.

[1973] H. Herrlich and G.E. Strecker .
*Category Theory* (Allyn and Bacon, Boston 1973).

[1930] A. Heyting .
*Die Regeln der intuitionistischen Logik*, Sitzungsbericht der Preuss. Akad. Wiss. Phys. Math. Kl. (**1930**), 42–56.

[1973] D. Higgs .
*A category approach to Boolean-valued set theory*, preprint Waterloo 1973 (unpublished).

[1984] ———.
*Injectivity in the topos of complete Heyting algebra valued sets*, Canad. J. Math. **36** (1984), 550–568.

[1991A] U. Höhle .
*Editorial of the Special Issue : Mathematical Aspects of Fuzzy Set Theory*, Fuzzy Sets and Systems **40** (1991), 253–256.

[1991B] ———.
*Monoidal closed categories, weak topoi and generalized logics* , Fuzzy Sets and Systems **42** (1991), 15–35.

[1992] ———.
*M-valued sets and sheaves over integral cl-monoids*, in "Applications of catgeory Theory to Fuzzy Subsets" , p. 33 – 72 , (Eds. S.E. Rodabaugh et al.) (Kluwer , Bosten 1992).

[1991] U. Höhle and L.N. Stout .
*Foundations of fuzzy sets*, Fuzzy Sets and Systems **40** (1991), 257–296.

[1992] C.S. Hoo.
*Unitary extensions of MV and BCK-algebras*, Math. Japon. **37** (1992), 585–590.

[1977] P.T. Johnstone .
*Topos Theory* (Academic Press 1977).

[1979] ———.
*On a topological topos*, Proc. London Math. Soc. (3) **38** (1979), 237–271.

[1982] ———.
*Stone Spaces* (Cambridge University Press, Cambridge 1982).

[1991] G. Khatcherian .
*Projectales*, Journal of Pure and Applied Algebra **74** (1991), 177–195.

[1965] D. Klaua .
*Über einen Ansatz zur mehrwertigen Mengenlehre*, Monatsb. Deutsch. Akad. Wiss. (Berlin) **7** (1965), 859–867.

[1967] ———.
*Einbettung der klassischen Mengenlehre in die mehrwertige*, Monatsb. Deutsch. Akad. Wiss. (Berlin) **9** (1967), 258–272.

[1988] G.J. Klir and T.A. Folger .
*Fuzzy Sets, Uncertainty, and Information* (Prentice Hall, Englewood Cliffs, N.J., 1988).

[1971] A. Kock and G.C. Wraith .
*Elementary Toposes*, Aarhus Lecture Note Series **30**, Aarhus 1971.

[1975] L. Kohout .
*Generalised topologies and their relevance to general systems*, Internat. Journal of General Systems **2** (1975), 25–34.

[1988] ———.
*Theories of possibility: Meta-axiomatics and semantics*, Fuzzy Sets and Systems **25** (1988), 357–367.

[1990] ———.
*A Perspective on Intelligent Systems: A Framework for Analysis and Design* (Chapman and Hall & Van Nostrand, London & New York, 1990).

[1991] ———.
*Quo vadis fuzzy systems: A critical evaluation of recent methodological trends*, Internat. J. of General Systems **19** (1991), 395–424.

[1957] G. Kreisel and H. Putnam .
*Eine Unableitbarkeitsbeweismethode für den intuitionistischen Aussagen-kalkül*, Arch. Math. Logic (Archiv für Math. Logik und Grundlagenforsch.) **3** (1957), 74–78.

[1991] R. Kruse, J. Gebhardt, F. Klawonn .
*Reasoning with mass distributions and the context model*, in: "Symbolic and Quantitative Approaches to Uncertainty", 81–85 (eds. R. Kruse and P. Siegel) (Springer–Verlag, Berlin 1991).

[1987] F. Lacava .
*Sulle classi delle L-algebre e degli ℓ-gruppi abeliani algebricamente chiusi*, Boll. Un. Mat. Ital. B (7) **1** (1987), 703–712.

[1989] ———— .
*Sulle L-algebre iniettive*, Bolletino UMI (7) 3-A (1989), 319 - 324.

[1964] F.W. Lawvere .
*An elementary theory of the category of sets*, Proc. Nat. Acad. Sci. USA **51** (1964), 1506–1510.

[1969] ———— .
*Adjointness in foundations*, Dialectica **23** (1969), 281–296.

[1970] ———— .
*Quantifiers and sheaves*, in : "Actes des Congrés International des Mathématiques" 1970, tome 1, 329–334.

[1987] J.W. Lloyd .
*Foundations of Logic Programming* (2nd ed.) (Springer–Verlag, Berlin, New York 1987).

[1920] J. Lukasiewicz .
*O logice trójwartościowej*, Ruch Filozoficzny **5** (1920),170-171.
English Translation : On three–valued logic, in "Jan Lukasiewicz Selected Works", page 87–88 (North–Holland, Amsterdam 1970).

[1930] ———— .
*Philosophische Bemerkungen zu mehrwertigen Systemen des Aussagenkalküls*, C. R. Sci. et Lettres Varsovie **23** (1930), CL. III, 51–77. English Translation : Philosophical remarks on many–valued systems of propositional logic, in "Jan Lukasiewicz Selected Works", page 153–178 (North–Holland, Amsterdam 1970).

[1930] J. Lukasiewicz and A. Tarski .
*Untersuchungen über den Aussagenkalkül*, C.R. Sci. et Lettres Varsovie **23** (1930), CL. III, 30–50.

[1971] S. MacLane .
*Categories for the Working Mathematician* (Springer–Verlag, Berlin New York 1971).

[1992] S. MacLane and I. Moerdijk . *Sheaves in Set Theory and Logic. A first introduction to topos theory* (Springer-Verlag, Berlin New York 1992).

[1976] E.G. Manes .
*Algebraic Theories*, Graduate Texts in Mathematics **26** (Springer–Verlag, Berlin, New York 1976).

[1883] T.G. Masaryk .
*Dav. Hume's Skepsis und die Wahrscheininlichkeitrechnung* (Carl Konegen, Wien, 1883).

[1887] ———— .
*Versuch einer concreten Logik*, (Carl Konegen, Vienna, 1887).
(Translation of Masaryk (1885); enlarged and revised. Reprinted 1969).

[1988] M.M. Mawanda .
*On a categorical analysis of Zadeh generalized subsets of sets I*, in : "Categorical Algebra and its Applications", Lecture Notes in Mathematics **1348** (1988), 257–269 (Springer–Verlag, Berlin New York).

[1962] JV. Medvedev .
*Finite problems*, Dokl. Sov. Acad. **142**, 5 (1962), 1015–1018.

[1966] K. Menger .
*Geometry and positivism – a probabilistic microgeometry*, in : "Selected Papers in Logic and Foundations, Didactics, Economics" (Reidel, Dordrecht, 1979).

[1986] G.P. Monro .
*Quasitopoi, logic and Heyting-valued models*, Jour. Pure Applied Algebra **42** (1986), 141–164.

[1974] R. Montague .
*Formal Philosophy*, Selected Papers of Richard Montague, edited and with an introduction by R. H. Thomason (Yale University Press, New Haven, 1974).

[1948] A. Mostowski .
*Proofs of non–deducibility in intuistionistic functional calculus* , Journal of Symbolic Logic **13** (1948), 204 – 207 .

[1983] C.J. Mulvey .
&, Tagungsbericht, Category Theory Meeting, Oberwolfach, 1983.

[1986] ———.
&, Rendiconti Circ. Mat. Palermo **12** (1986), 99-104.

[1986] D. Mundici .
*Interpretation of AF C\*-algebras in Łukasiewicz sentential logic*, Functional Analysis **65** (1986), 15–63.

[1992] ———.
*Normal forms in infinite-valued logic: the case of one variable*, Lecture Notes in Computer Science **626** (1992), 272–277.

[1993] ———.
*Ulam games, Łukasiewicz logic, and AF C\*-algebras*, Fundamenta Informaticae 18 (1993), 151–161.

[1986] N.J. Nilsson .
*Probabilistic logic*, Artificial Intelligence **28** (1986), 71–87.

[1960] I. Nishimura .
*On formulas on one variable in intuitionistic propositional calculus*, Journal of Symbolic Logic **25** (1960), 327–331.

[1989] V. Novák .
*Fuzzy Sets and Their Applications* (Adam–Hilger, Bristol, 1989).

[1990] ———.
*On the Syntactico-Semantical Completeness of First–Order Fuzzy Logic. Part I — Syntactical Aspects; Part II — Main Results.* Kybernetika **26** (1990), 47–66; 134–154.

[1991] ———.
*Fuzzy sets, fuzzy Logic and natural language*, Int. J. of General Systems **20** (1991), 83–97.

[1988] V. Novák and W. Pedrycz .
*Fuzzy sets and t-norms in the light of fuzzy logic*, Int. J. Man.-Mach. Stud. **29**(1988), 113–127.

[1977] H. Ono .
*On some intuitionistic modal logic*, Publication of the Research Institute for Mathematical Sciences (Kyoto University) **13** (1977), 687–722.

[1964] A.B. Paalman de Miranda .
*Topological Semigroups* (Amsterdam : Math. Centrum 1964).

[1979] J. Pavelka .
*On fuzzy logic I, II, III,* Zeit. Math. Logik u. Grundl. Math. **25** (1979), 45–52, 119–134, 447–464.

[1978] V. Pinkava .
*On a class of functionally complete multi-valued logical calculi,* Studia Logica **2** (1978), 201–212.

[1982] A.M. Pitts .
*Fuzzy sets do not form a topos,* Fuzzy Sets and Systems **8** (1982), 101–104.

[1973] J. Penon .
*Quasitopos,* C.R. Acad. Sci. Paris **276** (1973), Séries A, 237–240.

[1977] J. Penon .
*Sur les quasitopos,* Cahiers Top. Géom. Diff. **18** (1977), 181–218.

[1971] W.A. Pogorzelski .
*Structural completeness of the propositional calculus,* Bull. Acad. Polon. Sci. Ser. Math. Astr. Phys. **19** (1971), 349–351.

[1988] D. Ponasse .
*Categorical studies of fuzzy sets,* Fuzzy Sets and Systems **28** (1988), 235–244.

[1921] E.L. Post .
*Introduction to a general theory of elementary propositions,* Amer. J. Math. **43** (1921), 163–185.

[1972] T. Prucnal .
*On the structural completeness of some pure implicational propositional calculus,* Journal of Symbolic Logic **30** (1972), 45–52.

[1976] _____ .
*Structural completeness of Medvedev's propositional calculus,* Reports on Math. Log. **6** (1976).

[1979] _____ .
*On two problems of Harvey Friedman,* Journal of Symbolic Logic **38**, 3, (1979), 247–262.

[1976] A. Pultr .
*Fuzzy mappings and fuzzy sets,* Commentarii Mathematicae Universitae Carolinae **17** (1976), no. 3.

[1950]  H. Rasiowa and R. Sikorski .
*A proof of the completeness theorem of Gődel*, Fundamenta Math. **37**
(1950), 193–200.

[1970]  ———— .
*The Mathematics of Metamathematics* , third edition (Polish Scientific
Publishers , Warszawa 1970).

[1969]  N. Rescher .
*Many-Valued Logic* (McGraw-Hill, New York, 1969).

[1990]  K.I. Rosenthal .
*Quantales and Their Applications*, Pitman Research Notes in mathematics **234** (Longman, Burnt Mill, Harlow 1990).

[1958]  A. Rose and J.B. Rosser .
*Fragments of many-valued statement calculi*, Trans. Amer. Math.
Soc. **87** (1958), 1–53.

[1989]  J. Rosicky .
*Multiplicative Lattices and C\*-algebras*, Cahiers de Top. et Géom. Diff.
Cat. **XXX**-2 (1989), 95-110.

[1952]  J.B. Rosser and A.R. Turquette .
*Many-valued Logics* (North–Holland, Amsterdam 1952).

[1959]  A. Salomaa .
*On many-valued systems of logic*, Ajatus **22** (1959), 115–159.

[1962]  B. Scarpellini .
*Die Nichaxiomatisierbarkeit des unendlichwertigen Prädikatenkalküls
von Lukasiewicz*, J. of Symbolic Logic **27**(1962), 159–170.

[1983]  B. Schweizer and A. Sklar .
*Probabilistic metric spaces* (North Holland, New York, 1983).

[1967]  D. Scott .
*A proof of the independence of the continuum hypothesis*, Math. Systems Theory **1** (1967), 89-111.

[1974]  ———— .
*Completeness and axiomatizability in many-valued logic*, in : "Proceedings of the Tarski Symposium" (eds. Henkin et al.), pages 411–
435 (Proceedings of Symposia in Pure Mathematics Vol. XXV, Amer.
Math. Soc. 1974).

[1979] ———.
*Identity and existence in intuitionistic logic*, in: "Applications of Sheaves", Lecture Notes in Mathematics **753**, 660–696 (Springer–Verlag 1979).

[1964] R. Sikorski .
*Booelan Algebras* (Springer–Verlag, Berlin 1964).

[1991] L.N. Stout .
*A survey of fuzzy set and topos theory* Fuzzy Sets and Sytsems **42** (1991), 3–14 .

[1992] ———.
*The logic of unbalanced subobjects in a category with two closed structures*, in "Applications of Category Theory to Fuzzy Subsets" (eds. S.E. Rodabaugh et al.) (Kluwer, 1992).

[1989] N.I. Suzuki .
*An algrbraic approach to intuitionistic modal logic in connections with Intermediate predicate logic*, Journal of Symbolic Logic **48** (1989), 141–155.

[1923–1938] A. Tarski .
*Logic, Semantics and Meta-Mathematics* (Hacket Publishing Company, Indianapolis, 1983).
Papers from 1923 to 1938, translated by J.H. Woodger (2nd edition edited and introduced by J. Corcoran).

[1944] ———.
*The semantic conception of truth and the foundations of semantics*, Philosophy and Phenomenological Research **4** (1944), 341–376.

[1973] A. Urquhart .
*Free Heyting algebras*, Algebra Universalis **3**,1, (1973), 94–97.

[1976] O. Wyler .
*Are there topoi in topology ?* , in : "Categorical Topology, Mannheim 1975", Lecture Notes in Mathematics **540** (1976), 699–719 (Springer–Verlag, Berlin , Heidelberg, New York).

[1991] ———.
*Lecture Notes on Topoi and Quasitopoi* (World Scientific, Singapore 1991).

[1965] L.A. Zadeh .
*Fuzzy sets*, Information and Control **8** (1965), 338–353.

[1978] ———.

Fuzzy sets as a basis for a theory of possibility, Fuzzy Sets and Systems
1 (1978), 3–28.

# Index

# THEORY AND DECISION LIBRARY

## SERIES B: MATHEMATICAL AND STATISTICAL METHODS
*Editor*: H. J. Skala, *University of Paderborn, Germany*

The manufacturer's authorised representative in the EU is Springer
Nature Customer Service Centre GmbH, Europaplatz 3, 69115 Heidelberg,
Germany. If you have any concerns regarding our products, please
contact ProductSafety@springernature.com

Printed and bound by CPI Group (UK) Ltd, Croydon, CR0 4YY
24/04/2026
02096308-0012